Informed Decisionmaking for Sustainability

This Springer book series – Informed Decisionmaking for Sustainability – offers a roadmap for humankind to address issues, impacts and resources within, across and beyond the boundaries of nations. Informed decisions operate across a 'continuum of urgencies,' extending from security time scales (mitigating risks of political, economic or cultural instabilities that are immediate) to sustainability time scales (balancing economic prosperity, societal well-being and environmental protection across generations) for nations, peoples and our world. Moreover, informed decisions involve governance mechanisms (laws, agreements and policies as well as regulatory strategies, including insurance, at diverse jurisdictional levels) and built infrastructure (fixed, mobile and other assets, including communication, research, observing, information and other systems that entail technology plus investment), which further require close coupling to achieve progress with sustainability. International, interdisciplinary and inclusive (holistic) engagement in this book series involves decisionmakers and thought leaders from government, business, academia and society at large to reveal lessons about common-interest building that promote cooperation and prevent conflict. The three initial volumes utilize the high north as a case study, recognizing that we are entering a globally significant period of trillion-dollar investment in the new Arctic Ocean. Additional case studies are welcome and will be included in the book series subsequently. Throughout, to be holistic, science is characterized as 'the study of change' to include natural sciences, social sciences and Indigenous knowledge, all of which reveal trends, patterns and processes (albeit with different methods) that become the bases for decisions. The goal of this book series is to apply, train and refine science diplomacy as an holistic process, involving informed decisionmaking to balance national interests and common interests for the benefit of all on Earth across generations.

More information about this series at https://link.springer.com/bookseries/16420

Paul Arthur Berkman
Alexander N. Vylegzhanin • Oran R. Young
David A. Balton • Ole Rasmus Øvretveit
Editors

Building Common Interests in the Arctic Ocean with Global Inclusion

Volume 2

Editors
Paul Arthur Berkman
Science Diplomacy Center
EvREsearch LTD
Falmouth, MA, USA

Science Diplomacy Center
MGIMO University
Moscow, Russian Federation

United Nations Institute for Training and Research (UNITAR)
Geneva, Switzerland

Program on Negotiation
Harvard Law School
Cambridge, MA, USA

Alexander N. Vylegzhanin
International Law School
MGIMO University
Moscow, Russian Federation

David A. Balton
Polar Institute
Wilson Center and U.S. Arctic Steering Committee
Washington, DC, USA

Ole Rasmus Øvretveit
Initiative West Bergen
Bergen, Norway

Oran R. Young
Bren School of Environmental Science & Management
University of California Santa Barbara
Santa Barbara, CA, USA

ISSN 2662-4516 ISSN 2662-4524 (electronic)
Informed Decisionmaking for Sustainability
ISBN 978-3-030-89311-8 ISBN 978-3-030-89312-5 (eBook)
https://doi.org/10.1007/978-3-030-89312-5

Cover illustration: Holistic (international, interdisciplinary and inclusive) integration is represented in this figure with the eight Arctic states and six Indigenous Peoples' Organizations in the Arctic Council; biogeophysical features of the Arctic Ocean represented by the 2012 sea-ice mininum; and boundary of the Central Arctic Ocean High Seas established under the international framework of the law of the sea to which all Aprctic states and Indigenous Peoples' Organizations "remain committed." Details of this cover illustration are elaborated in the first figure in Chapter 1.

This Springer imprint is published by the registered company Springer Nature Switzerland AG
The registered company address is: Gewerbestrasse 11, 6330 Cham, Switzerland

Preface

This preface complements Chapter 1 ("Introduction: Building Common Interests with Informed Decisionmaking for Sustainability") for this book, *Building Common Interests in the Arctic Ocean with Global Inclusion*. The purpose of the preface is to provide background on the process to assemble as well as navigate this second edited volume in the Informed Decisionmaking for Sustainability book series published by Springer. Assembly of this volume also reinforces the international, interdisciplinary and inclusive (holistic) spirit envisioned in the book title, highlighting next-generation capacities with both early career and seasoned practitioners from diverse backgrounds to achieve progress with sustainable development (Figure).

The first three volumes in the book series represent a trilogy with the Arctic Ocean as a common feature, framed by the *Arctic Options* and *Pan-Arctic Options* projects to address *"Holistic Integration for Arctic-Coastal Marine Sustainability"* (Berkman et al. 2020):

Volume 1 – *Governing Arctic Seas: Regional Lessons from the Bering Strait and Barents Sea*
Volume 2 – *Building Common Interests in the Arctic Ocean with Global Inclusion*
Volume 3 – *Pan-Arctic Implementation of Coupled Governance and Infrastructure*

The *Arctic Options* and *Pan-Arctic Options* projects also supported production of the *Baseline of Russian Arctic Laws* (Berkman et al. 2019), as a complementary contribution to the book series, recognizing Russia's governance history in the Arctic since the early nineteenth century with sovereignty across much of the Arctic today.

The first volume in this book series and the *Baseline of Russian Arctic Laws* were coordinated through the Science Diplomacy Center (with P.A. Berkman as Founding Director) in the Fletcher School of Law and Diplomacy at Tufts University. In 2018, on behalf of the book series editors (Paul Arthur Berkman, Alexander N. Vylegzhanin and Oran R. Young), the Science Diplomacy Center also signed a Memorandum of Understanding (MoU) with Arctic Frontiers (with Ole Øvretveit as Director) to co-produce the second and third volumes in the book series (Arctic Frontiers 2020a). The Science Diplomacy Center is now part of EvREsearch LTD, which is coordinating production of this second volume with continued funding through the National Science Foundation for the *Pan-Arctic Options* project. The editorial team of this second volume includes these four individuals with the important addition of Ambassador David A. Balton, who had been a key contributor to the binding Arctic agreements that have emerged since 2009.

The MoU enabled the first volume in the book series (Young et al. 2020) to be launched at Arctic Frontiers in Tromsø in January 2020 and simultaneously to initiate this second volume, synchronizing potential chapter contributions just before global lockdowns due to the COVID-19 pandemic. The theme of that annual meeting was Knowledge to Action (Arctic Frontier 2020b), complementing the integration of research and action to produce informed decisions, which are the focus of this book series (please see Chap. 1).

Moreover, research and action underlie the general format of Arctic Frontiers 2020, which paralleled the "science" and "plenary" sessions, respectively:

- **Research:** Science sessions with open abstract submissions that are reviewed by committees who determine their acceptability for oral or poster presentations.
- **Action:** Plenary sessions with invited experts and high-level decisionmakers from governments as well as industry with international and interdisciplinary inclusion.

As identified in the Table of Contents, this book includes "research" (6000 to 10,000 words with figures, tables and references) and "action" (750 to 1200 words without figures, tables and references from written interventions or presentation transcripts) contributions from the Arctic Frontiers 2020 sessions. These written contributions also reflect general characteristics of presentations with the research and action communities, underlying different skills and methods involved to be helpful with their target audiences.

With Arctic Frontiers 2020, the option was introduced for submitted science session abstracts to be considered for inclusion in this second volume, which involved an additional review process with the editorial team. If the abstracts were acceptable by both the Arctic Frontiers and editorial teams, the authors from the science sessions were invited to submit draft chapters for further consideration. Among the 60 abstracts that were received for consideration as part of this book, just over 30 were invited as possible chapters, including contributions across the professional spectrum from graduate students to renown experts.

However, to ensure rigor and quality control, contributions from the early-career scientists involved additional interactions with the editorial team. Each young scientist iterated their contributions with a designated member of the editorial team, who was responsible to ensure possible chapter contributions were review-ready before soliciting input from anonymous reviewers. Each research chapter in this book was anonymously reviewed by at least two experts and revised accordingly before acceptance by the full editorial team.

Action chapters were invited from all of the plenary speakers to capture decisionmaking snapshots and emphases at the moment of the Arctic Frontiers 2020 conference. Recognizing that plenary speakers commonly read their statements, often carefully vetted by their institutions, the written interventions or transcripts from the plenary presentations are included in this book by self-selection of the authors without review or revision. The action chapters are grouped with reference to their respective sessions.

The research and action chapters in this volume together represent an holistic approach, as reflected by the interdisciplinary contributions from nearly 20 nations, ranging from a high-school student to foreign ministers along with graduate students, postdoctoral scholars, professors, organizational leaders, Indigenous peoples, subnational-national-international officials and community representatives. During the course of this book's production, there were two virtual author-editor meetings to further integrate the research chapters in view of informed decisionmaking (Berkman et al. 2017; Berkman et al. 2020; Berkman 2020a,b). The editorial team also had bi-monthly meetings to produce a book that has touch points for diverse readers who are interested to contribute with informed decisions that operate short-to-long term. The goal of this second volume in the INFORMED DECISIONMAKING FOR SUSTAINABILITY book series is to be helpful with Arctic sustainability in a local-global context with common-interest building as an inclusive process that is responsive to ever-changing circumstances with hope and inspiration across generations.

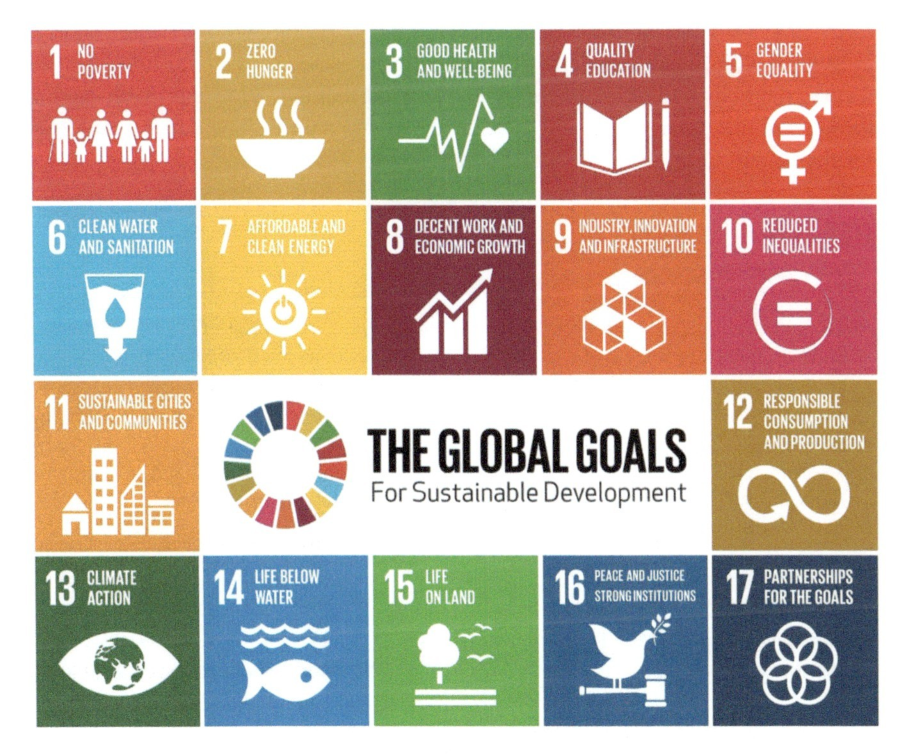

FIGURE: UNITED NATIONS SUSTAINABLE DEVELOPMENT GOALS (United Nations 2015) with international, interdisciplinary and inclusive relevance at local-global levels, balancing environmental protection, economic prosperity and societal well-being for the benefit of all on Earth across generations.

Falmouth, MA, USA — Paul Arthur Berkman
Santa Barbara, CA, USA — Oran R. Young
Moscow, Russian Federation — Alexander N. Vylegzhanin
Washington, DC, USA — David A. Balton
Tromsø, Norway — Ole Rasmus Øvretveit

Editors' Addendum We are living through an important moment in human history, which relates to the aspirations of this book. The COVID-19 pandemic erupted just after work began on this book at the Arctic Frontiers 2020 Conference, with devastating consequences worldwide. Now, just as this book is being published after being completed, events in Ukraine have created new peril that has the potential to cascade into world war, something we all have responsibilities to prevent forever after the twentieth century. Circumstances in Ukraine have already created significant instability in international relations, including for the Arctic region, jeopardizing dialogue, cooperation and progress among the eight Arctic states, six Indigenous Peoples' Organizations and international community of observers participating in the Arctic Council. This book and its journey to production reveal common interest building as a necessary complement to conflict resolution, utilizing science diplomacy as a "language of hope" to operate across a "continuum of urgencies" short to long term with direct application to current circumstances. It is our hope that the theory, methods and skills of informed decision-making – reflected with lessons throughout this book – will help to balance national interests and common interests for the benefit of all on Earth across generations.

References

Arctic Frontiers. (2020a). *Arctic frontiers 2020. Annual report: The power of knowledge.* Tromsø, Norway. https://www.arcticfrontiers.com/conference/2020-the-power-of-knowledge/

Arctic Frontiers. (2020b). *Arctic frontiers 2020. Conference summary. The power of knowledge.* Tromsø, Norway. https://www.arcticfrontiers.com/wp-content/uploads/2020/10/plenaryreport2020.pdf

Berkman, P. A. (2020a). Science diplomacy and its engine of informed decisionmaking: Operating through our global pandemic with humanity. *The Hague Journal of Diplomacy, 15,* 435–450. https://brill.com/view/journals/hjd/15/3/article-p435_13.xml

Berkman, P. A. (2020b). 'The pandemic lens': Focusing across time scales for local-global sustainability. *Patterns, 1*(8), 13 November 2020. 4p. https://pubmed.ncbi.nlm.nih.gov/33294877/

Berkman, P. A., Kullerud, L., Pope, A., Vylegzhanin, A. N., & Young, O. R. (2017). The Arctic science agreement propels science diplomacy. *Science, 358,* 596–598. https://science.sciencemag.org/content/358/6363/596

Berkman, P. A., Vylegzhanin, A. N., & Young, O. R. (2019). Baseline of Russian Arctic Laws. Springer. https://www.springer.com/gp/book/9783030062613

Berkman, P. A., Young, O. R., & Vylegzhanin, A. N. (2020). Preface for the book series on informed decisionmaking for sustainability. In: Young, O. R., Berkman, P. A., & Vylegzhanin

(Eds.) Volume 1: Governing Arctic Seas: Regional Lessons from the Bering Strait and Barents Sea. Springer. https://link.springer.com/content/pdf/bfm%3A978-3-030-25674-6%2F1.pdf

United Nations. (2015). *Transforming Our World: The 2030 Agenda for Sustainable Development.* Res. A/RES/70/1 (25 September 2015). United Nations General Assembly, New York. https://sustainabledevelopment.un.org/post2015/transformingourworld/publication

Young, O. R., Berkman, P. A., & Vylegzhanin, A. N. (Eds.). (2020). Informed Decisionmaking for Sustainability. Volume 1. Governing Arctic Seas: Regional Lessons from the Bering Strait and Barents Sea. Springer. https://link.springer.com/book/10.1007/978-3-030-25674-6

Acknowledgements

We are honoured to acknowledge the contributions of numerous individuals, teams, partner organizations and funding agencies that allowed us to carry out the work captured in this volume and to continue developing ideas about informed decisionmaking for sustainability in view of environmental, societal and economic considerations, short-to-long term, across our globally interconnected civilization.

This is the second volume in a trilogy with the Arctic as a global case study to balance national interests and common interests for the benefit of all on Earth across generations. On this journey, we are grateful to the residents of the Arctic, especially the Indigenous peoples, who have demonstrated resilience over millennia with lessons for humanity into the twenty-second century and beyond in the face of accelerating global changes.

This volume was initiated as part of the *Arctic Options* and *Pan-Arctic Options* projects (2013–2021), both addressing *"Holistic Integration for Arctic Coastal-Marine Sustainability"* with funding from the United States National Science Foundation (NSF-OPP 1263819 NSF-ICER 1660449 NSF-OPP 1719540 and NSF-ICER 2103490). The holistic (international, interdisciplinary and inclusive) characteristics of these intertwined projects were represented from the start with co-funding from the Centre national de la recherche scientifique (CNRS) and collaboration with the Université Pierre-et-Marie-Curie in France, considering options (without advocacy) with colleagues from the University of California Santa Barbara (UCSB) and the University of Alaska Fairbanks (UAF) in the United States.

Elements of the *Arctic Options/Pan-Arctic Options* projects are elaborated in the book series preface for Informed Decisionmaking for Sustainability published in *Volume 1. Governing Arctic Seas: Regional Lessons from the Bering Strait and Barents Sea* (2020). The options for Arctic sustainability further included colleagues from the Russian Federation, acknowledging MGIMO University, which enabled the first formal dialogue between the North Atlantic Treaty Organization (NATO) and Russia as part of a NATO Advanced Research Workshop project through the University of Cambridge in 2010, memorialized in the Springer book on

Environmental Security in the Arctic Ocean (2012). We thank Springer for their continuing support of science diplomacy, as reflected also by the *Baseline of Russian Arctic Laws* (2019), which accompanies this book series.

Accelerating from the 2009 *Antarctic Treaty Summit*,[1] science diplomacy has been the framework to address options with *Arctic Options/Pan-Arctic Options* projects, expanding with inclusion through the Belmont Forum. We thank the Agence Nationale de la Recherche (ANR) for additional funding in France along with new collaborations involving national funding agencies in Canada, China, Norway and Russia. In particular, we acknowledge the strong and ongoing financial support from the Research Council of Norway (RCN), representing durable and longstanding alignment with the NSF funding. We also acknowledge coordination with the Social Sciences and Humanities Research Council of Canada (SSHRC), National Science Foundation of China (NSFC) and the Russian Foundation for Basic Research (RFBR).

Moving from UCSB, the *Arctic Options/Pan-Arctic Options* projects were coordinated from 2015 to 2020 through the Fletcher School of Law and Diplomacy at Tufts University with additional research collaboration through the University of Colorado, Norwegian Polar Institute, Ocean University of China and Carleton University. We thank these institutions for enabling research and education to be triangulated with leadership, which evolved into ambassadorial panels in 2015 and 2016 – *"Arctic High Seas: Building Common Interests in the Arctic Ocean"* – at Reykjavik University in collaboration with the Arctic Circle, conceived and co-hosted by the *Arctic Options/Pan-Arctic Options* projects. These two ambassadorial panels along with earlier collaborations among the book series editors (P.A. Berkman, A.N. Vylegzhanin and O.R. Young) are the origin of this volume *Building Common Interests in the Arctic Ocean with Global Inclusion*.

We thank Ari Kristinn Jónsson, Rector of Reykjavik University, for enabling the 2015 and 2016 panel dialogues with ambassadors to Iceland from China, France, Germany, Norway, Russia, Sweden, the United Kingdom and the United States, along with other diplomats from Canada, Finland, Greenland, and Iceland, and thank especially the Foreign Minister of Iceland for her written remarks.[2] It is special honour to thank former President of Iceland, Ólafur Ragnar Grímsson, for his leadership and keynote presentations with both of these dialogues to build common interests in the Arctic Ocean, notably in the Central Arctic Ocean High Seas as an area beyond national jurisdiction where global lessons are emerging to promote cooperation and prevent conflict with a precautionary approach short-to-long term.

[1]Berkman, P.A., Lang, M.A., Walton, D.W.H. and Young, O.R. (eds.). Science Diplomacy: Antarctica, Science and the Governance of International Spaces. Smithsonian Institution Scholarly Press, Washington, DC. (https://repository.si.edu/handle/10088/16154).

[2]H.E. Lilja Alfreðsdóttir (Minister for Foreign Affairs of Iceland). *Opening Address. Second Annual Ambassadorial Panel on the Arctic High Seas. Pan Arctic Options and the Reykjavik University*. 6 October 2016 (https://www.stjornarradid.is/library/04-Raduneytin/Utanrikisraduneytid/PDF-skjol/Arctic-pre-event%2D%2D-raeda-ra%CC%81dherra.pdf).

From the *Arctic Options/Pan-Arctic Options* project, the Science Diplomacy Center was founded in 2017 to coordinate rapidly progressing co-production of knowledge, including development of this Informed Decisionmaking for Sustainability book series. We thank the collaborating scholars, decisionmakers, patrons, next-generation leaders and friends who have shared insights, observations and ideas to transform research into action across the data-evidence interface with informed decisionmaking as the engine of science diplomacy.

We thank many foreign ministries (including from Algeria, Armenia, Canada, Colombia, Costa Rica, Ethiopia, Hungary, India, Indonesia, Japan, Norway, Russian Federation and United States) and the United Nations Institute for Training and Research (UNITAR) for enabling science diplomacy to be shared in a scalable manner that builds on lessons from the *Arctic Options/Pan-Arctic Options* projects. Since 2020, with continued NSF funding from the *Arctic Options/Pan-Arctic Options* projects, the Science Diplomacy Center has been operating through EvREsearch LTD, coordinating production of this book to address issues, impacts and resources across all kinds of boundaries (ecopolitical, geographic, generational, population, cultural, legal, historic and other dimensions) inclusively in view of Arctic and Earth systems.

On behalf of the book series editors (Berkman, Young and Vylegzhanin), in 2018, the Science Diplomacy Center (Berkman) signed a Memorandum of Understanding (MOU) with Arctic Frontiers (Øvretveit), which had assembled *"the first global scientific conference on economic, societal and environmental sustainable growth in the north"* in 2006, becoming an annual international forum in Tromsø, Norway. Production of this book has been enriched by the additional member of the editorial team (Balton), bridging research and action with contributions of a key diplomat behind the binding Arctic agreements that emerged over the past decade.

We specially thank Akvaplan-Niva for enabling knowledge co-production with this book, which emanated from Arctic Frontiers 2020, and Alexey Pavlov for his key assistance to engage a self-selected team of contributors inclusively with the research chapters (from the science sessions) and action chapters (from the plenary sessions) herein. We further thank Anu Fredrikson and her leadership with co-production of *Volume 3. Pan-Arctic Implementation of Coupled Governance and Infrastructure*, which will arise from Arctic Frontiers 2022 with the Woodrow Wilson Center and Mike Sfraga also as co-editors.

Most importantly, we thank the chapter contributors to this book, who have enabled us to compile an unique collection of observations about the Arctic from diverse vantages, ranging from a high-school student to foreign ministers along with graduate students, postdoctoral scholars, professors, organizational leaders, Indigenous peoples, subnational-national-international officials and community representatives across a dozen nations. We appreciate their interdisciplinary breath across the natural sciences, social sciences and Indigenous knowledge – inclusively as the "study of change" – in view of patterns, trends and processes (albeit with different methodologies) that become the bases for decisions. We further appreciate their transdisciplinary efforts to address questions of common concern, integrating data into evidence in view of the decisionmaking institutions that produce governance mechanisms and built infrastructure as well as their coupling with sustainable development. In these efforts, we thank the research-chapter authors especially for the passion and creative contributions delivered with enthusiasm in the two editor-author team teleconferences during our global pandemic.

The rigor, quality and legacy of this book is influenced further by the community of experts who have shared their time and candid comments with the chapter authors, as part of a process with two anonymous reviews and subsequent revisions, leading to consensus decisions by the co-editors. We are grateful to these reviewers for their essential input, recognizing their voluntary contributions, and thank: Steinar Andresen; Raymond V. Arnaudo; Betsy B. Baker; Michael S. Bank; Mia M. Bennett; Lawson W. Brigham; Robert W. Correll; William Eichbaum; Gail Fondahl; Jeffrey T. Fong; Jon L. Fuglestad; Jacqueline M. Grebmeier; Peter Harrison; Alf Håkon Hoel; Gunhild Gjørv Hoogensen; Noor Johnson; Randy "Church" Key; Timo Koivurova; Yekaterina Y. Kontar; Valeriy A. Kryukov; James H. Lever; Molly McCammon; Erik J. Molenaar; Frode Nilssen; Andreas Østhagen; Andrey N. Petrov; Svein Vigeland Rottem; Jon-Arve Røyset; Scott Stephenson; D.R. Fraser Taylor; and David L. VanderZwaag.

Additional acknowledgements are included in the chapters by the design of the authors, and we are grateful for these important sources of support to enable this book to be produced. Behind the scenes, in a crucial supporting role with diverse editorial contributions, we thank Julie Anne Hambrook Berkman with deep appreciation.

As an editorial team, it has been an honour as well as pleasure working together with all involved to produce this book, especially during the COVID-19 pandemic when there once again is a common interest in survival of humankind at local to global levels, unlike any period since the Second World War.

Building Common Interests in the Arctic Ocean with Global Inclusion is an ongoing journey, involving a "continuum of urgencies" to address across generations with informed decisions. We are grateful to all contributors in this international, interdisciplinary and inclusive process with hope for humanity.

Paul Arthur Berkman

Science Diplomacy Center,
EvREsearch LTD, Falmouth, MA, USA

Science Diplomacy Center,
MGIMO University,
Moscow, Russian Federation

United Nations Institute for Training and
Research (UNITAR), Geneva,
Switzerland

Program on Negotiation, Harvard Law
School, Cambridge, MA, USA

Bren School of Environmental Science & Management, University of California Santa Barbara, Santa Barbara, CA, USA — Oran R. Young

International Law School, MGIMO University, Moscow, Russian Federation — Alexander N. Vylegzhanin

Polar Institute, Wilson Center and U.S. Arctic Steering Committee, Washington, DC, USA — David A. Balton

Initiative West Bergen, Bergen, Norway — Ole Rasmus Øvretveit

Contents[3]

[3]"Research" and "Action" chapters are described in the Preface (see also Figure 1.6)

About the Contributors

Dina Abdel-Fattah is an Associate Professor in the Department of Technology and Safety at UiT – The Arctic University of Norway and a Senior Lecturer in the Department of Computer and Systems Science at Stockholm University. She has an interdisciplinary background in natural and social science as well as decision and risk analysis. Her research is primarily focused on cryospheric and natural hazards and their impacts on local communities in the Arctic and sub-Arctic.

Mathias Albert is Professor of Sociology at Bielefeld University. His research covers the history and sociology of world politics, youth studies, and, more recently, polar politics. His latest book publications in English include: *A Theory of World Politics* (Cambridge: Cambridge University Press 2016); *The Politics of International Political Theory* (ed. with A.F. Lang. London: Palgrave 2019); *What in the World? Understanding Global Social Change* (ed. with T. Werron. Bristol: Bristol University Press 2021).

Ekaterina S. Anyanova completed her LLM and Dr.iur studies at Hamburg University and 3 years later received her Candidate of legal science degree from Institute of State and Law in Moscow. During her studies she conducted legal research and analysis in the law of the sea area. At present she is working on her Doctor of Legal Science thesis on the maritime security issue. She also works as a Legal adviser.

Eda Ayaydin is a lecturer at Sciences Po Bordeaux and PhD candidate at the University of Paris-Saclay in Paris, France. Her research interests include history of Arctic exploration, Arctic geopolitics, science diplomacy and governance.

David A. Balton is a Senior Fellow with the Woodrow Wilson Center's Polar Institute. He previously served as the U.S. Ambassador for Oceans and Fisheries, with a portfolio that included managing U.S. foreign policy issues relating to the Arctic and Antarctica. During the U.S. Chairmanship of the Arctic Council (2015–2017), he served as Chair of the Senior Arctic Officials. He also led negotiations that produced the 2011 *Agreement on Cooperation on Aeronautical and Maritime Search and Rescue in the Arctic*, the 2013 *Agreement on Cooperation on Marine Oil Pollution Preparedness and Response in the Arctic*, and the 2018 *Agreement to Prevent Unregulated High Seas Fisheries in the Central Arctic Ocean.* Ambassador Balton now serves as Executive Director of the U.S. Arctic Executive Steering Committee.

Manuel Barange is Professor and Director of the Fisheries and Aquaculture Policy and Resources Division at the FAO. Previously he was Deputy Chief Executive and Director of Science at the Plymouth Marine Laboratory, UK, and Chair of the Science Committee of the International Council for the Exploration of the Sea (ICES). He is a global expert on the impacts of Climate Change on marine ecosystems, fisheries and aquaculture, and on the interactions between natural and socio-economic drivers of change. He is a Review Editor of the forthcoming IPCC Special Report on Oceans and the Cryosphere in a Changing Climate (SROCC). In 2010 he was awarded the UNESCO-IOC Roger Revelle Medal for his contribution to ocean science.

Marnix van den Berg completed a MSc in Offshore and Dredging Engineering at Delft University of Technology in 2014 and received a PhD degree from Norwegian University of Science and Technology (NTNU) in 2019. His research focusses on numerical modelling of ice-structure and ice-ship interaction processes with a specific interest in the influence of discrete interactions of ice floes within a broken ice field on the ice loads experienced by structures and ships.

Paul Arthur Berkman is the Fulbright Arctic Chair 2021–2022. Professor Berkman is Director of the Science Diplomacy Center at EvREsearch LTD in the United States and at MGIMO University in Moscow, Russian Federation. He is an Affiliated Fellow of the United Nations Institute for Training and Research (UNITAR) and a Faculty Associate of the Program on Negotiation at Harvard Law School. Paul wintered in Antarctica in 1981 on a SCUBA research expedition when he was 22 and began teaching science diplomacy the following year as a Visiting Professor, leading to his doctorate in oceanography and lifetime journey to help balance national interests and common interests for the benefit of all on Earth across generations.

Jon Bjørnø received in 2016 his MSc in marine cybernetics at the department of Marine Technology, Norwegian University of Science and Technology (NTNU), Trondheim, Norway. Today, he is working towards a PhD degree within the same field at the Marine Technology department, NTNU. His research interests include motion control for single or multiple vessels, guidance and control, and Arctic ice management operations. The PhD work is a part of the Center for Research-Driven Innovation, Sustainable Arctic Marine and Coastal Technology (SAMCoT).

Aileen Campbell Cabinet Secretary for Communities and Local Government, Scottish government, was first elected to the Scottish Parliament in 2007 as a list member for the South of Scotland. She was re-elected as the Member of the Scottish Parliament for Clydesdale in 2011 and appointed as Minister for Children and Young People. She was appointed Minister for Public Health and Sport in 2016. Ms. Campbell started a career in publishing as an editor for the construction magazine Keystone in 2003 and has been an editorial assistant on the newspaper, the Scottish Standard. Before her election to parliament, Aileen also worked as a Parliamentary Assistant. Aileen Campbell has an MA joint honours degree in Politics with Economic and Social History at the University of Glasgow.

Colleen J. Charles is a Woodland Cree First Nation from the Lac La Ronge Indian Band in northern Saskatchewan, Canada. She is a single mother of three children and has a granddaughter. She is a certified Antiracism Education Facilitator. Colleen has her Master of Education (MEd) from the University of Saskatchewan. Her research interest includes, Indigenous land-based education, Indigenous water rights, critical anti-racist learning and practice.

Joel Clement is an Arctic Initiative Senior Fellow at the Harvard Kennedy School's Belfer Center for Science and International Affairs with a background in climate and energy issues, resilience and climate change adaptation, landscape-scale conservation and management, and Arctic social-ecological systems. Mr Clement has priorly served as an executive at the U.S. Department of the Interior. In 2017, he was awarded The Joe A. Callaway Award for Civic Courage and has received multiple awards for ethics, courage, and his dedication to the role of science in public policy. In addition to his role at the Harvard Kennedy School, he is an Associate with the Stockholm Environmental Institute and a Senior Fellow with the Union of Concerned Scientists.

Ranjan Datta Canada Research Chair in Community Disaster Research at Indigenous Studies, Department of Humanities, Mount Royal University, Calgary. Alberta, Canada. Ranjan's research interests include advocating for Indigenous environmental sustainability, Indigenous food sovereignty, Indigenous energy management, decolonization, Indigenous reconciliation, community-based research, and cross-cultural community empowerment.

Yulia Ermilova is a leading GIS Specialist in the Lomonosov Moscow State University Marine Research Center. Yulia's main activities are selection, editing and systematizing of spatial data and GIS-support of engineering and environmental surveys. She takes part in the creation of ecological atlases for Russian energy companies (Gazprom, Rosneft) and develops map designs for scientific publications.

Nadezhda Filimonova is a PhD Candidate at the University of Massachusetts Boston. Her research addresses the topics of Arctic governance, Russian Arctic policy and Arctic urban climate governance. She holds two master's degrees: one in Political Science and International Studies from Uppsala University, and a second in International Relations from St. Petersburg State University. She held visiting fellowships at several foreign universities including Aalborg University, University of Granada, Ludwig-Maximilians-Universität München and the University of Missouri.

Greg Fiske is a geographer whose background includes extensive use of geographic techniques including Geographic Information Systems (GIS), Cartography, Remote Sensing and relational database management to conduct and communicate environmental science. He has broad expertise in cloud-based spatial data analysis and data management and administers technical aspects of Woodwell Climate Research Center's geospatial activities. He has designed and developed complex spatial analysis workflows and has taught scientists, students and researchers on the use of both open-source and proprietary geospatial software. Prior to joining Woodwell, he worked at the Lawrence Livermore National Laboratory in California. Mr Fiske earned his BS from Plymouth State University in New Hampshire and his MS from Oregon State University.

Anna Gebruk is a PhD candidate in the School of GeoSciences at the University of Edinburgh. Anna is a Head of international collaboration of the Lomonosov Moscow State University Marine Research Center and a President of the UK Polar Network 2020–2021. Anna's main interests are in functioning of benthic ecosystems and the challenges they are facing including invasive species and pollutants, as well as international scientific collaboration in Arctic observing networks.

Jacqueline M. Grebmeier is a Professor at the University of Maryland Center for Environmental Sciences, Chesapeake Biological Laboratory in the United States. She holds a BA in Zoology (1977), MS in Biology (1979), MS in Marine Affairs (1983) and a PhD in Biological Oceanography (1987). She completed a postdoctoral position at the University of Southern California (1988) and was a research faculty member at the University of Tennessee before joining the Chesapeake Bay Laboratory in 2008. Her oceanographic research interests include pelagic-benthic coupling, benthic carbon cycling and benthic faunal population structure in relation to ecosystem structure in polar marine systems, primarily in the Arctic.

Markus Haraldsvik Leader, Student Council, Bardufoss upper secondary school. Markus (16) was born in Namsos in Trøndelag County, Norway and spent his first years in Trælnes, Brønnøysund Municipality. He graduated from Brøstadbotn secondary school and is now a pupil at the Bardufoss upper secondary school in Tromsø, specializing in sports. Markus has held various leading positions in youth councils at local and regional levels in Tromsø, including deputy head of Tromsø County Youth Council.

Sian F. Henley is an interdisciplinary marine scientist based at the University of Edinburgh (UK), focusing primarily on the Polar Oceans and their importance in the Earth System. She specializes in marine biogeochemistry and ecosystem functioning in the Arctic and Antarctica, and the way these are changing in response to climate and environmental change and in turn influencing these changes. In addition to primary research, Sian is actively involved in science-policy interactions, public engagement and science education.

Lea-Anne Henry is a marine ecologist and Chancellor's Fellow in the School of GeoSciences at, the University of Edinburgh in Scotland, United Kingdom. Her research seeks to understand the spatial and temporal drivers of marine biodiversity and biogeographical patterns over ecological and geological timescales, including effects of man-made activities and climate change. Much of her research is in collaboration with the international offshore industry community, governments and non-governmental organizations.

Alf Håkon Hoel is professor of Ocean Law and Policy at UiT – the Arctic University of Norway in Tromsø. His research addresses the governance of oceans and the management of natural resources. He is also affiliated with the Norwegian Institute of Marine Research and the Polar Institute of the Wilson Center in Washington DC.

Jens Frølich Holte State Secretary in the Norwegian Ministry of Foreign Affairs has been State Secretary to the Minister for EEA and EU Affairs, acting State Secretary to the Minister of Transport and Communications, and political adviser in the Ministry of Climate and Environment. Prior to his current position he was State Secretary for International Development in the Ministry of Foreign Affairs. Frølich Holte has a master's degree in economics and business administration from the Norwegian School of Economics and Business Administration and a master's degree in economic history from the London School of Economics. He became politically active in the Young Conservatives at the age of seventeen.

Aili Keskitalo is the President of the Sámi Parliament in Norway. She is the first female President of any Sámi Parliaments. She served as President of the Sámi Parliament in Norway in 2005 for the first time. Keskitalo has a master's degree in Public Administration from Copenhagen Business School, focusing on higher education in Norway and Greenland. Keskitalo is married and has three daughters. She lives in Guovdageaidnu / Kautokeino, Norway.

Nathan Kettle is a Research Assistant Professor in the International Arctic Research Center at the University of Alaska Fairbanks and Co-Investigator for the Alaska Center for Climate Assessment and Policy. His research in the Arctic centres on knowledge co-production to inform decisionmaking, climate adaptation, decision support tools and evaluation. He received his PhD in Geography from the University of South Carolina in 2012.

Elena V. Kienko Candidate of Science (International Law), has LLM in International Economic Law from MGIMO University, Russia. She is currently working at the Primakov National Research Institute of World Economy and International Relations, Russian Academy of Sciences. She specializes in legal aspects of cooperation between Russia and Non-Arctic States in the Arctic Region (primarily between Russia and Japan, between Russia and the People's Republic of China and between Russia and the Republic of Korea). She has published a wide range of articles on International Law, International Maritime Law and Arctic Law (mainly in Russian).

Yunjin Kim is a former researcher at the Korea Legislation Research Institute. She participated in Arctic Frontiers as a student forum participant, young ambassador and research presenter for 3 years (2018–20). She did her internship at the International Tribunal for the Law of the Sea and has been involved in research related to marine protected areas and high seas. She expects to see younger generations gain hope to lead the efforts in making the Arctic sustainable.

Sebastian Knecht is a postdoctoral researcher at the Faculty of Sociology at Bielefeld University. He was a fellow at the Berlin Graduate School for Transnational Studies (BTS) at Freie Universität Berlin from which he received his PhD in 2019. His research is primarily concerned with international cooperation in the polar regions, the design and stratification of membership systems in international organizations, and science-policy interactions. He is co-author of the German-language textbook *Internationale Politik und Governance in der Arktis: Eine Einführung* (Springer 2018) and co-editor of *Governing Arctic Change: Global Perspectives* (Palgrave Macmillan 2017).

Sveinung Løset is professor of Arctic Marine Technology at the Norwegian University of Science and Technology (NTNU) since 1995. He is Adjunct Professor at the University Centre in Svalbard (UNIS), and in 2005, he was appointed honorary doctor of St. Petersburg State Polytechnical University, St. Petersburg, Russia. Currently Dr Løset is dean of research and innovation at the Faculty of Engineering, NTNU. Professor Løset has been a pioneer within Arctic Marine Engineering. He established this research field in Norway.

Wenjun Lu received a joint MSc degree in Coastal and Marine Engineering from NTNU (Norway), TU Delft (the Netherlands) and University of Southampton (UK) in 2010; and then in 2014 the PhD degree in Arctic Technology from NTNU. His thesis was awarded the "Chorafas foundation prize" for the best doctoral thesis of NTNU for that year. Since 2015, he has been engaged in the development of SAMS, especially the analytical ice fracture module.

Raed Lubbad is Associate Professor at NTNU. He holds an MSc Degree in Coastal Engineering and PhD degree in Arctic Marine Civil Engineering from NTNU. He has a broad knowledge and a wide range of experience related to engineering challenges in the Arctic. Dr Lubbad has extensive experience with leading research projects and he has good experience with tech start-ups (he is a co-founder and the managing director of the NTNU spin-off company: ArcISo).

Andrew Mahoney is a Research Associate Professor of Geophysics at the Geophysical Institute at the University of Alaska Fairbanks. His research focuses on the physics of sea ice and its relationship to the climate, ecology and human systems of the Polar Regions.

Alexandra Middleton is a researcher with a PhD in Economics and Business Administration from the University of Oulu, Finland. Her research has focused on sustainability, corporate social responsibility reporting and Arctic business development. Alexandra has published scientific articles on climate change accountability in the Arctic, Arctic sustainable business development, demographics, human capital, innovations and connectivity solutions in the Arctic. As part of her Arctic work, Alexandra has been a Fellow in Science In/for Diplomacy EU funded programme. She is actively engaged in science diplomacy to raise awareness of the Arctic economic, social and environmental challenges.

Bjørn Inge Mo is chair of the Troms and Finnmark County Government in Norway, representing the Labour Party. Mr Mo was deputy mayor for 4 years before serving as mayor for 12 years in the Municipality of Kåfjord. In 2017 he was elected to the Sami Parliament and in 2019 to the Troms and Finnmark County Parliament. In addition, Mo has been leader of the Troms Labour Party, member of the Labour Party's national board as well as member of the party's executive board for 4 years. Mr Mo has served as a board member of the Council of the Torne Valley for 17 years and chairing the Council for 2 years.

Vadim Mokievsky (PhD, D.Sci) is a Chief scientist, head of the Laboratory of Coastal benthic ecology in P.P. Shirshov Institute of Oceanology, Moscow, Russia. Zoologist and ecologist. Expert in taxonomy of marine nematodes. Primary interests in ecology dealing with dynamics of natural communities and ecosystems with special reference to benthic communities of Arctic seas. Leader or participant of a number of expeditions and research cruises in coastal zone and high seas, mainly of the Northern hemisphere.

Annika E. Nilsson is researcher at KTH Royal Institute of Technology, Sweden (Division of History of Science, Technology and Environment) and professor-II at the Nordland Research Institute, Norway. Her research focuses on the politics of Arctic change, environmental governance and communication at the science-policy interface. She is currently engaged in the Nordic Centre of Excellence 'Resource Extraction and Sustainable Arctic Communities' where she looks at participatory scenario methods and indicators for sustainable development as potential tools for improving assessments of the long-term impacts of mining.

Annika Olsen has been the Mayor of Tórshavn Faroe Islands since 2017. She was elected in the City Council in Tórshavn for the People's Party at the age of 28. Since first elected she has been Minister of the Interior and Minister of Social Affairs in the Faroe Islands. Annika Olsen was a Member of Parliament before becoming Mayor of Tórshavn. Between 2011 and 2015 she was the Deputy Prime Minister of the Faroe Islands. Annika Olsen holds a master's degree in Danish and Religion from the University of Southern Denmark. She has also been a university teacher as well as teaching at upper secondary schools.

Alexander Osadchiev is a senior researcher in Shirshov Institute of Oceanology, Moscow, Russia. His research focuses on coastal oceanography with the special emphasis on river plumes and estuarine processes. His main results describe, first, structure of large-scale river plumes formed by the Great Siberian Rivers in the Arctic Ocean and, second, structure, dynamics and variability of small river plumes in many World's coastal areas.

Anders Oskal Executive Director International Centre for Reindeer Husbandry (ICR) is the Secretary General of the Association of World Reindeer Herders (WRH) and the Executive Director of the International Centre for Reindeer Husbandry. Oskal is a reindeer herding Sámi from northern Norway, with a master's degree in Business. Throughout his career, Oskal has worked with reindeer herding, Indigenous issues, policy, innovation and business development. Oskal represents WRH in the Arctic Council, is a Member of the Arctic Economic Council and was a co-author of the 5th IPCC report. He is currently leading the Arctic Council EALLU project, focusing on traditional Indigenous food knowledge as a source of innovation, value added and, in a wider sense, adaptation to Arctic change.

Ole Rasmus Øvretveit is the director of Initiative West, a think tank focusing on the sustainable ocean economy, societal growth and the green transition operating from the west coast of Norway. Prior to his current position Øvretveit was Director of Science to Policy for SDGs at the university of Bergen and was the director of Arctic Frontiers for 8 years. Mr Øvretveit has an MA in Comparative Politics from the University of Bergen. He has served on several boards and committees. Mr Øvretveit has broad insight in ocean issues and the connection between science, business and politics.

Jay-Kwon James Park is a former graduate researcher and a teaching assistant at the University of Washington, who served as an Arctic Research Fellow at the International Policy Institute. He participated in Korea Arctic Academy (2016–17), Arctic Council's Ministerial Meeting (2017), and Korea Arctic Partnership Week (2017) as a participant and a presenter.

Anita L. Parlow Principal A.L. Parlow & Associates, LLC, Anita Parlow, Esq., MSt. Oxford University is a recent Fulbright Scholar to Iceland, was advisor to the

Harvard-MIT Arctic Fisheries Project and Founding Team Lead for the D.C.-based Woodrow Wilson International Center for Scholars' Program on the Polar Code. Parlow, who has spoken in symposia in the United States, Canada, Europe and China on Arctic issues, has authored numerous articles on Arctic shipping, energy and infrastructure and subsistence topics. She is currently consultant to the Port of Nome, Alaska.

Alexander Pilyasov Doctor of Economic Geography (1995), Professor in Regional Economics (1998), Professor at Lomonosov Moscow State University, General Director of the Autonomous Non-commercial Organization "Institute of Regional Consulting". Major research themes: 1) Economic development of the Russian and Circumpolar North and Arctic; 2) Regional science as the interdisciplinary field (European experience of regional economic development); 3) Innovative and small business development in the Russian regions. Author of more than 175 scientific published papers, including 41 monographs.

Mikhail Pogodaev was born on 7 June 1978 in Topolinoe, Sakha Republic (Yakutia). In 2001, he graduated from Saint-Petersburg State University of Economics and Finance with a Master's degree in Economics. In 2007, at the same university, he defended his PhD dissertation in Economics. He was the Senior Advisor for Economic and International Relations at the Permanent Representation of Sakha Republic under the President of the Russian Federation from 2005 to 2014. He served as Chair of the World Reindeer Herders Association from 2009 to 2019 and Executive Director of the Northern Forum from 2015 to 2019. Since 2019, Dr. Pogodaev is the Vice-minister for Arctic Development and Indigenous Peoples Affairs of the Sakha Republic.

Rolf Rødven (PhD, MBA) is Executive Secretary of the Arctic Monitoring and Assessment Program (AMAP). AMAP is a working group to the Arctic Council, responsible for monitoring and assessing the state of the Arctic region with respect to pollution and climate change issues, and their impacts on ecosystems and human health. Rødven holds a PhD in the Northern Populations and Ecosystems programme at UiT – the Arctic university of Norway, focusing of ecology of reindeer, as well as an MBA in strategic leadership and finance. He is author on several scientific papers on Arctic ecology often related to impacts of climatic change.

Hide Sakaguchi is a President of the Ocean Policy Research Institute, the Sasakawa Peace Foundation (OPRI-SPF). Prior to OPRI-SPF, he served as Executive Director for Research at JAMSTEC from 2018 to 2021. He joined JAMSTEC in 2003 and his roles included Group Leader and Program Director of the Institute for Research on Earth Evolution, Director of the Department of Mathematical Science and Advanced Technology, and Assistant Executive Director. Prior to JAMSTEC, he served as Research Associate at Kobe University; Senior Research Scientist at the Division of Exploration and Mining of the Commonwealth Scientific and Industrial Research

Organisation of Australia; and Adjunct Professor at the Earthquake Research Institute of the University of Tokyo. He received a PhD from the Graduate School of Agriculture, Kyoto University in 1995.

Raquel Soto Sanchez Scholar of the National Council of Science and Technology (CONACYT) of Mexico, conducting her doctoral research on the protection and preservation of the marine environment under the visionary umbrella of the "Common Heritage of Mankind" with applications of science diplomacy. She is author of the book "The right of innocent passage in Mexico" (in the Spanish) published in 2015 by the Legal Research Institute of her Alma Mater the National Autonomous University of Mexico (UNAM). Raquel has a high sense of learnability, multidisciplinary research experience, creative problem-solving capacity with professionals of different backgrounds in multicultural international scenarios, and excellent interpersonal skills.

Geir Seljeseth is the Head of Europe Office of the Norwegian trade union "Industri Energi" and located in Brussels. Mr Seljeseth is born, raised and have lived in the Arctic for most of his life. For many years he was a journalist in Nordlys, the biggest newspaper in North-Norway. He also has spent some years as a journalist in Moscow as a correspondent for Norway's biggest newspaper Aftenposten. Geir Seljeseth was for 8 years regional manager for Norwegian Oil & Gas in Northern Norway.

Igor Semiletov is a Research Professor, University of Alaska Fairbanks and a Head of the Laboratory of Arctic Research of the V. I. Iliyechev Pacific Oceanology Institute of the Far Eastern Branch of the Russian Academy of Sciences (Russia). His main research interests are Polar regions, biogeochemistry, Arctic Ocean, carbon cycle, permafrost degradation, greenhouse gas flows, climate change. Expedition lead of many research expeditions in the Russian Arctic.

Mike Sfraga is the founding Director of the Polar Institute and serves as the Director of the Global Risk and Resilience Program at the Woodrow Wilson International Center for Scholars. His work focuses on the changing geography of the Arctic and Antarctic landscapes, Arctic policy, and the impacts and implications of a changing climate on political, social, economic, environmental, and security regimes in the Arctic. Sfraga served as distinguished co-lead scholar for the Fulbright Arctic Initiative from 2015 to 2017 and from 2017 to 2019. He serves as the co-director of the University of the Arctic's Institute for Arctic Policy previously served in a number of academic, administrative and executive positions. Sfraga earned a PhD in geography and northern studies from the University of Alaska Fairbanks. Dr. Sfraga has been appointed by President Biden as Chair of the United States Arctic Research Commission from 2021–2024.

Nikolay Shabalin is a marine ecologist, graduated from the Lomonosov Moscow State University, department of invertebrate zoology in 2006 and specializing in the ecology of marine macrobenthos in the Arctic. Nikolay is an executive director of the Lomonosov Moscow State University Marine Research Center from 2014. Key interests: Arctic marine ecosystems functioning and biodiversity, technologies of collecting and analysing environmental data, marine conservation. Nikolay is an author of over 40 patents and 10 publications.

Roger Skjetne has BSc and MSc degrees from University of California Santa Barbara and PhD from the Norwegian University of Science and Technology (NTNU) within control engineering, with Exxon Mobil prize for best PhD thesis in 2005. He is a certified electrician, earlier working for Aker Elektro on structures for North Sea. In 2004–2009 he worked in Marine Cybernetics on Hardware-In-the-Loop testing of marine control systems. From 2009 he has held the Kongsberg Maritime Professor chair in Marine Cybernetics at NTNU Department of Marine Technology.

Yeona Son is a researcher at the Korea Legislation Research Institute. She has a bachelor's degree in Law and a master's in Development Studies. In particular, she has keen interests in human rights issues, sustainable development, legislation and public policy. In this regard, she had internships at the Vandeventer Black LLP (Virginia, USA), the United Nations Global Compact(Korea Network) and the BKL Dongcheon Foundation. She was honoured to be part of the Arctic Frontiers 2020 to share bright thoughts and broaden the horizon of knowledge.

Ine Eriksen Søreide has been Minister of Foreign Affairs for Norway since 2017. Prior to that she was Minister of Defence since 2013. A member of the Conservative Party, she was elected in 2005 as a member of the Norwegian Parliament, *Stortinget*, for Oslo.

Vassily Spiridonov (PhD, Dr.habil) studied in Kazan and Moscow Universities, worked in All Union Research Institute of Fishery and Oceanography, Zoological Museum of Moscow University, as a visiting scientist in the Alfred Wegener Institute and Senckenberg Institute (Germany). In 1999–2004 organized Marine Programme of WWF Russia. Since 2005 working in Shirshov Institute of Oceanology with focus on Arctic and Antarctic ecology, biological invasions and crustacean research, having been consultant to United Nations Development Programme, Food and Agriculture Organization, Convention on Biological Diversity and the Marine Stewardship Council.

Sam Tan is currently Minister of State for Ministry of Foreign Affairs and Ministry of Social and Family Development of Singapore. He also serves as Chairman of Government Feedback Unit REACH since 2015. He has previously served as Minister of State for the Prime Minister's Office for Culture, Community and Youth and the Ministry of Manpower. In 2006, Mr Tan was elected Member of Parliament and subsequently appointed Parliamentary Secretary and later Senior Parliamentary Secretary for various portfolios and served as Senior Parliamentary Secretary for various ministries. He received the Public Service Medal (PBM) in

2002. Mr Tan graduated with a Bachelor of Arts (Hons) from the National University of Singapore in 1983.

Sarah Trainor is an Associate Professor of Sustainability Science in the Department of Natural Resources and Environment and at the International Arctic Research Center both of the University of Alaska, Fairbanks. She co-directs the Alaska Center for Climate Assessment and Policy, a NOAA RISA program, and is Director of the Alaska Fire Science Consortium, a member of the Fire Exchange Network of the Joint Fire Science Program. Her research centres on the process and evaluation of boundary spanning and knowledge co-production as well as on elements of building capacity in Indigenous communities for climate adaptation.

Alexander N. Vylegzhanin Doctor of Law, Professor, is a Head of the Program of International Law, Moscow State Institute of International Relations (*MGIMO-University*) and Editor-in-Chief of the *Moscow Journal of International Law*. He is elected as a Vice-president of the Russian Association of International Law. He is elected as a Member of the Committee on the Arctic and Antarctic of the Council of Federation (the upper chamber of the Russian Parliament). He has publicized a wide array of books and papers on international law, mainly in Russian.

Simon Wilson is an environmental scientist with wide experience in science project management including the production of regional science-based assessments of pollution and climate-related issues. His work incorporates coordination of international monitoring activities, management and interpretation of environmental data, data visualization and communication of scientific results to policymakers. He is Deputy Executive Secretary to the Arctic Council's Arctic Monitoring and Assessment Programme (AMAP) and previously worked at the International Council for the Exploration of the Sea (ICES). He holds a PhD in environmental chemistry from Lancaster University, United Kingdom.

Evgeniy Yakushev (PhD, D.Sci) is a Senior Researcher in the Norwegian Institute for Water Research (NIVA) with an experience in chemical oceanography and biogeochemical modelling. His field of activity concerns the implementation of field and modelling techniques to investigate how biogeochemical cycles, chemical and biological processes affect nutrients and redox metals, with particular emphasis on oxygen depletion processes and anoxia. His current interests are connected with benthic-pelagic coupling modelling, Arctic ocean acidification and microplastics.

Oran R. Young is Professor Emeritus at the Bren School of Environmental Science and Management at the University of California (Santa Barbara). He has devoted his career to analysing the roles institutions play in meeting needs for governance, with applications to the Arctic going back to the 1970s.

Nadezhda Zamyatina PhD in geography (2001), leading researcher of the Faculty of Geography of Lomonosov Moscow State University, Deputy General Director of ANO (Institute for Regional Consulting) since 2017. Specialist in human geography, in particular of Arctic and Northern regions, Arctic urbanization, economic and cultural development of frontier regions. Deputy director of project groups developing strategies for socio-economic development of a number of Russian cities and regions. Author of over 180 scientific articles, a series of online journalistic materials on the Arctic, co-author of 14 books.

Acronyms

AAC	Arctic Athabaskan Council
AACA	Adaptation Actions for a Changing Arctic
ABNJ	Areas Beyond National Jurisdiction
AC	Arctic Council
ACAP	Arctic Contaminants Action Programme – Working Group (Arctic Council)
ACGF	Arctic Coast Guard Forum
ACIA	Arctic Climate Impact Assessment
ACOM	Advisory Committee
ADAC	Arctic Domain Awareness Center
AEC	Arctic Economic Council
AEPS	Arctic Environmental Protection Strategy
AGDC	Alaska Gasline Development Corporation
AHDR	Arctic Human Development Report
AI	Artificial intelligence
AIA	Aleutian International Association
AIP	Arctic Investment Protocol
AIS	Automatic Identification System
AMAP	Arctic Monitoring and Assessment Program – Working Group (Arctic Council)
AMSA	Arctic Marine Shipping Assessments
AMSP	Arctic Marine Strategic Plan
ANO	Institute for Regional Consulting at Lomonosov Moscow State University
ANR	Agence Nationale de la Recherche
ANTHC	Alaska Native Tribal Health Consortium
AO	autonomous okrug (Yamal Nenets)
K-AOOS	Korea-Arctic Ocean Observing System
ArcISo	Arctic Integrated Solutions
ASTD	Arctic Ship Traffic Database

BBNJ	Biodiversity Beyond National Jurisdiction
BIN	Business Index North
BIT	bilateral investment treaty
BKL	Berg Kaprow Lewis Foundation
BPA	bisphenol A
BTS	Berlin Graduate School for Transnational Studies
CAFF	Conservation of Arctic Flora and Fauna – Working Group (Arctic Council)
CAO	Central Arctic Ocean
CAOF	Central Arctic Ocean Fisheries
CAOFA	Central Arctic Ocean Fisheries Agreement
CBD	Convention on Biological Diversity
CETA	Comprehensive and Economic Trade Agreement
CIC	China Investment Corporation
CLEO	Circumpolar Local Environmental Observer
CLRTAP	Convention on Long-Range Transboundary Air Pollution
CMTS	U.S. Committee on the Marine Transportation System
CNPC	China National Petroleum Corporation
CNRS	Centre national de la recherche scientifique
CONACYT	Consejo Nacional de Ciencia y Tecnología
CSR	corporate social responsibility
DEM	discrete element method
DIKW	Data, Information and Knowledge to Wisdom
DUI	Doing, Using and Interacting
DWT	deadweight tonnage
EA EG	Ecosystem Approach Expert Group
EBM	ecosystem-based management
ECCHR	European Center for Constitutional and Human Rights
EEA	European Economic Area
EEZ	exclusive economic zone
EBM EG	Expert Group on Ecosystem-Based Management
EGBCM	Expert Group on Black Carbon and Methane
EITI	Extractive Industries Transparency Initiative
EPPR	Emergency Prevention, Preparedness and Response – Working Group (Arctic Council)
EU	European Union
FAIR	Findability, Accessibility, Interoperability, Reusability
FAO	Food and Agriculture Organization of the United Nations
FDI	foreign direct investment
FIRRMA	Foreign Investment Risk Review Modernization Act
FOEI	Friends of the Earth International
FTA	Free Trade Agreement
GA	General Assembly
GAC	Global Agenda Council

GACA	Global Agenda Council on the Arctic
GCI	Gwich'in Council International
GDP	gross domestic product
GESAMP	Group of Experts on the Scientific Aspects of Marine environmental Protection
GIS	Geographic Information Systems
GMT	Greenwich Mean Time
GPS	Global Positioning System
GRI	Global Reporting Initiative
GRID	Global Resource Information Database
HBC	Hudson Bay Company
HELCOM	Baltic Marine Environment Protection Commission (Helsinki Commission)
HF	high frequency
HIOMAS	High-Resolution Ice-Ocean Modeling and Assimilation System
HSE	health, safety and environment
HSVA	Hamburg Ship Model Basin
IASC	International Arctic Science Committee
ICC	Inuit Circumpolar Council
ICER	Integrative and Collaborative Education and Research
ICES	International Council for the Exploration of the Sea
ICR	International Centre for Reindeer Husbandry
ICT	Information and Communications Technology
IEA	Integrated Ecosystem Assessment
IFAW	International Fund for Animal Welfare
IFC	International Finance Corporation
IGO	international intergovernmental organization
IIA	International Investment Agreement
IIS	Institutional Investment Services
IK	Indigenous Knowledge
ILO	International Labour Organization
IM	Ice management
IMO	International Maritime Organization
IMS	industrial methylated solution
IOC	Intergovernmental Oceanographic Commission
IPCC	Intergovernmental Panel on Climate Change
IPY	International Polar Years
IRAB	Investment Review Advisory Board
IS	innovation systems
ISO	International Organization for Standardization
ITLOS	International Tribunal for the Law of the Sea
IUCN	International Union for the Conservation of Nature
IWGIA	International Work Group for Indigenous Affairs
JAMSTEC	Japan Agency for Marine-Earth Science and Technology

JAPEX	Japan Petroleum Exploration Co., Ltd
KASCO	Korea Arctic Science Council
KMA	Korea Meteorological Administration
KOPRI	Korea Polar Research Institute
KPI	key performance indicator
LKL	last known location
LME	large marine ecosystem
LNG	liquefied natural gas
LOSC	United Nations Convention on the Law of the Sea
MARPOL	International Convention for the Prevention of Pollution from Ships
MGIMO	Moscow State Institute for International Relations
MMSI	Maritime Mobile Service Identity
MOE	Ministry of Environment
MOF	Ministry of Oceans and Fisheries
MOFA	Ministry of Foreign Affairs
MOLIT	Ministry of Land, Infrastructure and Transport
MOPP	Agreement on Cooperation on Marine Oil Pollution Preparedness and Response in the Arctic
MOTIE	Ministry of Trade, Industry and Energy
MOU	Memorandum of Understanding
MPA	Marine Protected Areas
MSC	Maritime Safety Committee
MSIP	Ministry of Science, ICT and Future Planning
MSY	maximum sustainable yield
NACGF	North Atlantic Coast Guard Forum
NAFTA	North American Free Trade Agreement
NAMMCO	North Atlantic Marine Mammals Commission
NATO	North Atlantic Treaty Organization
NAVCEN	Navigation Center
NDC	Nationally Determined Contribution
NDEM	non-smooth discrete element method
NEAFC	North-East Atlantic Fisheries Commission
NEFU	North-Eastern Federal University
NEP	Northeast Passage
NGO	non-governmental organization
NIVA	Norwegian Institute for Water Research
NLFEA	Nonlinear Finite Element Analysis
NOAA	National Oceanic and Atmospheric Administration
NSF	National Science Foundation
NSFC	National Science Foundation of China
NSIDC	National Snow and Ice Data Center
NSR	Northern Sea Route
NTNU	Norwegian University of Science and Technology
NWP	Northwest Passage

OECD Organisation for Economic Co-operation and Development
OGA Oil and Gas Activities
OPP Office of Polar Programs
OPRI-SPF Ocean Policy Research Institute, the Sasakawa Peace Foundation
OSPAR North Atlantic Marine Environment Organization
PAME WG Protection of the Arctic Marine Environment Working Group
PBM Public Service Medal
PICES North Pacific Marine Science Organization
PINRO Russian Federal Research Institute of Fisheries and Oceanography
PNAS Proceedings of the National Academy of Sciences
PWP Plastic Waste Partnership (Basel Convention on the Control of Transboundary Movements of Hazardous Wastes and Their Disposal)
RAIPON Russian Association of Indigenous Peoples of the North
RCN Research Council of Norway
REACH Reaching Everyone for Active Citizenry@Home (Department under the Ministry of Communications and Information of Singapore)
RFBR Russian Foundation for Basic Research
RFMO Regional fisheries management organisation
RISA Regional Integrated Sciences and Assessments program (National Oceanographic and Atmospheric Administration of the United States)
SADC Southern African Development Community
SAMS Simulator for Arctic Marine Structures
SAO Senior Arctic Officials
SAON Sustaining Arctic Observations Network
SAPEA Science Advice for Policy by European Academies
SAR search and rescue
SAROPS Search and Rescue Optimal Planning System
SCUBA Self-Contained Underwater Breathing Apparatus
UNTS United Nations Treaty Series
UN SC United Nations Security Council
SCAR Scientific Committee on Antarctic Research
SCICOM Science Committee
SDEM smooth discrete element method
SDG Sustainable Development Goal
SDWG Sustainable Development Working Group (Arctic Council)
SAMCoT Sustainable Arctic Marine and Coastal Technology
SMM SAO-based Marine Mechanism
SOE state-owned enterprises
SOLAS Safety of Life at Sea Convention
OPRI-SPF Ocean Policy Research Institute, the Sasakawa Peace Foundation
SROCC Special Report on Oceans and the Cryosphere in a Changing Climate
SSHRC Social Sciences and Humanities Research Council of Canada

STCW	Standards of Training, Certification and Watchkeeping
SUA	Suppression of Unlawful Acts
SWIPA	Snow, Water, Ice and Permafrost in the Arctic
TAC	total allowable catch
TIPS	Trade and Industrial Policy Strategies
TLK	Traditional and Local Knowledge
TTNU	territories of traditional nature use
UAF	University of Alaska Fairbanks
UCSB	University of California Santa Barbara
UN	United Nations
UNAM	National Autonomous University of Mexico
UNCBD	United Nations Convention on Biological Diversity
UNCED	United Nations Conference on Environment and Development
UNCLOS	United Nations Convention on the Law of the Sea
UNCTAD	United Nations Conference on Trade and Development
UNDOS	United Nations Decade on Ocean Science for Sustainable Development
UNEP	United Nations Environment Programme
UNESCO	United Nations Educational, Scientific and Cultural Organization
UNFCCC	United Nations Framework Convention on Climate Change
UNFSA	United Nations Fish Stocks Agreement
UNIS	University Centre in Svalbard
UNITAR	United Nations Institute for Training and Research
USCG	U.S. Coast Guard
USGS	United States Geological Survey
USSR	Union of Soviet Socialist Republics
WEF	World Economic Forum
WGICA	Working Group on Integrated Ecosystem Assessment for the Central Arctic Ocean
WRH	World Reindeer Herders
WWF	World Wide Fund for Nature
XFEM	eXtended Finite Element Method

Part I
Introduction

Chapter 1
(Research): Introduction: Building Common Interests with Informed Decisionmaking for Sustainability

Paul Arthur Berkman, Oran R. Young, Alexander N. Vylegzhanin, David A. Balton, and Ole Rasmus Øvretveit

Abstract This chapter introduces conceptual threads woven within and between the chapters, applying the book title as the organizing framework and the Arctic as a case study with global relevance. The book focuses on science diplomacy and its engine of informed decisionmaking together with the theory, methods and skills introduced in view of BUILDING COMMON INTERESTS. As an exemplar, the ARCTIC OCEAN highlights holistic (international, interdisciplinary and inclusive) integration with marine and surrounding terrestrial systems interacting with humanity at local-global levels, especially in relation to Earth's changing climate. The importance of this book is WITH GLOBAL INCLUSION, recognizing challenges to engage diverse stakeholders, rightsholders and other actors, as illustrated with special respect for the Indigenous

---

P. A. Berkman (✉)
Science Diplomacy Center, EvREsearch LTD, Falmouth, MA, USA

Science Diplomacy Center, MGIMO University, Moscow, Russian Federation

United Nations Institute for Training and Research (UNITAR), Geneva, Switzerland

Program on Negotiation, Harvard Law School, Harvard University, Cambridge, MA, USA
e-mail: paul.berkman@unitar.org; pberkman@law.harvard.edu

O. R. Young
Bren School of Environmental Science & Management, University of California Santa Barbara, Santa Barbara, CA, USA
e-mail: oran.young@gmail.com

A. N. Vylegzhanin
International Law School, MGIMO University, Moscow, Russian Federation
e-mail: danilalvy@mail.ru

D. A. Balton
Polar Institute, Wilson Center and U.S. Arctic Steering Committee, Washington, DC, USA
e-mail: davebalton@comcast.net

O. R. Øvretveit
Initiative West Bergen, Bergen, Norway
e-mail: ole@initiativvest.no

P. A. Berkman et al. (eds.), *Building Common Interests in the Arctic Ocean with Global Inclusion, Volume 2*, Informed Decisionmaking for Sustainability,
https://doi.org/10.1007/978-3-030-89312-5_1

peoples who have inhabited the Arctic for millennia with resilience across ice ages and past periods of global warming. The goal of this book on Building Common Interests in the Arctic Ocean with Global Inclusion (involving contributions from graduate students to foreign ministers at the *Arctic Frontiers* 2020 conference) is to help produce informed decisions that operate short-to-long term at local-global levels *for the benefit of all on Earth across generations.*

1.1 Building Common Interests

1.1.1 Science as the 'Study of Change'

We are living during a complex period in human history, reflecting our evolution as a **globally-interconnected civilization.**[1] The book series on Informed Decisionmaking for Sustainability is conceived to offer lessons that have **local-global** relevance *"for the benefit of all on Earth across generations"* (Berkman et al., 2017, 2020a; Berkman 2018, 2019, 2020a, b; Young et al., 2020a, b).

This volume on Building Common Interests in the Arctic Ocean with Global Inclusion is second in the book series and linked to others with **holistic** (international, interdisciplinary and inclusive) integration. On this journey, Volume 1. Governing Arctic Seas: Regional Lessons from the Bering Strait and Barents Sea introduced the concept of **'ecopolitical'** to elevate the focus on our homes ('eco') above the geopolitical fray of nations (Table 1.1).

In the book series, the first three volumes are a trilogy with the Arctic Ocean as a global case-study to elaborate decisionmaking with holistic integration (Fig. 1.1). The Arctic Ocean is international unlike the surrounding national territories on land, involving areas within and beyond sovereign jurisdictions with impacts, issues and resources in constant motion. Dynamics of the Arctic Ocean system as an integral part of our globally-interconnected civilization are a portrait of change with interdisciplinary analogues elsewhere on Earth in view of diverse time and space scales, revealed with natural and social sciences along with Indigenous knowledge. Importantly, the Arctic Ocean represents an inclusive journey of common-interest building, considering most immediately the eight Arctic States and six Indigenous Peoples Organizations that established the Arctic Council in 1996 as high-level forum (Ottawa Declaration 1996), progressively engaging non-Arctic States and other observers.

This book connects to the third volume (Pan-Arctic Implementation of Coupled Governance and Infrastructure) with the goal of contributing to decisionmaking for sustainable development in the Arctic, where progress is measured across generations on a Pan-Arctic scale. The intergenerational feature of sustainable development underlies the quest to operate short-to-long term inclusively. Moreover, like the Earth system with its local-global connections, Pan-Arctic progress involves

[1] **Highlighted terms** in Chapter 1 involve definitions to avoid jargon with concepts that are threaded through this book series.

Table 1.1 Holistic characteristics of ecopolitical regions with informed decisionmaking to couple governance mechanisms and built infrastructure for sustainable development[a]

Ecopolitical region characteristics	Bering Strait Region (BeSR)	Barents Sea Region (BaSR)
International	Russian Federation and United States	Russian Federation and Norway
Local-global connections	Maritime ship traffic and marine living resources	Maritime ship traffic, marine living resources and mineral resources
Cultural and historical heritage	Small predominantly Indigenous communities	Large populations with settler majorities and close links to national governments
System dimensions for comparisons over time and space	Regions defined explicitly within polygon boundaries that are mapped with geographic information systems	
Operating across a 'Continuum of urgencies'	Informed decisionmaking at security to sustainability time scales with skills, methods and theory that are being applied, trained and refined	
Common-interest building	Skill to facilitate inclusive dialogues among allies and adversaries alike	
Holistic integration	Skill to be international, interdisciplinary and inclusive	

[a]This table is adapted from the concluding chapter in Governing Arctic Seas: Regional Lessons from the Bering Strait and Barents Sea. Volume 1. Informed Decisionmaking for Sustainability (Young et al., 2020a)

cooperation and coordination among stakeholders, rightsholders and other actors within as well as between regions inclusively (Fig. 1.1). Inclusion is the biggest challenge, considering the temporal and spatial scope for sustainability from diverse perspectives that ultimately translate into actions. With contributions from the 2020 *Arctic Frontiers* conference in Tromsø, Norway (Arctic Frontiers 2020a, b; Steinveg 2020), this book seeks to be inclusive, exploring cooperation and coordination to achieve Arctic sustainability from diverse perspectives with global relevance (please see the Conclusions, Chapter 35).

Organization of this introductory chapter corresponds with phrases in the book title – Building Common Interests / In the Arctic Ocean / With Global Inclusion – to elaborate the theory, methods and skills that are intertwined across the contributions (please see the *Preface*). What do ecology, ecosystems and economics have in common (Table 1.1)? Practical answers to such questions emerge with science, which starts with curiosity and inquiry, elaborated into ways of knowing. With this objective, the natural sciences, social sciences and Indigenous knowledge all reveal patterns, trends and processes (albeit with different methodologies) that become the bases for decisions across our globally-interconnected civilization (Fig. 1.2a-d).

As an umbrella concept across knowledge systems, **science** can be characterized broadly as the 'study of change' (symbolized by the Greek letter delta Δ) to be holistic (Berkman et al., 2020; Berkman, 2020a, b), considering biophysical and socioeconomic factors as well as their intersections. This umbrella characterization of science builds on Socratic methods of learning that are stimulated by questions with the human quest for knowledge to understand our world and its relative motions

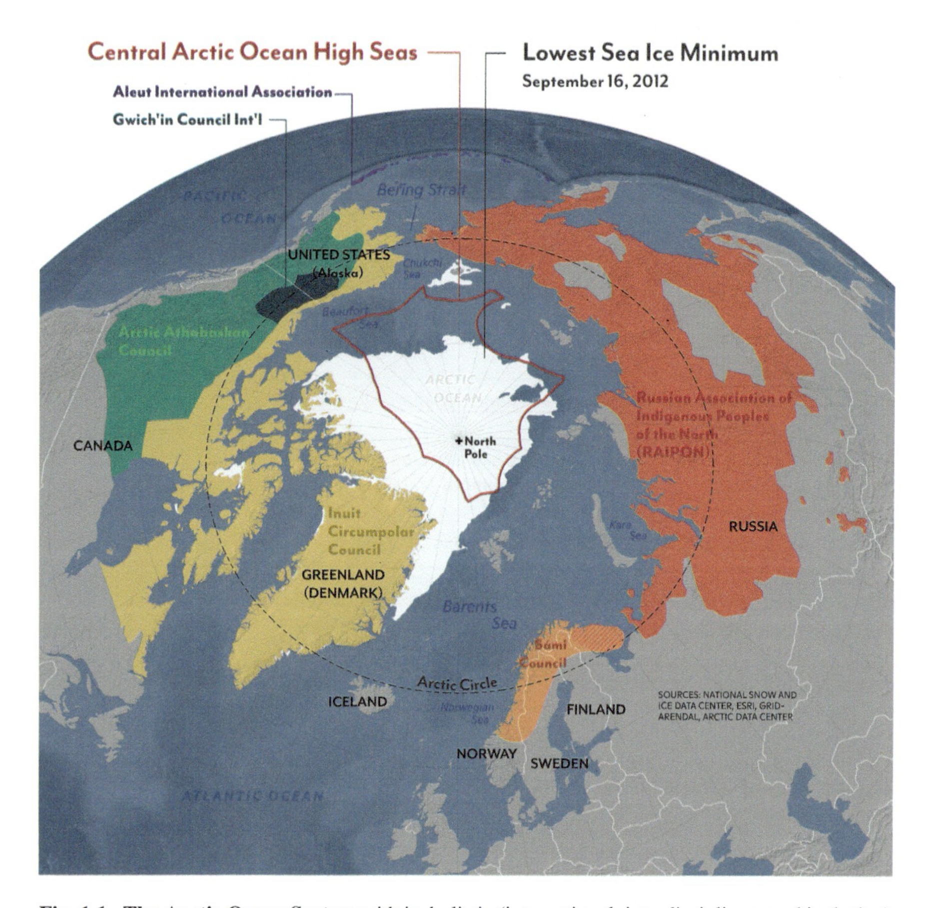

Fig. 1.1 The Arctic Ocean System with its holistic (international, interdisciplinary and inclusive) dimensions surrounding the North Pole is being transformed with climate warming as its surface boundary changes from permanent sea-ice cover to seasonally open water between the North Pacific and North Atlantic, as illustrated with the lowest sea-ice minimum observed during the satellite era (NASA 2012). The superimposed legal boundary of the high seas in the Central Arctic Ocean (CAO) illustrates connections between biogeophysical and socioeconomic dynamics with the maritime Arctic associated with *"sustainable development and environmental protection"* as *"common Arctic issues"* among all signatories of the *Ottawa Declaration* (1996) that established the Arctic Council. To be inclusive, the eight Arctic states (north of the Arctic Circle) and six Indigenous Peoples Organizations as Permanent Participants in the Arctic Council are shown together. Mapping of the Indigenous Peoples' Organizations has been co-produced iteratively with feedback from the Indigenous Peoples' Secretariat (2021a, b) and in cooperation with GRID-Arendal (2021). The resulting polygon shape files have been deposited with the Arctic Data Center for open access (Fiske 2021)

(Lucretius 55 BCE) with implications for societal development at all levels (Ruffert & Steinecke, 2011). In this book and in the series on Informed Decisionmaking for Sustainability, science is illustrated as an unifying framework to operate with inclusion across time and space, especially in view of urgencies.

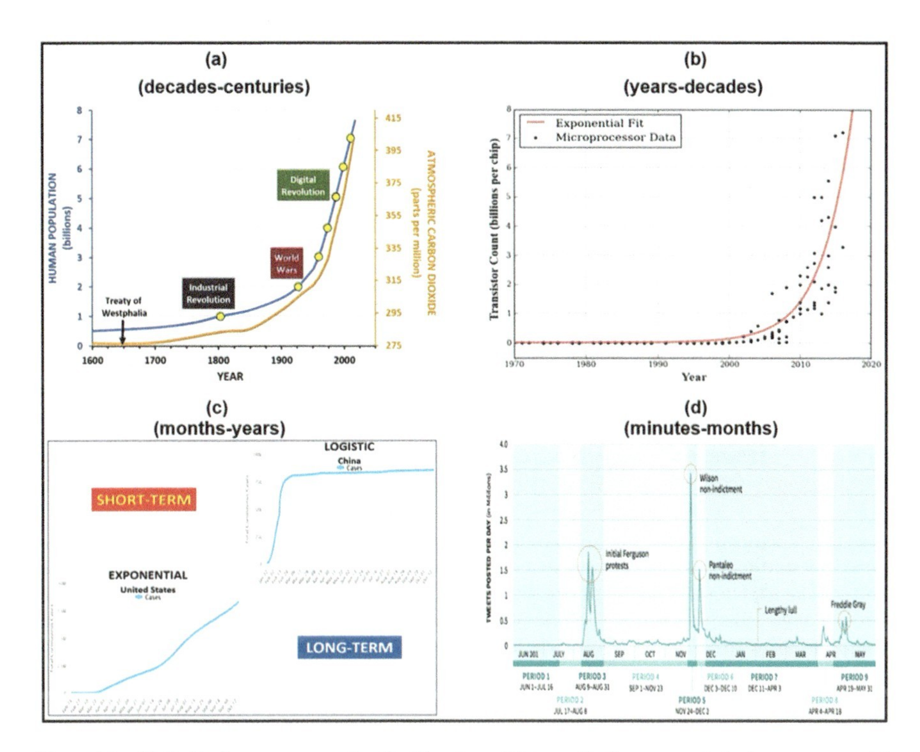

Fig. 1.2 Globally-Interconnected Civilization Times Scales revealed by exponential changes with: (**a**) Climate and human-population change over decades to centuries; (**b**) High-technology change over years to decades illustrated by "Moore's Law" with transistors on a chip; (**c**) Global pandemic change over months to years with COVID-19 cases accelerating across the Earth, illustrated by the United States (scale of 10,000,000–10^7) in contrast to China (scale of 100,000–10^5) through 12 October 2020, as recorded by *Worldometer*; and (**d**) Social-media change over minutes to months in relation to specific events, illustrated by 2014–2015 tweets about "Black Lives Matter", posted per day (in millions), as reported in *Mother Jones* on 13 March 2016. Adapted from Berkman et al. (2020) and Berkman (2020b)

Moreover, to be objective requires understanding about the dynamics of issues, impacts and resources in relation to defined systems, which is the reason the first volume in this book series focused on regional lessons. In this second volume, there is emphasis to further consider the temporal domain in view of the Arctic Ocean, especially to make **informed decisions**, operating across a **'continuum of urgencies'** (Vienna Dialogue Team, 2017; Berkman et al., 2017, 2020; Berkman, 2019, 2020a, b).

How can we study change to characterize a 'continuum of urgencies'? How can we place the present in context of the past and future? Answers to both of these questions are revealed with 'The Pandemic Lens' (Fig. 1.3), placing our world today in context across embedded time scales that all operate on a planetary scale with the common driver of our global human population (Fig. 1.2a-d). With science as the 'study of change,' we can describe as well as respond to the biophysical and socioeconomic dynamics that influence our home on Earth.

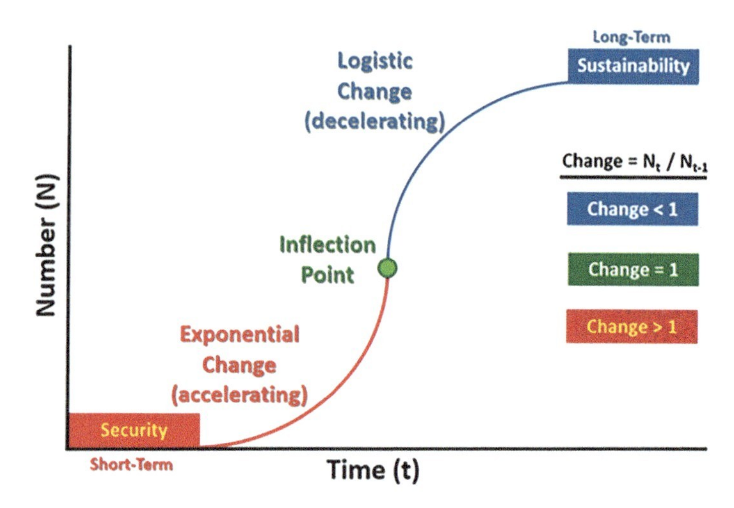

Fig. 1.3 'Pandemic Lens' for Sustainability, highlighting exponential change across an inflection point toward logistic (S-shaped, sigmoid) change, as described by numbers (N) changing per unit of time (t). Informed decisions operate across a "continuum of urgencies" (Fig. 1.4) – before-through-after inflection points to 'bend the curve' short-to-long term with scalability across embedded time scales of our globally-interconnected civilization (Figs. 1.2a-d). (Adapted from Berkman (2020a, b))

1.1.2 Operating Across a 'Continuum of Urgencies'

Working from first principles in view of humankind on a planetary scale, the shape of urgent change involves exponential rates at the time scales of minutes-months, month-years, years-decades and decades-centuries (Fig. 1.2a-d). Informed decisionmaking operates across these embedded time scales, addressing changes that impact our **sustainable development** (United Nations 1987, 2015), balancing environmental protection, economic prosperity and societal well-being at local-global levels with stability (Hardin 1968) as well as resilience (Berkes et al. 2000, 2008; Arctic Council 2016) in the face of change.

At the shortest global periods of minutes-months, the decisionmaking is largely reactive, especially when communications are contributed in view of self-interests without consideration of perspectives, drivers and consequences over time. The unfortunate outcomes of thinking short-term or long-term only are **uninformed decisions**, as illustrated in the United States during the global pandemic (Fig. 1.2c), pulling out of the World Health Organization with exponential change of COVID-19 infections and mortality continuing unchecked in the absence of foresight and leadership (Berkman 2020a, b).

Operating across a 'continuum of urgencies' (Fig. 1.4) to produce informed decisions involves short-to-long term thinking, distinguishing **exponential** and **logistic** rates of change (Krebs 1972) bounded by an **inflection point** (Figs. 1.2c and 1.3) – when the past, present and future converge with clarity about common interests. We are living during such a moment with the global pandemic making survival a common interest at local-global levels.

Fig. 1.4 Informed Decisionmaking as a scalable proposition operating across a 'continuum of urgencies,' illustrated for peoples, nations and our world from security to sustainability time scales. In parallel, there are negotiation strategies that contribute to the decisionmaking – short-term in view of conflicts to resolve and long-term in view of common interests to build – balancing issues, impacts and resources. (Adapted from Berkman et al. (2020) and Berkman (2020a, b))

The relevance of an inflection point is immediate and long-term, noting there will be a global inflection point with the COVID-19 pandemic with certainty, either because the plague runs its course or because we have vaccines with effective distribution channels around the Earth. In this moment, leadership involves setting expectations correctly. Such intervention to 'bend the curve' – which is a source of hope – is exhibited today among some nations over months-years during the global pandemic (1.2c), underscoring the imagination and capacity of humanity to address issues and impacts over longer time scales: across years-decades with advanced technologies (Fig. 1.2b) as well as decades-centuries with our global human population and the Earth's climate (Fig. 1.2a).

The challenge is to recognize the inflection points, which are few and far between, and then to capitalize on those rare moments as levers for transformation. This application of informed decisionmaking is scalable, as there are inflection points in each of our lives, sometimes together at local-global levels across different time scales (Figs. 1.2a-d). The **theory** of informed decisionmaking scales from an individual to humanity, like driving a car constantly adjusting to the immediacies on the left and right while maneuvering in view of future urgencies with red lights ahead and circumstances to consider in the rear.

Speaking to humanity with lessons, the COVID-19 pandemic is the *"most challenging crisis we have faced since the Second World War,"* as stated in March 2020 by the Secretary-General of the United Nations (Guterres 2020). The end of the Second World War in August 1945 was another global inflection point, educating future generations about how to operate across a 'continuum of urgencies' (Fig. 1.4) from:

Security Time Scales (mitigating risks of political, economic, cultural and environmental instabilities that are immediate); to

Sustainability Time Scales (balancing economic prosperity, environmental protection and societal well-being across generations).

The *Bretton Woods Conference* in New Hampshire in July 1944 produced a vision of a stronger international governance regime that included establishment of the *International Monetary Fund* and the *International Bank for Reconstruction and Development* that became the *World Bank* (Steil, 2013). The *United Nations Conference on International Organization* in San Francisco from April to June 1945, produced the *Charter of the United Nations and Statute of the International Court of Justice* (United Nations, 1945), symbolised for the ages with the California redwoods, where Franklin Delano Roosevelt, the '*chief architect of the United Nations, and apostle of lasting peace for all mankind', was memorialised in May 1945 (National Park Service,* 2020*).* The new international architecture created in the post-War years and supplemented by a burgeoning array of international institutions in the decades that followed has manifestly reduced the risk of another global conflict on the scale of the two World Wars. That said, rising nationalism and political polarization combined with advanced technologies has generated new challenges to peace and security around the globe.

At the level of peoples, nations and our world, the 'continuum of urgencies' extends from security to sustainability time scales (Vienna Dialogue Team, 2017). However, knowing the time span of a 'continuum of urgencies' is a research exercise unless there are methods and skills to apply with decisions that commonly involve negotiations, enabling actions that operate short-to-long term (Fig. 1.4).

1.1.3 Science Diplomacy to Negotiate Transformation

With informed decisionmaking before-through-after inflection points (Figs. 1.2, 1.3 and 1.4), the opportunity is to turn science fiction into science reality with inspiration and hope for humanity as in the case of travelling from the Earth to the Moon with progress across a century (Verne, 1865). In this quest, informed decisionmaking is the engine of **science diplomacy** as an holistic process to *"balance national interests and common interests for the benefit of all on Earth across generations"* (Berkman et al., 2011, 2017, 2020; Berkman, 2009, 2020a, b). However, balancing national interests first requires common interests, as reflected across the twentieth century, contrasting periods of conflict and cooperation (Fig. 1.5), which mirror the capacities of humankind to operate short-to-long term on a planetary scale (Fig. 1.4).

As a skill, common-interest building promotes cooperation among allies and adversaries alike, without the conflict that would persist otherwise. For example, throughout the Cold War, there was continuous cooperation between the United States and Soviet Union regarding both Antarctica and outer space. How was this continuous cooperation facilitated with these international spaces (Berkman, 2009) in the face of geopolitical confrontation among these superpowers everywhere else on Earth? What are the Cold War lessons of common-interest building in the Arctic today (Berkman, 2013, Nature, 2020)?

Both the Antarctic Treaty (1959) and Outer Space Treaty (1967) were built around the *"common interest of all mankind."* Asking what was the umbrella interest that enabled continuous cooperation between superpower adversaries in these international spaces during the Cold War, the answer is the same today: survival in the face of mutually assured destruction, which can happen quickly with global conflict or more slowly without planetary action, as required in the cases of Earth's climate (Fig. 1.2a) and human population (Erlich & Holdren, 1971;

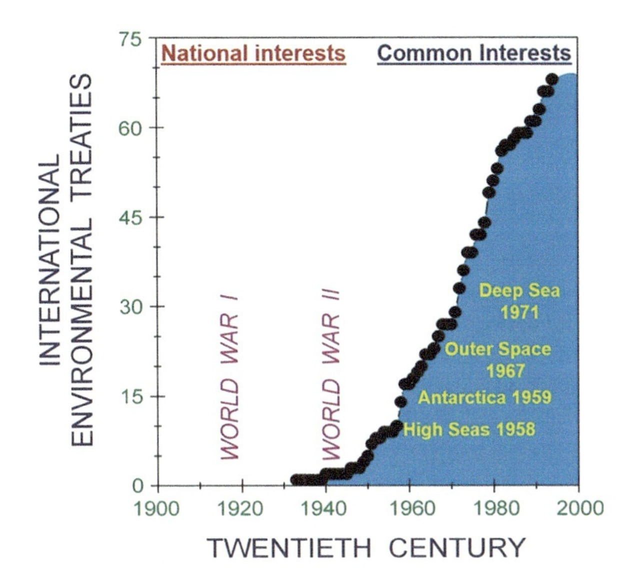

Fig. 1.5 Balancing National and Common Interests on a planetary scale during the twentieth century with international environmental treaties to address sustainability questions in our globally-interconnected civilization (Fig. 1.2a). (Adapted from Berkman (2002), including legal establishment of areas beyond national jurisdictions (yellow), which are international spaces (Kish, 1973, Berkman et al., 2011) to build common interests and minimize risks of conflict over jurisdictional boundaries across the Earth (Berkman, 2009))

Holdren, 2008). Notably, the *Antarctic Treaty* (1959) – the first nuclear arms agreement (Berkman, 2011) – also emphasizes *"facilitation of scientific research"* and *"international scientific cooperation."*

With science diplomacy in Antarctica, superpower adversaries and allies alike have been *"consulting together on matters of common interest"* across more than sixty years with continuous cooperation. As complementary negotiation strategies, conflict resolution and common-interest building (Fig. 1.4) both have the same end objectives to promote cooperation and prevent conflict, but the journeys are entirely different, depending on the starting point. The Cold War lesson with superpower adversaries in Antarctica as well as outer space is the starting point determines the journey with negotiations, continuously resolving conflicts or continuously cooperating based on common interests. Like with a glass half empty or half full, the starting point is a choice affecting the course of subsequent negotiations, emphasizing there is great scope in our world to increase capacities with common-interest building.

These ecopolitical lessons at the local-global scales of our home planet – "all mankind" – were carried from Antarctica to the Arctic before-through-after the end of the Cold War inflection point. Heralded with the Murmansk speech by Soviet President Mikhail Gorbachev (1987), an *"Arctic Research Council"* was conceived to *"let the North Pole be a pole of peace,"* leading to the formation of the Arctic

Council under the terms of the *Ottawa Declaration on the Establishment of the Arctic Council* (1996) to focus on *"common Arctic issues,"* particularly *"sustainable development and environmental protection."* The evolving international law of common interests (Berkman, 2012) is highlighted in the polar regions (Kish, 1973, Berkman, 2020c), underscoring theory, methods and skills with informed decisionmaking to apply, train and refine in a scalable manner. Moreover, the polar regions reveal science diplomacy as a process (Berkman et al., 2011), complementing science into policy as a product (Berkman, 2002), ultimately to develop **options** (without advocacy) that can be used or ignored explicitly, contributing to informed decisions beyond short-term political agendas.

As an example, the scalability of science diplomacy is reflected by two university professors convening the first formal dialogue between the North Atlantic Treaty Organization (NATO) and Russia regarding security in the Arctic (Berkman & Vylegzhanin, 2012a, b). Implications of this high-level dialogue among allies and adversaries alike continue to evolve, including with the Ambassadorial Panels (2015, 2016) on *"Building Common Interests in the Arctic Ocean"* (Berkman & Vylegzhanin, 2012b) that are the conceptual origin of this book. The underlying methods and skills that enable such dialogues are framed with the *Pyramid of Informed Decisionmaking* (Fig. 1.6), recognizing synergies exist between **research** and **action** like connections between *"the internal and the external"* realms of the human spirit (King, 1964).

With informed decisionmaking theory, methods and skills, it also becomes possible to train science diplomacy in a scalable manner, as reflected with the joint courses at universities in the United States and Russian Federation since 2017 (Berkman & Vylegzhanin, 2020), extending across the University of the Arctic with the *Science Diplomacy Thematic Network* (UArctic 2017). More broadly, science-diplomacy training is applicable across the diplomatic corps of foreign ministries and with the United Nations Institute for Training and Research (UNITAR, 2019a, b; 2020a, b; 2021). In all these venues, the objective is to enhance capacities with informed decisionmaking (Figs. 1.3, 1.4 and 1.6), involving both:

Governance Mechanisms (laws, agreements and policies as well as regulatory strategies, including insurance, at diverse jurisdictional levels); and

Built Infrastructure (fixed, mobile and other assets, including communication, research, observing, information and other systems that require technology plus investment).

Coupling governance mechanisms and built infrastructure underlies progress toward sustainable development, which will be elaborated in the third volume in the INFORMED DECISIONMAKING FOR SUSTAINABILITY book series, considering PAN-ARCTIC IMPLEMENTATION OF COUPLED GOVERNANCE AND INFRASTRUCTURE.

The underlying research that was translated into these education and leadership initiatives with science diplomacy emerged from the intertwined *Arctic Options / Pan-Arctic Options* projects from 2013–2021 with participants from Canada, China, France, Norway, Russian Federation and United States addressing *"Holistic Integration for Arctic Marine-Coastal Sustainability"* (Berkman et al., 2020a, b; Young

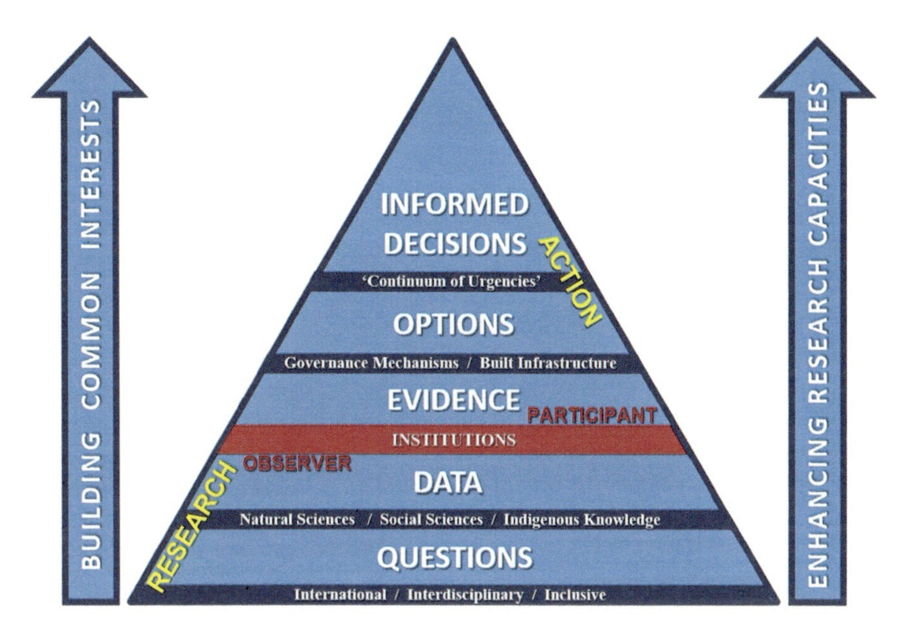

Fig. 1.6 Pyramid of Informed Decisionmaking as an holistic methodology with science diplomacy to apply, train and refine across a 'continuum of urgencies' (Fig. 1.4), characterizing the scope of an **informed decision**, as the apex goal with governance mechanisms and built infrastructure as well their coupling for sustainable development (Figs. 1.2a-d and 1.3). With holistic integration, **questions of common concern** reveal the methods of science to study change, generating the necessary **data** to produce answers in a transdisciplinary manner. These stages of research are transformed into action with **evidence** for decisions, involving institutions and their decisionmakers. Across the data-evidence interface, the diplomacy with science simply is in revealing **options** (without advocacy), which can be used or ignored explicitly, respecting the institutions. Starting with questions among allies and adversaries underlies the skill to build common interests. The engine of informed decisionmaking operates with common-interest building, enhancing research capacities as a positive feedback with individuals contributing as observers and participants inclusively. (Elaborated progressively from Berkman et al. (2017))

et al., 2020b). Holistic questions with science diplomacy (Table 1.2), at the base of the *Pyramid of Informed Decisionmaking,* emerged from the *Antarctic Treaty Summit* (2009) and were memorialised in the first book on Science Diplomacy (Berkman et al., 2011).

Holistic questions to build common interests and enhance research capacities (Fig. 1.6) also introduce a comparative framework to map the user "research" and "action" chapters of this book with diverse stakeholders, rightsholders and actors inclusively(Table 1.2) beyond listing them in the Table of Contents (please also see the *Preface*). As an illustration, this mapping also provides an inclusive framework to introduce all of the book contributions in a balanced manner.

Striving to achieve balance, science diplomacy can be viewed as a **language of hope** because of its international and interdisciplinary inclusion with common-interest building. Like any language, there are definitions for words and phrases (which are bolded and defined above) with syntax to connect meanings from paper to

Table 1.2 Mapping of chapters in this book to categories of holistic (international, interdisciplinary and inclusive) questions with science diplomacy and its engine of informed decisionmaking

Categories of Questions for Decisionmaking[a,b] Involving Science as…	Chapters in VOLUME 2 (see *Table of Contents and Preface*)[c,d]	
	Research	Action
An essential gauge of changes over time and space.	1, 5, 6, 8, 9, 10, 15, 16, 17, 23, 24, 25, 26, 35	2, 3, 4, 11, 22, 26, 29, 32, 33, 34
An instrument for Earth system monitoring.	1, 6, 9, 15, 16, 23, 24, 25, 35	2, 3, 11, 13, 22, 29, 34
An early warning system.	1, 6, 9, 10, 15, 17, 23, 24, 25, 35	2, 3, 4, 11, 12, 14, 22, 29, 30
A determinant of public policy agendas.	1, 5, 6, 8, 9, 16, 17, 18, 19, 23, 24, 25, 26, 27, 35	2, 3, 4, 13, 21, 22, 29, 30, 31, 32, 33, 34
An element of international legal institutions.	1, 5, 6, 8, 15, 18, 23, 24, 25, 26, 35	33, 34
A source of invention and commercial enterprise.	1, 7, 10, 15, 17, 18, 19, 24, 35	11, 13, 14, 20, 22, 28, 33
An element of continuity in our global society.	1, 5, 6, 7, 8, 9, 15, 16, 19, 24, 26, 35	2, 3, 4, 11, 22, 28, 29, 32, 33, 34
A tool of diplomacy to build common interests. (chapter mentions "common")[e]	1, 5, 6, 7, 8, 9, 15, 16, 18, 19, 23, 24, 26, 35	3, 4, 11, 29, 30

[a]Decisions involve governance mechanisms and built infrastructure, coupled for sustainability
[b]Elaborated from Berkman et al. (2011)
[c]Stages of research and action are elaborated in Fig. 1.6
[d]Appendix regarding the United Nations Decade for Ocean Science and Sustainable Development (UNDOS) is included in all categories of questions
[e]Searched and integrated comprehensively with the KnoHow™ knowledge bank (https://knohow.co) for VOLUME 2. BUILDING COMMON INTERESTS IN THE ARCTIC OCEAN WITH GLOBAL INCLUSION, using the final drafts of PDF files for the research and action chapters as well as the Appendix

practice. As puzzle pieces, the language starts with science and diplomacy, empowering synergies with these processes to facilitate holistic integration with research and action that together enable informed decisionmaking with governance mechanisms and built infrastructure as well as their coupling for sustainable development in a scalable manner with the Arctic Ocean as a case-study (Figs. 1.1, 1.2, 1.3, 1.4, 1.5 and 1.6; Tables 1.1 and 1.2).

With holistic integration, the biggest challenge is to be inclusive, recognizing the prevalence and problems with systemic exclusion that exist worldwide. On our shared journey as a globally-interconnected civilization (Figs. 1.2a-d and 1.5), science diplomacy promotes informed decisions: not good or bad decisions, not right or wrong decisions, but decisions that optimize the available data in view of the underlying questions inclusively (Table 1.2). All the chapters in this book touch on science diplomacy and its engine of informed decisionmaking, providing open-ended starting points (Fig. 1.6) for readers to consider skills, methods and theory to build common interests with lifelong learning.

1.2 The Arctic Ocean

1.2.1 Interconnected Home Systems

The Arctic Ocean is a case study with global relevance, conceptually and in practice as an holistic system (Fig. 1.1). International and interdisciplinary questions operate inclusively (Table 1.2) in view of the Arctic Ocean at diverse time and space scales (Vörösmarty et al., 2010, Petrov et al., 2017), across regions and jurisdictions (Gad and Strandsbjerg, 2019), which are discussed throughout this book. Moreover, the diminishing sea-ice boundary of the Arctic Ocean is like removing the ceiling to your room, exposing all to the outside conditions with inherent risks of instabilities. Operating at this security time scale (Fig. 1.4) – especially considering the Arctic Ocean could become a $1 trillion arena for investment (Roston, 2016, World Economic Forum, 2016) – reflects the challenge to achieve progress with sustainable development as a "common" Arctic issue.

When **boundaries** change, so do the associated and dependent systems, which is why the world has been introduced to a "new" Arctic Ocean with the diminishing sea-ice boundary (Berkman & Vylegzhanin, 2012a, b; Carmack et al., 2015). Applications of bounded regions to characterize inflows and outflows as well as system dynamics in the Arctic Ocean are the focus of the first volume of this book series in view of GOVERNING ARCTIC SEAS: REGIONAL LESSONS FROM THE BERING STRAIT AND BARENTS SEA (Young et al., 2020b). The international, interdisciplinary and inclusive focus of this second volume is on a Pan-Arctic scale in view of diverse biogeophysical, socioeconomic and institutional boundaries associated the Arctic Ocean (Berkman, 2015), underscoring holistic considerations with science as the study of change that are necessary to generate knowledge for Arctic sustainability (Greybill & Petrov, 2020).

In a general sense, progressing southward from the North Pole, the Arctic Ocean is bounded by the sea floor and sea surface with the surrounding continents (Jakobsson et al., 2004). Defining the southern boundary is where the ambiguities arise, underscoring the diverse interests of the associated stakeholders, rightsholders and other actors with this Pan-Arctic system. Nonetheless, while there is no fully agreed definition of the Arctic Ocean applicable in all situations, boundary characterizations of the Arctic Ocean system and its subsystems become essential to position the research and observations that are necessary to interpret change (Lee et al., 2019). These interconnections within the Earth system become especially important to manage the resulting data products in view of desired stakeholder outcomes (Eicken et al., 2016).

For some purposes, using the Arctic Circle as the southern limit of the Arctic Ocean has a number of advantages. The Arctic Circle boundary allows changes to be assessed in the Arctic Ocean with consistency over time, even back to the time when Indigenous peoples were able to walk across the Bering Strait with sea-level 120 meters lower than today more than 11,000 years ago (Hopkins, 1967, Jakobsson et al., 2017). Moreover, the Arctic Circle at 66.5° North latitude reflects the seasonality of the Arctic system in relation to the Sun as the primary external driver of Earth's climate, as with the climates of other planets in our Solar system (Kondratyev & Hunt, 1982), further highlighting the embedded and interacting nature of systems. In addition, the Arctic Circle demarcates the eight Arctic States, which established the Arctic Council, along with six Indigenous Peoples' Organizations (Ottawa Declaration, 1996), highlighting extensions of the Pan-Arctic region across lower latitudes (Fig. 1.1).

Dynamics of the Arctic Ocean system (Fig. 1.1) also can be characterized by diverse inflows and outflows across its boundaries, including from the North Pacific and North Atlantic as well as from adjacent land masses and across land-air-sea interfaces (Fig. 1.7). Among the many projects and programs, these diverse biogeophysical interactions are illustrated recently in view of sea-ice with research from the *Multidisciplinary drifting Observations for the Study of Arctic Climate* (MOSAiC) project (Shupe et al., 2020) and its complementary *Terrestrial Multidisciplinary distributed Observations for the Study of Arctic Connections* (T-MOSAiC) project (Vincent et al., 2019).

Amplified warming (Holland Bitz, 2003, Stuecker et al., 2018) and feedbacks with Earth's climate further illustrate geophysical connections with the Arctic Ocean (Merideth et al., 2019), especially with reduced albedo from diminishing sea ice (Winton, 2006, Pistone et al., 2014) as well as with increased greenhouse gases in the atmosphere from devolving methane (Shakhova et al., 2010, Sultan et al., 2020, James et al., 2016). Additionally, melting from the Greenland Ice Sheet is raising global sea level (Briner et al., 2020). The biogeophysical dynamics of the Arctic Ocean system (Falardeau & Bennett, 2020) are represented importantly by species interactions across associated and dependent ecosystems, involving humans as the primary internal system driver of changes across the Earth during the Anthropocene (Ehlers & Krafft, 2006, National Research Council, 2014), beyond the external drivers associated with changes in Solar radiation and Earth's orbital geometries (Berger, 1988, Eddy, 2009). System perspectives that center on the Arctic Ocean are reflected in the organization of this book (please see the Preface) with mapping of the research and action chapters (Fig. 1.6 and Table 1.2) in each of the sections:

Section 1. Introduction (Chapters 1–4);
Section II. The Arctic Ocean: Evolving Ecological and Sustainability Challenges (Chapters 5–14);
Section III. The Broader Arctic Setting (Chapters 15–22)
Section IV. Informed Decisionmaking Tools and Approaches for the Arctic (Chapters 23–32);
Section V. Conclusion (Chapters 33–35).

Like any natural system, the Arctic Ocean is represented by boundaries that depend on the eye of the beholder and the questions being addressed (Table 1.2). Boundaries also are conceptual features of learning systems widely considered in view of stages from Data, Information and Knowledge to Wisdom across the 'DIKW Pyramid' (Ackoff, 1989, 1999; Rowley, 2007). Although inquiry has been considered essential to gain knowledge and wisdom since Socrates, an innovation with the Informed Decisionmaking for Sustainability book series is that questions underlie

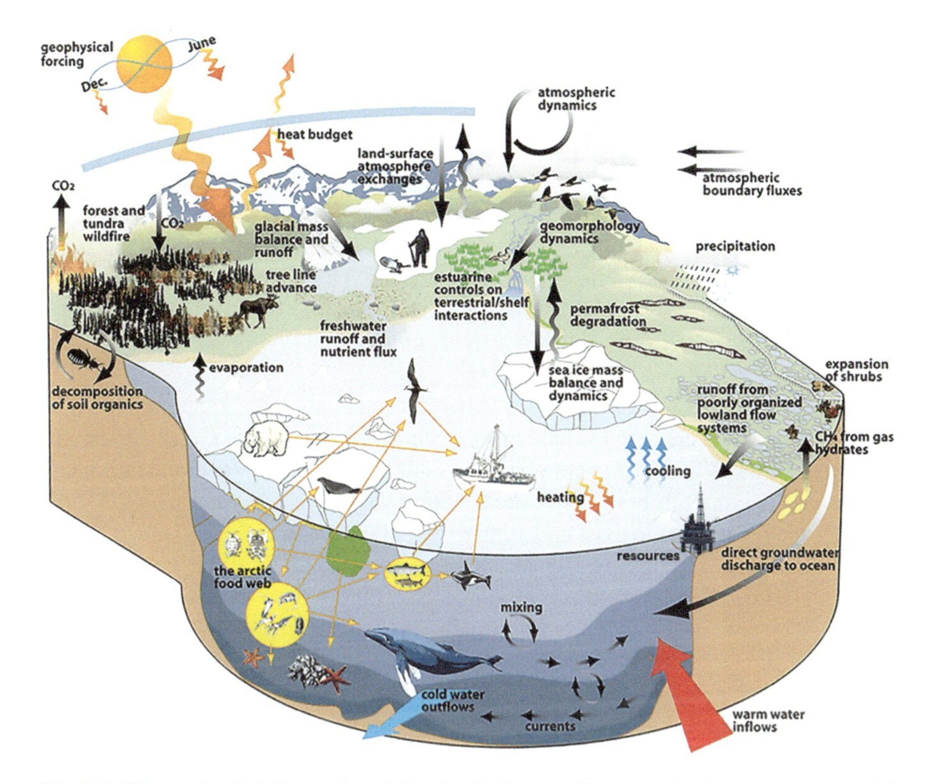

Fig. 1.7 Biogeophysical Dynamics of the Arctic System illustrated in view marine-terrestrial coupling with external and internal forcing. In the Arctic Ocean system (Fig. 1.1) – with its sea floor, sea surface and terrestrial boundaries – there are: (**a**) geophysical features with water masses, currents and sea ice with land-air-sea exchanges; (**b**) biological features with organisms, including humans, interacting with their dependent and associated ecosystems over diverse time and space scales; and (**c**) socioeconomic features with human uses of the maritime system along with its living and mineral resources, involving short-to-long term impacts (Fig. 1.3). (Modified from Roberts et al. (2010))

data as a fundamental feature of learning systems, evolving with the 'Pyramid of Informed Decisionmaking' (Fig. 1.6). Moreover, starting with questions inclusively is the essence of being transdisciplinary (beyond interdisciplinary, multidisciplinary or disciplinary) to achieve progress with knowledge democracy (Bunders et al., 2010), creating the opportunity to build common interests among allies and adversaries alike (Fig. 1.4).

The aspiration of this book is to be practical, helping readers to overlay ecopolitical regions (Table 1.1; Figs. 1.1 and 1.7) with the methodologies of science that contribute to informed decisions (Vörösmarty et al. 2018) with governance mechanisms and built infrastructure as well as their coupling for sustainable development in the Arctic Ocean (Pongrácz et al., 2020). This journey across the Pyramid of Informed Decisionmaking recognizes there are stages of research and action, which are distinguished across the data-evidence interface, where science diplomats can contribute as both observers and participants by serving as brokers of dialogues among stakeholders, rightsholders and actors inclusively (Fig. 1.6).

1.2.2 Interconnected Governance Systems

In the Arctic Ocean, just as on a planetary scale, managing ecopolitical regions (Table 1.1; Figs. 1.1 and 1.7) involves common-interest building with research-action connections that operate short-to-long term (Figs. 1.3, 1.4, 1.5 and 1.6). In this local-global context, the "***Law of the Sea***" provides "*an extensive international legal framework*" to which the Arctic States and Indigenous Peoples' Organizations "*remained committed*" (Arctic Council Secretariat, 2013), including the five Arctic coastal states (Canada, Denmark, Norway, Russia and the United States) that have declared their "***sovereignty***, *sovereign rights and jurisdiction in large areas of the Arctic Ocean*" (*Ilulissat Declaration* 2008).

Law of the sea and international environmental law (like other branches of international law) are applied universally, either as a binding system under the *Law of Treaties* (Vienna Convention, 1969) or as a matter of international customary law among nations. In both cases, areas within, across and beyond national jurisdictions are recognized to exist under international law. However, it is **international spaces** (Kish, 1973; Berkman et al., 2011; Berkman, 2020c) that best illustrate inclusive frameworks to build common interests (Figs. 1.4, 1.5 and 1.6), promoting peace (Berkman, 2009) with lessons of unity beyond sovereign jurisdictions.

Law of the sea provides lessons about international spaces in the high seas and deep sea that are different than with Antarctica and outer space (Fig. 1.5), considering the **jurisdictional zonation** from national boundaries into Areas Beyond National Jurisdiction (ABNJ) that cross a gradient of roles and responsibilities among nations (Fig. 1.8). The law of the sea also distinguishes ecopolitical regions that are bounded by the sea floor and superjacent waters, which are related to sovereign jurisdictions differently. As a prominent example (Berkman & Young, 2009), the deep sea floor to the North Pole still could be delineated as continental shelf attached to national jurisdictions, whereas the overlying high seas (Fig. 1.1) in the Central Arctic Ocean (CAO) are recognized universally as an international space (Fig. 1.5) where: "*No State may validly purport to subject any part of the high seas to its sovereignty*" (United Nations, 1982).

As the first ABNJ in human history (*Convention on the High Seas*, 1958), the high seas with its freedoms reflects the evolution of our globally-interconnected civilization since Grotius' crafting *Mare Liberum* in the early seventeenth century (Bull et al. 1995), when humankind was formulating the legal prerogatives of the nation-state with the *Treaty on Westphalia* (Fig. 1.2a). The journey ahead with international spaces includes the "*common heritage of mankind*" (United Nations, 1982) in the deep sea floor as a visionary concept with equitability for the benefit of humanity. As with all international spaces (Fig. 1.5), in the deep sea, complications to balance national interests and common interests are highlighted by accelerating commercial developments with short-term considerations only (Banet, 2020, Tunnicliffe et al., 2020), illustrating the precursors for uninformed decisions without formulation across a 'continuum of urgencies' (Fig. 1.4).

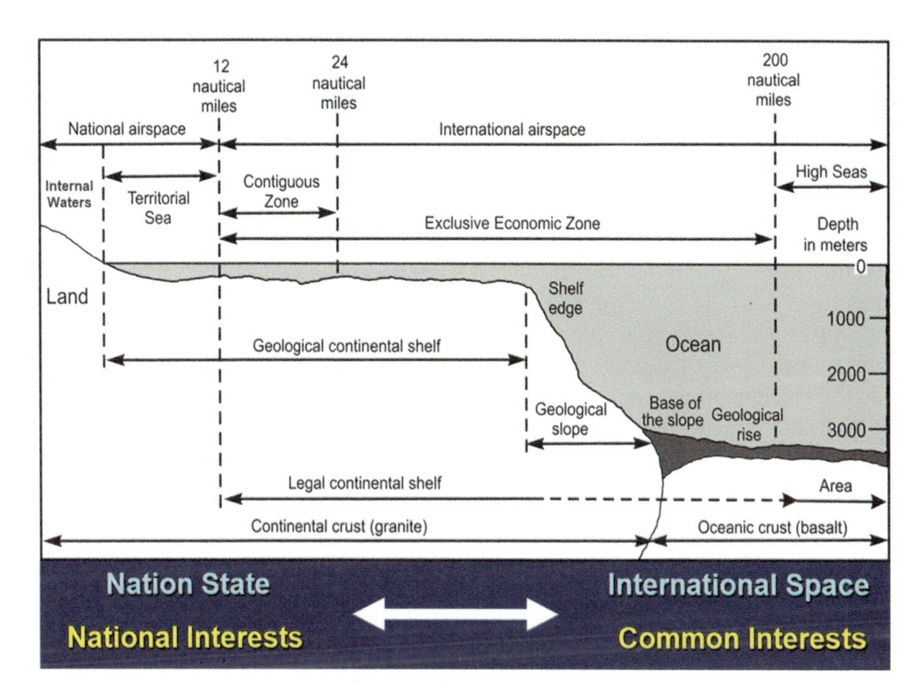

Fig. 1.8 Law of the Sea Zonation from boundary baselines of coastal nations into international spaces (Figs. 1.1 and 1.5), applied under customary international law (as by the United States) and through the 1982 *United Nations Convention on the Law of the Sea* (UNCLOS) with provisions for *"strengthening of peace, security, co-operation and friendly relations among all nations"* with keystone contributions from "*Marine Scientific Research*" (United Nations, 1982). The law of the sea illustrates the challenge of humanity forever after the Second World War – as long as there are nations – to balance national interests and common interests on a planetary scale. (Adapted from Berkman et al. (2020) as a core feature of the INFORMED DECISIONMAKING FOR SUSTAINABILITY book series)

With the 2018 *Agreement to Prevent Unregulated High Seas Fisheries in the Central Arctic Ocean* that entered into force on 25 June 2021, the CAO high seas (Fig. 1.1) have become a common interest of Arctic and non-Arctic states with Canada, China, Denmark (on behalf of the Faroe Islands and Greenland) Iceland, Japan, Norway, Russia, South Korea and the United States as well as the European Union (Vylegzhanin et al., 2020, Balton, 2020). Importantly, this binding Agreement brings home the point about the North Pole as a *"pole of peace"* (Gorbachev, 1987), considering especially the relations of superpowers who have signed this historic agreement.

The high seas of the CAO (Fig. 1.1.) also are a focal region to assess the interconnected biogeophysical and socioeconomic dynamics of associated and dependent ecosystems in the changing Arctic with **transboundary** governance considerations (Platjouw, 2019). Such assessment involves the cross-cutting features of science (Table 1.2), which are integrated across thematic, institutional and jurisdictional boundaries in a Pan-Arctic context (Figs. 1.1 and 1.7) with the 2017 *Agreement on Enhancing International Arctic Scientific Cooperation*.

Conceptually, the Arctic Science Agreement (2017) can be viewed as a key piece of the governance complex in the Arctic Ocean, complementing the *Marine Scientific Research* provisions of UNCLOS (United Nations 1982) that are central to informed decisionmaking (Figs. 1.3, 1.4 and 1.6) with law of the sea (Figs. 1.1 and 1.8). For example, the needs and applications of information are illustrated with the Polar Code (2017), which refers to: *"chart information;" "current information;" "hydrographic information;" "ice information;" "information available;" "information exchange;" land-based support information;" "meteorological information;" "positioning information;" "reference information;" "statistical information;" "sufficient information;" "supporting information;" "up-to-date information;"* and *"weather information."* Similarly, data and information needs are identified in other agreements that have come into force in the past decade (Arctic Search and Rescue Agreement, 2011; Arctic Marine Oil Pollution Preparedness Agreement, 2013) as well as earlier agreements (Berkman et al., 2019).

The process to produce these recent governance mechanisms are represented with the *Arctic Ocean Review* (PAME, 2013), implementing informed decisions from the *2005–2015 Arctic Marine Strategic Plan* (Arctic Council 2004), which is continuing with the *2015–2025 Arctic Marine Strategic Plan* (Arctic Council, 2015) across decadal time scales (Fig. 1.2a, b). Together, coupling of the scientific and governance products in the Arctic Ocean illustrates the pathway of science diplomacy from research to action, integrating data into evidence for informed decisionmaking (Figs. 1.3, 1.4 and 1.6).

1.3 With Global Inclusion

1.3.1 Local-Global Considerations

The premise of this book is that we live in a globally-interconnected civilization (Figs. 1.2a-d; 1.5 and 1.8), which includes all of us today and across generations (Table 1.1). What does it mean to be inclusive? How can we operate with inclusion, recognizing that institutions as well as systems have boundaries? Answering these questions is a matter of lifelong learning, which is the journey with informed decisionmaking (Figs. 1.3, 1.4 and 1.6; Table 1.2) that is illustrated with focus on the Arctic Ocean as a global case study (Figs. 1.1, 1.7, 1.8 and 1.9).

The dimensions of inclusion are local-global (Figs. 1.2a-d), starting with questions (Fig. 1.6, Table 1.2) that involve transboundary perspectives and transdisciplinary capacities that apply across institutions and jurisdictions (Figs. 1.5, 1.8 and 1.9) as well as across ecosystems that have dynamic geospatial dimensions. The Arctic Ocean system (Figs. 1.1 and 1.7) illustrates such holistic integration to achieve progress with sustainable development (Table 1.3) as a Pan-Arctic issue, building common interests among allies and adversaries alike while enhancing research capacities (Fig. 1.6) to produce informed decisions (Figs. 1.3 and 1.4) that will operate across generations in the high north.

Fig. 1.9 **Holistic Integration** of interests in the Arctic Ocean with all humankind represented, recognizing there are concentric stewardship roles and responsibilities with respect to the Arctic residents who are most immediately impacted by changes in this region. At the center of the Arctic Ocean, surrounding the North Pole as a *"pole of peace"* (Gorbachev, 1987), is the CAO high seas (Figs. 1.1 and 1.8), which is an ABNJ that lends itself to building common interests with global relevance. (Adapted from Berkman and Vylegzhanin (2012b))

1.3.2 Precaution Across Generations

With the Arctic Ocean system (Figs. 1.1, 1.7, 1.8 and 1.9), how can we maintain the high north as region of low tension (Støre 2010), continuously promoting cooperation and preventing conflict (Table 1.3)? How can we balance economic prosperity, environmental protection and societal well-being at the heart of sustainable development in the Arctic as well as elsewhere on Earth across generations? Answers to these questions involve common-interest building (Fig. 1.4 and 1.6), which is reflected with the CAO high seas (Figs. 1.1 and 1.8) as an international space with inclusion among allies and adversaries alike.

Table 1.3 Attributes and local-global characteristics of sustainability

Attributes	Local-Global Characteristics
Balance	Environmental Protection + Economic Prosperity + Societal Well-Being
	National Interests + Common Interests
Resilience	Present Generations + Future Generations
	Governance Mechanisms + Built Infrastructure
Stability	Promoting Cooperation + Preventing Conflict
	Peace + Survival

The CAO High Seas Fisheries Agreement (2018) highlights informed decisionmaking (Figs. 1.3, 1.4 and 1.6) under international law with the principle of precaution (Pan & Huntington, 2016, Hoag, 2017, Schatz et al., 2019). In specific, the CAO High Seas Fisheries Agreement (2018) considers the "*application of precautionary conservation and management measures as part of a long-term strategy*" to address "*potential adverse impacts*" – safeguarding "*healthy marine ecosystems*" and ensuring "*sustainable use of fish stocks*" as the specific focus of this historic agreement among Arctic and non-Arctic states. With inclusive relevance, considering examples and lessons to manage activities short-to-long term (Figs. 1.3 and 1.4), related international legal instruments with the "precautionary" principle and approaches are compiled as an Appendix to this chapter.

Applying precaution underscores the challenge we face collectively to operate as a globally-interconnected civilization. The solutions we seek are not magic bullets, but processes that operate with scalability over time and space in the face of changing circumstances. With inclusion as the biggest challenge to achieve scalability, science diplomacy introduces an holistic process with informed decisionmaking to enhance integration skills in an unbounded fashion independent of language, location and culture, complementing the seventeen United Nations Sustainable Development Goals (United Nations, 2015) at local-global levels (Figs. 1.8, 1.9 and 1.10). In view of international and interdisciplinary inclusion, the scalability with informed decisionmaking is a testable proposition, potentially with lifelong learning in an intergenerational context.

How can individuals, institutions and governments be inclusive? Part of the answer is to think beyond self-interests, which are most urgent now with greatest immediacy in the present. The future also is urgent, requiring present considerations to anticipate and prepare for eventualities, which are clearly evident in view of exponential change across diverse time scales (Figs. 1.2a-d). The notion of precaution places responsibility on present and future generations to inform the decisionmaking with governance mechanisms and built infrastructure as well as their coupled contributions to sustainable development. Importantly, thinking with precaution has the benefit of empowering diverse stakeholders, rightsholders, and other actors to contribute in an inclusive manner, building common interests with continuous iteration of questions, methods and capacities to address change.

Since the *Treaty of Westphalia* (Fig. 1.2a), nations have been the primary jurisdictional unit across the Earth. However, with "world" wars in the twentieth century, it became necessary for nations to create international institutions that also operate on a planetary scale. To be inclusive today involves subnational as another step in our evolution as a globally-interconnected civilization, noting the jurisdictional spectrum on Earth is like meters aggregated into kilometers and divided into centimeters (Fig. 1.10). How subnational fits into international legal frameworks with nations itself is a question to be resolved with informed decisionmaking.

Nonetheless, subnational jurisdictions do operate on planetary scale, as with the mayors of forty major cities considering their shared responses to climate change (World Mayors Summit, 2019) or with California as the fifth largest economy on Earth (CBS New 2018). Addressing the COVID-19 pandemic now over months to

Fig. 1.10 Spectrum of Jurisdictions on Earth, representing an inclusive framework for humankind to address impacts, issues and resources with informed decisionmaking (Figs. 1.3, 1.4, 1.5 and 1.6; Tables 1.1, 1.2 and 1.3) across our globally-interconnected civilization (Figs. 1.2a-d) at subnational-national-international levels. With such integration, the Arctic Ocean system (Figs. 1.1, 1.7, 1.8 and 1.9) provides a global case-study with timeless lessons for humanity to operate on a planetary scale. (Modified from Berkman (2019))

years (Fig. 1.2c) – and with precaution for global pandemics to come in the future – requires informed decisionmaking across the jurisdictional spectrum (Fig. 1.10) with local-global implementation strategies before-through-after the inflection point (Fig. 1.3) that has yet to arrive after two years.

All humans share a common interest in survival now with the COVID-19 pandemic, just as with "world" wars of the twentieth century, considering decades to centuries into the future with the passion of our world entering each *Conference of the Parties* ever after the 1992 *United Nations Framework Convention on Climate Change* (United Nations, 1992). Earth's climate brings into renewed focus that our global human population has accelerated from 1 billion people living around 1800 to 8 billion alive within the next five years (Fig. 1.2a). **The challenge we collectively face is one of common-interest building on a planetary scale.**

The Arctic Ocean offers holistic lessons about decisionmaking, both informed and uninformed, involving governance mechanisms and built infrastructure as well as their coupling for sustainability. The Arctic Ocean also is part of the decisionmaking with global issues, especially climate, operating across decades and centuries to come. In this sense, it is important to note that young adults living today will be alive in the twenty-second century, which brings great responsibility of those reading this book to consider how each of us can enhance next-generation capacities. The perspectives about time – past, present and future – are what guide us individually and collectively throughout our lives inclusively.

Research connects the present, past and the future, which is the essence of science diplomacy to negotiate with time (Fig. 1.4), turning questions of common concern into informed decisions (Fig. 1.6). Stimulated by curiosity, research skills are the most basic feature to make an informed decision, operating short-to-long term before-through-after inflections points (Fig. 1.3). In this realm of imagination, children are innately curious, emphasizing responsibilities to develop skills that begin with questions across a lifetime.

With science fiction into reality (Verne, 1865), the synergies of informed decisionmaking will contribute to lifelong learning, triangulating education, research and leadership with common-interest building and compassion. Learning lessons of global inclusion from the Arctic Ocean and elsewhere with common-interest building to produce informed decisions underscores the opportunity to act *"for the benefit of all on Earth across generations."*

Acknowledgements This chapter and those that follow emerged with *Arctic Frontiers* 2020, building on the 2018 *Memorandum of Understanding* with the Science Diplomacy Center on behalf of the editors for the book series on Informed Decisionmaking for Sustainability. This chapter is a product of the Science Diplomacy Center through EvREsearch LTD, coordinating *Arctic Options/Pan-Arctic Options* and related projects with support from the United States National Science Foundation (Award Nos. NSF-OPP 1263819, NSF-ICER 1660449, NSF-OPP 1719540 and NSF-ICER 2103490) along with the Fulbright Arctic Chair 2021-2022 awarded to P.A. Berkman by the United States Department of State and Norwegian Ministry of Foreign Affairs with funding from the United States Congress. These international projects include support also from national science agencies in Canada, China, France, Norway, Russia and United States from 2013 to 2022 in coordination with the Belmont Forum, gratefully acknowledging the collaboration with the University of California Santa Barbara, MGIMO University, Université Pierre et Marie Curie, Norwegian Polar Institute, University of Alaska Fairbanks, University of Colorado Boulder, Carleton University, Ocean University of China and University of the Arctic among other institutions throughout this period. We also thank the Polar Institute with the Wilson Center for their leadership and support of this contribution. Knowledge-discovery application of KnoHow™ (http://knohow.co) was supported by the National Science Foundation project on "*Automated Discovery of Content-in-Context Relationships from a Large Corpus of Arctic Social Science Data*" (Award No. NSF-OPP 1719540).

Appendix

The Precautionary Principle or Approach

Possible Citation: Soto Sanchez, R. 2022. *Appendix to Chapter 1. Building Common Interests with Informed Decisionmaking for Sustainability*. IN: Berkman, P.A., Young, O.R., Vylegzhanin, A.N., Balton, D.A. and Øvretveit, O. (eds). Building Common Interests in the Arctic Ocean with Global Inclusion. Volume 2. Informed Decisionmaking for Sustainability. Springer, Dordrecht.

Appendix: Table 1 The precautionary principle or approach in the 1982 United Nations Convention on the Law of the Sea (LOSC)

The 1982 United Nations Convention on the Law of the Sea (LOSC) provides the widely agreed basic international law framework for balancing the rights and duties of coastal States, including protecting and preserving the marine environment in the different maritime zones, with the rights and duties of all States, including to freedom of navigation. The LOSC applies to the Arctic Ocean as it applies to other parts of the seas and oceans. Although the LOSC does not expressly refer to the precautionary principle or precautionary approach, a number of its provisions, highlighted in this table nevertheless give effect to the basic concept of precaution. The LOSC is not comprehensive in the sense of providing detailed rules for the regulation of all marine operations and shipping at sea, especially in areas beyond national jurisdiction (ABNJ). Other international instruments, included in Appendix: Table 2, *build on or supplement the provisions of the LOSC relating to precaution an evolving concept that must be interpreted in accordance with the full complement of relevant international instruments.*[2]

Year adopted	Instrument	Type	Provision(s)	Textual quotation
1982	LOSC.[3]	Multilateral, International.	Articles 194 para(s). 1 to 3; and 195.	*Article 194* *Measures to prevent, reduce and control pollution of the marine environment* *"1. States shall take, individually or jointly as appropriate, all measures consistent with this Convention that are necessary to prevent, reduce and control pollution of the marine environment from any source, using for this purpose the best practicable means at their disposal and in accordance with their capabilities, and they shall endeavour to harmonize their policies in this connection.* *2. States shall take all measures necessary to ensure that activities under their jurisdiction or control are so conducted as not to cause damage by pollution to other States and their environment, and that pollution arising from incidents or activities under their jurisdiction or control does not spread beyond the areas where they exercise sovereign rights in accordance with this Convention.* *3. The measures taken pursuant to this Part shall deal with all sources of pollution of the marine environment. (...)"*

(continued)

[2] *1969 Vienna Convention on the Law of Treaties* (Vienna, 23 May 1969, in force 27 January 1980), Article 31, para. 3, lit. c. *1982 United Nations Convention on the Law of the Sea*, opened for signature 10 December 1982, (entered into forced 16 November 1994) Article 311, para. 2.

[3] United Nations Convention on the Law of the Sea (opened for signature 10 December 1982, entered into force 16 November 1994) 1833 UNTS 3 (LOSC).

				Article 195 *Duty not to transfer damage or hazards or transform one type of pollution into another* "In taking measures to prevent, reduce and control pollution of the marine environment, States shall act so as not to transfer, directly or indirectly, damage or hazards from one area to another or transform one type of pollution into another."
			Articles 207 to 212.	*Article 207* *Pollution from land-based sources* *"1. States shall adopt laws and regulations to prevent, reduce and control pollution of the marine environment from land-based sources, including rivers, estuaries, pipelines and outfall structures, taking into account internationally agreed rules, standards and recommended practices and procedures.* *2. States shall take other measures as may be necessary to prevent, reduce and control such pollution.* *3. States shall endeavour to harmonize their policies in this connection at the appropriate regional level.* *4. States, (. . .), shall endeavour to establish global and regional rules, standards and recommended practices and procedures to prevent, reduce and control pollution of the marine environment from land-based sources, (. . .). Such rules, standards and recommended practices and procedures shall be re-examined from time to time as necessary.* *5. Laws, regulations, measures, rules, standards and recommended practices and procedures referred to in paragraphs 1, 2 and 4 shall include those designed to minimize, to the fullest extent possible, the release of toxic, harmful or noxious substances, especially those which are persistent, into the marine environment."*

(continued)

				Article 208 *Pollution from seabed activities subject to national jurisdiction* *"1. Coastal States shall adopt laws and regulations to prevent, reduce and control pollution of the marine environment arising from or in connection with seabed activities subject to their jurisdiction and from artificial islands, installations and structures under their jurisdiction, (...).* *2. States shall take other measures as may be necessary to prevent, reduce and control such pollution.* *3. Such laws, regulations and measures shall be no less effective than international rules, standards and recommended practices and procedures.* *5. (...) Such rules, standards and recommended practices and procedures shall be re-examined from time to time as necessary."*
				Article 209 *Pollution from activities in the Area* *"1. International rules, regulations and procedures shall be established in accordance with Part XI to prevent, reduce and control pollution of the marine environment from activities in the Area. Such rules, regulations and procedures shall be re-examined from time to time as necessary.* *2. Subject to the relevant provisions of this section, States shall adopt laws and regulations to prevent, reduce and control pollution of the marine environment from activities in the Area undertaken by vessels, installations, structures and other devices flying their flag or of their registry or operating under their authority, as the case may be. The requirements of such laws and regulations shall be no less effective than the international rules, (...)."*
				Article 210 *Pollution by dumping* *"1. States shall adopt laws and regulations to prevent, reduce and control pollution of the marine environment by dumping.*

(continued)

				2. States shall take other measures as may be necessary to prevent, reduce and control such pollution. *3. Such laws, regulations and measures shall ensure that dumping is not carried out without the permission of the competent authorities of States.* *4. (…) Such rules, standards and recommended practices and procedures shall be re-examined from time to time as necessary.* *5. Dumping within the territorial sea and the exclusive economic zone or onto the continental shelf shall not be carried out (…) after due consideration of the matter with other States which by reason of their geographical situation may be adversely affected thereby.* *6. National laws, regulations and measures shall be no less effective in preventing, reducing and controlling such pollution than the global rules and standards."*
				Article 211 *Pollution from vessels* *"1. States, (…), shall establish international rules and standards to prevent, reduce and control pollution of the marine environment from vessels (…). Such rules and standards shall, in the same manner, be re-examined from time to time as necessary.* *2. States shall adopt laws and regulations for the prevention, reduction and control of pollution of the marine environment from vessels flying their flag or of their registry. Such laws and regulations shall at least have the same effect as that of generally accepted international rules and standards established through the competent international organization or general diplomatic conference. (…)"*
				Article 212 *Pollution from or through the atmosphere* *"1. States shall adopt laws and regulations to prevent, reduce and control pollution of the marine environment from or through the atmosphere,*

(continued)

				applicable to the air space under their sovereignty and to vessels flying their flag or vessels or aircraft of their registry, taking into account internationally agreed rules, standards and recommended practices and procedures and the safety of air navigation. *2. States shall take other measures as may be necessary to prevent, reduce and control such pollution.* *(…)"*
			Article 234.	*Article 234* *Ice-covered areas* *"Coastal States have the right to adopt and enforce non-discriminatory laws and regulations for the prevention, reduction and control of marine pollution from vessels in ice-covered areas within the limits of the exclusive economic zone, where particularly severe climatic conditions and the presence of ice covering such areas for most of the year create obstructions or exceptional hazards to navigation, and pollution of the marine environment could cause major harm to or irreversible disturbance of the ecological balance. Such laws and regulations shall have due regard to navigation and the protection and preservation of the marine environment based on the best available scientific evidence."*

Table 2 Existing instruments that embody the precautionary principle or approach

This table sets forth key provisions from international instruments that relate to the precautionary principle or precautionary approach, and that build on or supplement the provisions of the LOSC contained in Table 1.[4]

Year adopted	Instrument	Type	Provision(s)	Textual quotation
1969	OPRC Convention.[5]	Multilateral, International.	Article V.	*Article V.* *"1. Measures taken by the coastal State in accordance with Article I shall be proportionate to the damage actual or threatened to it.* *2. Such measures shall not go beyond what is reasonably necessary to achieve the end mentioned in Article I and shall cease as soon as that end has been achieved; they shall not unnecessarily interfere with the rights and interests of the flag State, third States and of any persons, physical or corporate, concerned.* *3. In considering whether the measures are proportionate to the damage, account shall be taken of:* *(a) the extent and probability of imminent damage if those measures are not taken; and* *(b) the likelihood of those measures being effective; and* *(c) the extent of the damage which may be caused by such measures."*

(continued)

[4] *1982 United Nations Convention on the Law of the Sea*, opened for signature 10 December 1982, (entered into forced 16 November 1994) Article 311, para.3.

[5] *1969 International Convention relating to intervention on the high seas in cases of oil pollution casualties*, opened for signature 29 November 1969, (entered into forced 06 May 1975) UNTS 970 (p.211).

1973	CITES.[6]	Multilateral, International.	Articles VIII, XIII and XIV.	*Article VIII* *Measures to be Taken by the Parties* *"1. The Parties shall take appropriate measures to enforce the provisions of the present Convention and to prohibit trade in specimens in violation thereof.* *(...)"* *Article XIII* *International Measures* "1. When the Secretariat in the light of information received is satisfied that any species included in Appendix I or II is being affected adversely by trade in specimens of that species or that the provisions of the present Convention are not being effectively implemented, it shall communicate such information to the authorized Management Authority of the Party or Parties concerned. 2. When any Party receives a communication as indicated in paragraph 1 of this Article, it shall, as soon as possible, inform the Secretariat of any relevant facts insofar as its laws permit and, where appropriate, propose remedial action. Where the Party considers that an inquiry is desirable, such inquiry may be carried out by one or more persons expressly authorized by the Party. (...)" *Article XIV* *Effect on Domestic Legislation and International Conventions*

(continued)

[6]Convention on International Trade in Endangered Species of Wild Fauna and Flora, 03 March 1973,, (entered into force 01 July 1975), 993 U.N.T.S. 243 (CITES).

				"1. The provisions of the present Convention shall in no way affect the right of Parties to adopt: (a) stricter domestic measures regarding the conditions for trade, taking, possession or transport of specimens of species included in Appendices I, II and III, or the complete prohibition thereof; or (b) domestic measures restricting or prohibiting trade, taking, possession or transport of species not included in Appendix I, II or III. (...) 6. Nothing in the present Convention shall prejudice the codification and development of the law of the sea by the United Nations Conference on the Law of the Sea convened pursuant to Resolution 2750 C (XXV) of the General Assembly of the United Nations nor the present or future claims and legal views of any State concerning the law of the sea and the nature and extent of coastal and flag State jurisdiction."
1980	CAMLR Convention.[7]	Multilateral, Regional.	Article II, para. 3, item c).	*Article II.* *"1. The objective of this Convention is the conservation of Antarctic marine living resources.* *(...)* *3. Any harvesting and associated activities in the area to which this Convention applies shall be conducted in accordance with the*

(continued)

[7] *1980 Convention for the Conservation of Antarctic Marine Living Resources*, opened for signature 20 May 1980, (entered into forced 7 April 1982) 1329 UNTS.

				provisions of this Convention and with the following principles of conservation: *(...)* *(c)revention of changes or minimisation of the risk of changes in the marine ecosystem which are not potentially reversible over two or three decades, taking into account the state of available knowledge of the direct and indirect impact of harvesting, the effect of the introduction of alien species, the effects of associated activities on the marine ecosystem and of the effects of environmental changes, with the aim of making possible the sustained conservation of Antarctic marine living resources."*
1982	World Charter for Nature.[8]	United Nations' General Assembly Resolution.		*GENERAL PRINCIPLES* *"1. Nature shall be respected and its essential processes shall not be impaired.* *2. (...) habitats shall be safeguard.* *3. All areas of the earth, both land and sea, shall be subject to these principles of conservation; special protection shall be given to unique areas, (...) to the habitats of rare or endangered species.* *(...)* *5. Nature shall be secured against degradation caused by (...) hostile activities."*

(continued)

[8]UN General Assembly, *World Charter for Nature.*, 28 October 1982, A/RES/37/7, available at: https://digitallibrary.un.org/record/39295?ln=es[accessed 06 September 2020]

1985	Vienna Convention.[9]	Multilateral.	Article 2, para.1.	*Article 2: General obligations* *"1. The Parties shall take appropriate measures in accordance with the provisions of this Convention and of those protocols in force to which they are party to protect human health and the environment against adverse effects resulting or likely to result from human activities which modify or are likely to modify the ozone layer.* *(...)"*
1987	Declaration on the Second International Conference on the Protection of the North Sea.[10]	Regional Declaration.	Articles VII; XV item(ii); and XVI para. (1).	*Ministerial Declaration* *"(...)* *VII. Accepting that, in order to protect the North Sea from possibly damaging effects of the most dangerous substances, a precautionary approach is necessary which may require action to control inputs of such substances even before a causal link has been established by absolutely clear scientific evidence;* *(...)* *XV. Decide to:* *(...)* *(ii) accept that by combining, simultaneously and complementarily, approaches based on emission standards and environmental quality objectives, a more precautionary approach to dangerous substances will be established;*

(continued)

[9] *1985 Vienna Convention for the Protection of the Ozone Layer*, opened for signature 22 march 1985, (entered into forced 22 September 1988) UNTS 1513, (p.293).

[10] Second International Conference on the Protection of the North Sea, London, 27 November 1987.

				(...) *XVI. Therefore agree to:* *(...)* *I. accept the principle of safeguarding the marine ecosystem of the North Sea by reducing polluting emissions of substances that are persistent, toxic and liable to bioaccumulate at source by the use of the best available technology and other appropriate measures. This applies especially when there is reason to assume that certain damage or harmful effects on the living resources of the sea are likely to be caused by such substances, even where there is no scientific evidence to prove a causal link between emissions and effects ("the principle of precautionary action"); (...)"*
1987	The Montreal Protocol.[11]	Protocol, Multilateral.	Preamble, para (s). 6 and 8.	*Preamble.* *"(...) Determined to protect the ozone layer by taking precautionary measures to control equitably total global emissions of substances that deplete it, with the ultimate objective of their elimination on the basis of developments in scientific knowledge, taking into account technical and economic considerations and bearing in mind the developmental needs of developing countries,* *(...)*

(continued)

[11] Montreal Protocol on substances that deplete the ozone layer, opened for signature 16 September 1987, (entered into forced 01 January 1989) UNTS 26369.

				Noting the precautionary measures for controlling emissions of certain chlorofluorocarbons that have already been taken at national and regional levels, (...)"
1992	United Nations Framework Convention on Climate Change.[12]	Multilateral.	Article 3.	*Article 3. PRINCIPLES. "(...) 3. The Parties should take precautionary measures to anticipate, prevent or minimize the causes of climate change and mitigate its adverse effects. Where there are threats of serious or irreversible damage, lack of full scientific certainty should not be used as a reason for postponing such measures, taking into account that policies and measures to deal with climate change should be cost-effective so as to ensure global benefits at the lowest possible cost. To achieve this, such policies and measures should take into account different socio-economic contexts, be comprehensive, cover all relevant sources, sinks and reservoirs of greenhouse gases and adaptation, and comprise all economic sectors. Efforts to address climate change may be carried out cooperatively by interested Parties. (...)"*

(continued)

[12] United Nations Framework Convention on Climate Change, opened for signature 09 May 1992, (entered into forced 21 March 1994) UNTS 1771 (p.107).

1992	The Water Convention.[13]	Regional.	Article 2, para. 5, item (a).	*Article 2. General Provisions. "(...)5. (...) the Parties shall be guided by the following principles: (a) The precautionary principle, by virtue of which action to avoid the potential transboundary impact of the release of hazardous substances shall not to be postponed on the ground that scientific research has not fully proved a causal link between those substances, on the one hand, and the potential transboundary impact, on the other hand; (...)"*
1992	Rio Declaration.[14]	International.	Principle 15.	*Principle 15. "In order to protect the environment, the precautionary approach shall be widely applied by States according to their capabilities. Where there are threats of serious or irreversible damage, lack of full scientific certainty shall not be used as a reason for postponing cost-effective measures to prevent environmental degradation."*
1992	The Maastricht Treaty.[15]	Regional.	Article 130r, para. (2).	*Title XVI Environment Article 130r "(...) 2. Community policy on the environment shall aim at a high level of protection taking into account the diversity of*

(continued)

[13]*Convention on the Protection and Use of Transboundary Watercourses and International Lakes*, adopted on 17 March 1992, (entered into forced 06 October 1996).

[14]Rio Declaration on Environment and Development, 14 June 1992, UN Doc. A/CONF.151/26/ Rev.1 Report of the UNCED Vol.1 (New York).

[15]Treaty on European Union, adopted on 07 February 1992, (entered into forced 01 November 1993) UNTS 298, (p.11).

				situations in the various regions of the Community. It shall be based on the precautionary principle and on the principles that preventive action should be taken, that environmental damage should as a priority be rectified at source and that the polluter should pay. Environmental protection requirements must be integrated into the definition and implementation of other Community policies. (...)"
1992	Convention on the Transboundary Effects of Industrial Accidents.[16]	Regional.	Article 3, para. 1.	*Article 3* *General provisions* *"1. The Parties shall, taking into account efforts already made at national and international levels, take appropriate measures and cooperate within the framework of this Convention, to protect human beings and the environment against industrial accidents by preventing such accidents as far as possible, by reducing their frequency and severity and by mitigating their effects. To this end, preventive, preparedness and response measures, including restoration measures, shall be applied. (...)"*
1992	Helsinki Convention.[17]	Regional.	Article 3, para. 2.	*Article 3.* *Fundamental principles and obligations.* *"(...)*

(continued)

[16]Convention on the Transboundary Effects of Industrial Accidents, adopted on 17 March 1992, (entered into forced 19 April 2000) UNTS 2105, (p. 457), with Amendments as Adopted in 2015.

[17]Convention on the Protection of the Marine Environment of the Baltic Sea Area, opened for signature 17 March 1992, (entered into force 17 January 2000) 1507 UNTS.

				2. The Contracting Parties shall apply the precautionary principle, i.e., to take preventive measures when there is reason to assume that substances or energy introduced, directly or indirectly, into the marine environment may create hazards to human health, harm living resources and marine ecosystems, damage amenities or interfere with other legitimate uses of the sea even when there is no conclusive evidence of a causal relationship between inputs and their alleged effects. (...)"
1992	OSPAR Convention.[18]	Multilateral, Regional.	Article 2, para. 2, item a).	*Article 2.* *"(...) The Contracting Parties shall apply:* *a) the precautionary principle, by virtue of which preventive measures are to be taken when there are reasonable grounds for concern that substances or energy introduced, directly or indirectly, into the marine environment may bring about hazards to human health, harm living resources and marine ecosystems, damage amenities or interfere with other legitimate uses of the sea, even when there is no conclusive evidence of a causal relationship between the inputs and the effects; (...)"*

(continued)

[18]Convention for the Protection of the Marine Environment of the North-East Atlantic (opened for signature 22 September 1992, entered into force 25 March 1998) 2354 UNTS.

1994	Protocol to the 1979 Convention on Long-Range Transboundary Air Pollution on Further Reduction of Sulphur Emissions.[19]	Protocol, Multilateral.	Preamble.	*Preamble.* *"(...) Resolved to take precautionary measures to anticipate, prevent or minimize emissions of air pollutants and mitigate their adverse effects, Convinced that where there are threats of serious or irreversible damage, lack of full scientific certainty should not be used as a reason for postponing such measures, taking into account that such precautionary measures to deal with emissions of air pollutants should be cost-effective, (...)"*
1994	Energy Charter Treaty.	Multilateral.	Article 19, para. (1).	*Article 19: Environmental Aspects* *"(1) In pursuit of sustainable development and taking into account its obligations under those international agreements concerning the environment to which it is party, each Contracting Party shall strive to minimise in an economically efficient manner harmful Environmental Impacts occurring either within or outside its Area from all operations within the Energy Cycle in its Area, taking proper account of safety. In doing so each Contracting Party shall act in a Cost-Effective manner. In its policies and actions each Contracting Party shall strive to take precautionary measures to prevent or minimise environmental degradation.(...)"*

(continued)

[19]Protocol to the 1979 Convention on Long-Range Transboundary Air Pollution on Further Reduction of Sulphur Emissions, Oslo, 14 June 1994, in force 05 August 1998, UNTS 2030, (p. 122).

1995	UN Fish Stocks Agreement.[20]	Multilateral, Regional.	Article 6.	*Article 6. Application of the precautionary approach. "1. States shall apply the precautionary approach widely to conservation, management and exploitation of straddling fish stocks and highly migratory fish stocks in order to protect the living marine resources and preserve the marine environment.* *2. States shall be more cautious when information is uncertain, unreliable or inadequate. The absence of adequate scientific information shall not be used as a reason for postponing or failing to take conservation and management measures.* *3. In implementing the precautionary approach, States shall:* *(a) improve decision-making (...) by obtaining and sharing the best scientific information available and implementing improved techniques for dealing with risk and uncertainty; (...)* *(c) take into account, inter alia, uncertainties (...) and the impact of fishing activities on non-target and associated or dependent species, as well as existing and predicted oceanic, environmental and socio-economic conditions;*

(continued)

[20]Agreement for the Implementation of the Provisions of the United Nations Convention on the Law of the Sea of 10 December 1982 Relating to the Conservation and Management of Straddling Fish Stocks and Highly Migratory Fish Stocks (UN Fish Stocks Agreement), 2167 UNTS 3.

				(d) develop data collection and research programmes to assess the impact of fishing on non-target and associated or dependent species and their environment, and adopt plans which are necessary to ensure the conservation of such species and to protect habitats of special concern. *(...)* *6. For new or exploratory fisheries, States shall adopt as soon as possible cautious conservation and management measures, including, inter alia, catch limits and effort limits. Such measures shall remain in force until there are sufficient data to allow assessment of the impact of the fisheries on the long-term sustainability of the stocks, whereupon conservation and management measures based on that assessment shall be implemented. The latter measures shall, if appropriate, allow for the gradual development of the fisheries.* 7. If a natural phenomenon has a significant adverse impact on the status of straddling fish stocks or highly migratory fish stocks, States shall adopt conservation and management measures on an emergency basis to ensure that fishing activity does not exacerbate such adverse

(continued)

				impact. States shall also adopt such measures on an emergency basis where fishing activity presents a serious threat to the sustainability of such stocks. (...)"
1996	London Protocol.[21]	Protocol, Multilateral.	Article 3, para. 1.	*Article 3.* *General Obligations* *"1. In implementing this Protocol, Contracting Parties shall apply a precautionary approach to environmental protection from dumping of wastes or other matter whereby appropriate preventative measures are taken when there is reason to believe that wastes or other matter introduced into the marine environment are likely to cause harm even when there is no conclusive evidence to prove a causal relation between inputs and their effects. (...)"*
1999	Convention on the Protection of the Rhine.[22]	Regional.	Article 4, item (a).	*Article 4* *Principles* *"To this end, the Contracting Parties shall be guided by the following principles:* *(a) precautionary principle; (...)"*
2000	Cartagena Protocol on Biosafety.[23]	Multilateral.	Preamble, and Articles 1; 10 para. 6; and 11 para. 8.	*Preamble* *"(...) Reaffirming the precautionary approach contained in Principle*

(continued)

[21] 1996 Protocol to the Convention on the Prevention of Marine Pollution by Dumping of Wastes and Other Matter (London Convention), London, 7 November 1996, in force 24 March 2006, 2006 ATS 11 (London Protocol).

[22] Convention on the Protection of the Rhine, adopted on 12 April 1999, (entered into forced 01 January 2003).

[23] Protocol to the 1992 Convention on Biological Diversity (Cartagena Protocol on Biosafety), Cartagena de Indias, adopted on 29 January 2000, (entered into force on 11 September 2003) UNTS 2226, (p.208).

				15 of the Rio Declaration on Environment and Development,(. . .)" Article 1 OBJECTIVE In accordance with the precautionary approach contained in Principle 15 of the Rio Declaration on Environment and Development, the objective of this Protocol is to contribute to ensuring an adequate level of protection in the field of the safe transfer, handling and use of living modified organisms resulting from modern biotechnology that may have adverse effects on the conservation and sustainable use of biological diversity, taking also into account risks to human health, and specifically focusing on transboundary movements. Article 10 DECISION PROCE DURE "(. . .) 6. Lack of scientific certainty due to insufficient relevant scientific information and knowledge regarding the extent of the potential adverse effects of a living modified organism on the conservation and sustainable use of biological diversity in the Party of import, taking also into account risks to human health, shall not prevent that Party from taking a decision, as appropriate, with regard to the import of the living modified organism in question as referred to in

(continued)

				paragraph 3 above, in order to avoid or minimize such potential adverse effects. (...)" *Article 11* *PROCEDURE FOR LIVING MODIFIED OR GANISMS INTENDED FOR DIRECT USE AS FOOD OR FEED, OR FOR PROCESSING* *"(...) 8. Lack of scientific certainty due to insufficient relevant scientific information and knowledge regarding the extent of the potential adverse effects of a living modified organism on the conservation and sustainable use of biological diversity in the Party of import, taking also into account risks to human health, shall not prevent that Party from taking a decision, as appropriate, with regard to the import of that living modified organism intended for direct use as food or feed, or for processing, in order to avoid or minimize such potential adverse effects. (...)"*
2001	Stockholm Convention on Persistent Organic Pollutants.[24]	Multilateral.	Preamble para. 8; and Articles 1, and 8 para. 9.	*Preamble* *"(...) Acknowledging that precaution underlies the concerns of all the Parties and is embedded within this Convention, (...)"* *Article 1* *Objective* *Mindful of the precautionary approach as set forth in Principle 15 of*

(continued)

[24] Stockholm Convention on Persistent Organic Pollutants, adopted on 22 May 2001, (entered into force on 17 May 2004) UNTS 2256, (p.119).

				the Rio Declaration on Environment and Development, the objective of this Convention is to protect human health and the environment from persistent organic pollutants. *Article 8* *"(...)* *9. The Committee shall, based on the risk profile referred to in paragraph 6 and the risk management evaluation referred to in paragraph 7 (a) or paragraph 8, recommend whether the chemical should be considered by the Conference of the Parties for listing in Annexes A, B and/or C. The Conference of the Parties, taking due account of the recommendations of the Committee, including any scientific uncertainty, shall decide, in a precautionary manner, whether to list the chemical, and specify its related control measures, in Annexes A, B and/or C."*
2003	Framework Convention for the Protection of the Marine Environment of the Caspian Sea.[25]	Regional.	Article 5, para. (a).	*Article 5. Principles* *"In the actions for goal achievement of this Convention and accomplishment of its provisions Contracting Parties are guided, including:* *(a) the principle of taking measures of precaution according to which, in the presence of threat of serious or irreversible*

(continued)

[25] Framework Convention for the Protection of the Marine Environment of the Caspian Sea, adopted on 04 November 2003, (entered into force on 12 August 2006).

				damage for the marine environment of the Caspian Sea, references to lack of complete scientific confidence are not used as the reason for delay of cost-efficient measures for the prevention of similar damage; (...)"
2014	The Polar Code.[26]	Multilateral, Regional (Polar waters).	Part II-A and Part II-B Pollution Prevention Measures.	See the Chapter 1 text with reference to "information" needs and applications from the Polar Code (2017)
2018	The CAO Fisheries Agreement.[27]	Multilateral, Regional.	Article 5, para. 1, item c).	Article 5. Review and Further Implementation. "(...) c) on the basis of the scientific information derived from the Joint Program of Scientific Research and Monitoring, from the national scientific programs, and from other relevant sources, and taking into account relevant fisheries management and ecosystem considerations, including the precautionary approach and potential adverse impacts of fishing on the ecosystems, consider, inter alia, whether the distribution, migration and abundance of fish in the Agreement Area would support a sustainable commercial fishery and, (...)."

[26] International Code for Ships Operating in Polar Waters (Polar Code), adopted on 94th Session of IMO's Maritime Safety Committee (MSC) in November 2014, entered into force 01 January 2017.

[27] 'Agreement to Prevent Unregulated High Seas Fisheries in the Central Arctic Ocean', 12.6.2018, COM (2018) 454 final, available at https://eur-lex.europa.eu/legal-content/EN/TXT/PDF/?uri=CELEX:52018PC0453&from=EN; last accessed 05 May 2020.

References

Ackoff, R. L. (1989). From data to wisdom. Presidential address to ISGSR, June 1988. *Journal of Applied Systems Analysis, 16*, 3–9. http://www-public.imtbs-tsp.eu/~gibson/Teaching/Teaching-ReadingMaterial/Ackoff89.pdf

Ackoff, R. L. (1999). *Ackoff's best: his classic writings on management*. John Wiley.

Ambassadorial Panel. (2015). *Building common interests in the Arctic Ocean*. University of Reykjavik. https://en.ru.is/news/arctic-high-seas-oct15

Ambassadorial Panel. (2016). *Building common interests in the Arctic Ocean*. University of Reykjavik. https://en.ru.is/news/building-common-interests-in-the-arctic-ocean-1

Arctic Council. (2004). *2005–2015 Arctic marine strategic plan*. Protection of the Arctic Marine Environment (PAME) Working Group. https://www.pame.is/document-library/amsp-documents/173-amsp-2005-2015/file

Arctic Council. (2015). 2015–2025 Arctic marine strategic plan. 9th Arctic Council Ministerial Meeting, Iqaluit. https://www.pame.is/document-library/amsp-documents/174-amsp-2015-2025/file

Arctic Council. (2016). In M. Carson & G. Peterson (Eds.), *Arctic resilience report*. Stockholm Environment Institute and Stockholm Resilience Centre. https://oaarchive.arctic-council.org/handle/11374/1838

Arctic Council Secretariat (2013) Vision for the Arctic. Arctic Council Secretariat, Kiruna, Sweden 15 May 2013

Arctic Frontiers (2020a) Arctic Frontiers 2020. Annual Report*: The Power of Knowledge*. Tromsø, Norway. (https://www.arcticfrontiers.com/conference/2020-the-power-of-knowledge/)

Arctic Frontiers (2020b) Arctic Frontiers 2020. Conference Summary. *The Power of Knowledge*. Tromsø, Norway. (https://www.arcticfrontiers.com/wp-content/uploads/2020/10/plenaryreport2020.pdf)

Arctic Marine Oil Pollution Preparedness Agreement (2013) Agreement on cooperation on marine oil pollution preparedness and response in the Arctic. Signed: 15 May 2013, Kiruna, Sweden Entry into Force: 25 March 2016. https://oaarchive.arctic-council.org/handle/11374/529

Arctic Science Agreement (2017) *Agreement on Enhancing International Arctic Scientific Cooperation*. Signed Fairbanks, Alaska, United States, 11 May 2017. Entry into Force, 23 May 2018. (https://www.state.gov/e/oes/rls/other/2017/270809.htm)

Arctic Search and Rescue Agreement (2011) Agreement on cooperation on aeronautical and maritime search and rescue in the Arctic. Signed: 12 May 2011, Nuuk, Greenland Entry into Force: 19 January 2013. (https://oaarchive.arctic-council.org/handle/11374/531)

Balton, D. A. (2019). What will the BBNJ agreement mean for the Arctic fisheries agreement? *Marine Policy, 109*, 103745. https://www.sciencedirect.com/science/article/abs/pii/S0308597X1930449X

Balton, D. A. (2020). Chapter 21 implementing the new Arctic fisheries agreement. In T. Heidar (Ed.), *New knowledge and changing circumstances in the law of the sea* (pp. 429–445). Brill Nkjhoff. https://doi.org/10.1163/9789004437753_023

Banet, C. (Ed.). (2020). *The law of the seabed: access, uses, and protection of seabed resources*. Brill Nijhoff.

Berger, A. (1988). Milankovitch theory and climate. *Reviews of Geophysics, 26*(4), 624–657. https://www.researchgate.net/publication/230890888_Milankovitch_Theory_and_Climate

Berkes, F., Folke, C., & Colding, J. (Eds.). (2000). *Linking social and ecological systems: management practices and social mechanisms for building resilience*. University of Cambridge Press.

Berkes, F., Colding, J., & Folke, C. (2008). *Navigating social-ecological systems: building resilience for complexity and change*. Cambridge University Press.

Berkman, P. A. (2002). *Science into policy: global lessons from Antarctica*. Academic.

Berkman, P. A. (2009). International spaces promote peace. *Nature, 462*, 412–413. https://www.nature.com/articles/462412a

Berkman, P. A. (2011). President Eisenhower, the Antarctic treaty and origin of international spaces. In P. A. Berkman, M. A. Lang, D. W. H. Walton, & O. R. Young (Eds.), *Science diplomacy: antarctica, science and the governance of international spaces* (pp. 17–28). Smithsonian Institution Scholarly Press. https://repository.si.edu/handle/10088/16154

Berkman, P. A. (2012). 'Common interests' as an evolving body of international law: applications for Arctic Ocean stewardship. In R. Wolfrum (Ed.), *Arctic Marine Science, International Law and Climate Protection. Legal Aspects of Future Marine Science in the Arctic Ocean* (pp. 155–174). Springer.

Berkman, P. A. (2013) Preventing and Arctic Cold War. New York Times 12 March 2013. https://www.nytimes.com/2013/03/13/opinion/preventing-an-arctic-cold-war.html

Berkman, P. A. (2015). Institutional dimensions of sustaining Arctic observing networks (SAON). *Arctic, 68*(Suppl. 1). https://doi.org/10.14430/arctic4499

Berkman, P. A. (2019). Evolution of science diplomacy and its local-global applications. Special issue, 'broadening soft power in EU-US relations'. *European Foreign Affairs Review, 24*, 63–79. https://www.ingsa.org/wp-content/uploads/2019/09/Evolution-of-Science-Diplomacy-and-its-Local-Global-Applications_23JUL19.pdf

Berkman, P. A. (2020a). Science diplomacy and it engine of informed Decisionmaking: operating through our global pandemic with humanity. *The Hague Journal of Diplomacy, 15*, 435–450. https://brill.com/view/journals/hjd/15/3/article-p435_13.xml

Berkman, P. A. (2020b) 'The pandemic lens': focusing across time scales for local-global sustainability. *Patterns* 1(8): 13 November 2020. 4p. https://pubmed.ncbi.nlm.nih.gov/33294877/

Berkman, P. A. (2020c). Chapter 6. Polar science diplomacy. In K. N. Scott & D. VanderZwaag (Eds.), *Research Handbook on Polar Law* (pp. 105–123). Edward Elgar.

Berkman, P. A., & Vylegzhanin, A. N. (Eds.). (2012a). *Environmental security in the Arctic Ocean. NATO Science for Peace and Security Series* (p. 459p). Springer. https://www.springer.com/gp/book/9789400747128

Berkman, P. A., & Vylegzhanin, A. N. (2012b). Conclusion: building common interests in the Arctic Ocean. In P. A. Berkman & A. N. Vylegzhanin (Eds.), *Environmental security in the Arctic Ocean. NATO Science for Peace and Security Series* (pp. 371–404). Springer. https://www.springer.com/gp/book/9789400747128

Berkman, P. A., & Vylegzhanin, A. N. (2020). Training Skills with Common-Interest Building. *Science Diplomacy Action* 4:1–65. [Student-Ambassador Declarations (2016–2020) from the Joint Video-Conferencing Course on *Science Diplomacy: Environmental Security and Law in the Arctic Ocean* with the Fletcher School of Law and Diplomacy at Tufts University (United States) and International Law Programme at MGIMO University (Russian Federation)]. https://scidiplo.org/wp-content/uploads/2020/11/Synthesis_4.pdf

Berkman, P. A., & Young, O. R. (2009). Governance and environmental change in the Arctic Ocean. *Science, 324*, 339–340. https://science.sciencemag.org/content/324/5925/339

Berkman, P. A., Lang, M. A., Walton, D. W. H., & Young, O. R. (Eds.). (2011). *Science diplomacy: Antarctica, science and the governance of international spaces*. Smithsonian Institution Scholarly Press. https://repository.si.edu/handle/10088/16154

Berkman, P. A., Kullerud, L., Pope, A., Vylegzhanin, A. N., & Young, O. R. (2017). The Arctic science agreement propels science diplomacy. *Science, 358*, 596–598. https://science.sciencemag.org/content/358/6363/596

Berkman, P. A., Vylegzhanin, A. N., & Young, O. R. (2019). *Baseline of Russian Arctic Laws*. Springer. https://www.springer.com/gp/book/9783030062613

Berkman, P. A., Young, O. R., & Vylegzhanin, A. N. (2020). *Preface for the book series on informed Decisionmaking for sustainability. IN: Young, O.R., Berkman, P.A. and Vylegzhanin (eds.) Volume 1: Governing Arctic Seas: Regional Lessons from the Bering Strait and Barents Sea*. Springer. https://link.springer.com/content/pdf/bfm%3A978-3-030-25674-6%2F1.pdf

Briner, J. P., Cuzzone, J. K., Badgeley, J. A., Young, N. E., Steig, E. J., Morlighem, M., Schlegel, N.-J., Hakim, G. J., Schaefer, J. M., Johnson, J. V., Lesnek, A. J., Thomas, E. K., Allan, E., Bennike, O., Cluett, A. A., Csatho, B., de Vernal, A., Downs, J., Larour, E., & Nowicki,

S. (2020). Rate of mass loss from the Greenland ice sheet will exceed Holocene values this century. *Nature, 586*, 70–74. https://www.nature.com/articles/s41586-020-2742-6

Bull, H., Kingsbury, B., & Roberts, A. (1995). *Hugo Grotius and international relations*. Clarendon Press.

Bunders, J. F. G., Broerse, J. E. W., Keil, F., Pohl, C., Scholz, R. W., & Zweekhorst, M. B. M. (2010). *How can transdisciplinary research contribute to knowledge democracy? IN: in 't veld, Roel (Ed.) Knowledge Democracy: Consequences for Science, Politics, and Media* (pp. 125–152). Springer. https://www.researchgate.net/publication/274239011_Knowledge_Democracy

CAO High Seas Fisheries Agreement. (2018). Agreement to Prevent Unregulated High Seas Fisheries in the Central Arctic Ocean. Signed: Ilulissat, 3 October 2018. Entry into Force: pending remaining ratification by China. https://eur-lex.europa.eu/legal-content/EN/TXT/?uri=COM:2018:453:FIN

Carmack, E., Polyakov, I., Padman, L., Fer, I., Hunke, E., Hutchings, J., Jackson, J., Kelley, D., Kwok, R., Layton, C., Melling, H., Perovich, D., Persson, O., Ruddick, B., Timmermans, M.-L., Toole, J., Ross, T., Vavrus, S., & Winsor, P. (2015). Toward quantifying the increasing role of oceanic heat in sea ice loss in the new Arctic. *Bulletin of the American Meteorological Society, 96*(12), 2079–2105. https://journals.ametsoc.org/downloadpdf/journals/bams/96/12/bams-d-13-00177.1.xml

CBS News. (2018). California now has the world's 5th largest economy. CBS News, 4 May 2018. https://www.cbsnews.com/news/california-now-has-the-worlds-5th-largest-economy/

Convention on the High Seas. (1958). Convention on the high seas. Signed: Geneva, 29 April 1958 Entry into Force: 30 September 1962. https://treaties.un.org/pages/ViewDetails.aspx?src=TREATY&mtdsg_no=XXI-2&chapter=21

De Lucia, V. (2019). The BBNJ negotiations and ecosystem governance in the Arctic. *Marine Policy, 110*, 103756. http://refhub.elsevier.com/S0308-597X(20)30860-5/sref32)

DeEicken H., Lee OA, Lovecraft AL (2016) Evolving roles of observing systems and data co-management in Arctic Ocean governance. *OCEANS 2016, Marine Technology Society*. pp. 1-8 (https://ieeexplore.ieee.org/document/7761298)

Eddy, J. A. (2009). *The sun, the earth and the near-earth space: A guide to the sun-earth system*. National Aeronautics and Space Administration. https://lwstrt.gsfc.nasa.gov/images/pdf/john_eddy/SES_Book_Interactive.pdf

Ehlers, E., & Thomas Krafft, T. (Eds.). (2006). *Earth system science in the Anthropocene*. Springer.

Ehrlich, P. R., & Holdren, J. P. (1971). Impact of population growth. *Science, 171*, 1212–1217. https://science.sciencemag.org/content/171/3977/1212

Falardeau, M., & Bennett, E. M. (2020). Towards integrated knowledge of climate change in Arctic marine systems: a systematic literature review of multidisciplinary research. *Arctic Science, 6*, 1–23. https://cdnsciencepub.com/doi/10.1139/as-2019-0006

Fiske, G. (2021). Indigenous Peoples of the Arctic boundary polygons, 2021. *Arctic Data Center*. https://doi.org/10.18739/A24746S61

Freestone, D. (Ed.). (2019). *Conserving biodiversity in areas beyond National Jurisdiction*. Brill Nijhoff.

Gad, U. P., & Strandsbjerg, J. (Eds.). (2019). *The politics of sustainability in the Arctic. Reconfiguring identity, space, and time*. Routledge.

Greybill, J. K., & Petrov, A. N. (Eds.). (2020). *Arctic sustainability, key methodologies and knowledge domains*. Routledge.

GRID-Arendal. (2021). Global Resource Information Database (GRID) Partner of the United Nations Environmental Programme in Arendal, Norway. https://www.grida.no/

Guterres, A. (2020). 'Transcript of UN Secretary-General's virtual press encounter to launch the report on the socio-economic impacts of COVID-19'. United Nations Secretary-General, 31 March 2020. https://www.un.org/sg/en/content/sg/press-encounter/2020-03-31/transcript-of-un-secretary-general%E2%80%99s-virtual-press-encounter-launch-the-report-the-socio-economic-impacts-of-covid-19

Hardin, G. (1968). The tragedy of the commons. *Science, 162*, 1243–1248. https://science.sciencemag.org/content/162/3859/1243

Hoag, H. (2017). Nations put science before fishing in the Arctic. *Science, 358*, 1235. https://science.sciencemag.org/content/sci/358/6368/1235.full.pdf

Holdren, J. P. (2008). Science and technology for sustainable well-being. *Science, 319*, 424–434. https://science.sciencemag.org/content/319/5862/424

Holland, M. M., & Bitz, C. M. (2003). Polar amplification of climate change in coupled models. *Climate Dynamics, 21*, 221–232. https://link.springer.com/article/10.1007/s00382-003-0332-6

Hopkins, D. M. (1967). *The Bering land bridge*. Stanford University Press.

Indigenous Peoples Secretariat. (2021a). Indigenous Peoples Secretariat of the Arctic Council. https://www.arcticpeoples.com/

Ilulissat Declaration. (2008). *Declaration from the Arctic Ocean Conference*. Declared by Canada, Denmark, Norway, Russian Federation and United States. Ilulissat, 28 May 2008

Indigenous Peoples Secretariat. (2021b). *Arctic indigenous languages and revitalization: an online educational resource*. Indigenous Peoples Secretariat. https://www.arcticpeoples.com/arctic-languages#feedback

Jakobsson, M., Grantz, A., Kristoffersen, Y., Macnab, R., MacDonald, R. W., Sakshaug, E., Stein, R., & Jokat, W. (2004). The Arctic Ocean: boundary conditions and background information. In R. Stein & R. W. MacDonald (Eds.), *The organic carbon cycle in the Arctic Ocean*. Springer. https://doi.org/10.1007/978-3-642-18912-8_1

Jakobsson, M., Pearce, C., Cronin, T. M., Backman, J., Anderson, L. G., Barrientos, N., Björk, G., Coxall, H., de Boer, A., Mayer, L. A., Mörth, C.-M., Nilsson, J., Rattray, J. E., Stranne, C., Semiletov, I., & O'Regan, M. (2017). Post-glacial flooding of the Bering land bridge dated to 11 cal ka BP based on new geophysical and sediment records. *Climate of the Past, 13*, 991–1005. https://doi.org/10.5194/cp-13-991-2017

James, R. H., Bousquet, P., Bussmann, I., Haeckel, M., Kipfer, R., Leifer, I., Niemann, H., Ostrovsky, I., Piskozub, J., Rehder, G., Treude, T., Vielstädte, L., & Greinert, J. (2016). Effects of climate change on methane emissions from seafloor sediments in the Arctic Ocean: A review. *Limnology and Oceanography, 61*, S283–S299. https://doi.org/10.1002/lno.10307

King, M. L. (1964). The quest for peace and justice. Nobel Peace Prize Lecture, 11 December 1964, Oslo. https://www.nobelprize.org/prizes/peace/1964/king/lecture/

Kish, J. (1973). *The law of international spaces*. A. W. Sijthoff.

Kondratyev, K. Y., & Hunt, G. E. (1982). *Weather and climate on planets*. Elsevier.

Krebs, C. J. (1972). *Ecology: the experimental analysis of distribution and abundance*. Harper & Row.

Lee, C.M., Starkweather, S., Eicken, H., Timmermans, M.L., Wilkinson, J., Sandven, S., Dukhovskoy, D., Gerland, S., Grebmeier, J., Intrieri, J.M., Kang, S.Hj., McCammon, M, Nguyen, A.T., Polyakov, I., Rabe, B., Sagen, H., Seeyave, S., Volkov, D., Beszczynska-Möller, A., Chafik, L., Dzieciuch, M., Goni, G., Hamre, T., King, A.L., Olsen, A., Raj, R.P, Rossby, T., Skagseth, Ø., Søiland, H., & Sørensen, K. (2019). A framework for the development, design and implementation of a sustained Arctic Ocean observing system. *Frontiers in Marine Science* 6: (19 August 2019):1–21. https://www.frontiersin.org/articles/10.3389/fmars.2019.00451/full

Lucretius, 55 BCE. On the Nature of Things (De Rerum Natura). Rome. [Translated by Stallings, A.E. 2007. Penguin Classics]

Meredith, M., Sommerkorn, M., Cassotta, S., Derksen, C., Ekaykin, A., Hollowed, A., Kofinas, G., Mackintosh, A., Melbourne-Thomas, J., Muelbert, M. M. C., Ottersen, G., Pritchard, H., & Schuur, E. A. G. (2019). Chapter 3. Polar regions. In H.-O. Pörtner, D. C. Roberts, V. Masson-Delmotte, P. Zhai, M. Tignor, E. Poloczanska, K. Mintenbeck, A. Alegría, M. Nicolai, A. Okem, J. Petzold, B. Rama, & N. M. Weyer (Eds.), *IPCC special report on the ocean and cryosphere in a changing climate* (pp. 203–320) https://www.ipcc.ch/srocc/

NASA. (2012). *Arctic sea ice hits smallest extent in satellite era*. National Aeronautics and Space Administration. https://www.nasa.gov/topics/earth/features/2012-seaicemin.html

National Research Council. (2014). *The Arctic in the Anthropocene. Emerging research questions*. National Academies Press. http://nap.edu/18726

Nature. (2020). Editorial. Arctic Science Cannot Afford a New Cold War. *Nature* 30 September 2020. https://www.nature.com/articles/d41586-020-02739-x

Ottawa Declaration. (1996). *Declaration on the Establishment of the Arctic Council*. 19 September 1996, Ottawa: Foreign Affairs and International Trade Canada. https://oaarchive.arctic-council.org/handle/11374/85

PAME. (2013). The Arctic Ocean Review Project, Final Report, (Phase II 2011–2013). Protection of the Arctic Marine Environment (PAME) Secretariat, Akureyri. https://www.pame.is/projects/arctic-marine-shipping/the-arctic-ocean-review-aor

Platjouw, F. M. (2019). Dimensions of transboundary legal coherence needed to foster ecosystem-based governance in the Arctic. *Marine Policy, 110*, 103666. https://www.sciencedirect.com/science/article/pii/S0308597X19303185

Polar Code. (2017). *International Code for Ships Operating in Polar Waters (Polar Code). Marine Environmental Protection Committee, MEPC 68/21/Add. 1, Annex 10. International Maritime Organization*. Entry into Force 1 January 2017. http://www.imo.org/en/MediaCentre/HotTopics/polar/Pages/default.aspx

National Park Service. (2020). *The United Nations Memorial Service at Muir Woods*. United States National Park Service. https://www.nps.gov/articles/the-united-nations-memorial-service-at-muir-woods.htm

Pan, M., & Huntington, H. P. (2016). A precautionary approach to fisheries in the Central Arctic Ocean: policy, science, and China. *Marine Policy, 63*, 153–157. https://doi.org/10.1016/j.marpol.2015.10.015

Petrov, A. N., BurnSilver, S., Chapin, F. S., Fondahl, G., Graybill, J. K., Keil, K., Nilsson, A. E., Riedlsperger, R., & Schweitzer, P. (2017). *Arctic sustainability research past, present and future*. Routledge.

Pistone, K., Eisenman, I., & Ramanathan, V. (2014). Observational determination of albedo decrease caused by vanishing Arctic Sea ice. *Proceedings of the National Academy of Sciences, 111*(9), 3322–3326. https://doi.org/10.1073/pnas.1318201111

Pongrácz, E., Pavlov, V., & Hänninen, N. (Eds.). (2020). *Arctic marine sustainability Arctic maritime businesses and the resilience of the marine environment*. Springer.

Roberts, A., Hinzman, L., Walsh, J. E., Holland, M., Cassano, J., Döscher, R., Mitsudera, H., & Sumi, A. (2010). A science plan for regional Arctic system modeling. A report to the National Science Foundation from the International Arctic Science Community. International Arctic Research Center, University of Alaska Fairbanks

Roston, E. (2016). The world has discovered a $1 Trillion Ocean. *Bloomberg* 21 January 2016

Rowley, J. (2007). The wisdom hierarchy: representations of the DIKW hierarchy. *Journal of Information Science, 33*(2), 163–180. https://doi.org/10.1177/0165551506070706

Ruffert, M., & Steinecke, S. (2011). The global administrative law of science. In *Beiträge zum ausländischen öffentlichen Recht und Völkerrecht (Veröffentlichungen des Max-Planck-Instituts für ausländisches öffentliches Recht und Völkerrecht), Volume 228*. Springer. https://doi.org/10.1007/978-3-642-21359-5_2

Schatz, V. J., Proelss, A., & Liu, N. (2019). The 2018 Agreement to prevent unregulated high seas fisheries in the Central Arctic Ocean: A critical analysis. *International Journal of Marine and Coastal Law, 34*(2), 195–244. https://brill.com/view/journals/estu/34/2/article-p195_2.xml

Shakhova, N., Semiletov, I., Salyuk, A., Yusupov, V., Kosmach, D., & Gustafsson, Ö. (2010). Extensive methane venting to the atmosphere from sediments of the east Siberian Arctic shelf. *Science, 327*, 1246–1250. https://science.sciencemag.org/content/327/5970/1246

Shupe, M. D., Rex, M., Dethloff, K., Damm, E., Fong, A. A., Gradinger, R., Heuzé, C., Loose, B., Makarov, A., Maslowski, W., Nicolaus, M., Perovich, D., Rabe, B., Rinke, A., Sokolov, V., & Sommerfeld, A. (2020). The MOSAiC expedition: a year drifting with the Arctic sea ice. In *Arctic Report Card, Update for 2020*. National Oceanic and Atmospheric Administration. https://repository.library.noaa.gov/view/noaa/27898

Steil, B. (2013). *The Battle of Bretton woods: John Maynard Keynes, Harry Dexter white, and the making of a New World order*. Princeton University Press.
Steinveg, B. (2020). The role of conferences within Arctic governance. *Polar Geography, 43*. https://doi.org/10.1080/1088937X.2020.1798540
Støre, J. G. (2010). The Norwegian high north policy. *Foreign Minister* Speech, Norwegian Ministry of Foreign Affairs, Oslo 7 June 2010. https://www.regjeringen.no/en/aktuelt/m/id609025/
Stuecker, M. F., Bitz, C. M., Armour, K. C., Proistosescu, C., Kang, S. M., Xie, S. P., Kim, D., McGregor, S., Zhang, W., Zhao, S., Cai, W., Dong, Y., & Jin, F. F. (2018). Polar amplification dominated by local forcing and feedbacks. *Nature Climate Change, 8*, 1076–1081. https://doi.org/10.1038/s41558-018-0339-y
Sultan, N., Plaza-Faverola, A., Vadakkepuliyambatta, S., Buenz, S., & Knies, J. (2020). Impact of tides and sea-level on deep-sea Arctic methane emissions. *Nature Communications, 11*, 5087. (2020). https://doi.org/10.1038/s41467-020-18899-3
Tunnicliffe, V., Metaxas, A., Jennifer Le, J., Ramirez-Llodra, E., & Levin, L. A. (2020). Strategic environmental goals and objectives: setting the basis for environmental regulation of deep seabed mining. *Marine Policy, 114*, 103347. https://www.sciencedirect.com/science/article/pii/S0308597X1830321X
UArctic. (2017). *Science diplomacy thematic network*. University of the Arctic. https://www.uarctic.org/organization/thematic-networks/science-diplomacy/
UNITAR. (2019a). *Retreat for directors of Arab diplomatic academies and institutes*. United Nations Institute for Training and Research. https://www.unitar.org/event/full-catalog/retreat-directors-arab-diplomatic-academies-and-institutes
UNITAR. (2019b). *Diplomacy 4.0 training programme*. Geneva: United Nations Institute for Training and Research. https://unitar.org/about/news-stories/news/united-nations-diplomacy-40-training-programme
UNITAR. (2020a). *Science diplomacy and informed decision-making during our global pandemic*. United Nations Institute for Training and Research. https://www.unitar.org/event/full-catalog/science-diplomacy-and-informed-decision-making-during-our-global-pandemic
UNITAR. (2020b). *Executive diploma on international law*. United Nations Institute for Training and Research. https://www.unitar.org/event/full-catalog/executive-diploma-international-law-online
UNITAR. (2021). *Executive summer programme on innovations in science diplomacy*. United Nations Institute for Training and Research. https://www.unitar.org/event/full-catalog/unitar-executive-summer-programme-innovations-science-diplomacy
United Nations. (1945). *United Nations Archive: United Nations Conference on International Organization* (UNCIO) (1945) – AG-012. https://search.archives.un.org/united-nations-conference-on-international-organization-uncio-1945
United Nations. (1982). *United Nations Convention on the law of the sea*. (Signed: Montego Bay, 10 December 1982; entry into force: 16 November 1994). https://www.un.org/depts/los/convention_agreements/texts/unclos/unclos_e.pdf
United Nations. (1987). Our common future: from one earth to one world. Report Transmitted to the General Assembly as an Annex to Resolution A/RES/42/187. World Commission on Environment and Development, New York. https://sustainabledevelopment.un.org/content/documents/5987our-common-future.pdf
United Nations. (1992). *United Nations Framework Convention on Climate Change*. (Signed: Rio de Janeiro, 9 May 1992; Entry into Force: 21 March 1994). https://unfccc.int/sites/default/files/convention_text_with_annexes_english_for_posting.pdf
United Nations. (2015). *Transforming Our World: The 2030 Agenda for Sustainable Development*. Res. A/RES/70/1 (25 September 2015). United Nations General Assembly, New York. https://sustainabledevelopment.un.org/post2015/transformingourworld/publication
Verne, J. (1865). *De la Terre á la Lune*. Pierre-Jules Hetze.

Vienna Convention (1969) Vienna convention on the law of treaties. Signed: 23 May 1969, Vienna, Austria; Entry into Force: 27 January 1980

Vienna Dialogue Team. (2017). A global network of science and technology advice in foreign ministries. *Science Diplomacy Action, 1*, 1–20. https://scidiplo.org/wp-content/uploads/2020/11/Synthesis_1.pdf

Vincent, W. F., Canário, J., & Boike, J. (2019). Understanding the terrestrial effects of Arctic Sea ice decline. *Transactions of the American Geophysical Union (Eos)* 100, 17 July 2019. https://doi.org/10.1029/2019EO128471

Vörösmarty, C. J., McGuire, A, D., & Hobbie, J. E. (2010) Scaling studies in Arctic system science and policy support. *A Call-to-Research.* U.S. Arctic Research Commission. Anchorage. https://digital.library.unt.edu/ark:/67531/metadc949504/

Vörösmarty, C., Rawlins, M., Hinzman, L., Francis, J., Serreze, M., Liljedahl, A., McDonald, K., Piasecki, M., & Rich, R. (2018). *Opportunities and challenges in Arctic system synthesis: A consensus report from the Arctic research community*. City University of New York. https://www.arcus.org/publications/28459

Vylegzhanin, A. N., Young, O. R., & Berkman, P. A. (2020). The Central Arctic Ocean fisheries agreement as an element in the evolving Arctic Ocean governance complex. *Marine Policy, 118*, 1–10. https://www.sciencedirect.com/science/article/abs/pii/S0308597X20301780

Winton, M. (2006). Amplified Arctic climate change: what does surface albedo feedback have to do with it? *Geophysical Research Letters, 33, L03701*. https://doi.org/10.1029/2005GL025244

World Mayors Summit. (2019). C40 world mayors Summit. Copenhagen, 9–12 October 2019. https://c40summit2019.org/agenda/

World Economic Forum. (2016). *Arctic investment protocol. Guidelines for responsible investment in the Arctic*. World Economic Forum. http://www3.weforum.org/docs/WEF_Arctic_Investment_Protocol.pdf

Young, O. R., Berkman, P. A., & Vylegzhanin, A. N. (2020a). Chapter 15: Conclusions. In O. R. Young, P. A. Berkman, & A. N. Vylegzhanin (Eds.), *Informed decisionmaking for sustainability. Volume 1. Governing Arctic Seas: Regional Lessons from the Bering Strait and Barents Sea* (pp. 341–353). Springer. https://link.springer.com/chapter/10.1007/978-3-030-25674-6_15

Young, O. R., Berkman, P. A., & Vylegzhanin, A. N. (Eds.). (2020b). *Informed decisionmaking for sustainability. Volume 1. Governing Arctic seas: regional lessons from the Bering Strait and Barents Sea*. Springer. https://link.springer.com/book/10.1007/978-3-030-25674-6

Chapter 2
(Action): Welcome to Arctic Frontiers 2020, Plenary Introductory Remarks, Opening Speech

Aili Keskitalo

Welcome greetings to Sápmi, the homeland of the Sámi People.

Buorre iđit, ráhkis árktalaš guovllu verddet!

Good morning, dear friends of the Arctic!

It is a pleasure to welcome all of you to Sápmi, the homeland of the Sámi people. On behalf of the Sámi Parliament, I especially like to welcome our Indigenous sisters and brothers to join us her in Romsa this week. According to the Sami origin story, we are the sons and daughters of the sun. The return of our father, the sun, and the transition towards the nightless nights of the Arctic summer gives reassurance. It is a time to be grateful, and hopeful for the future.

It is so important to never forget where we came from. Where our roots belong. I would like to share with you some lyrics from the Sámi rap artist Áilu Valle.

Mun in dárbbaš dáid sániid go oainnán duoddára. (I need no words when I see the tundra.)

His song *Suotnjárat Beaivváža* (The Rays of the Sun) explores the links between Indigenous knowledge, culture and nature. The sight of our ancestral lands can be overwhelming and humbling, and yet our languages tie everything together.

Our tundra is changing. It is growing trees where there should be none. Sometime in the future, we might not recognize our treeless mountain plains anymore. The only memory we might have left, is the word itself duottar, the tundra. The memory of our lands embedded in the languages. It is both sad and beautiful to reflect on how our languages are mirrors of the world we live in.

The themes in this year program go straight to the heart of the future of our unique cultures, languages and communities. The word we use for Sámi knowledge is *árbediehtu*. Literally it means "inherited knowledge". *Árbediehtu* is the collective wisdom and skills of the Sámi people used to enhance our livelihood for centuries. It has been passed down from generation to generation both orally and through work

A. Keskitalo (✉)
Sámediggi/Sámi Parliament of Norway, Norway, Kárášjohka/Karasjok
e-mail: aili.keskitalo@samediggi.no

P. A. Berkman et al. (eds.), *Building Common Interests in the Arctic Ocean with Global Inclusion, Volume 2*, Informed Decisionmaking for Sustainability,
https://doi.org/10.1007/978-3-030-89312-5_2

and practical experience. Through this continuity, the concept of *árbediehtu* ties the past, present and future together.

We have also the concept of *birgejupmi,* a Sámi term for life sustenance, livelihood, in the spheres of economy and social life. *Birgejupmi* is to be understood as survival capacity, how to maintain yourself in a certain area with its respective resources. It requires know-how, skills, resourcefulness, reflexivity, professional and social competence. It ties together people and communities, landscape and natural environment, the ecosystem, healthy social and spiritual development, and identity. It is also a term that connects to a core Sami value – to not overindulge.

That is truly a message for the future.

More rapid changes in society-at-large, the environment and the climate have led to more pressure on the planet, while increasing the demand for Indigenous knowledge. We still have the power of knowledge to live in harmony with the nature.

With these words I wish us the best with Arctic Frontiers 2020.

Ollu giitu – Thank You!

Chapter 3
(Action): Welcome to Arctic Frontiers 2020, Plenary Introductory Remarks

Bjørn Inge Mo

Your Excellencies, dear fellow speakers, guests and friends of the Arctic Frontiers,

It is a great pleasure for me to welcome you all to the Arctic Frontiers 2020 – and to Tromsø. For our international guests I would like to add that you are now visiting Norway's by far largest regional administrative unit, Troms and Finnmark. Since January 1 the former Troms and Finnmark County Councils are merged, and the new one comprises almost 75,000 km^2, which is a little less than the size of Ireleand. So welcome to all of you also to the Arctic region of Norway.

Dear friends, knowledge is the answer to all questions on the Arctic. It was the answer 14 years ago when the Arctic Frontiers was established, it is the answer today, and will be the answer in the time to come as well.

It is an alarming fact that the temperature in the Arctic has increased *twice as fast* as in the rest of the world during the last 50 years. And in this context it is safe to say that what happens in the Arctic, does not stay in the Arctic. Disappearing snow, sea ice, glaciers and permafrost mean massive changes also for the rest of the planet.

To meet these challenges, knowledge is our most valuable resource. We need more research, focusing on sustainability, value creation and living conditions in the High North. And we need it more than ever. It is not necessarily so that yesterday's solutions are the answers to tomorrow's challenges. However, the fact that so many of the world's centres of competence every year gather in Tromsø, gives me hope that we will find a basis for a common understanding and an answer to the development of the Arctic.

For us who live in the North, it is important to underline that we do want also our children and grandchildren to be able to live good lives here. It means that we have to draw the line between exploitation of natural resources and preservation, between the interests of industries and environmental protection. In order to strike the

B. I. Mo (✉)
Tromsø and Finnmark County Government, Oslo, Norway
e-mail: bente.k.helland@tffk.no

P. A. Berkman et al. (eds.), *Building Common Interests in the Arctic Ocean with Global Inclusion, Volume 2*, Informed Decisionmaking for Sustainability,
https://doi.org/10.1007/978-3-030-89312-5_3

balance, there has to be a close dialogue between regional, national and international decisionmakers and the scientific and business communities.

The Arctic Frontiers provides us with a platform for this kind of dialogue. In order to address the challenges and opportunities facing the Arctic region, knowledge has always been one of the corner stones of the Arctic Frontiers. At this year's conference knowledge is also the overarching topic, which will be discussed from many perspectives today and in the days to come.

Dear friends, I wish you interesting days here in Tromsø: Listen, discuss – agree and disagree – and continue discussing, hopefully, also after you have gone home.

And now I would like to give the floor to Markus, a representative of the young Arctic, our future. Markus is also a member of our regional Youth Council.

Chapter 4
(Action): Welcome to Arctic Frontiers 2020, Plenary Introductory Remarks

Markus Haraldsvik

Thank you, Bjørn Inge.

Dear Ministers, Mayors, dear participants,

A warm welcome to the Arctic region of Troms and Finnmark.

First, I would like to thank you for this opportunity to address this session. Young people and *the voice* of young people are important to fulfil the potential of the Arctic region. I stand here today as a representative of Arctic youth, but also as a representative of the Indigenous peoples of the Arctic.

It is important for me to underline what Bjørn Inge just said, we need to strike the balance between sustainability and growth. The Indigenous peoples have lived close to nature for hundreds of years, and we feel a strong sense of responsibility to take care of it. Living close to nature also means being grateful for its abundant resources, which have given us the possibility to live in the North. But just like the rest of the world, the Arctic region is entitled to growth – in a sustainable way.

This is an issue that concerns us young people a lot. We do want to stay in the High North, we want to contribute and to live good lives here. A good life means, among other things, to have interesting jobs and live in communities offering social and cultural services relevant to us. As I am sure you are all aware of, without young people the Arctic will, over time, not be able to develop.

Perhaps the most important element in order to make young people stay in the Arctic, is education. Today many young people leave the region to study, and many of them do not return after their final exams. In a time where the focus on the Arctic and its resources is increasing, it is more important than ever to make the young

M. Haraldsvik (✉)
Bardufoss Upper Secondary School, Bardufoss, Norway
e-mail: bente.k.helland@tffk.no

P. A. Berkman et al. (eds.), *Building Common Interests in the Arctic Ocean with Global Inclusion, Volume 2*, Informed Decisionmaking for Sustainability,
https://doi.org/10.1007/978-3-030-89312-5_4

people stay to contribute to shaping our common Arctic future. We need our educational institutions to offer relevant studies – at local and regional levels.

In my opinion, the best investment decisionmakers can make, is to invest in education and research, because it is also an investment in the future, in development, and in young people.

Thanks for your attention!

Part II
The Arctic Ocean: Evolving Ecological and Sustainability Challenges

Framing Questions

1. How are anthropogenic forces affecting the biophysical systems of the Arctic Ocean?
2. What is the role of knowledge in understanding the requirements for sustainable uses of Arctic marine resources?
3. What institutional arrangements are needed to secure a sustainable Arctic Ocean?

Chapter 5 (Research): Preventing Unlawful Acts Against the Safety of Navigation in Arctic Seas

Alexander N. Vylegzhanin and Ekaterina S. Anyanova

Abstract With the on-going environmental state-change in the Arctic, security issues are becoming more urgent for increasing Arctic shipping. This chapter examines how the existing rules of international customary and treaty law on repression of piracy and other unlawful acts against the safety of maritime navigation might be applicable to the waters of the Arctic Ocean. The main characteristics of such rules – as they are provided in the *1982 UN Convention on the Law of the Sea* and the *1988 Convention on the Suppression of Unlawful Acts against the Safety of Maritime Navigation* – are scrutinized in this chapter in the specific context of the developing Arctic law. This chapter also considers the relevant provisions of other international agreements and soft-law instruments in the light of a broader framework of international customary law.

While addressing the applicability of these universal legal rules to the Arctic Ocean, this chapter summarizes the challenges we face in preventing crimes in the Arctic Region. In this context, this chapter considers the lessons learned from attempts to repress unlawful acts against the safety of shipping in the regions of Somalia, the Strait of Malacca and South-East Asia in general. The chapter suggests the development of a "precautionary regional approach" for anti-criminal activity along the Arctic shipping routes, including transpolar routes.

In sum, this chapter addresses two key questions: how do we build (within the rules of international law which are applicable to the waters of the Arctic Ocean) common interests among Arctic and non-Arctic states to combat unlawful acts against the safety of navigation in those waters and prevent such acts from expanding? To this end, how can we adopt a precautionary anti-criminal approach that takes into account urgencies arising from the warming of the Arctic region and the recession and thinning of ice in the Arctic Ocean and the discovery of new islands previously covered by ice, as well as the relevant increase in maritime economic activities in the Arctic?

A. N. Vylegzhanin
International Law School, MGIMO University, Moscow, Russian Federation

E. S. Anyanova (✉)
Candidate of Legal Science, Legum Magister, Doctor iuris, Kaliningrad, Russia

P. A. Berkman et al. (eds.), *Building Common Interests in the Arctic Ocean with Global Inclusion, Volume 2*, Informed Decisionmaking for Sustainability,
https://doi.org/10.1007/978-3-030-89312-5_5

5.1 Introduction

Potential instabilities are emerging along with potential opportunities in the Arctic Ocean, which is undergoing the largest environmental state-change on Earth (Berkman & Vylegzhanin, 2013). This chapter examines one such potential instability – the risk of emerging piracy and other maritime crimes in the Arctic Ocean. *In concreto* the chapter examines how the existing rules of international customary and treaty law on prevention of piracy and other criminal acts against the security of maritime navigation apply to northern polar waters, taking into account the particular legal regime that governs the Arctic Ocean as well as options as to how the Arctic states might prevent such unlawful acts.

Piracy is not a totally new question for the Arctic region, as will be shown further. The maritime Arctic has also experienced acts of terrorism, including two terrorist incidents in the Russian Arctic that took place in the Arkhangelsk coastal community (in 1998 – when a hostage was taken in a local school and in 2006 – when a bomb blew up near a municipal building). Other potential objects of terrorism or other criminal attacks exist in the Arctic Ocean, such as Russian drilling rigs on the Arctic.

With the rising average temperature in the Arctic, commercial shipping and other uses of the area are growing; criminal communities across the continents are certainly monitoring these trends. The growth in the intensity of navigation (same as in "warm" ocean areas) will very likely be accompanied by an increase of piracy, terrorism and other forms of organized criminal actions.

The main goals of this chapter are to provide not only a comprehensive theoretical assessment of international law rules that are applicable to combating unlawful acts against the security of navigation in the Arctic waters but also options as to how the Arctic states might prevent such acts. However, this chapter does not address "technical matters" affecting maritime safety which are dealt with by the International Maritime Organization (IMO) including such IMO conventions as: *the International Convention on the Safety of Life at Sea* (SOLAS 1974); *the International Convention for the Prevention of Pollution from Ships*, 1973, as modified by the *Protocol of 1978* relating thereto (MARPOL 1978); *the International Convention on Standards of Training, Certification and Watchkeeping for Seafarers* (STCW 1978); and *the Torremolinos Protocol* of 1993 to the 1977 *Torremolinos International Convention for the Safety of Fishing Vessels* (Torremolinos 1993). Nor does this chapter address the activity of vessels involved in military security activities in the Arctic, including the use of naval force. The specific objective of this chapter is to interpret customary and treaty rules of international law that are applicable to combating and preventing piracy and other maritime crimes in the particular legal environment of the Arctic Ocean and its adjacent seas.

To be sure, the conditions that exist in the Arctic Ocean are different from those in warm ocean areas. But experience from the other regions has demonstrated that pirates and other criminals can operate in coastal waters and in the high seas without using vessels that do not meet IMO safety standards. Moreover, piracy is not a totally new question for the Arctic region. Even in the Middle Ages, pirates threatned Finnic

Karelian people and Russians navigating to Grumant (Spitsbergen). Two of the most famous pirates of that era were Phillipus Defos and Jan Mandhaus. These pirates were eventually captured, bringing an end to Arctic piracy in those days (Perabo, 2015). The threat of resumed piracy in the Arctic Ocean is quite real, however. In addition to piracy, illegal actions of Greenpeace and other environmental activists are most likely to increase with the off-shore oil and gas development on the Arctic shelf.

5.2 Climate Change in the Arctic: New Possibilities for Navigation and New Risks Due to Piracy and Other Unlawful Acts Against Shipping in the Arctic Ocean

Between the Pacific and Atlantic Oceans there are three main Arctic routes: 1) the most navigable is the *Northeast Passage (NEP)*, of which the *Northern Sea Route (NSR)* called in the Russian legal documents Sevmorput is the longest part[1]; 2) the *Northwest Passage (NWP)* crossing the Canadian Arctic archipelago; and 3) the *High North ('nordområdene') Arctic Ocean Route* (or *Transpolar Sea Route*) which crosses the Central Arctic Ocean (beyond the exclusive economic zones of the Arctic coastal States). However, due to global warming the Arctic sea ice is dramatically melting. In the past 30 years approximately 10 percent of the Arctic sea ice has melted.

Nowadays spring freshwater ice breakup in the Arctic region occurs nine days earlier than 150 years ago, and the autumn freeze-up occurs ten days later. In Alaska the ground has subsided more than 15 feet (4.6 meters) due to the melting permafrost (Glick, 2018). By the 2050s most of the Arctic sea ice is expected to melt for at least part of the year (National Geographic, 2018).

Less ice means more economic activity: first and foremost, the growth of Arctic shipping and of oil and gas development on the Arctic shelf. The licensing areas of Rosneft in the Arctic Zone of the Russian Federation alone cover the Arctic shelf in the Barents, Kara, the East Siberian Sea, the Laptev Sea, the Chukchi Sea. Thirty significant oil fields are already discovered in the Arctic Zone of the Russian Federation (Arctic and Antarctic Council..., 2019, p. 422) as this Zone is defined by the Russian Legislation (Berkman et al., 2019b). By 2030, geological exploration and large-scale development are expected to be conducted in the Arctic, including

[1] According to the Federal Law of the Russian Federation of 31 July 1998 (titled *"Internal Sea Waters, Territorial Sea and Contiguous Zone of the Russian Federation"*), "Navigation on the seaways of the Northern Sea Route, a historically established national uniform transport communication of the Russian Federation in the Arctic, including through the Vilkitsky, Shokalsky, Dmitry Laptev, Sannikov straits, is carried out according to the present federal law, other federal laws, international treaties of the Russian Federation and Regulations for navigation on the seaways of the Northern Sea Route, approved by the federal enforcement organ authorized by the Government of the Russian Federation".

the Pechora Sea, the Barents Sea, the Kara Sea, the Gulf of Ob and the East Siberian Sea, the Laptev Sea, and the Chukchi Sea. The continental shelf of Greenland and Norway is also expected to be extensively developed in future.

Recent events confirm new opportunities for using the NSR (along the Russian Arctic coasts) as the shortest sea route from northern ports of the Pacific Ocean (including ports of Japan, South and North Korea and China) to European ports and in the reverse direction:

- new ice-breaking oil and gas tankers are able to operate year-round, although they consume considerable fuel an produce approximately 33 percent of carbon emissions in the Russian Arctic. For example, the Arc7 ice-class LNG tanker "Vladimir Rusanov" navigates the Northern Sea Route ("NSR") passage from the port of Sabetta to China without ice-breaking support, taking less than twice the time required to transport a cargo than via the Suez Canal and Strait of Malacca (National Geographic, 2018);
- in July 1990, the motor vessel *"Kola"* owned by Murmansk Shipping Company ("Norilsk" type, 20,000 DWT with Russian highest ice class) transited from Hamburg to Tokyo in 19 days, crossing the NSR in only 8 days;
- in September 2010, the tanker "SCF Baltica" (owned by NOVATEK Company) transited from Murmansk to Ningbo (China) in 23 days, crossing the NSR in 10 days;
- in August 2011, the vessel "Vladimir Tikhonov" became the largest supertanker (162,300 tons deadweight) to transit the NSR, in a record 7.4 days. The following month, the tanker "Palva" (74,940 tons deadweight) eclipsed the NSR transit record in 6.5 days with an average speed about 14 knots;
- in September 2011, the "M/V Sanko Odyssey" became the first Japanese tanker and largest bulk carrier (74,800 tons deadweight) to cross the NSR (Bunik & Mikhaylichenko, 2013).

Moreover, shipping in the Arctic Ocean as a whole is increasing, as recently revealed on the basis of satellite records providing continuous pan-Arctic coverage of individual ships. As noted, "ship traffic is increasing across the entire Arctic Ocean at a faster rate" than in Barents Sea Region, most of which "is ice-free throughout the year" (Berkman et al., 2019a). Other consequences of warming in the Arctic are increasing coastal fishing activity in the exclusive economic zone (EEZ) of Russia and more Russian and foreign ships are used for Arctic tourism. By 2021–2023 more icebreaker ships are expected for polar cruises (National Geographic, 2018). People are also interested in tourism to see the changing environment in the Arctic (WWF, 2020) and to get acquainted with the cultural heritage of Greenland (Ren & Chimirri, 2018). At present commercial fisheries are prohibited in the high seas part of the Central Arctic Ocean by the 2018 CAO international agreement (Vylegzhanin et al., 2020), However, the states expect that commercial fishing will be possible in future in the EEZ of the USA to the north of Alaska at some point in the future (U.S. Department of State, 2019).

To sum up, widespread melting of Arctic sea ice and expanding numbers of ships transiting the Arctic Ocean have implications for the regional maritime security.

To date, most international Arctic shipping has involved the export of minerals and fish from the Arctic region, and the worldwide delivery of goods, including those which are destined for ports on the Arctic coasts.

At present there are already several months a year when safe navigation is feasible via NSR without ice-breaker escorts, though with risks of collision with floating ice. Although modern ice-class vessels (Norilsk Nickel class, LNG-carriers) are able to operate in winter without icebreaker escort, the icebreaking and ice pilotage services will be still needed in the coming years, especially for emergency situations (Icebreaking needs, 2017). In summer and autumn, however, piracy and other unlawful acts at sea may present more risks for the development of future Arctic shipping, as the experience in other parts of the world ocean demonstrates.

5.3 A Review of Piracy in Other Regions and What It May mean for the Arctic Shipping

Piracy at sea is one of the oldest international crimes, and has been characterized as the most dangerous threat to maritime shipping (Mejia et al., 2008). Legal qualification of piracy as an international crime is primarily based on the "Lotus case" considered by the Permanent Court of International Justice in 1927. In this case, Judge Moore described piracy as "an offence against the law of nations" and pirates as "the enemy of mankind – *hostis humani generis* – whom any nation may in the interest of all capture and punish" (Lotus, 1927). The word "punish" was not limited, according to the Court's position, so in practice pirates were put to death at sea, even were hanged on yardarm, by order of an officer. This legal practice was confirmed as legitimate in 1934 by the Judicial Committee of the Privy Council in Great Britain. According to the Committee, a person guilty of piracy committed on the high seas "has placed himself beyond the protection of any State. He is no longer a national, but '*hostis humani generis*' and as such he is justiciable by any State anywhere" (Dixon & McCorquodale, 2003: 148). On the basis of this legal principle, since pirates are "*hostis humani generis*", one can argue that they are not protected by human rights conventions.

Piracy is one of the few crimes that is subject to universal jurisdiction, i.e., the prosecution of piracy is not limited to the jurisdiction of a particular state (Reuland, 1989). Piracy is a crime *erga omnes*, violating the rules of the *law of nations* and interests of *all states*, and jurisdiction over piracy is based first and foremost on *international customary law* (Lotus, 1927).

In 1926 the League of Nations considered the first draft of a Convention on combating piracy (Golitsyn, 2012). The draft was not universally supported, so for a long period customary rules relevant to measures against pirates remained the sole universal legal basis for taking action against *hostis humani generis* (Sidorchenko, 2004).

As a result of the first United Nations Conference the Law of the Sea, the 1958 Convention on the High Seas was adopted. The Convention provides for obligations of States Parties *to cooperate to the fullest possible extent in the repression of piracy* (art. 14). In fact, article 14 is identical to article 38 of the Draft Articles on the Law of the Sea, prepared in 1956 by the United Nations (UN) International Law Commission. The conventional definition of piracy limits the area in which a crime could qualify as piracy: only within the high seas.

During the Third UN Conference on the Law of the Sea a trend towards a more "humanitarian attitude" toward pirates (in contrast to customary rules qualification of pirates as *hostis humani generis*) seemed to prevail, as is reflected in articles 100–107 of 1982 UN Convention on the Law of the Sea (UNCLOS). Given that more than 170 States are parties to UNCLOS, some scholars consider these articles as a modern universal and sole legal regime applicable to piracy. In addition to UNCLOS, the 1988 *Convention on the Suppression of Unlawful Acts against the Safety of Maritime Navigation* (the SUA Convention*) along with its 2005 Protocol* is regarded as the specific "relevant treaty" for the repression of piracy (Nordquist, 1995). Some authors consider that the SUA Convention goes even one step further than UNCLOS by regarding additional offenses at sea (not limited to those committed in the high seas) as piracy (Qureshi, 2017).

Based on these customary and treaty rules for combating piracy, States conduct anti-piracy operations in different areas of the world's ocean. For example, States have sought to address piracy in the Strait of Malacca, a global strategic waterway in the contemporary world transportation system. Despite the coordinated measures of control by Singapore and other states since July 2004 in the Strait of Malacca, the piracy problem still exists.

The piracy and armed robbery phenomenon as a threat to the common interests of civilized nations is also addressed in the 2004 *Regional Cooperation Agreement on Combating Piracy and Armed Robbery against Ships in Asia (ReCAAP).* This document was adopted on 29 January 2009 at the intergovernmental meeting *on maritime security, piracy and armed robbery against ships for Western Indian Ocean, Gulf of Aden and Red Sea States* (26–29 January 2009).

Another example of international law measures against piracy is those addressing the outbreak of criminal activities off the coast of Somalia. In 2006 10 attacks by pirates were reported in the Somalia region; from 2007 the region became home to a sort of organized criminal operation (Transnational piracy, 2011): 78 attacks in 2007 took place near Somalia coasts; in 2009 – 44 attacks, in 2010 – 51, in 2011 – 125 (!). Piracy off the coast of Somalia was addressed in a number of resolutions of the *United Nations Security Council* and the *General Assembly* (United Nations Security Council (UN SC) resolutions 1816 (2008), 1838 (2008), 1846 (2008) and 1851 (2008) and of UN General Assembly (GA) resolution 63/111). According to these resolutions, adopted with the consent of the Transitional Federal Government of Somalia, the States combating piracy could extend their operations *from the high seas to the territorial sea* of Somalia. International legal and political measures (taken afterwards) provided positive results: in 2017 – only 2 attacks were reported in this area; in 2018 – only 1; in 2019 – 0; in 2020 (as of 31 March) – 0.

As noted above, conditions in warm ocean areas such as the Indian Ocean and the Strait of Malacca differ in several respects from conditions in the Arctic Ocean. The extreme cold and harsh environment of the Arctic may make it difficult for pirates to operate there. But those same difficult conditions will make it more difficult to pursue pirates in the Arctic. Newly discovered islands or inhabited islands and rocks in the Arctic Ocean may also present additional advantages to pirates.

In this context, an example of international legal efforts against piracy in the format of "hot spot activity" might be of some utility in the context of the Arctic Ocean. The waters of the *Guinea Gulf* run along the coasts of Cote d'Ivoire, Ghana, Togo, Benin, Nigeria, Cameroon, Equatorial Guinea, the Republic of the Congo, and the Democratic Republic of the Congo. In 2011 piracy in the waters off these coasts became another international crisis, especially off the coast of Nigeria. For example, piracy and other attacks in these waters were responsible for *up to a 70%* (!) *decrease* in trade through the port of Cotonou in 2012 (Piracy, Report for the period 1 January – 30 June 2016; Piracy, Report for the period 1 January – 31 December 2015; Piracy, Report for the period 1 January – 31 December 2011). In the Gulf of Guinea, however, most attacks occurred in marine areas under the jurisdiction of the Gulf's coastal states. The facts reveal a shaky success story here. In 2011, piracy in the Gulf triggered an international crisis of shipping off the coast of Nigeria. Coordinated measures of the international community aimed at enforcing law order and maritime security in the Gulf produced positive results. While more than 40 attacks were committed in Gulf of Guinea in 2012 (31 – off the coast of Nigeria), the number of attacks in the Gulf decreased after 2013. In 2016 the number of registered attacks was 10, in 2017 – 7, in 2018 – 30, in 2019 – 20, while in 2020 – 16 (Piracy, Report for the period 1 January – 31 March 2020).

This brief summary reveals that the international community has already achieved positive results in battling piracy and other armed attacks on vessels in marine areas adjacent to Somalia and in South-Eastern Asia and other problematic regions of the world by means of the international cooperation.

Although at present the Arctic is free of piracy, the average temperature in Arctic is rising, while commercial shipping and other uses of the area are growing (McGrath M, 2017), so temptations of piracy and other criminal attacks in the Arctic waters might be also growing (Vylegzhanin & Anyanova, (2020). Besides, the world statistics remain quite alarming: 201 attacks in 2018, 162 in 2019 and 47 in 2020. More specifically, in 2018 in the sea areas adjacent to Nigeria 22 attacks against ships were committed; and in waters adjacent to Indonesia – 9 such illegal acts took place and in 2019 – 14 attacks occurred in the sea waters adjacent to Nigeria and 3 – near the coast of Indonesia, in 2020 – 11 and 5 correspondingly. In 2019 in Latin America's waters 8 attacks took place (1 in Brazil, 1 in Colombia, 1 in Dominican Republic, 1 in Peru, 4 in Venezuela), in 2020 – 6 attacks (1 in Brazil, 1 in Colombia, 1 in Haiti, 3 in Peru). In short, incidents of piracy and other armed attacks against ships or persons or property on board such ships continue to take place in waters near Eastern Africa and Latin America and seas of Indian Ocean and criminal organization of piracy is successfully based on the coasts. So a question arises whether inhabited coasts of the numerous islands in the Arctic might be similarly used.

5.4 Is Piracy a Threat to the Growing Economic Activities in the Arctic Ocean?

As the era of intensive international Arctic navigation is beginning, maritime security issues related to terrorism and piracy are already arising in the Arctic Ocean: as observed, new ice-free northern passages need more security control (Bekkers, 2019). New and different challenges in the Arctic region will arise with melting ice and opening of new sea lanes, and this development definitely impacts not only Arctic but also non-Arctic states (Sousa, 2019). Although there have been some positive technological and regulatory developments, the possibility of new risks to maritime safety remains high for the Arctic shipping (Klepikov et al., 2005).

The risks of attacks on tankers and other merchant vessels in the Arctic Ocean are usually connected to the developing production of offshore oil and gas resources and possible attacks not only on tankers but also on offshore rigs and other permanent structures on the Arctic shelf, as well as to the increasing navigation that takes place in the Arctic areas: from 1298 ships in 2013 to 1628 at present, that is a 25% increase (The increase in Arctic shipping, 2020).

Today there exists a high possibility for vessels to encounter icebergs in these waters, especially near the Svalbard (Spitsbergen) archipelago, the Novaya Zemlya islands, and the Franz Josef Archipelago (Shaposhnikov et al., 2017, 76–77). In expanding ice-free sea routes there still remains a threat of changing winds, currents and ice flows. However, existing ice along the Arctic sea routes partly "protects" Arctic shipping from the attention of pirates and other criminal groups.

Criminal communities across the continents (Elgsaas, 2017, p. 112) are certainly monitoring the on-going decrease of sea areas permanently covered by ice in Arctic and the resulting increase in Arctic ship traffic. The growth in the intensity of navigation in "warm" ocean areas (not covered by ice) is often accompanied by an increase of piracy, smuggling, illegal migration, drug trafficking, terrorism and other forms of organized crime. Recent criminal activities show the ability and readiness of terrorists and other criminals to conduct attacks in unexpected areas not commonly associated with terrorist threats (Elgsaas, 2017, p. 110).

Will the Arctic states, especially those with the long coastlines (Russia and Canada) be vulnerable to piracy, terrorist attacks, and other criminal activity by midcentury?

Some analysts argue that the maritime routes in the Arctic will be protected by military bases in USA (Alaska), in Russia, in Canada, in Norway and in Greenland (Vyatkin 2019), as well as by the law enforcement vessels (both naval and coast guard vessels) of the Arctic States. We do not share this optimistic view. A combination of a number of factors might create new threats to the security of the vessels transiting the Arctic waters. These factors include the increased amount of shipping and oil and gas development in the region, as was noted earlier, and the existence of a number of inhabited islands or without permanent population in the Arctic Ocean from which pirates might operate. In addition to that, many vast areas on the Arctic mainland coast with the rapidly diminishing sea ice cover lack proper infrastructure to effectively protect Arctic shipping from criminal acts even within the state territory, in particular, to protect crews of merchant vessels and even coastal communities from armed attacks. Following the collapse of the Soviet Union, the number of criminals in the Russian Federation has increased, in part as a result of the greater disparities in wealth between the rich and the poor. As noted above, two terrorist incidents took place in the Arkhanglesk region in the past 30 years, suggesting that other incidents are possible. Russian drilling rigs the Arctic seem to be also highly attractive as potential objects of criminal attacks. Terrorists may also take into account current tensions between Russia and USA, etc. (Elgsaas, 2017, p. 119).

A terrorist attack on an offshore oil or gas installation on the Arctic coast or on a permanent facility at sea may cause a huge environmental disaster. The scale of security threats might increase if even one act of piracy or one armed robbery proves to be successful, thereby attracting more "criminal investors". The Arctic states are obliged to protect the security of navigation in the Arctic, as well as to protect local people who live and work in the Arctic, including not only the coastal communities but also persons on board ships and the personnel of offshore oil and gas installations.

In short, we need to keep the Arctic waters protected from any growth in criminal activity. At present these threats are most hypothetical – often relating to Greenpeace, an environmental organization concerned with preservation of the marine environment, sometimes not by legitimate means. For example, in 2011 Greenpeace activists approached the "Leiv Eriksson" platform and were arrested by the Danish officers. In light of this incident, Cairn Energy filed for an injunction against Greenpeace, which a Dutch court granted. According to the injunction, any violation of the safety zone established around the platform "Leiv Eriksson" was subject to a fine, the amount of which was established between 50,000 and 1 million euro (Cairn Energy, 2011). In 2012 Greenpeace activists also embarked on drill ships "Noble Discoverer" and "Kulluk". Royal Dutch Shell, owner of these vessels, filed a petition seeking a temporary restraining order from the District Court of Alaska "with the purpose of preventing Greenpeace from interfering with Shell's drilling plans in the Arctic Ocean" (Memorandum, 2012). The court granted this petition, which meant that Greenpeace activists were prohibited from trespassing on these ships; from interfering with the operation or movement of both platforms; from barricading, blocking, or preventing access or egress from both platforms; and from creating dangers or threatening Shell's employees or its visitors/affiliates who are present on, or as they enter or exit, the two ships (Order granting plaintiff's motion, 2012). Taking into account the geography of Shell's drilling operations off the coast of Alaska, the District Court of Alaska issued two orders: one relates to ports and territorial waters of the United States (Order granting motion for preliminary injunction); and the other – to the U.S. exclusive economic zone north of Alaska (Order pending motions, 2012). Due to the oil and gas deposits development in the Arctic, eco-terrorists might target oil and gas facilities as a part of their campaign against economic projects damaging the vulnerable Arctic environment (Elgsaas, 2017. p. 118).

In contrast to these "soft" environmental actions of protest against oil and gas installations at sea, potential criminal acts against any facilities at sea might be devastating and the Coast Guards of the Arctic states should be preparing to address such crimes in the Arctic Ocean.

The evolving security threats in the Arctic region (which are increasing with the economic development of the Arctic and melting ice) require common reactions first and foremost from the Arctic states. Counter-terrorist exercises started in the Russian state territory in the Arctic (Elgsaas, 2017, p. 130–131). However, the vast Arctic waters beyond the state boundaries of the Arctic States do not seem to be protected around the clock. The common reaction of the Arctic states to the maritime security issues remains to be properly developed.

5.5 The Northern Sea Route: Towards an Initial "Precautionary Anti-piracy Approach"?

The shortest way from Northern Europe to Siberia or further to Asian States (such as Japan and China) is via the Northern Sea Route (the NSR), along the Russian coast in the Arctic, currently functioning under the support of Russian ice-breakers and pilots and under the protection of Russian Coast Guard vessels. The marine environment within the NSR is also protected under the standards of the Russian legislation (Jekspertnyj Sovet, 2017).

The Northern Sea Route is defined in the Russian Law as "the national transport communication" which was formed in Russia "historically", in the Bering Strait from the Kara Gate to Providence Bay (Arctic and Antarctic Council…, 2019, p. 198). During the USSR period, the NSR was considered as a route for domestic navigation only within the ports along the country's Arctic coast. Today, the NSR is also regarded as an international sea route of increasing importance, primarily because more Russian Arctic natural products are being shipped out of the Russian Arctic to global markets. A relatively small number of other foreign ships are conducting trans-Artic voyages along the NSR. Some non-Arctic States plan to cooperate with Russia in developing additional transport routes to the North of the Russian EEZ (Arctic and Antarctic Council…, 2019, p. 191). Considering the continued retreat of the Arctic sea ice due to climate change, as well as the fact that the NSR is approximately 40% shorter than the Suez Canal Route, the NSR could become a prominent alternative to the "southern" sea routes – particularly for vessels travelling from China, Japan and Korea to major European ports (Lee and Song 2014) during at least the summer and early autumn months. Use of the NSR can shorten transit times considerably. For example, a transit from Murmansk to Yokohoma via the NSR can be completed in 18 days, as compared to 37 days via the Suez Canal (Rosatomflot, 2020).

Developing the viability of the Northern Sea Route is one of the national priorities of Russia in the Arctic. Russia is seeking to develop the entirety of its Arctic Zone, as is confirmed in the recently promulgated "Fundamentals of the State Policy of the Russian Federation in the Arctic for the Period up to 2035" (Fundamentals 2020). Based on such governmental priorities, Russian legislation on encouraging international shipping via the NSR is developing (Berkman et al., 2019b).

Law-abiding people as well as pirates and their sponsors are certainly taking note of the evolving international shipping potential of the NSR (Arctic and Antarctic Council..., 2019, p. 422):

- Today the existence of ice along the NSR during the winter still limits the navigable season in these waters, but such a limitation is diminishing.
- Effective oil and gas development on the Arctic shelf requires development of maritime infrastructure of the NSR (Pryahina et al., 2018). Oil and gas fields discovered on the Arctic shelf of the Barents, Pechora and Kara Seas are partly licensed and initially developed. Maritime transport makes modest progress in delivering hydrocarbons to markets outside the region (Kryukov, Poudel, 2020, 153–154).
- Slowly increasing traffic volumes are generated for the NSR by the carriage of Russian fish products, of Korean electronic and bulk goods, by the export of Arctic minerals to global markets, and especially the transport of Russian natural gas (Yamal LNG and Arctic LNG projects) to foreign markets. International shipping companies operate on the NSR, approximately 40% of them are non-Russian (mostly Norwegian) (Maritime Transportation in the North, 2018, 101). China transports a lot via the NSR; by 2025 China plans to conduct up to 20% of foreign trade via the NSR on its own (Arctic and Antarctic Council..., 2019, p. 399). Supplying local Arctic communities also comprises an important part of shipping through the NSR, as well as the rise of cruise tourism, of marine research, and of fisheries in the Russian EEZ. The cabotage as well as destination traffic with non-Russian ports constitute most of the NSR transportation – in total, approximately 10 million tons per year (Maritime Transportation in the North, 2018, 101).
- Russia plans to build new ice-breakers to make the specific lanes of the NSR navigable all the year round. The Russian government invests in the use of this route, including by building additional search and rescue (SAR) infrastructure, e.g. opening of the new SAR centers in Dudinka and Naryan-Mar, Pevek and Anadyr, as well as in Tiksi. Still, most shipping through the NSR at present is "intra-Arctic" in nature, as is the case elsewhere in the Arctic, for example along the arctic coasts of Canada and Greenland.

Current environmental trends have led people to regard the NSR as a more and more attractive alternative to the southern routes from Europe to Asia (via Strait of Malacca, for example). But if the risk of attacks of pirates and other criminals on the NSR becomes a reality, the situation might drastically change – in spite of the fact that the use of the NSR may reduce the distance of a voyage by a third, and may reduce vessel's freight and the costs of the delivery by 30–35% (Bekkers, 2019); which means also reducing the environmental pollution relevant to the same economic result.

Today there is essentially no risk of piracy along the NSR, in contrast to such risks during a voyage of ships between Europe and Northeast Asia via the Indian Ocean. But will the NSR remain an attractive and safe alternative for maritime shippers in the long run? Approximately two-thirds of all trade presently transported via southern shipping routes (through the Suez Canal) could ultimately be rerouted to the shorter NSR, if: (a) relevant new (or additional) port infrastructure is created in the NSR, especially a regular container line between West Europe and Japan and China, and (b) the NSR can remain immune from piracy and other unlawful acts against shipping. Most surveys and studies show minimal future trans-Arctic navigation use of the NSR except for a short summer navigation season. Further development of this route in the nearest future will not impact the world transportation significantly: around 300 vessels in Arctic per year transport 7 479 000 tons total cargo cannot be compared with approximately 17 000 vessels via the Suez Canal transporting 974 million tons of cargo. But the global warming could cause the increase of the commercial traffic on the NSR (Maritime Transportation in the North, 2018, 98–103).

The reality is that although the NSR is considered as "a national transport communication of the Russian Federation" (regulated on the basis of stringent Russian environmental legislation), the NSR is becoming more and more in demand for international merchant shipping, especially for destinational shipping of natural (raw) products, which is in the current commercial interests of Russia. One expects that the transport of hydrocarbon products by ships in the Russian Arctic will expand in 2021- 2025, up to 65,000,000 tons per year. And that may also be a reason for pirates and other criminals at sea to focus on shipping on the NSR. Polar days and polar nights, fogs, barren coastal areas without permanent population – these might be additional factors to be considered by "criminal investors".

Neither the new "Fundamentals of the State Policy of the Russian Federation in the Arctic" (2020) nor any other legal act of the Russian Federation on the NSR addresses such threats as piracy and other unlawful acts at sea.

Hundreds of vessels have already been escorted by the Russian nuclear ice-breakers via the NSR. The annual shipping volume via the NSR at present constitutes 19 600 000 tons without transit freight, transported mostly from the Gulf of Ob: Sabetta port (gas) and Kamenny Cape (oil). Such shipments require strong ice-breaking support during 8–9 months a year (Rosatomflot, 2020). Today, even with climate change in the Arctic, enhanced icebreaking capabilities are needed to ensure safe Arctic navigation. Before its collapse, the USSR had the biggest fleet of icebreakers in the world. The capabilities of Russia today are much more modest, though nuclear ice-breakers of a new generation still promote the economic interests of Russia in the Arctic. Some of new icebreakers are reportedly armed with anti-aircraft missile systems, artillery mount, radar and hydro-meteorological stations (Gudev, 2016). In this context such ice-breakers might be used to combat piracy or other criminal acts not only along the NSR, but potentially in other areas of the Arctic Ocean if appropriate arrangements between Russia and other Arctic coastal States are reached. It is extremely doubtful that countries such as the USA and Canada are going to involve Russian ice-breakers in the law enforcement activities in their EEZ, but these countries might cooperate on such activities in the Central

Arctic Ocean, beyond their EEZ. In this case icebreakers (beside their main function of escorting ships through the ice) may have also other functions, from patrolling sea routes to towage to the nearest port of detained vessels of criminals.

The Arctic coastal states might agree to arrange patrols and to establish additional border outposts, to bolster the resources of their Coast Guards and SAR centers. Relevant coastal states might agree to broaden patrolling by air and maritime space surveillance. The huge Arctic spaces can only be effectively controlled through cooperation among the Arctic states. Some authors propose the use unmanned flying objects and satellites to monitor the Arctic sea lanes (Konyshev & Sergunin, 2013), which might also be an option as a part of a regional arrangement of the Arctic states.

5.6 Towards a Regional Maritime Security Regime in the Arctic?

As was noted, the problem of keeping the Arctic maritime routes secure has already arisen. The unlawful acts that may take place in the Arctic Ocean are to be addressed on the basis of relevant provisions of *UNCLOS* and the *SUA Convention*, taking into account the 1958 *Geneva Convention on the High Seas* and the *2005 Protocols* to the SUA Convention. The general legal bases are applicable *customary rules* as formulated first in the "*Lotus case*" by the *Permanent Court of International Justice* in 1927.

Of the eight States lying within the Arctic Circle – Canada, the Kingdom of Denmark (Greenland), Finland, Iceland, Norway, the Russian Federation, Sweden and the United States (Alaska) – only the United States has not ratified UNCLOS. However, the United States regards UNCLOS rules on navigation as reflecting customary law and abides by them.

In its territorial sea, each of the Arctic coastal State exercises sovereignty subject to relevant rules of international law (article 2 of UNCLOS). In the contiguous zone, UNCLOS provides for the rights of the coastal states in order to prevent infringement of its customs, fiscal, immigration or sanitary laws and regulations. These rights could be also used for prevention of the drug trafficking, smuggling of people, and other crimes at sea.

Some Arctic routes are situated partly (the NSR) or fully (the North West Passage) in the internal waters of the relevant coastal state. However, there is a dispute between Canada and the United States regarding the status of the NWP: Canada maintains that the NWP is a part of its internal waters (without the right of innocent passage), whereas the United States believes the NWP is a strait used for international navigation with the right of transit passage (Rowe, 2019). This dispute, however, should present no barrier to the possibility that the United States, Canada and other Arctic states could collaborate in preventing unlawful acts in the Arctic waters, including those which are adjacent to the Canadian Arctic coast. The same is true for U.S.-Russia disagreements regarding the status of some parts of the NSR including Vilkitskiy and other straits. Russia qualified these coastal waters under its sovereignty since the 18th century and this qualification was confirmed in 1985 Governmental Decree on the Arctic baselines (Zhuravleva, 2019:110–113).

Indeed, some maritime security issues are already addressed in IMO instruments including the *1974 Safety of Life at Sea Convention (SOLAS). The International Shipping and Port Facility Security*, the anti-terrorist amendment to *SOLAS (chapter XI-2)*, and the *SUA Convention* with two *2005 Protocols* address crimes committed at sea (Vylegzhanin & Anyanova, 2020). The provisions of these documents might be applicable to the Arctic Ocean, with proper adaption.

A reasonable option for the Arctic states is not only to rely upon the applicable universal rules of international law on combating piracy and other unlawful acts but also to create *lex specialis* in this area for the Arctic region. This might become another important contribution to the development of the Arctic regional legal regime. Under the auspices of the Arctic Council, legally binding regional agreements strengthening cooperation, coordination and mutual assistance among Arctic nations have already been negotiated (the 2017 *Agreement on Enhancing International Arctic Scientific Cooperation;* the 2013 *Agreement on Cooperation on Marine Oil Pollution Preparedness and Response in the Arctic;* and the 2011 *Agreement on Cooperation on Aeronautical and Maritime Search and Rescue in the Arctic*). Creating a new regional agreement on preventing piracy and other unlawful acts in Arctic waters could be the next challenge. However, it is not obvious that the development of a new Arctic agreement on maritime security should become the responsibility of the Arctic Council, since the latter is not a "classical", "operational" organization formed by States, but an intergovernmental forum, albeit one with a permanent Secretariat. As an option this task could be addressed by special agencies of the Arctic states whose functions are targeted on the security matters at sea, like the Coast Guards, first and foremost.

5.7 The Content of the Regional Arctic Anti-criminal Agreement: Further Options

It is becoming more and more obvious that with the increasing maritime activity in the Arctic in recent years the risk of large-scale criminal acts in the Arctic waters have also increased (ACGF, 2019). Commercial shipping – whether in the territorial sea, in the EEZ or on the high seas – is not isolated: security issues are interrelated with shipping. As was noted above, a priority option might be to develop a regional agreement among the Arctic states, aiming at coordinated measures against pirates and other criminals in the Arctic Seas. The Arctic States could also consider opening such an agreement for accession by non-Arctic states. In this format, Arctic and non-Arctic states may have a mechanism for combating any crimes in the Arctic waters, both in the EEZ and on the high seas, including drug smuggling, illegal migration, illegal shipping of weapons, etc.

Today there exists in the Arctic region an impressive level of cooperation among the border guards, coast guards, and SAR services of the Arctic states.

In 2000 the North Pacific Coast Guard Forum was established and in 2007 the North Atlantic Coast Guard Forum (NACGF) was formed as informal platforms for encouraging the exchange of the best practices among agencies with coast guard missions. They serve also as fora to conduct joint exercises addressing transnational threats such as drug smuggling and illegal migration. One of the NACGF subcommittee working groups deals with "maritime security".

On October 30, 2015 the Coast Guard agencies of the eight Arctic states signed a Joint Statement formally establishing the Arctic Coast Guard Forum (ACGF). The ACGF Joint Statement avoids addressing delicate legal issues (such as overlapping claims to the Arctic Ocean seabed or territorial boundary disputes). However, this document does provide a basis for the ACGF to address maritime security issues on an operational level. The ACGF coordinates effective use of the assets of the eight Arctic coast guards.

The Forum holds two annual meetings per year organized by the chair State. Chairmanship duties of the ACGF rotate every two years in concert with the Chairmanship of the Arctic Council. Iceland occupies Forum's chair in 2019–2021 (Finland in 2017–2019; the United States – in 2015–2017). Russia will chair the Arctic Council and the ACGF for the two coming years, starting in 2021.

Decisions are made in the Forum on the basis of consensus (ACGF, 2019). As a new forum for pan-Arctic security cooperation, the ACGF has a mandate to strengthen multilateral, bilateral and regional cooperation in the specific areas, including the coordination of functions of coast guard services within the region. The ACGF deals with the maritime security by sharing information about the Arctic region and integrating scientific research in support of Coast Guard operations and joint practical measures on maintaining security at sea. Some authors question why the ACGF has not yet taken a high-profile role in the region, in particular, in connection with naval exercises (Regehr, 2017). It is also noted that the ACGF establishes basis for confidence-building between the Arctic states including the matters of cooperative law enforcement and other maritime security Arctic mechanisms (Regehr, 2017).

One should mention, however, that the ACGF has certain limitations. This forum has never been used for international law-making. However, it could be used as a forum for the development by the Arctic States of a regional binding agreement on Arctic maritime security. Usually governments use other fora to develop international maritime agreements, at the regional level or within the IMO. However, the ACGF offers an attractive option as a venue in which to develop a new Arctic anti-criminal agreement, considering that it has successfully adopted joint statements and protocols for emergency response and combined operations in the Arctic Ocean. The fact that the ACGF also has experience in promoting collaboration by sharing information and best practices also supports this idea.

Working together, the Arctic states might develop an effective legal framework for ensuring appropriate anti-criminal security measures in the Arctic Ocean. Taking into account that IMO has failed to help the littoral nations in the Straits of Malacca to develop an effective anti-criminal security regime, and that the success in combating piracy off the coast of Somali was achieved primarily with little IMO involvement, we do not consider that the IMO should be the primary forum for developing a special regional regime of anti-criminal security in the Arctic Ocean. On the contrary, the necessity of strengthening regional cooperation of the Coast Guard services through the Arctic Coast Guard Forum (ACGF) – specifically targeted on the security matters – seems a better option. A "compromise option" might be as following: to give the ACGF primary responsibility in developing such a regional agreement with support from the Arctic Council.

The "common Arctic issues" mentioned in the 1996 Ottawa Declaration that established the Arctic Council require development of more "smart" regulations in the law of the sea, both at the universal and regional levels, to balance national interests and common interests in the Arctic Ocean (Berkman et al., 2017). Within this trend, the ACGF might play a leading role in constructing professional mechanisms for combating crimes in the Arctic waters.

The problem of the anti-criminal Arctic security *de lege ferenda* is not limited to cooperation among the coast guards of the Arctic coastal states, or to the necessity to build more patrol ships or to use some of the military ships of the Arctic states for this purpose. In such a regional format the Arctic maritime security legal network (based on inter-state cooperation) may include the development of common patrolling marine surveillance aircrafts for anti-piracy control; jointly building new patrol icebreakers; developing SAR and surveillance capabilities in the process of monitoring of the Arctic navigational routes; international patrol icebreakers may be properly equipped, including the best available navigational means, reliable SAR systems, etc.

Concluding such a regional anti-criminal agreement for the Arctic waters before such crimes *de facto* occur will in a very real sense constitute a precautionary approach to strengthening the regional legal order and stability in the Arctic. This approach (aimed at preventing crimes at sea before they actually take place on the Arctic shipping routes) seems also a better option for non-Arctic flag states. Such an agreement would be a wonderful contribution in the common-interest building in the inter-state relations in the Arctic.

5.8 Conclusion

The threats of piracy and other unlawful acts at sea are a common concern for the Arctic states and for other states that are involved (or will become involved) in Arctic shipping. Cooperative measures to prevent piracy and other unlawful acts in the Arctic waters is a concrete area in which the Arctic states are able to work together even at a time of growing tensions elsewhere in the world.

The necessity of "smart" regional policies in the high north is one of the future challenges for the Arctic states to develop for combating and even preventing new crimes in the Arctic waters.

The ice-free Northern routes require strategic mobility throughout the Arctic for the security purposes based on common interests of the Arctic and non-Arctic states. The international attention to the issues of the maritime security in the Arctic has intensified in the recent times. The common interest in the maritime security in the Arctic Ocean must be built with the consideration of the perspectives of diverse stakeholders (Arctic and non-Arctic States, international organizations, state authorities, and local communities, etc.) in an inclusive manner. The challenge for the preventive regulation in this special area is again the balance between existing national interests (sovereignty and sovereign rights) and common needs (to ensure security in the Arctic waters). An increased and coordinated "security presence" of the Arctic states in the region as a short term international solution (with greater ice-breaking capability and ice-strengthened patrol ships and the relevant best available technologies) may contribute to preventing new long-term maritime security threats.

This might be accompanied with infrastructure improvements: harbor and dock reconstruction, installation of modern aids to navigation, including satellite navigational systems and up-dated charts, and also traffic separation systems for the evolving high intensity traffic sea areas, with better SAR capabilities, effective communications networks, etc. Some of these are already in place, including those on the ground and along the Arctic coasts, some just need to be enhanced, and some need to be created – especially on the islands in high latitudes.

In this context, the Arctic states might take into account the difficult experience of repression of unlawful acts in other regions, described above –not to copy those efforts exactly but to help informed decisionmaking for the Arctic region. The list of maritime security measures of the Arctic States might include diverse international legal measures at national, bilateral and regional levels. While for the area of Somalia the UN General Assembly (GA) resolutions were in use and for the area of the Gulf of Guinea the coordinated enforcing law order operations took place, it is shown in this chapter that different international mechanisms to prevent unlawful acts in the Arctic are available. In addition to a possible regional agreement regarding such a preventive measures, it might be advisable to organize anti-terrorism training in the Arctic States in order to work out different scenarios of possible terrorist attacks. It might also be a good option to build additional patrol ships with ice-breaking functions or to use on a coordinated basis some of the military ships of the Arctic states for this purpose. Another optional measure might be to organize common patrolling marine surveillance aircrafts for anti-piracy control; and to develop SAR and surveillance capabilities in the process of monitoring of the Arctic navigational routes. The Arctic States might organize international tenders for declaring "the best available navigational means" for the "most reliable SAR systems", etc. In this way they might develop preventive maritime security mechanisms, before the first criminal incident occurs in the Arctic.

Optional suggestions in this specific area of knowledge presented in this chapter may serve as a starting point for formulating relevant coordinated Arctic policy and corresponding legal instruments guiding further cooperation of the Arctic States. As an optional priority, this chapter emphasizes the importance of preventive measures through regional collaboration to ensure the security of shipping in the Arctic Ocean well before the first criminal attack on ships or on permanent oil and gas facilities at sea occurs. The Arctic and non-Arctic States will benefit if such a precautionary anti-criminal approach aimed at prevention of piracy and other unlawful acts becomes a part of "smart" governance of the Arctic Ocean and its seas.

References

ACGF. (2019). *Finland's Chairmanship of the ACGF. 2017–2019* (Chairmanship report). https://www.raja.fi/download/77927_ACGF_raportti_web.pdf?6dcb41d342f9d688. Accessed 15 Feb 2020.

Arctic and Antarctic Council at the Council of Federation of the Federal Assembly of the Russian Federation. (2019). Annual report. Moscow, 2019. On the status and problems of the legislative support of the realization of the strategy of the developments of the Arctic zone of the Russian Federation and ensuring of the national security for the period till 2020.

Bekkers, F. (2019). *Geopolitics and maritime security. A broad perspective on the future capability portfolio of the Royal Netherlands Navy*. https://hcss.nl/sites/default/files/files/reports/07-Geopolitics%20and%20Maritime%20Security-web.pdf. Accessed 25 Feb 2020.

Berkman, P. A., & Vylegzhanin, A. N. (2013). Preface: International, interdisciplinary and inclusive perspectives. In *Environmental security in the Arctic Ocean* (pp. XIX–XXV). Springer.

Berkman, P. A., Vylegzhanin, A. N., & Young, O. R. (2017). Application and interpretation of the agreement on enhancing International Arctic Scientific Cooperation. *Moscow Journal of International Law, 3*, 6–17.

Berkman, P. A., Fiske, G., Royset, J. A., Brigham, L. W., & Lorenzini, D. (2019a). Next-generation Arctic marine shipping assessments. Governing Arctic Seas: Regional lessons from the Bering Strait and Barents Sea (Vol. 1, pp. 241–268). Springer.

Berkman, P. A., Vylegzhanin, A. N., & Young, O. R. (2019b). *Baseline of Russian Arctic Laws*. Springer.

Bunik, I. V., & Mikhaylichenko, V. V. (2013). Legal aspects of navigation through the Northern Sea Route. In P. Berkman & A. Vylegzhanin (Eds.), *Environmental security in the Arctic Ocean* (pp. 231–239). Springer.

Cairn Energy wins injunction against Greenpeace. (2011). https://www.bbc.com/news/uk-scotland-13719485 Accessed 15 Feb 2020.

Crépin, A.-S., Karcher, M., & Gascard, J.-C. (2017). Arctic Climate Change, Economy and Society (ACCESS): Integrated perspectives. *Ambio, 46*(Suppl 3), 341–354.

Dixon, M., & McCorquodale, R. (2003). *Cases and materials on international law*. Oxford University Press.

Elgsaas, I. M. (2017). *Counterterrorism in the Russian Arctic: Legal framework and central actors*. Arctic and North. 2017. No. 29. pp. 110–132. Fundamentals of the State Policy of the Russian Federation in the Arctic for the Period up to 2035 approved by Presidential Decree 05.03.2020 N 164. http://www.consultant.ru/document/cons_doc_LAW_347129/ Accessed 15 Aug 2020.

Glick, D. (2018). *The Big Thaw*. https://www.nationalgeographic.com/environment/global-warming/big-thaw/ Accessed 16 Aug 2020.

Golitsyn, V. (2012). Maritime security (case of piracy). In H. Hestermeyer et al. (Eds.), *Coexistence, cooperation and solidarity. Lieber Amoricum Rüdiger Wolfrum* (Vol. II, pp. 1157–1176). Martinus Nijhoff Publishers.

Gudev, P. J. (2016). Nevoennye ugrozy bezopasnosti v Arktike. *Mirovaja jekonomika i mezhdunarodnye otnoshenija, 60*(2), 72–82.

Icebreaking needs and plans on the Northern Sea Route (2017, March) Aker Arctic Technology Inc Newsletter. https://akerarctic.fi/app/uploads/2019/05/arctic_passion_news_1_2017_Icebreaking-needs-and-plans-on-the-Northern-Sea-Route.pdf Accessed 13 Aug 2020.

International Chamber of Commerce International Maritime Bureau (ICC IMB). (2016). Piracy and armed robbery against ships. Report for the period 1 January – 31 December 2015. London: ICC IMB. http://www.icc-ccs.org/. Accessed 2 Apr 2019.

International Chamber of Commerce International Maritime Bureau (ICC IMB). (2020). *Piracy and armed robbery against ships*. Report for the period 1 January – 31 March 2020. ICC IMB, London. http://www.icc-ccs.org/. Accessed 28 Apr 2020.

International Chamber of Commerce International Maritime Bureau (ICC IMB) (2016) Piracy and armed robbery against ships. Report for the period 1 January – 30 June 2016. ICC IMB, London. http://www.icc-ccs.org/. Accessed 02 Apr 2019.

Jekspertnyj Sovet pri Pravitel'stve RF. Rabochaja gruppa «Razvitie Arktiki i Severnogo morskogo puti». Severnyj morskoj put' – glavnaja transportnaja arterija Rossii. (2017). http://будущее-арктики.рф/severnyj-morskoj-put-glavnaya-transportnaya-arteriya-rossii/. Accessed 30 Apr 2017.

Klepikov, A., Danilov, A., & Dmitriev, V. (2005). Consequences of Rapid Arctic Climate Changes. In M. Nordquist, J. Moore, & A. Skaridov (Eds.), *International energy policy, the Arctic and the law of the sea* (pp. 277–281). Martinus Nijhoff Publishers.

Konyshev, V. N., & Sergunin, A. A. (2013). Strategii inostrannyh gosudarstv v Arktike: obshhee i osobennoe. In I. S. Ivanov (Ed.), *Arkticheskij region: Problemy mezhdunarodnogo sotrudnichestva: Hrestomatija v 3 tomah* (Ros. sovet po mezhd. delam, Vol. 1) (pp. 112–144). Aspekt Press.

Kryukov, V., & Poudel, D. (2020). Economies of the Barents Sea Region. In O. R. Young, P. A. Berkman, & A. N. Vylegzhanin (Eds.), *Governing Arctic Seas: Regional lessons from the Bering Strait and Barents Sea* (Vol. 1, pp. 143–164). Springer.

Lee, S. W., & Song, J. M. (2014). *Economic possibilities of shipping though Northern Sea Route*, p. 418.

"Lotus"-case (Fr. v. Turk.). (1927). P.C.I.J. Ser. A. No. 10; Collection of judgments. Leyden: A.W. Sijthoff's Publishing Company.

Maritime Transportation in the North. (2018). Maritime activity on the Northern Sea Route. Section (06). Issue #02:96–111. https://businessindexnorth.com/site/plugins/Article/download.php?url=06_maritime_activity.pdf. Accessed 16 Aug 2020.

McGrath M First tanker crosses northern sea route without ice breaker. (2017). https://www.bbc.com/news/amp/science-environment-41037071. Accessed 13 Aug 2020.

Mejia, M., Cariou, P., & Wolff, F. C. (2008). Ship piracy: Ship type and flag. In W. Talley (Ed.), *Maritime safety, security and piracy* (pp. 103–120). Informa.

Memorandum in support of motion for temporary restraining order and preliminary injunction. (2012) https://pugetsoundblogs.com/waterways/files/2012/06/Memorandum-in-support-of-motion.pdf. Accessed 13 Aug 2020.

National Geographic (eds.) OPEN FOR BUSINESS. Monthly Sea Ice Extent. (2018). https://www.nationalgeographic.com/environment/2019/08/map-shows-how-ships-navigate-melting-arctic-feature/. Accessed 16 Aug 2020.

Nordquist, M. H. (Ed.). (1995). *UN Convention on the Law of the Sea 1982. A Commentary* (Vol. III). Martinus Nijhoff Publishers.

Order granting motion for preliminary injunction (Within United States Territorial Waters and Ports). (2012). https://cases.justia.com/federal/district-courts/alaska/akdce/3:2012cv00042/47761/87/0.pdf?ts=1411516040. Accessed 13 Aug 2020.

Order granting plaintiff's motion for temporary restraining order and scheduling hearing. (2012). https://cases.justia.com/federal/district-courts/alaska/akdce/3:2012cv00042/47761/27/0.pdf?ts=1543542149 Accessed 13 Aug 2020.

Order re all pending motions. (2012). https://casetext.com/case/shell-offshore-inc-v-greenpeace-1 Accessed 13 Aug 2020.

Perabo, L. (2015). Arctic Pirates. BIVROST Stories 22.11.2015. https://www.bivrost.com/arctic-pirates/ Accessed 5 Aug 2020.

Prjahina, V. A., Hinkiladze, V., & Nikulina, A. (2018). Analiz i perspektivy razvitija Arkticheskogo goroda Pevek. In: Djachenko NG. Budushhee Arktiki nachinaetsja zdes'. Sbornik materialov II Vserossijskoj nauchno-prakticheskoj konferencii s mezhdunarodnym uchastiem. Izd-vo filiala MAGU v g. Apatity, Apatity (pp. 245–252). http://www.arcticsu.ru/. Accessed 15 Feb 2020.

Qureshi, W. A. (2017). The prosecution of pirates and the enforcement of counter-piracy laws are virtually incapacitated by law itself. *San Diego International Law Journal*, *19*, 95–126.

Regehr, E. (2017, November 16). Arctic security briefing papers. In: *ACGF forum – Cooperative security under construction*. http://www.thesimonsfoundation.ca/sites/default/files/Arctic%20Coast%20Guard%20Forum%20%E2%80%93%20Cooperative%20Security%20Under%20Construction%20-%20Arctic%20Security%20Briefing%20Paper%2C%20November%2016%202017_0.pdf. Accessed 15 Feb 2020.

Ren, C., & Chimirri, D. (2018). *Arctic Tourism – More than an Industry?* https://www.thearcticinstitute.org/arctic-tourism-industry/ Accessed 13 August 2020.

Reuland, R. C. F. (1989). Interference with non-national ships on the high seas peacetime exceptions to the exclusivity rule of flag-state jurisdiction. *Vanderbilt Journal of Transnational Law* *22*:1161–1229.

Rosatomflot (eds.). (2020). *Severnyy morskoy put*. http://www.rosatomflot.ru/o-predpriyatii/severnyy-morskoy-put/ Accessed 15 Aug 2020.

Rowe, R. (2019). Legal Article: The Northwest Passage – What is its status under the international law of the sea? UK P&I 27.02.2019. https://www.ukpandi.com. Accessed 12 Feb 2020.

Shaposhnikov, V. M., Aleksandrov, F. V., Matantsev, R. A., Ivanovskaya, O. D. (2017). Iceberg risk analysis for the Northern sea route: LNG carrier study case. *Arktika: ekologiya i ekonomika*, *2*(26), 76–81.

Sidorchenko, V. F. (2004). Morskoe piratstvo. Izdatel'skij Dom S.-Peterburgskogo gosudarstvennogo universiteta, Izdatel'stvo juridicheskogo fakul'teta S.-Peterburgskogo gosudarstvennogo universiteta, St. Petersburg.

Sousa, I. (2019). *Maritime territorial delimitation and maritime security in the Atlantic*. http://www.atlanticfuture.eu.files/325-ATLANTIC%20FUTURE_07_Maritime%20Security%20in%20the%20Atlantic.pdf. Accessed 15 Feb 2020.

The increase in Arctic shipping. (2020). 2013-2019. Arctic shipping status report (ASSR) #1 12 March 2020 https://storymaps.arcgis.com/stories/592bfe70251741b48b0a9786b75ff5d0 Accessed 13 August 2020.

Transnational piracy: to pay or to prosecute? (2011). American Society of International Law Proceedings *105*:543–554.

U.S. Department of state (eds.). (2019). The United States Ratifies Central Arctic Ocean Fisheries Agreement. Media note. Office of the spokesman. 27.08.2019. https://www.state.gov/the-united-states-ratifies-central-arctic-ocean-fisheries-agreement/ Accessed 13 Aug 2020.

Vyatkin, J. (2019). Voennye ledokoly berut Arktiku pod ohranu. *Argumenty Nedeli*, *47*(691). 04.12.19 https://argumenti.ru/army/2019/12/640219. Accessed 15 Feb 2020.

Vylegzhanin, A., & Anyanova, E. (2020). International Law versus Piracy: Issues in Legal Theory. *International Journal of Psychosocial Rehabilitation, 24*(1), 25–42.

Vylegzhanin, A., Young, O., & Berkman, P. (2020). The Central Arctic Ocean Fisheries Agreement as an element in the evolving Arctic Ocean governance complex. *Marine Policy, 118.*

WWF. (Eds.). (2020). *Arctic tourism.* https://arcticwwf.org/work/people/tourism/ Accessed 13 Aug 2020.

Zhuravleva, I. (2019). *International Law Doctrines of the Current Status of the Arctic.* University of Belgrade. Faculty of Law, 434 p.

Chapter 6
(Research): Microplastics in the Arctic Benthic Fauna: A Case Study of the Snow Crab in the Pechora Sea, Russia

Anna Gebruk, Yulia Ermilova, Lea-Anne Henry, Sian F. Henley, Vassily Spiridonov, Nikolay Shabalin, Alexander Osadchiev, Evgeniy Yakushev, Igor Semiletov, and Vadim Mokievsky

Abstract Microplastics have been declared a threat to ocean health and status under the United Nations Sustainable Development Goal (SDG) 14 Target 14.1. Microplastics are bioavailable for a wide range of marine organisms and may cause adverse physiological and biochemical effects, including decreased growth and energy intake, and impaired reproduction. Accumulation of microplastics in benthic (seafloor) fauna is of particular concern in commercially important species, as this poses threats to human health. A baseline assessment of microplastic ingestion by Arctic benthic fauna is of urgent necessity.

In this chapter, we present initial results on microplastics ingestion by nine species of benthic fauna from the Pechora Sea, south-eastern Barents Sea, including the snow crab *Chionoecetes opilio*, a commercially-exploited and invasive benthic crustacean. From a sample set of 154 specimens, we compare microplastics ingestion by snow crabs with that of the eight other species to assess the impact of

A. Gebruk (✉)
School of GeoSciences, The University of Edinburgh, Edinburgh, UK

Marine Research Centre, The Lomonosov Moscow State University, Moscow, Russia
e-mail: Anna.Gebruk@ed.ac.uk

Y. Ermilova · N. Shabalin
Marine Research Centre, The Lomonosov Moscow State University, Moscow, Russia

L.-A. Henry · S. F. Henley
School of GeoSciences, The University of Edinburgh, Edinburgh, UK

V. Spiridonov · A. Osadchiev · V. Mokievsky
Shirshov Institute of Oceanology, Russian Academy of Sciences, Moscow, Russia

E. Yakushev
Norwegian Institute for Water Research, Oslo, Norway

I. Semiletov
Il'ichov Pacific Oceanological Institute, Far Eastern Branch of the Russian Academy of Sciences, Vladivostok, Russia

P. A. Berkman et al. (eds.), *Building Common Interests in the Arctic Ocean with Global Inclusion, Volume 2*, Informed Decisionmaking for Sustainability,
https://doi.org/10.1007/978-3-030-89312-5_6

different feeding strategies on ingestion rates. Microplastic fibres were recorded in 35% of snow crab stomachs and 21% of stomachs of all species studied. Benthic omnivores (organisms with flexible feeding strategies) are shown to have more ingested microplastics (29%) than sessile filter-feeding organisms (17%).

A comprehensive and well-integrated monitoring program is needed in the Arctic for monitoring of microplastic pollution in both benthic and pelagic ecosystems, with consideration of regionally-specific features, such as seasonality of the ice cover, primary production, and riverine discharge. We believe that the Regional Action Plan on Marine litter in the Arctic currently under development by the Arctic Council's Protection of the Arctic Marine Environment (PAME) Working Group will constitute an internationally-recognised framework for investigation and mitigation of plastic pollution in the Arctic. More broadly, adding ingestion rates of microplastics by benthic fauna to the SDG indicator 14.1.1 as a globally-important indicator of the impacts of plastic pollution would greatly advance development of a more comprehensive understanding of ecosystem status and mitigation measures to reduce plastic pollution globally.

6.1 Microplastics in the Arctic Ocean

6.1.1 Distribution and Sources of Plastic Debris in the Arctic

Plastic litter has been reported globally as the most abundant form of marine debris accounting for up to 80% of the world's marine litter (UNEP, 2016). Plastics and microplastics (defined as plastic particles of 1 μm – 5 mm) cause adverse repercussions from direct impacts on marine biota and habitats, to aesthetic impacts of litter leading to loss of tourism and economic damage (costs for cleaning, research and monitoring) (Thompson, 2015; Avio et al., 2016). Marine plastics are recognised as a threat to marine ecosystems worldwide; particularly Target 14.1 of the United Nations Sustainable Development Goals (SDG) clearly identifies the need to remove marine debris from our oceans. The Report of the Secretary-General on the progress of SDGs in 2019 however did not indicate significant achievements in mitigating plastic pollution (E/2019/68). Only floating plastic debris are currently included in the global indicator framework (indicator 14.1.1), whereas other forms of marine plastics are not considered. The most recent scientific evidence demonstrates that only 1% of marine plastics are floating in the sea surface, whereas the remaining 99% likely accumulate in the seafloor ecosystems and in particular in the deep-sea (Kane et al., 2020). Clearly, plastic pollution of the oceans remains a significant problem and threat to the marine environment, and there is a great need for international cooperation to address this problem robustly. Lack of knowledge regarding the extent of plastic pollution in the Arctic Ocean, one of the world's most sensitive areas to environmental change due to Arctic amplification of climate change (Serreze & Francis, 2006), makes baseline research of critical importance.

Whilst data on plastic pollution in the Arctic Ocean remain scarcer than for other regions, in the last years the number of publications reporting microplastics in the Arctic waters (Lusher et al., 2015), sea ice (Kanhai et al., 2020), sediments (Bergmann et al., 2017), and biota (Bråte et al., 2018) have been increasing (reviewed for the Arctic region in Halsband and Herzke, 2019, and for the Barents Sea – in Grøsvik et al., 2018). Cozar et al. (2017) reported substantial accumulation of microplastics near the Novaya Zemlya archipelago in the eastern Barents Sea, with concentrations of hundreds of thousands of pieces of debris per square kilometre. Buoyant plastics sourced primarily from the US East Coast and northwest European shelf are transported by the North Atlantic drift to the Greenland and Barents Seas. Large-scale oceanographic circulation and sinking of water masses makes this region likely to accumulate transported plastics from distal sources in the water column and benthic (seafloor) environment (Cozar et al., 2017). Results of the recent Cruise AMK-78 to the Russian Arctic in autumn 2019 demonstrated consistent presence of microplastics in the surface and subsurface waters of the Kara, Laptev, and East Siberian Seas (Yakushev et al., 2021). The study has also suggested the Atlantic surface water inflowing from the North Atlantic and discharge plumes of the Great Siberian Rivers to be the two main sources of microplastic pollution in the Eurasian Arctic shelf (Yakushev et al., 2021).

With rivers serving as key vectors for terrestrial plastic runoff (Barrows et al., 2018), it is important to consider these as input to Arctic marine ecosystems. The Arctic Ocean receives 11% of the global freshwater discharge (Fichot et al., 2013). Large Arctic rivers, namely, the Northern Dvina, Pechora, Ob, Yenisei, Lena, Indigirka, Kolyma, Yukon, and Mackenzie, drain large areas of Europe, Asia, and North America (Fig. 6.1). Microplastics likely remain contained within large river plumes, whose dynamics then determine the spreading and accumulation of plastics in the Arctic Ocean. Transport of microplastics and larger plastic items in Arctic sea ice and its subsequent release during summertime ice melting is another important source of plastic litter to Arctic surface waters (Fichot et al., 2013; Obbard et al., 2014; Kanhai et al., 2020).

6.1.2 Interactions of Marine Fauna with Microplastics in the Arctic

The growing presence of plastic items in the marine environment leads to increasing exposure of marine biota to plastics, with a range of physiological, biochemical and ecological consequences from transportation of non-native species to blockages of digestive systems or death from entanglement (Gregory, 2009; Zettler et al., 2013; Avio et al., 2016). Microplastics correspond to the size of prey items of various marine biota and ingestion is the most common interaction of animals with microplastics, particularly when feeding mechanisms do not discriminate between plastic particles and food items (Courtene-Jones et al., 2018; La Beur et al., 2019).

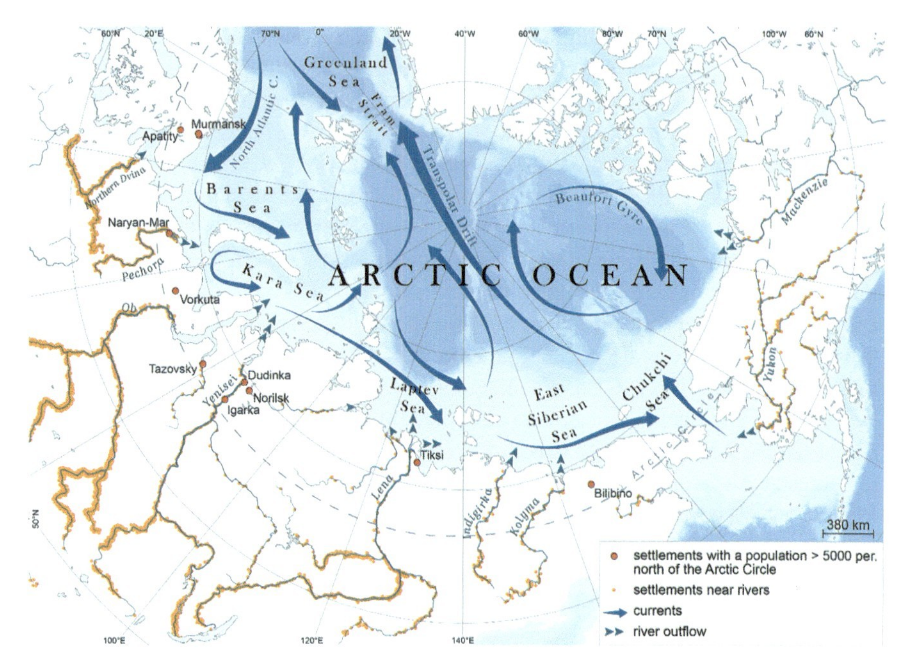

Fig. 6.1 Map showing the major Arctic rivers, ocean currents and human populations. Density of human populations in the Arctic coastal areas and near the rivers contributing to annual freshwater discharge to the Arctic Ocean is shown by the orange dots that represent settlements with a population of >5000 people north of the Arctic Circle. (Data from the Federal State Statistic Service, Russia); blue arrows show directions of prevailing surface currents and the blue vectors from the rivers show the riverine inflow

Emerging concerns have been raised regarding ecotoxicological impacts of ingested microplastics, including absorption and transport of environmental pollutants and leaching of toxic additives such as phthalates, bisphenol A (BPA), alkylphenols and others (Hirai et al., 2011; Rochman et al., 2014).

Presence of microplastics in zooplankton species is of particular ecological concern, since ingestion by organisms from the primary trophic levels (zooplankton) is a pathway for transfer of microplastics into the pelagic food web (Avio et al., 2016), though implications for Arctic marine ecosystem functioning are yet to be determined. Benthic ecosystems are predicted to constitute a global sink for marine microplastics as a result of direct sinking, biodegradation, biofouling or ingestion and transport via food webs (Avio et al., 2016). Particularly in the deep sea, complexity of topographic and hydrographic regimes acts to funnel marine litter and microplastics into the seafloor ecosystems (La Beur et al., 2019; Kane et al., 2020). Still, there remain very few studies dedicated to microplastics in deep-sea sediments, especially in the Arctic Ocean (Bergmann et al., 2017). Ingested microplastics have been identified in deep-sea macrobenthic organisms in the North Atlantic (Courtene-Jones et al., 2018; La Beur et al., 2019), mid-Atlantic and Indian Oceans (Taylor et al., 2016). Studies focusing on the ingestion of microplastics by benthic fauna in the Arctic are lacking, one of the few exceptions is a recent study in the Bering-Chukchi Seas (Fang et al., 2018).

To address the knowledge gap on microplastic ingestion by Arctic benthos, we conducted a case study, investigating stomach contents of the snow crab and other benthic invertebrates in order to evaluate the baseline level of microplastic ingestion by macrobenthos in the Pechora Sea.

6.2 Case Study: Ingestion of Microplastics by the Snow Crab in the Pechora Sea

6.2.1 The Pechora Sea – Key Features and Importance of Ecological Studies

The Pechora Sea lies in the south-eastern basin of the Barents Sea. Historically, the Pechora Sea had little anthropogenic pressure with a small coastal population, limited shipping, and no large-scale commercial fishery activities in the area (Denisenko et al., 2003). However, in recent decades vast oilfields have been discovered in the area resulting in increasing rates of offshore oil exploration and production (Malyutin et al., 2003; Kaminskii et al., 2011). Intensification of human activities, along with climate change, the introduction of invasive species, and release of contaminants are predicted to have cumulative impacts on the unique marine ecosystems of the Pechora Sea (Sukhotin et al., 2019; Semenova et al., 2019).

In this chapter, we measured microplastic ingestion by the commercially exploited invasive benthic decapod species, the snow crab *Chionoecetes opilio*, in the Pechora Sea. The snow crab is native to the Northern Pacific and was first recorded in the Barents Sea in fisheries by-catch in 1996, thereafter forming a self-maintaining population (Sokolov et al., 2016). Unlike another invasive benthic decapod in the Barents Sea, the red king crab *Paralithodes camtschaticus*, the snow crab was not deliberately introduced to the area and most likely was carried by ballast waters (Jørgensen & Spiridonov, 2013). The snow crab *Chionoecetes opilio* is an important commercial species, with fishery activity currently being carried out mainly by a Russian fleet in the Russian part of the Barents Sea shelf since 2016 and being under development in Norway (ICES, 2019). In 2020, the commercial stock of the snow crab in the Barents Sea was estimated as 523,000 tonnes with 984,000 tonnes total allowable catch (TAC) in the Russian EEZ (Bakanev and Pavlov, 2020, Table 6.1). In the Pechora Sea, snow crabs are most abundant near the Novaya Zemlya archipelago where the cold Arctic waters entering from the Kara Strait mix with Barents Sea water masses (Zalota et al., 2018) (Fig. 6.2).

Table 6.1 Estimate of the snow crab population and total allowable catch (TAC) based on PINRO reports (Bakanev and Pavlov, 2020)

Year	Snow crab stock in the Barents Sea, thousand tonnes	Snow crab TAC,[a] thousand tonnes
2020	523	9.84
2019	516	9.83
2018	601	9.84
2017	489	7.84
2016	436	1.6

[a]*TAC* total allowable catch

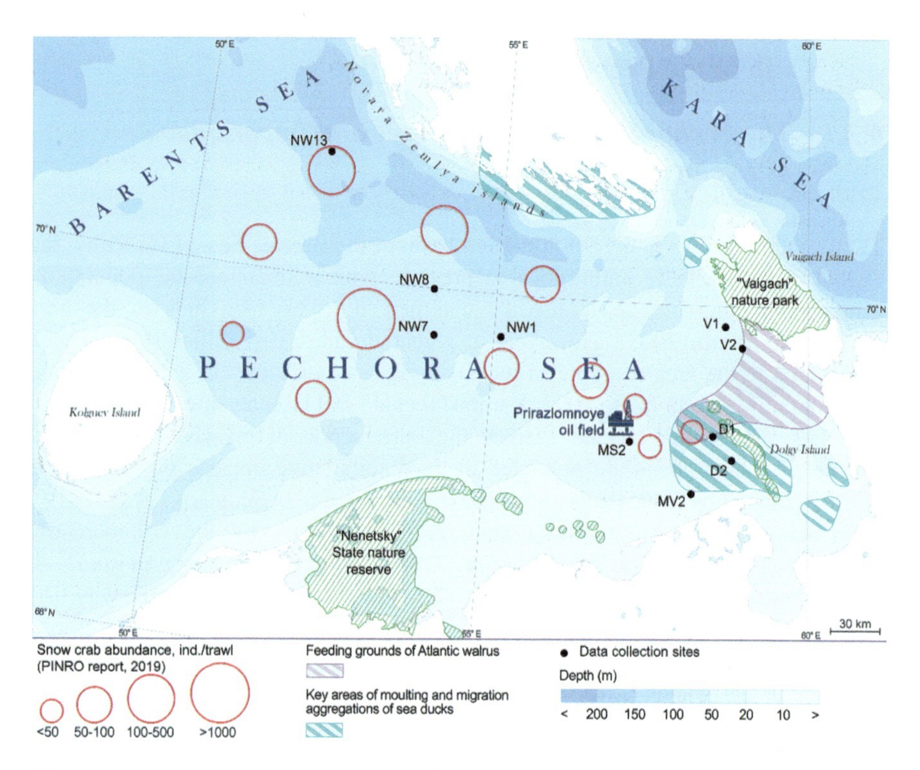

Fig. 6.2 Map of the key ecological and socio-economic features of the Pechora Sea. Feeding grounds of Atlantic walruses shown by purple dashing; moulting grounds of the sea ducks shown (dark green-blue dashing); abundance of snow crab, the invasive commercial species (red circles); shelf oil-field Prirazlomnoye position (black icon) and the land-based protected areas (light green shading)

The present case study in the Pechora Sea aimed to (1) determine the frequency of occurrence of microplastics in the digestive system of snow crabs and; (2) compare these ingestion rates with data on other benthic species from different feeding guilds, including non-commercial crustaceans and bivalves, in order to establish a baseline assessment of microplastic ingestion in the Arctic benthos.

6.2.2 *Material and Methods*

6.2.2.1 Sampling

Samples of benthic fauna were collected during the RV *Kartesh* cruises during the summers of 2017 and 2018 (Cruise reports Kartesh, 2017, 2018). In 2017, decapod specimens were collected from sites V1 and V2 near Vaigach Island using the *Sigsbee* bottom trawl. In 2018, additional benthic samples were taken from the *Okean-0.1* grab samples at 10 sites in the Pechora Sea (Fig. 6.2). Collected species represent different feeding guilds: filter feeders consisted of *Astarte borealis*, *Astarte montagui*, *Ciliatocardium ciliatum*, and *Serripes groenlandicus*; mixed filter feeders and surface deposit feeders consisted of *Macoma calcarea*, subsurface deposit feeders were represented by *Yoldia hyperborea,* and mobile omnivores consisted of three benthic decapods (*C. opilio*, *H. araneus*, and *P. pubescens*). A total of 154 specimens of Arctic benthic fauna were studied.

Bottom sediments from trawls and bottom grabs were washed with seawater over a 0.5 mm mesh, then decapods and bivalves were extracted manually for further analyses. Samples were preserved in buffered 4% formalin and then transferred to the 70% industrial methylated solution (IMS).

6.2.2.2 Dissection and Digestion

Tissues and organs of each specimen were examined for presence of microplastics with a combination of digestion and dissection techniques following protocols adapted from Courtene-Jones et al. (2018) and La Beur et al. (2019). In the laboratory, specimens were dissected, and stomach contents were placed in trypsin/deionised solution in 50 mL glass covered vials in a water bath at 40 °C to digest overnight (15–20 h). Samples were then washed over a 40 μm metal sieve to separate microparticles made of artificial synthetic polymers from digested organic matter of biological origin. Microparticles were then examined and photographed under a Zeiss SteREO Discovery V20 stereomicroscope (Appendix). For each specimen, presence/absence of microplastic particles was noted, as well as their abundance and location in the body. Only ingested particles from digestive systems were used for further analyses. A presence/absence data matrix of ingested microfibers was then used to calculate the frequency of occurrence of microplastics across sampling sites, species and feeding guilds.

6.2.2.3 Quality Assurance and Quality Control

Laboratory quality control measures to reduce artificial contamination of samples were adopted from Courtene-Jones et al. (2018) and included the following: only 100% cotton laboratory coats were used for the duration of the study; the working area was cleaned with IMS solution prior to any analysis; dissection kit and tools were cleaned with IMS then triple-rinsed in deionised water before used. A control petri dish was placed in the fume cupboard and under the microscope to identify potential airborne contaminants (Woodall et al., 2015).

6.2.3 Results and Discussion

In total 65 microparticles were found in 154 examined specimens, all classified as microfibers. A total of 34 specimens contained microplastics in their digestive systems. No microplastics were found in control petri dishes, therefore no particles were excluded from the further analyses. All animals with ingested microplastics were collected at the three sites between the Vaigach and Dolgy Islands: V1, V2 and D1 (Fig. 6.3). The site with the largest number of ingested particles corresponds to the area with the highest biomass of macrobenthic communities (Denisenko et al., 2003), an area that also likely serves as a foraging ground for the local population of Atlantic walrus (Semenova et al., 2019).

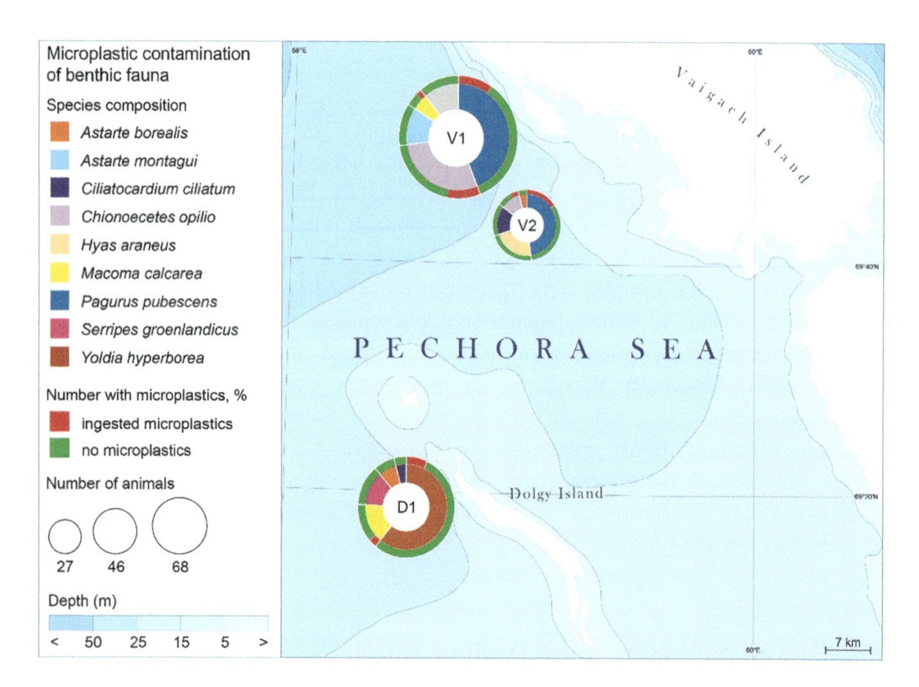

Fig. 6.3 Microplastic ingestion by benthic fauna in the Pechora Sea. Size of circles represents the total number of specimens studied per site; sectors of different colour represent species composition and the outer contour shows the proportion of animals (%) with ingested microplastics for each taxon

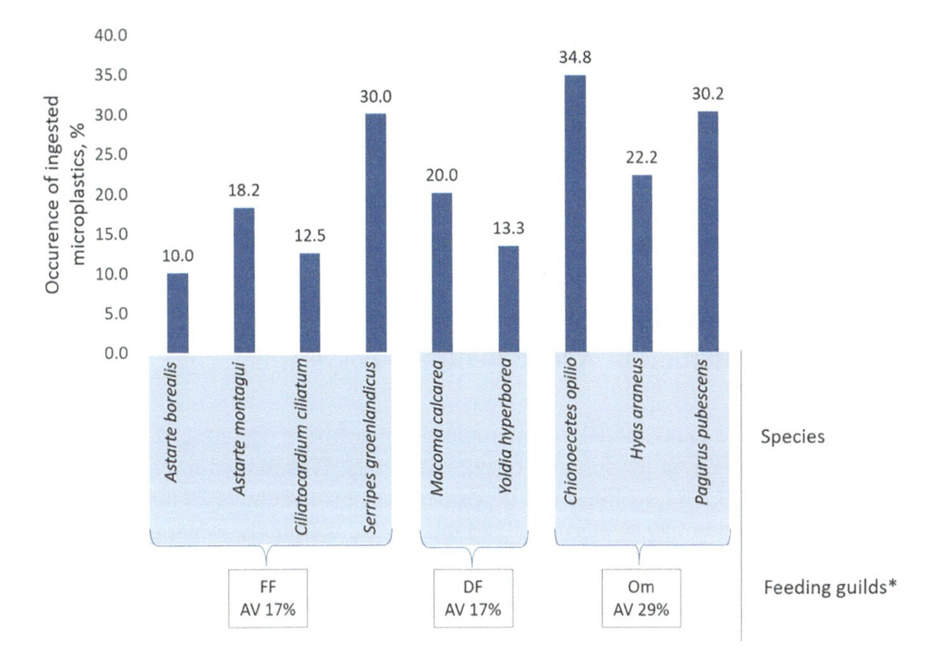

Fig. 6.4 Diagram of occurrence of ingested microplastics in benthic fauna from different feeding guilds
**FF* filter feeders, *DF* deposit feeders, *Om* omnivores, *AV* average

The average frequency of occurrence of ingested microplastics for all studied species was 21 ± 8% ranging from 10% in the bivalve *Astarte borealis* to 35% in the snow crab *Chionoecetes opilio* (Fig. 6.4). The observed differences between the species can be attributed to inter-specific differences in size, age and feeding behaviour. Previous studies showed contradicting results regarding differences of microplastic ingestion between the feeding guilds of macrobenthos, with some authors arguing that suspension feeders were more likely to ingest microplastics than filter feeders (Taylor et al., 2016). However, some studies did not find a statistically significant difference between feeding guilds (La Beur et al., 2019). Our results agree with the latter study, in that levels of microplastics ingestion were indistinguishable between filter feeders and deposit feeders (17% for both). Furthermore, the frequencies of microplastic ingestion by filter feeders and deposit feeders are also very close to those reported from the deep-sea North Atlantic biomes (La Beur et al., 2019) (16% and 15% respectively).

In benthic omnivores (the snow crab, the spider crab and hermit crabs) the ingestion rate of microplastics was noticeably higher at 29 ± 6%. This may be explained by food web dynamics. Bivalves, both deposit and filter-feeding, are primary consumers extracting organic matter from the water column and/or the detritus layer on sediments. Benthic decapods are higher-level consumers actively preying or scavenging on organisms at the seafloor and have complex diets. Therefore, in addition to microplastics from the substrate, benthic decapods also passively ingest particles accumulated in the prey (Taylor et al., 2016), which results in increased magnitude of microplastic occurrence compared to filter, deposit, or suspension feeders.

6.3 Conceptual Diagram Model of Microplastic Accumulation in Arctic Benthic Ecosystems

A conceptual diagram model of microplastic ingestion by benthic fauna in the Pechora Sea, based on the findings of our case study, is presented in Fig. 6.5. The scheme demonstrates how filter- and deposit-feeding bivalve molluscs that dominate macrobenthic assemblages ingest sinking microplastics from the water column and sediments, whereas mobile omnivores (the snow crab) passively consume microplastics from lower trophic levels. Therefore, larger quantities of microplastics

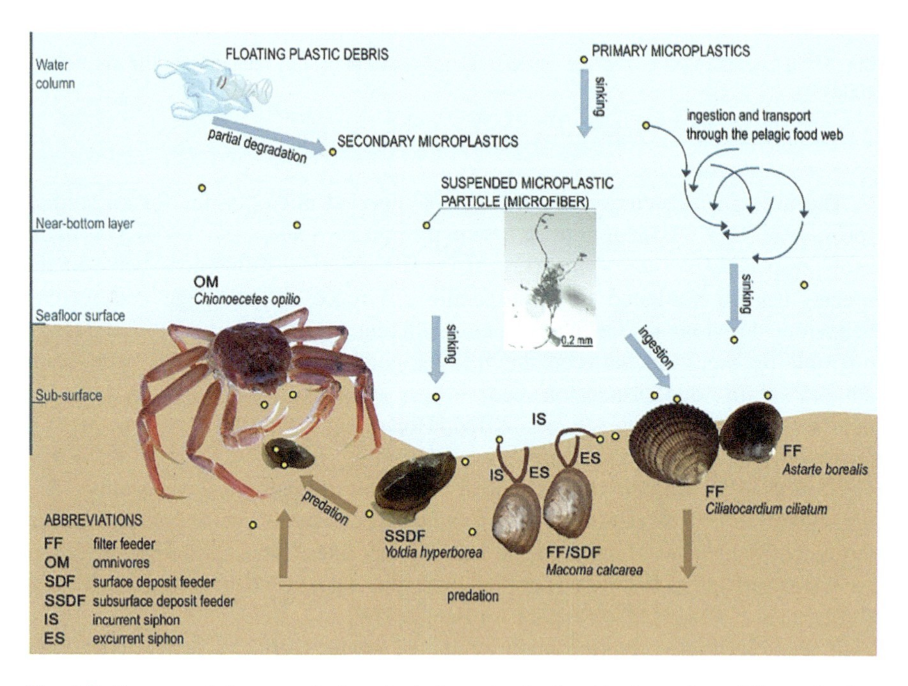

Fig. 6.5 Conceptual diagram of microplastic ingestion by benthic fauna from different feeding guilds in the Pechora Sea based on our findings

accumulate in the species of higher trophic levels as a result of biomagnification. The apex benthic predators in the research area include the Atlantic walrus and the common eider duck; there is currently no published evidence of microplastics ingestion for these species. However, presence of microplastics has been previously recorded in the Arctic and north Atlantic for marine birds, for example northern fulmars (Trevail et al., 2015), and marine mammals including fin whales (Sadove & Morreale, 1989) and bowhead whales (Finley, 2001). It is therefore likely that the lack of data on walruses and eiders is due to a lack of research rather than an absence of microplastics from the food webs in the Barents Sea.

The presented diagram illustrates the significant role of seafloor ecosystems and specifically benthic fauna in accumulating sinking microplastics. The model can also be used to demonstrate diversity of feeding strategies of benthic organisms and therefore importance of targeting species from different feeding guilds for monitoring of microplastic accumulation and (re)distribution.

6.4 Microplastics in the Context of Informed Decisionmaking

6.4.1 Existing Frameworks for Microplastics Studies

The legacy of marine plastic pollution has led governments, international organisations, non-governmental organisations, and other stakeholders to recognise the need for assessing and monitoring the magnitude, distribution, and sources of plastic pollution throughout the global oceans. As one example, the Basel Convention on the Control of Transboundary Movements of Hazardous Wastes and Their Disposal has recently adopted Plastic Waste Amendments that will come into force in January 2021 and add plastic polymers including polyethylene, polypropylene or polyethylene terephthalate to the list of hazardous wastes requiring removal from the oceans (BC-14/12). With a view to building common interests with global inclusion, of great interest is decision BC-14/21 on International cooperation and coordination that clearly mandates governments and other stakeholders to develop measures to ensure the effective implementation of the plan "Towards a Pollution-Free Planet". The UN has also placed microplastics high on its agenda within the framework of the UN Decade of Ocean Science for Sustainable Development, wherein 'a clean ocean where sources of pollution are identified and removed' is stated among the key societal outcomes of the Decade (UNDOS, 2020).

Data collection is a first step towards mitigation of plastic pollution, and at this stage developing, harmonising, and adopting international protocols for both pelagic and benthic plastics assessments on a regional scale is crucial to obtaining comparable and reliable data. Recent example is the *Guidelines for the Monitoring and Assessment of Plastic Litter in the Ocean* by the joint Group of Experts on the Scientific Aspects of Marine environmental Protection (GESAMP) (Kershaw et al., 2019). Guidelines produced by GESAMP (Kershaw et al., 2019) provide an extensive framework for microplastics studies in different components of marine ecosystems including the sea surface, water column, seafloor, and marine biota. However, these guidelines are not regionally specific and there is a lack of internationally-recognised protocols for marine microplastics assessments specifically for the Arctic.

The Arctic Council has also recognised plastic pollution in the Arctic marine environment as an emerging threat, as reflected in their Arctic Marine Strategic Plan, and initiated the first study on marine litter in the Arctic conducted by the Protection of the Arctic Marine Environment (PAME) Working Group in 2019. The PAME report identified multiple knowledge gaps in sources of plastic pollution in the Arctic, the drivers and pathways of distribution, interactions with marine fauna, socio-economic impacts, as well as the urgent need for developing monitoring programmes and regional action plans to underpin informed decisionmaking (PAME, 2020).

6.4.2 Limitations of this Study

The present case study contributes to baseline data on microplastic ingestion by Arctic benthic fauna, which is an essential first step for developing a more comprehensive understanding of ecosystem status and mitigation measures required to reduce plastic pollution in the Arctic. However, a large sampling size is needed to assess differences in microplastic accumulation between the different species, feeding guilds, and geographic areas.

The high levels of microplastic contamination in the snow crab revealed by this study offer concerning insights into plastics accumulation in mobile benthic omnivores, which highlight the need for further investigation of the drivers of accumulation and its ecological consequences. Regular surveys of microplastic contamination in commercially harvested benthic invertebrate species are needed in the Arctic; in the case of the Barents Sea, those are *P. camtschaticus*, *C. opilio* and *C. islandica.* Future studies should investigate the physiological mechanisms of plastic ingestion by different species and feeding guilds and determine the timescales over which ingested particles remain in the organism.

In addition to quantitative assessment, chemical characterisation of microplastics, for instance by Fourier transform infrared spectroscopy or Raman spectroscopy would be beneficial for understanding potential sources of contamination. Advantages and disadvantages of each method are discussed in the GESAMP report (Kershaw et al., 2019) and should be considered on a case-by-case basis, depending on the number and size of particles, budget, and timeframes of the project.

6.4.3 Further Research Questions for Monitoring of Microplastics in Arctic Benthic Ecosystems

For the Arctic Ocean, oceanographic features such as the seasonality of sea ice cover, primary production and riverine discharge need to be considered for effective and representative monitoring. In particular, the role of riverine inflow is crucial for understanding the distribution and accumulation of microplastics in the Arctic. Study of the delivery and fate of river-borne plastic litter in the Arctic Ocean requires an end-to-end system-scale understanding of its inflow with fluvial water, transformation in the estuarine and deltaic zones, transport by river plumes during ice-free periods and by sea ice during cold periods, settling to subjacent seawater below river plumes and accumulation at the seafloor. Collection of specific *in situ* data is essential to quantify these processes and determine the key factors that govern the dynamics and variability of transport and accumulation of marine plastic litter in the Arctic Ocean.

Incorporating the following research objectives into the agendas of regional and international programmes focused on microplastic studies in the Arctic, such as the Regional Action Plan on Marine Litter in the Arctic by PAME and Arctic Monitoring and Assessment Programme (AMAP) would address the key knowledge gaps in microplastic pollution in the Arctic:

1. Understanding the role and distribution of river-borne versus ice-borne microplastics in the marine environment.
2. Revealing the role of riverine plumes in governing the distribution of microplastics on the pan-Arctic scale.
3. Defining ecotoxicological consequences of microplastic ingestion by commercially valuable species as well as humans.
4. Investigating differences in physiological mechanisms of microplastic ingestion and inter-tissue translocation by fauna from different feeding guilds.
5. Identifying target species representative of Arctic benthic assemblages, habitats and feeding guilds for monitoring of microplastics ingestion.

Plastic litter is currently considered under SDG 14.1, with the density of floating plastic litter listed among key indicators of ocean pollution (Indicator 14.1.1). However, no other parameters of plastic and microplastic contamination are considered. Macrobenthos are a suitable subject for ecological monitoring because their tendency to accumulate pollutants enables them to demonstrate retrospectively the condition of the marine environment. We argue therefore that parameters such as the abundance of microplastic items in seafloor sediments and ingestion rates of microplastics by benthic fauna should be added to the SDG 14 as globally-important indicators of plastic pollution. Clearly, marine plastic pollution – including microplastics – remains a serious threat, with an increasing amount of scientific evidence revealing the global scale of the problem, and an ever-increasing amount of plastics entering the oceans annually (Kershaw et al., 2019). International collaboration with global inclusion is needed to address this problem by developing mitigation strategies on a global scale (such as the SDG targets, GESAMP reports and the PWP action plan of the Basel Convention) and implementation plans on a regional scale (such as those of the AMAP and PAME working groups of the Arctic Council).

6.5 Conclusions

Despite growing evidence of the magnitude of microplastic contamination in the Arctic and its impacts on the marine environment, sources of plastic pollution and ecological repercussions of ingested microplastics are poorly studied and potential harm to human health from ingestion of microplastics is yet to be determined.

The snow crab *Chionoecetes opilio* is an important commercial species in the Arctic region and our discovery of ingested microplastics in 35% of snow crabs suggests potential passive consumption of microplastics by humans from seafood. The present study has shown that ingestion of microplastics by benthic fauna in the Pechora Sea occurs commonly, with an average of 21 ± 8% of stomachs of all macrobenthos containing ingested microplastics. We have also shown that benthic omnivores, including the snow crab, demonstrate higher frequency of occurrence of ingested microplastics than sessile filter-feeding organisms.

Adding ingestion rates of microplastics by benthic fauna to the SDG 14 as a globally-important indicator of plastic pollution is an essential step towards developing a more comprehensive understanding of ecosystem status and the mitigation measures required to reduce plastic pollution and its impacts world-wide. In addition, a harmonized monitoring program is needed for monitoring microplastic pollution in the Arctic in both benthic and pelagic marine ecosystems with consideration of regional specificities such as seasonality of the ice cover, primary production, and riverine discharge.

Acknowledgements Authors express gratitude to the organising committee of the Arctic Frontiers conference for the opportunity to contribute to the edition. Authors are also grateful to the *Lomonosov Moscow State University Marine Research Centre* for organising the RV Kartesh cruises to the Pechora Sea and to the Captain and crew of RV Kartesh for their excellent work. Advice on methods of microplastic analyses by the Changing Oceans Research Group at the University of Edinburgh is gratefully acknowledged, in particular Dr. Seb Hennige and Laura La Beur. Authors are also grateful to Dr. Andrey Gebruk for helpful feedback on the first drafts and to Professor Paul Arthur Berkman for reviewing the paper.

Appendix: Photographs of Microfibers from Stomach Contents of Specimens of Macrobenthos (Examples)

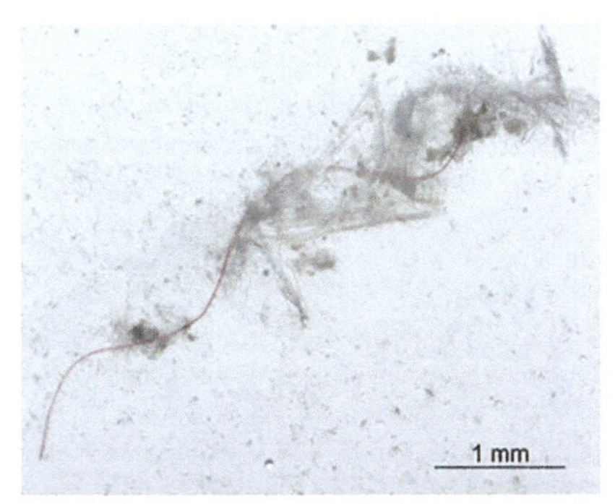

Yoldia hyperborea 36 (visceral mass)

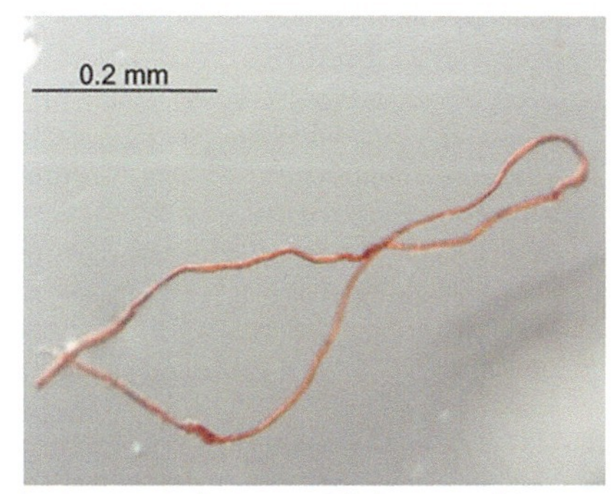

Pagurus pubescens 35 (stomach)

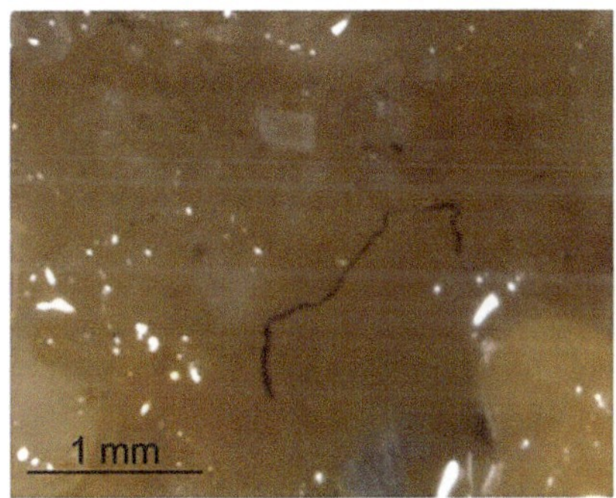

Chionoecetes opilio 20 (stomach)

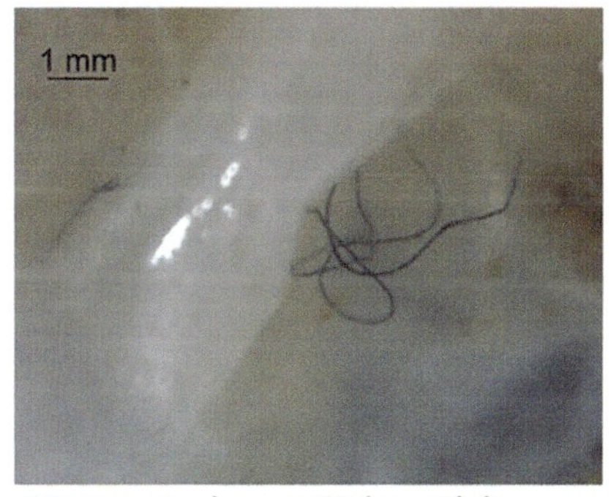

Macoma calcarea 42 (mantle)

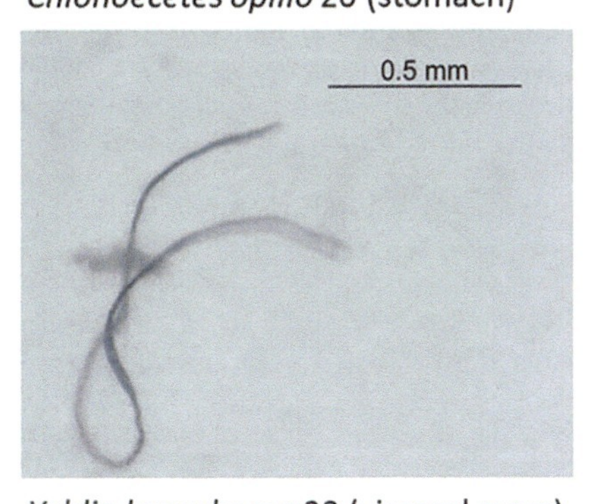

Yoldia hyperborea 32 (visceral mass)

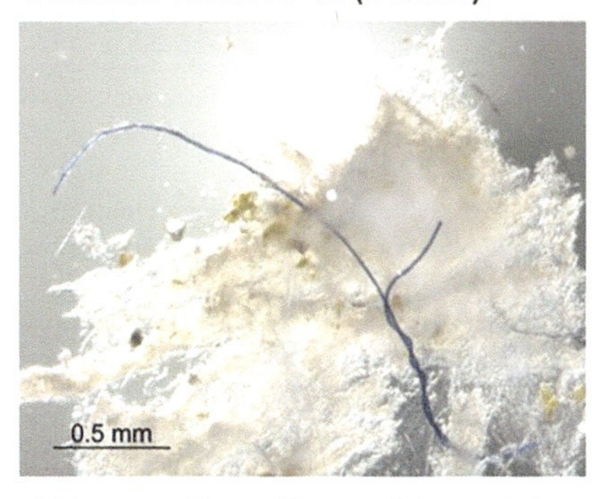

Ciliatocardium ciliatum 4 (mantle)

References

Avio, C. G., Gorbi, S., & Regoli, F. (2016). Plastics and microplastics in the oceans: From emerging pollutants to emerged threat. *Marine Environmental Research*. https://doi.org/10.1016/j.marenvres.2016.05.012.

Bakanev, S. V., & Pavlov, V. A. (2020). *Materials establishing the total allowable catch of the snow crab-opilio in the Russian EEZ in the Barents Sea in 2021*. In Preliminary materials of the total allowable catch in the area of extraction (catch) of aquatic biological resources in the internal waters, territorial seas, the continental shelf and the EEZ of the Russian Federation and the Caspian Sea for 2021 (with environmental impact assessment). PINRO, Murmansk, pp. 15–27 (In Russian).

Barrows, A. P. W., Cathey, S. E., & Petersen, C. W. (2018). Marine environment microfiber contamination: Global patterns and the diversity of microparticle origins. *Environmental Pollution, 237*, 275–284.

Basel Convention Decisions (BC-14/21). *International cooperation and coordination*. http://www.basel.int/Portals/4/download.aspx?d=UNEP-CHW-COP.14-BC-14-21.English.pdf.

Basel Convention Plastic Waste Amendments (BC-14/12). Amendments to Annexes II, VIII and IX to the Basel Convention. http://www.basel.int/Implementation/Plasticwaste/Amendments/tabid/8339/Default.aspx.

Bergmann, M., Wirzberger, V., Krumpen, T., Lorenz, C., Primpke, S., Tekman, M. B., & Gerdts, G. (2017). High quantities of microplastic in Arctic deep-sea sediments from the HAUSGARTEN observatory. *Environmental Science & Technology, 51*(19), 11000–11010.

Bråte, I. L. N., Hurley, R., Iversen, K., Beyer, J., Thomas, K. V., Steindal, C. C., … Lusher, A. (2018). Mytilus spp. as sentinels for monitoring microplastic pollution in Norwegian coastal waters: A qualitative and quantitative study. *Environmental Pollution, 243*, 383–393.

Courtene-Jones, W., Quinn, B., Ewins, C., Gary, S. F., & Narayanaswamy, B. E. (2018). Consistent microplastic ingestion by deep-sea invertebrates over the last four decades (1976–2015), a study from the north East Atlantic. *Environmental Pollution, 244*, 503–512. https://doi.org/10.1016/j.envpol.2018.10.090.

Cózar, A., Martí, E., Duarte, C. M., García-de-Lomas, J., Van Sebille, E., Ballatore, T. J., … Troublè, R. (2017). The Arctic Ocean as a dead end for floating plastics in the North Atlantic branch of the thermohaline circulation. *Science Advances, 3*(4), e1600582.

Denisenko, S. G., Denisenko, N. V., Lehtonen, K. K., Andersin, A. B., & Laine, A. O. (2003). Macrozoobenthos of the Pechora Sea (SE Barents Sea): Community structure and spatial distribution in relation to environmental conditions. *Marine Ecology Progress Series, 258*, 109–123.

Fang, C., Zheng, R., Zhang, Y., Hong, F., Mu, J., Chen, M., … Bo, J. (2018). Microplastic contamination in benthic organisms from the Arctic and sub-Arctic regions. *Chemosphere, 209*, 298–306.

Fichot, C. G., Kaiser, K., Hooker, S. B., Amon, R. M., Babin, M., Bélanger, S., … Benner, R. (2013). Pan-Arctic distributions of continental runoff in the Arctic Ocean. *Scientific Reports, 3*, 1053.

Finley, K. J. (2001). Natural history and conservation of the Greenland whale, or bowhead, in the Northwest Atlantic. *Arctic*, 55–76.

Gregory, M. R. (2009). Environmental implications of plastic debris in marine settings—Entanglement, ingestion, smothering, hangers-on, hitch-hiking and alien invasions. *Philosophical Transactions of the Royal Society B: Biological Sciences, 364*(1526), 2013–2025.

Grøsvik, B. E., Prokhorova, T., Eriksen, E., Krivosheya, P., Horneland, P. A., & Prozorkevich, D. (2018). Assessment of marine litter in the Barents Sea, a part of the joint Norwegian–Russian ecosystem survey. *Frontiers in Marine Science, 5*, 72.

Halsband, C., & Herzke, D. (2019). Plastic litter in the European Arctic: What do we know? *Emerging Contaminants, 5*, 308–318.

Hirai, H., Takada, H., Ogata, Y., Yamashita, R., Mizukawa, K., Saha, M., . . . Zettler, E. R. (2011). Organic micropollutants in marine plastics debris from the open ocean and remote and urban beaches. *Marine Pollution Bulletin, 62*(8), 1683–1692.

International Council for the Exploration of the Sea (ICES). (2019). *Fisheries overviews Barents Sea Ecoregion*. Published 29 November 2019. https://www.ices.dk/sites/pub/Publication%20 Reports/Advice/2019/2019/FisheriesOverview_BarentsSea_2019.pdf.

Jørgensen, L. L., & Spiridonov, V. (2013). *Effect from the king-and snow crab on Barents Sea benthos*. In Results and conclusions from the Norwegian-Russian Workshop in Tromsø (Vol. 2013).

Kaminskii, V. D., Suprunenko, O. I., & Suslova, V. V. (2011). The continental shelf of the Russian Arctic region: The state of the art in the study and exploration of oil and gas resources. *Russian Geology and Geophysics, 52*(8), 760–767.

Kane, I. A., Clare, M. A., Miramontes, E., Wogelius, R., Rothwell, J. J., Garreau, P., & Pohl, F. (2020). Seafloor microplastic hotspots controlled by deep-sea circulation. *Science*. https://doi.org/10.1126/science.aba5899.

Kanhai, L. D. K., Gardfeldt, K., Krumpen, T., Thompson, R. C., & O'Connor, I. (2020). Microplastics in sea ice and seawater beneath ice floes from the Arctic Ocean. *Scientific Reports, 10*(1), 1–11.

Kartesh. (2017). *Field report on the RV Kartesh cruise to Pechora Sea in summer 2017*. LMSU MRC.

Kartesh. (2018). *Field report on the RV Kartesh cruise to Pechora Sea in summer 2018*. LMSU MRC.

Kershaw, P., Turra, A., & Galgani, F. (2019). *Guidelines for the monitoring and assessment of plastic litter in the Ocean-GESAMP reports and studies No. 99*. GESAMP reports and studies.

La Beur, L., Henry, L.-A., Kazanidis, G., Hennige, S., McDonald, A., Shaver, M. P., & Roberts, J. M. (2019). Baseline assessment of marine litter and microplastic ingestion by cold-water coral reef benthos at the East Mingulay marine protected area (sea of the Hebrides, Western Scotland). *Front Marine Science, 6*, 80. https://doi.org/10.3389/fmars.2019.00080.

Lusher, A. L., Tirelli, V., O'Connor, I., & Officer, R. (2015). Microplastics in Arctic polar waters: The first reported values of particles in surface and sub-surface samples. *Scientific Reports, 5*, 14947.

Malyutin, A. A., Gintovt, A. R., Toropov, Y. Y., & Chernetsov, V. A. (2003). *Offshore platforms for oil and gas production on the Russian Arctic Shelf*. In Proceedings of the international conference on port and ocean engineering under Arctic conditions.

Obbard, R. W., Sadri, S., Wong, Y. Q., Khitun, A. A., Baker, I., & Thompson, R. C. (2014). Global warming releases microplastic legacy frozen in Arctic Sea ice. *Earth's Future, 2*(6), 315–320.

PAME. (2020). *Protection of the Arctic marine environment*. Arctic Council Working Group official website. Online. https://www.pame.is/.

Rochman, C. M., Kurobe, T., Flores, I., & Teh, S. J. (2014). Early warning signs of endocrine disruption in adult fish from the ingestion of polyethylene with and without sorbed chemical pollutants from the marine environment. *Science of the Total Environment, 493*, 656–661.

Sadove, S. S., & Morreale, S. J. (1989). *Marine mammal and sea turtle encounters with marine debris in the New York Bight and the northeast Atlantic*. In Proceedings of the second international conference on Marine Debris, Honolulu, Hawaii, 2–7.

Semenova, V., Boltunov, A., & Nikiforov, V. (2019). Key habitats and movement patterns of Pechora Sea walruses studied using satellite telemetry. *Polar Biology*, 1–12.

Serreze, M. C., & Francis, J. A. (2006). The Arctic amplification debate. *Climatic Change, 76*(3–4), 241–264.

Sokolov, K. M., Strelkova, N. A., Manushin, I. E., & Sennikov, A. V. (Eds.). (2016). Snow crab Chionoecetes opilio in the Barents and Kara seas. PINRO (in Russian). ISBN 978-5-86349-221.

Sukhotin, A., Denisenko, S., & Galaktionov, K. (2019). Pechora Sea ecosystems: Current state and future challenges. *Polar Biology, 42*(9), 1631–1645.

Taylor, M. L., Gwinnett, C., Robinson, L. F., & Woodall, L. C. (2016). Plastic microfibre ingestion by deep-sea organisms. *Scientific Reports, 6*, 33997.

Thompson, R. C. (2015). Microplastics in the marine environment: Sources, consequences and solutions. In *Marine anthropogenic litter* (pp. 185–200). Springer.

Trevail, A. M., Gabrielsen, G. W., Kühn, S., & Van Franeker, J. A. (2015). Elevated levels of ingested plastic in a high Arctic seabird, the northern fulmar (Fulmarus glacialis). *Polar Biology, 38*(7), 975–981.

UN Secretary-General (E/2019/68). *Progress towards the Sustainable Development Goals: report of the Secretary-General: Report of the Secretary-General*, May 2019. https://digitallibrary.un.org/record/3810131?ln=en.

UNDOS. (2020). *The science we need for the ocean we want brochure*. Online. https://www.oceandecade.org.

UNEP. (2016). *Marine plastic debris and microplastics – Global lessons and research to inspire action and guide policy change*. United Nations Environment Programme, Nairobi. Online. http://hdl.handle.net/20.500.11822/7720.

Woodall, L. C., Gwinnett, C., Packer, M., Thompson, R. C., Robinson, L. F., & Paterson, G. L. (2015). Using a forensic science approach to minimize environmental contamination and to identify microfibres in marine sediments. *Marine Pollution Bulletin, 95*(1), 40–46.

Yakushev, E., Gebruk, A., Osadchiev, A., Pakhomova, S., Lusher, A., Berezina, A. . . . & Semiletov I. (2021). Microplastics distribution in the Eurasian Arctic is affected by Atlantic waters and Siberian rivers. *Under review in Nature Commun Earth Environ 2* (23). https://doi.org/10.1038/s43247-021-00091-0.

Zalota, A. K., Spiridonov, V. A., & Vedenin, A. A. (2018). Development of snow crab Chionoecetes opilio (Crustacea: Decapoda: Oregonidae) invasion in the Kara Sea. *Polar Biology, 41*(10), 1983–1994.

Zettler, E. R., Mincer, T. J., & Amaral-Zettler, L. A. (2013). Life in the "plastisphere": microbial communities on plastic marine debris. *Environmental Science & Technology, 47*(13), 7137–7146.

Chapter 7
(Research): Sustainable Business Development in the Arctic: Under What Rules?

Alexandra Middleton

Abstract This chapter addresses the future of Arctic investment and sustainable business development by investigating available hard laws (frameworks and agreements that support sustainable development) and soft laws, especially by focusing on the Arctic Investment Protocol. Arctic Investment Protocol (AIP) was introduced in 2015 by the Global Agenda Council on the Arctic organised by the World Economic Forum. The chapter reviews currently available hard laws applicable to foreign direct investments in the Arctic states (National Investment Laws and International Investment Agreements) and AIP. The data comes from publicly available sources such as the World Economic Forum (WEF) publications and the UN Conference on Trade and Development database and interviews with WEF and Arctic Economic Council. Results demonstrate that hard investment laws and international investment treaties in the Arctic states do not support three pillars of sustainability and corporate social responsibility, and the adaption of soft law mechanisms is slow. The results of the analysis led to recommendations for developing future sustainable investment framework for the Arctic.

7.1 Introduction

The Arctic with its abundant natural resources such as oil and gas, bioresources and opening transport routes is becoming attractive for investment opportunities. Non-Arctic countries like France, Spain, Germany, and the UK and Scotland within the UK have developed their national policies on the Arctic. Moreover, China, South-Korea, Japan and the EU are joining this arena with their strategic documents and increased presence in the Arctic (Heininen et al., 2020). At the same time, investment in the Arctic remains challenging. Declining population, youth decrease and the lack of physical and digital infrastructure (BIN, 2017, 2018, 2019) combined

A. Middleton (✉)
University of Oulu, Oulu, Finland
e-mail: alexandra.middleton@oulu.fi

P. A. Berkman et al. (eds.), *Building Common Interests in the Arctic Ocean with Global Inclusion, Volume 2*, Informed Decisionmaking for Sustainability,
https://doi.org/10.1007/978-3-030-89312-5_7

with vulnerable nature and accelerated climate change impact create difficulties for business development and investment flows. The future development of the Arctic requires international investment, but how to balance protecting investors' rights and protecting the fragile Arctic environment while at the same time practising socially responsible investment remains unclear.

Sustainability is defined here through the prism of sustainable development that encompasses three pillars: economic, social and environmental. Economic pillar relates to growth, development, productivity and poverty alleviation, social pillar includes equity, participation, wellbeing, cultural identity and environmental pillar involves caring for natural environment and biodiversity (Kahn, 1995). The notion of sustainable development is at the core of the UN Sustainable Development Goals (UN SDGs) that incorporate all three pillars of sustainability to achieve 17 sustainable development goals that are a universal call to action to end poverty, protect the planet and improve the lives of peoples everywhere (UN SDGs, 2020). Closely related to sustainability is the notion of corporate social responsibility (CSR) that "*encompasses the economic, legal, ethical, and discretionary expectations that society has of organizations at a given point in time*" (Carroll, 1979; 500). Both sustainability and CSR concepts share the same vision, which intends to balance economic responsibilities with social and environmental ones (Montiel, 2008).

The development of the Arctic requires substantial investments in infrastructure such as roads, connectivity and civic infrastructure (Guggenheim Partners, 2019). Investments can come in the form of domestic and foreign direct investments. Foreign direct investment (FDI) is defined as an investment involving a long-term relationship and reflecting a lasting interest and control by a resident entity in one economy (foreign direct investor or parent enterprise) in an enterprise resident in an economy other than that of the foreign direct investor (FDI enterprise or affiliate enterprise or foreign affiliate) (UNCTAD, 2017). Analysis of investments in this chapter focuses solely on FDI and related policy framework for foreign investment.

Policy framework for FDI comprises a system of hard and soft laws. Hard law refers generally to legal obligations that are binding on the parties involved and which can be legally enforced before a court, while soft laws denote agreements, principles and declarations that are not legally binding (ECCHR, 2020). National Investment Laws and treaties such as International Investment Agreements are examples of hard law. Soft law is represented by various resolutions and declarations that are not legally binding.

This chapter addresses the future of sustainable Arctic investment by investigating currently available hard laws applicable to FDIs in the Arctic states (National Investment Laws and International Investment Agreements) and soft laws, by focusing predominantly on the Arctic Investment Protocol (Shaffer & Pollack, 2009).

The chapter provides a literature review and answers the following questions: (1) what hard laws related to investment are available in the Arctic states' policy frameworks for FDI?; (2) to what extent do they incorporate principles of responsible investment?; (3) what soft laws are available pertaining to sustainable investment in the Arctic?; (4) are the current mechanism adequate to guarantee sustainable investments in the Arctic? For analysis of the laws related to investment in the Arctic states, I perform hard laws content analysis using publicly available data from the UN Conference on Trade and Development database (Investment Policy Hub). For analysis of soft laws, i.e. Arctic Investment Protocol (AIP), I collect data through media and bibliographic searches, complemented by interviews with the World Economic Forum and the Arctic Economic Council. Theoretical lenses such as stakeholder theory analysis (Freeman, 1984; Garriga & Melé, 2004) and vested interest theory (Lehman & Crano, 2002) are used to analyse AIP.

The Chapter proceeds as follows. First, the scale of Arctic investment is reviewed, followed by a summary of the International Investment Framework and Sustainable Development and analysis of Arctic states' current National Investment laws and International Investment Agreements. The following subsection provides an analysis of soft law (Arctic Investment Protocol). The Chapter reviews the current status and provides possible solutions on how to design a sustainable investment framework in the Arctic.

7.2 Scale of the Arctic Investment

The scope of Arctic investments ranges from transport, fossil fuels, mining to civic ones. In general, Arctic investments can be categorised into two main categories: infrastructure and development and climate-change impact mitigation-related investments (Guggenheim Partners, 2019).

Figure 7.1 demonstrates that investments in the Arctic are required in mining, ports and related infrastructure, railways, roads, marine-related investments, aviation, power and renewable energy, telecommunications, and social targets (such as housing, schools, hospitals, etc.). According to Guggenheim Partners' estimates, infrastructure requirements in the Arctic region are expected to reach nearly $1 trillion over the next 15 years (Guggenheim Partners, 2019). The EU conducted Arctic stakeholders' consultation process and, in a report published in 2017 revealed that the major priorities for investment identified in the consultation are extension and improvement of digital infrastructure and development of internal and external transport connections.

The scale of the investments in the Arctic often requires foreign capital. For example, the Arctic telecommunication cable project establishing an alternative route between Asia and Europe over the Arctic Ocean is estimated at 0.7–1.1 billion USD and would most likely be funded by the contribution of private investment, primarily by Eurasian and Asian investors (Arctic Cable Initiative, 2019). Similarly, the Russian Arctic infrastructure along the Northern Sea Route is part of a $9.5 billion credit agreement signed with the China Development Bank (Staalesen, 2019).

The second category of investment is linked to climate-change impact mitigation-related investments. It has been estimated that in Russia alone, thawing permafrost would cost the Russian economy $2.3 billion a year (Fedorinova & Tanas, 2019), hence requiring investments in infrastructure reconstruction.

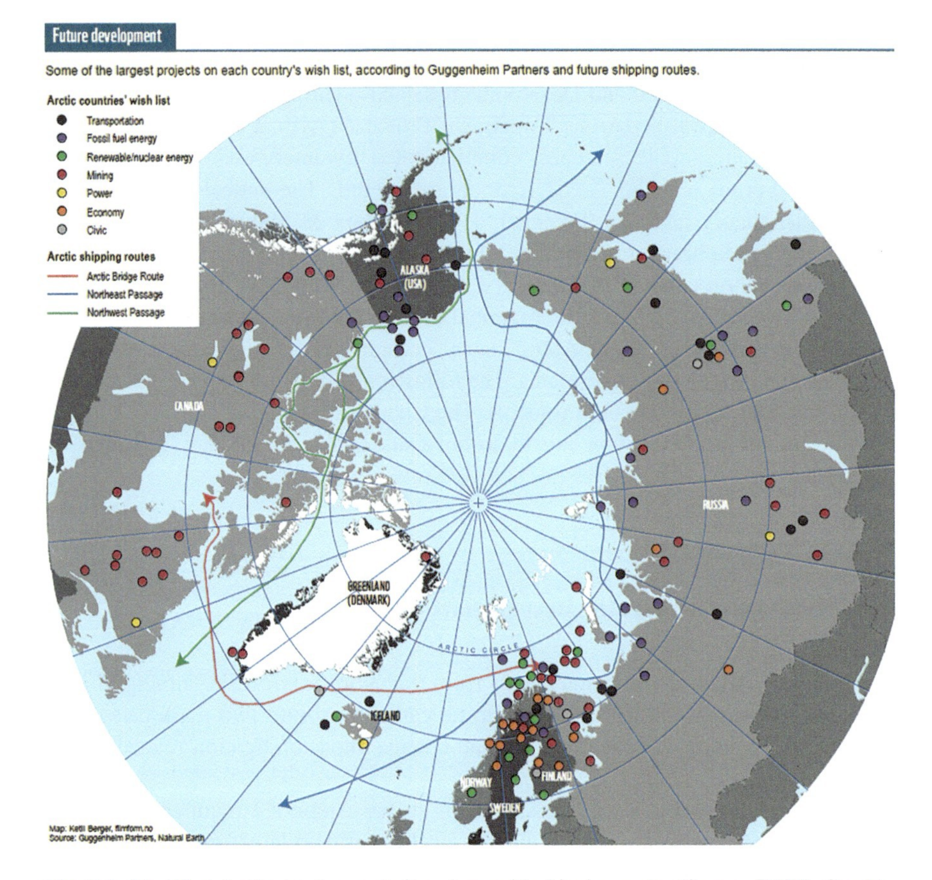

Fig. 7.1 Identified Arctic development sites along with shipping routes (Source: WWF. Graphic: © Ketill Berger, filmform.no – Source: © Guggenheim Partners, Natural Earth)

Unique characteristics of developing business and investment in the Arctic include: harsh climate, limited existing infrastructure, long project lead times, social, environmental, human rights and Indigenous peoples' rights impact concerns (Conley et al., 2013; Rosen & Thuringer, 2017). The exploitation of Arctic resources carries risks such as oil spills, ship causalities, chemical runoffs from mining and smelting activities (Rosen & Thuringer, 2017). Furthermore, Arctic investment is subject to uncertainty due to fluctuations in commodity prices, changing geopolitical forces and limited legal framework that concerns sustainable business development and sustainable investment in the Arctic. Development in Arctic areas is likely to be more expensive because of distance from consumption centres and increased transportation times and costs (Conley et al., 2013). Arctic investment should account for possible conflicts over land, environmental impacts, labour relations and Indigenous peoples' rights. Investment is regulated on one hand by international investment framework (national investment laws and international investment agreements) with a legally binding nature but not addressing Arctic specifics. On the other hand, guidelines and soft laws that are tailored for the Arctic are all voluntary and do not have international enforcement and arbitrage mechanisms.

7.3 International Investment Framework and Sustainable Development

7.3.1 Components of International Investment Framework

This section focuses on hard laws related to investments in the Arctic states. International investment rule-making is taking place at the bilateral, regional, interregional and multilateral levels. The process of international investment rule-making developed in two stages post WW2 period (1945–1989) and from 1989 to present (UNCTAD, 2008). In the first stage, customary international law established an international minimum standard of treatment to which foreign investors were entitled in the territory of the host country. In the second stage, growing FDI flows initiated the creation of more complex international investment agreements.

International investments are primarily regulated by National Investment Laws and International Investment Agreements (IIAs) (UNCTAD, 2020). National investment laws govern investments made by foreign nationals within a specific country boundary. The purpose of such law is to establish protection of investments, provide tax settings and incentives, (and) procedures for dispute settlements. In some instances, national investment laws limit foreign ownership of certain strategic industries. International investment agreements (IIAs) are designed to protect the investments of foreign investors in the state hosting the investment. IIAs reduce perceived risk and may result in an increased flow of FDI. The term "investment" must be defined by the state since only investment falling under the scope of the agreement would be guaranteed protection (Malik, 2011).

International investment agreements (IIAs) can be classified into two types. The first one, a bilateral investment treaty (BIT) is an agreement between two countries regarding the promotion and protection of investments made by investors from respective countries in each other's territory (UNCTAD, 2020). Most of IIAs are bilateral investment treaties. The second type is treaties with investment provisions or TIPs, they bring together various types of investment treaties that are not BITs. There are also Multilateral Investment Treaties (MITs)[1] that represent international investment agreements made between several countries and containing provisions to protect investments made by individuals and companies in each other's territories, e.g. North American Free Trade Agreement (NAFTA) and the Energy Charter Treaty

[1] https://uk.practicallaw.thomsonreuters.com/4-502-5545?transitionType=Default&contextData=%28sc.Default%29

The current International Investment Agreements (IIAs) regime is multi-layered and fragmented. As of 2020, there are 2335 BITs and 314 TIPs in force worldwide (UNCTAD, 2020). While investment law and investment agreements are efficient in bringing capital and technology, they fail to address social and environmental concerns (Gordon & Pohl, 2011). Most of the IIAs in force today were drafted in the 1990s supporting free markets (Spears, 2010). The debate about the current investment law regime evolves around "needs to strike a better balance between the rights of investors and the rights of host states to regulate" (Spears, 2010). Traditionally, IIAs have focused on protecting investment and included no mentioning of responsible conduct (Cotula, 2017).

The issues related to economic and social conduct are often addressed in national laws (Cotula, 2017). On a global scale addressing climate change and environmental problems has resulted in several International and Multilateral Environmental Agreements (MEAs). Some examples of these agreements include the Espoo Convention on Environmental Impact Assessment in a Transboundary Context (1997), the Stockholm Convention on Persistent Organic Pollutants (2004), and the Paris Agreement (2015). Moreover, many existing international instruments can be used to set standards for responsible investment, for example, human rights treaties, labour conventions and soft law instruments on respecting land rights (Cotula, 2017). The integration of responsible investment principles and reference to other relevant soft laws into IIAs can provide more clarity in case of disputes. But to what extent do International Investment Agreements incorporate or refer to available environmental and other sustainability-related agreements?

Addressing social and environmental concerns can be done via sustainability provisions in the text of the international investment treaties. Major types of these provisions include: general sustainable development provision, anti-corruption provision, environmental provision, labour rights and human rights provision, substantive transparency provision, procedural transparency provision, national security provision and responsible business practices (Chi, 2019).

However, analysis of the treaties demonstrates that only a minimal amount of treaties includes such provisions. When analysing a sample of 1623 IIAs, Gordon and Pohl (2011) find that only 8% of all treaties include references to environmental concerns. According to Chi (2019) the IIA regime requires a reform guided by sustainable development objectives, and IIAs should include a provision concerning not only economic but also social and environmental development. The UNCTAD Policy framework proposes a reform on the inclusion of such provisions in IIAs. Chi (2019) finds that recently a growing number of IIAs are starting to incorporate provisions that have binding obligations for foreign investors concerning corporate social responsibility.

The recent trend overall is while investment is encouraged, it should not be at the cost of the long-term environmental and social well-being of the host state. For instance, the EU-Canada Comprehensive and Economic Trade Agreement (CETA) refers to the Universal Declaration of Human Rights and OECD Guidelines for Multinational Enterprises, Morocco-Nigeria BIT 2016 stands out because of its innovative human rights approach (Zugliani, 2019) and commits states to ensure their laws, policies and actions are consistent with international human rights treaties. Moreover, the Dutch model BIT (2018) commits to promote equal opportunities and participation for women and men in the economy, it mentions the OECD Guiding Principles for Global Investment Policymaking, Paris Agreement, ILO Conventions, Universal Declaration of Human Rights, United Nations Guiding Principles on Business and Human Rights, OECD Guidelines for Multinational Enterprises (Cotula, 2017; Peacock et al., 2019). In the context of the EU, EU–Canada Comprehensive Economic and Trade Agreement (CETA) and the EU–Singapore Free Trade Agreement (FTA) represent a shift to adopt better practices that include sustainable and responsible behaviour by investors (Titi, 2015).

7.3.2 Analysis of Arctic States Investment Frameworks

7.3.2.1 Investment Laws

The analysis of investment laws of the Arctic countries is conducted by studying the inclusion of sustainability language in the text of the laws. This is done to see whether the laws address environmental and social concerns associated with the investment. The search was conducted by using NVivo software with the search words "environ*" and "social*" in the full texts of the investment laws and their official English translations. Altogether 14 unique investment laws were included in the analysis (see Table 7.1). Most of the laws in the Arctic states date back to the 1980s and 90s. Only two laws (the EU and Icelandic) to some extent address sustainability concerns. Recent additions such as the EU's, Norway's and the US laws directly address national security with the language determining registrations and authorisation, and procedures of investment screening.

In the case of the US investment law, the Foreign Investment Risk Review Modernization Act (FIRRMA) signed into law in 2018 is believed to address national security concerns largely related to particular Chinese investment trends (White & Case LLP, 2018). The law limits foreign investments, such as real estate acquisitions in sensitive areas and minority investments that might provide access to sensitive information or technology of the target US business (White & Case LLP, 2018).

7.3.2.2 IIAs of Arctic States

This subsection reviews the investment framework in force in the Arctic countries including IIAs, TIPs and an open-ended category of investment-related instruments (IRIs). IRIs encompasses various binding and not-binding instruments and includes, for example, model agreements and draft instruments, multilateral conventions on dispute settlement and arbitration rules, documents adopted by international organisations, and others. The degree of these binding agreements and treaties varies greatly from country to country. Table 7.2 demonstrates that countries that are part

Table 7.1 Investment Laws in the Arctic countries

Country	Investment Law	Year	Inclusion of sustainability/CSR language
Canada	National Security Review of Investments Regulations	2009	No
	Investment Canada Regulations	1985	No
	Investment Canada Act	1985	No
EU	Regulation (EU), 2019/452 of the European Parliament and of the Council of 19 March 2019 establishing a framework for the screening of foreign direct investments into the Union	2019	To some extent
Finland	Act on the Monitoring of Foreign Corporate Acquisitions in Finland	2012	No
Denmark	Lov om krigsmateriel/Act on War Material (The Danish Consolidated Act No. 1004 of October 22 2012)	2012	No
Iceland	Act on Investment by Non-residents in Business Enterprises	1991	To some extent
Norway	National Security Act	2018	No
Russia	Federal Law on Foreign Investments	1999	No
	Federal Law on the procedure for making foreign investments in business entities of strategic importance for ensuring the country's defense and state security	2008	No
USA	Section 721 of the Defense Production Act of 1950	1950	No
	Foreign Investment Risk Review Modernization Act of 2018	2018	No

Note: Data from UNCTAD, compiled by the author
Inclusion of sustainability language = inclusion on words "environ*" and "social*" in the text of investment laws

Table 7.2 Investment framework in force in the Arctic countries

Country	International Investment Agreements (IIAs)	Treaties with Investment Provisions (TIPs)	Investment Related Instruments (IRIs)	Total	Number of instruments including sustainability/ CSR language as /% of total (in force)
Canada	36	17	29	82	14 (17%)
Finland	65	56	29	150	6 (4%)
Greenland	0	1	–		–
Denmark	46	56	29	131	5 (4%)
Iceland	7	29	27	63	0
Norway	14	28	29	71	0
Russia	64	6	21	91	0
Sweden	63	56	29	148	3 (2%)
USA	39	50	33	122	0

Note: Data from UNCTAD, compiled by the author

of the EU have most instruments in use including IIAs, TIPs and IRIs. I use the mapping of IIA content[2] available as part of the UNCTAD IIAs mapping project to access how many of the instruments include corporate social responsibility (any mentioning in the text, except preamble).

Table 7.2 demonstrates that only a few countries have investment agreements incorporating corporate social responsibility and environmental protection, with the biggest share in Canada (17%), followed by Finland and Denmark where 4% of all instruments include sustainability language.

What kind of sustainability language can an investment instrument potentially have? An example from Agreement Between the Government of Canada and the Government of the Republic of Benin for the Promotion and Reciprocal Protection of Investments (2016) has the following article on corporate social responsibility "*Each Contracting Party should encourage enterprises operating within its territory or subject to its jurisdiction to voluntarily incorporate internationally recognized standards of corporate social responsibility in their practices and internal policies, such as statements of principle that have been endorsed or are supported by the Contracting Parties. These principles address issues such as labour, the environment, human rights, community relations and anti-corruption*".

While the analysis of investment agreements text provides only indicative results, the pattern confirms findings of Gordon and Pohl (2011) that only a few of them include corporate social responsibility (CSR) concerns. The ones that include were drafted starting from 2010. Furthermore, the language of CSR is advisory, not a binding one. To conclude, while the Arctic states have a wide arsenal of IIAs, TIPs and IRIs the issues of CSR are addressed in them at a low degree only.

A reform of the IIAs incorporation sustainability pillars such as protection of the environment, Indigenous people and their livelihoods by the Arctic states while applicable to the whole Arctic state would benefit the Arctic territories. Another way to move forward would be an introduction of a multilateral Arctic investment agreement applicable to all Arctic states based on the best practices incorporating environmental and social issues that would cater to specific needs of protecting the Arctic environment and traditional livelihoods.

The Sustainable Development Working Group (SDWG) of the Arctic Council has in its mandate to propose and adopt steps to be taken by the Arctic States to advance sustainable development in the Arctic. This includes pursuing opportunities to protect and enhance the environment and the economies, culture and health of Indigenous peoples and Arctic communities. The results of SDWG work are presented in reports, but so far SDWG has not produced any reports or guidelines specifically on responsible investment in the Arctic. The report "Environmental Impact Assessment and Meaningful Engagement in the Arctic" produced by SDWG in 2019 report contains some examples of sustainable business practices including models for meaningful

[2]The IIA Mapping Project is a collaborative initiative between UNCTAD and universities worldwide to map the content of IIAs. The IIA Mapping Project is an ongoing effort that aims to map all IIAs for which texts are available (about 3000), 2577 or 86% of IIAs have been mapped already.

engagement of Indigenous peoples. This document, however, does not mention investments or investors' rights and responsibilities in the Arctic.

7.4 Arctic Investment Protocol (AIP)

7.4.1 Purpose and Content

Arctic Investment Protocol is an example of a soft law that directly deals with investment in the Arctic. Arctic Investment Protocol (AIP) was introduced in 2015 by the Global Agenda Council on the Arctic organised by the World Economic Forum. The World Economic Forum is the international organisation for public-private cooperation addressing key global challenges, such as climate change, poverty, investments, by engaging the foremost political, business and other leaders of society to shape global, regional and industry agendas (WEF, 2020).

The Global Agenda Council (GAC) network, latterly known as the Global Future Council Network, is the world's foremost interdisciplinary knowledge network, facilitated by the World Economic Forum and comprised several expert Councils that sit for two-year terms with the objective of address specific issues. The Global Agenda Council on the Arctic (GACA) was established in 2012 and sat for two terms (2012–2014 and 2014–2016). The Arctic Investment Protocol was created by the Global Agenda Council on the Arctic during the 2014–2016 term and released in 2015. The composition of experts on the Global Agenda Council on the Arctic was determined by the World Economic Forum that provided a platform for the facilitation of the process. Stakeholders from academia, states, media, business and Indigenous people organisation took part in the GACA. The protocol was developed within the network of experts constituting GACA (Interview with WEF 18.5.2018).

AIP falls under the category of soft law as it is not legally binding and is advisory. In his 2016 statement Scott Minerd, Global Chief Investment Officer of Guggenheim Partners said "*Building a sustainable future for the Arctic region and its people requires long-term capital that carefully weighs environmental and societal impact. The Arctic Investment Protocol represents a new approach to sustainable development that the world needs today.*[3]"

7.4.2 Analysis of AIP

First, stakeholder theory (Freeman, 1984; Garriga & Melé, 2004) is applied in the analysis of the AIP. A stakeholder is "any group or individual who can affect, or is

[3] https://www.guggenheimpartners.com/firm/news/guggenheim-partners-endorses-world-economic-forums

affected by, the achievement of a corporation's purpose" (Freeman, 1984:46). Each of these groups has a stake in the modern corporation, hence, the term "stakeholder". Stakeholder theory recognises the importance of balancing between aligning potentially converging interests of various groups.

A stakeholder analysis of the AIP is built taking Arctic Council, representing Arctic stakeholders. The Arctic Council established in 1996, represents an intergovernmental forum for promoting cooperation, between the eight Arctic nations, Indigenous peoples represented by six Permanent Participants, and other Arctic stakeholders (non-Arctic states, intergovernmental and interparliamentary organizations and nongovernmental organizations). Altogether 22 people contributed to the development of the AIP, out of them the USA was represented seven times, followed by Norway (6), Russia (2) and Canada (1). Observer non-Arctic countries were represented by the UK (2), China (1), Japan (1) and South Korea (1). At the same time, Arctic countries such as Iceland, Denmark, Finland and Sweden did not have a representation in the drafting of the AIP. Indigenous people were represented by Sami Reindeer Herders' Association of Norway and the National Union of the Swedish Sami People. Analysed from an organisational background perspective the AIP creators were dominated by private industry (28%), followed by finance industry (18%), government officials (18%), academia (18%), public organisation (9%) and Indigenous people organisations (5%) (see Table 7.3).

Stakeholder analysis demonstrates that Indigenous people are underrepresented (5% of creators), so are half of the Arctic states (Iceland, Denmark, Finland and Sweden), the composition of experts is skewed towards private sector and finance sector representatives. Additionally, some important NGOs (e.g. World Wide Fund for Nature, Arctic Programme and intergovernmental organizations (e.g. International Union for the Conservation of Nature (IUCN)) that are Arctic Council observers were not included either.

In order to study how widely AIP is used I apply vested interest theory lenses that suggest that vested interest is an important moderator of consistency between attitudes and policy endorsement (Lehman & Crano, 2002). By 2020, 5 years since its launch AIP has received 16 official endorsements,[4] 40% of them coming from the same organisations involved in drafting the protocol. In the 2017 Arctic Economic Council (AEC) Annual Meeting endorsed the intent of the AIP and took over the ownership of the protocol. AEC represents an independent international organization established in 2014 with the aim to promote sustainable business development in the Arctic. The AEC membership comprises over 40 members, amongst others including Confederation of Danish Industry, Confederation of Icelandic Enterprise and Confederation of Norwegian Enterprise representing their member organizations After taking up the ownership of the AIP, the AEC consulted

[4]American Geographical Society, Eykon Energy, Guggenheim Partners, High North Center, Graduate School of Business, Nord University, The Icelandic Arctic Chamber of Commerce, Lapland Chamber of Commerce (Finland), Lapland Chamber of Commerce (Sweden), The Northern Chamber of Commerce and Industry, Murmans; Norwegian Shipowners' Association, Norwegian Shipowners' Association, The Polar Journal, Pt Capital, Statoil, Tschudi Shipping

Table 7.3 Composition of AIP creators

Sector	Number of people	Percentage from Total
Academia	4	18%
Finance sector	4	18%
Indigenous people organization	1	5%
Private	5	23%
Private Consulting	1	5%
Public	2	9%
Public-private partnership	1	5%
State	4	18%
Total	22	100%

all its members from all Arctic states and Permanent Participant if they would like any amendment or update to the AIP. The members of the AEC considered the AIP to still have the right language and tone (Interview with AEC 19.03.2021). However, the constituents of AEC represent mostly business community stakeholders and Arctic Indigenous peoples; to access the need to update the AIP a consultation with a broader Arctic stakeholder base (e.g. policy-makers, local communities, academia, etc.) would be advisable. According to experts, the AEC has the potential to become an effective international law development mechanism of socially and environmentally responsible business in the Arctic region; however, this organization has not yet become an effective dialogue platform for the development of economic activities in the region (Vylegzhanin et al., 2020).

Low recognition and endorsement of the protocol in the world can also be attributed to the fact that the drafting process did not go through a process of open public stakeholder consultation. When asked about the procedure WEF commented as follows according to the World Economic Forum "*The Forum worked to ensure the Council was representative (by region, sector, industry, gender, etc.). The development of the Arctic Investment Protocol drew on informal discussions led by the Arctic GAC members amongst their own professional networks, representing different stakeholder groups.*"[5]

Content analysis of the AIP shows that the protocol consists of a text body of the protocol (672 words or 23% of the text) and an appendix totalling 2554 words (77%). The appendix has references to 11 CSR and other frameworks: Aspen Principles of Arctic Governance, The Equator Principles, The Extractive Industries Transparency Initiative (EITI), International Council on Mining and Metals (2013), Indigenous Peoples and Mining Position Statement, International Maritime Organization (IMO), International Finance Corporation (IFC) Performance Standards on Environmental and Social Sustainability, National Science Foundation (US), Principles for the Conduct of Research in the Arctic, Organisation for Economic Cooperation and Development Guidelines for Multinational Enterprises, Guiding

[5]interview with WEF 18.5.2018.

Principles on Business and Human Rights, UN Principles for Responsible Investment and World Bank Group Environmental, Health and Safety Guidelines.

The Protocol does not explain why these standards were selected. Some major CSR standards (e.g. GRI, ISO 14001) and the UN Declaration on the Rights of Indigenous Peoples are missing, these are particularly important in the light of protecting Arctic Indigenous peoples' rights. The lack of Russian representatives represents a concern, especially since Russia abstains from voting for the UN Declaration on the Rights of Indigenous Peoples (IWGIA). The Protocol heavily relies on CSR standards, which are voluntary by nature (Fransen, 2012). Additionally, the AIP does not include provisions for independent monitoring and dispute settlement mechanism.

There are examples of investment protocols that serve as a departure from BITs and multilateral investment laws. For instance, the Southern African Development Community (SADC) Protocol on Finance and Investment was enacted in 2006 to facilitate foreign investment and protect it from, e.g. political risks (SADC, 2006). The protocol features harmonised tax policies throughout the region that allow for ease and equity of investment through SADC, protection of all investors regardless of their nationality, and prohibits the nationalisation of the investment property (Kotuby et al., 2014). In 2016 the protocol was amended to include new sustainability provisions on domestic health, safety and environmental measures, investor responsibility and the right of a state party to regulate (SADC, 2016).

The ownership of the Arctic Investment Protocol was handed over to the Arctic Economic Council in 2017. Consequently, the AEC initiated a process for stakeholders to provide input and concrete examples of best practices to achieve sustainable investments in the Arctic and elsewhere. These examples will be incorporated into the Arctic Investment Protocol Annex listing best practices (AEC website). The idea is to collect investment cases from the region taking the AIP as a starting point (Interview with AEC 15.03.2021). The results of stakeholder input are not yet available as it stands.

7.5 Future of Sustainable Investments in the Arctic

Different models have been suggested for introducing sustainable and responsible investment principles in the Arctic. Zhang (2020) proposed an Investment Review Advisory Board (IRAB) sitting within the Arctic Economic Council consisting of advisors from each of the eight Arctic Council states. The role of IRAB would be to apply the Arctic Investment Protocol's guidelines to guide and advise Arctic states on the multi-faceted, long-term impact of large-scale investments proposed for the region (Zhang, 2020). Alternative solutions include establishing a set of multilateral Arctic FDI review criteria directed by each nation, an Arctic Development Code, and setting up the Arctic Development Bank (Rosen & Thuringer, 2017). Another solution includes the introduction of a more holistic approach to infrastructure development through strong institutions and the establishment of an Arctic accreditation scheme to focus development within the High North, consistent with the UN Sustainable Development Goals (Scherwin, 2020).

Sustainable business development and responsible investment in the Arctic requires functioning laws, agreements, and guidelines. Therefore, the work can be done in parallel on both hard and soft law fronts. International investment agreements are predominantly bilateral and at their current state do not accommodate the demand for enhanced corporate social responsibility. One option to address this challenge is the restructuring of IIAs by the Arctic states following UNCTAD IIS reform that places inclusive growth and sustainable development while benefiting from investments. One option to initiate the discussion about the need to update national investment laws and IIAs would be to include responsible investor conduct in light of the Arctic development needs. This preliminary discussion can potentially take place at the Arctic Council and Arctic Economic Council level. While each country decides on the national laws, it would be beneficial to outline common interests and provision clauses relevant for the Arctic at a collegial Arctic state level.

Updating the Arctic Investment Protocol Arctic Investment Protocol provides a framework perspective but lacks some essential elements discussed further. First, it would be beneficial to define the Arctic investments, followed by objectives, definitions, and parties to who the AIP applies. The AIP can be built upon the best corporate social responsibility practices available in International Investment Agreements and Treaties with Investment Provisions and on already existing finance and investment protocols, e.g., SADC protocol. The functioning AIP would require governance mechanisms and built-in infrastructure, such as, for instance, suggestions for investor-state dispute settlement. A report "Business Financing in the Arctic" (2018) as part of the Kingdom of Denmark's work aiming to promote sustainable economic development in the Arctic recommends developing the Arctic Investment Protocol to make it more practically applicable.

The options for the update of the AIP could include cooperation between the Arctic Economic Council and the Arctic Council, especially taking into consideration Arctic Council's efforts resulting in three binding international agreements. In 2019 the Arctic Economic Council signed a Memorandum of Understanding (MoU) with the Arctic Council aiming to provide a framework for cooperation and to facilitate collaboration between the Arctic Council and the Arctic Economic Council. This cooperation in drafting an updated version of the AIP including guidelines on its practical implementation can benefit from the input of the Arctic Council's constituents.

So far, the AIP received a rather weak endorsement, as by 2020, 5 years since its launch AIP has received 16 official endorsements, 40% of them coming from the same organisations involved in drafting the protocol. Additionally, the AIP is endorsed by all AEC members. Several options exist to alleviate this challenge. The creation of an updated investment protocol that is aimed to be useful for sustainable and responsible investment in the Arctic, in the long run, could include the following elements. As demonstrated by analysis of the AIP, parties that were involved in drafting are likely to endorse it and promote it among their network. The drafting of the protocol through an open, inclusive and transparent process would be beneficial for future protocol endorsement (workable examples of wide stakeholder consultation processes exist, for example, in the EU stakeholder dialogue on the Arctic). Additionally, standard setters have well-functioning processes of stakeholder involvement (e.g. GRI, ISO). Hence, the draft of the protocol can be made available for an open consultation process, including collecting feedback from all Arctic stakeholders. Consequently, the Arctic states would need to demonstrate their commitment to recommend adherence to the AIP by investors.

7.6 Conclusions

This Chapter provides an overview of available hard and soft law mechanisms for sustainable business development and investment in the Arctic states. The analysis demonstrates that currently the international investment framework presented by hard investment laws and international investment treaties in the Arctic states is fragmented and does not support the three pillars of sustainability and corporate social responsibility. The hard law related to investment in the Arctic states does not incorporate sustainable investment provisions.

Concerning soft law, the Arctic Investment Protocol that was specifically designed for the Arctic still receives little attention and endorsement from business. To proceed, the AEC has been actively promoting the Arctic Investment Protocol at conferences, public events and high-level meetings with stakeholders. So, there is an ongoing discussion about applying it. Nonetheless, it is important to understand that AIP represents only a set of guidelines similar to the Sustainable Development Goals (interview with AEC 15.03.2021). One important step towards wider recognition of the protocol is a mention of it in Summary of the results of the public consultation on the EU Arctic policy (2021), whereby "*several contributors suggest that the EU incorporate sustainability criteria in its Arctic investments and that it promotes utilisation of the Arctic Investment Protocol. Various contributors also propose that the EU takes action on sourcing Arctic natural resources, including through certification, p.1*" However, in the Joint Communication on a stronger EU engagement for a peaceful, sustainable and prosperous Arctic released in October 2021, the AIP is not mentioned (Joint Communication, 2021).

The way forward for decreasing uncertainty related to the Arctic investment is outlined that would include defined roles of the Arctic Council, the Arctic Economic Council and the Arctic states in updating hard law instruments and the creation of the updated Arctic Investment Protocol through an open, inclusive and transparent process. Greater involvement of local communities and Indigenous peoples in the design and consultation phases of the AIP could potentially result in a higher rate of acceptance and implementation of this instrument. The financial community, e.g. European Investment Bank, can potentially step-up and demand that funding is only granted to the projects that abide by the AIP or similar sustainable investment criteria applicable to the Arctic.

The Arctic is expected to receive large infrastructural investments that need to be sustainable and respectful of people and the environment. Cooperation within Arctic Council that marked 25th anniversary has been built on achieving sustainable development. Further, joint efforts of the Arctic Council and the Arctic Economic Council can be targeted towards clear rules and guidelines on sustainable investment in the Arctic.

References

Act on the Monitoring of Foreign Corporate Acquisitions in Finland. (2012). https://investmentpolicy.unctad.org/investment-laws/laws/72/finland-act-on-the-monitoring-of-foreign-corporate-acquisitions-in-finland. Accessed 20 Dec 2019.

Arctic360. (2020). https://www.arctic360.org/. Accessed 15 Dec 2020.

Agreement Amending Annex 1 – Cooperation on investment – On the Protocol on Finance & Investment – English – 2016. https://www.sadc.int/documents-publications/show/4999.

Agreement Between the Government of Canada and the Government of the Republic of Benin for the Promotion and Reciprocal Protection of Investments. (2016). https://www.international.gc.ca/trade-commerce/trade-agreements-accords-commerciaux/agr-acc/benin/fipa-apie/index.aspx?lang=eng.

Arctic Cable Initiative. (2019). https://www.cinia.fi/en/archive/arctic-telecom-cable-initiative-takes-major-step-forward.html. Accessed 20 Dec 2019.

Arctic Council Sustainable Development Working Group. https://www.sdwg.org/activities/project-reports-from-completed-sdwg-projects-1998-to-2015/. Accessed 20 Dec 2019.

Arctic Economic Council (AEC). (2020). https://arcticeconomiccouncil.com/about/arctic-investment-protocol/. Accessed 1 Feb 2021.

Business Index North (BIN). (2017). *A periodic report with insights to business activity and opportunities in the Arctic*. https://businessindexnorth.com/reports/?Article=37. Accessed 20 Dec 2019.

Business Index North (BIN). (2018). *A periodic report with insights to business activity and opportunities in the Arctic*. https://businessindexnorth.com/reports/?Article=61. Accessed 20 Dec 2019.

Business Index North (BIN). (2019). *A periodic report with insights to business activity and opportunities in the Arctic*. https://businessindexnorth.com/reports/?Article=67. Accessed 20 Dec 2019.

Business Financing in the Arctic. (2018). https://um.dk/en/foreign-policy/the-arctic/business-financng-in-the-arctic/. Accessed 20 Mar 2021.

Canada National Security Review of Investments Regulations. (2009). https://laws-lois.justice.gc.ca/eng/regulations/sor-2009-271/page-1.html. Accessed 12 Aug 2020.

Canada's Arctic and Northern Policy Framework. (2019). https://www.rcaanc-cirnac.gc.ca/eng/1560523306861/1560523330587. Accessed 10 Dec 2020.

Carroll, A. B. (1979). A three-dimensional conceptual model of corporate performance. *AMR, 4*, 497–505. https://doi.org/10.5465/amr.1979.4498296.

Chi, M. (2019). *Sustainable development provisions in investment treaties*. UNESCAP. https://repository.unescap.org/bitstream/handle/20.500.12870/973/Sustainable%20Development%20Provisions%20in%20Investment%20Treaties.pdf?sequence=1&isAllowed=y.

Conley, H., David, L., Pumphrey, D., Toland, T., & David, M. (2013). *Arctic economics in the 21st century: The benefits and costs of cold*. Report of the CSIS Europe Program, July 2013. https://csis-prod.s3.amazonaws.com/s3fs-public/legacy_files/files/publication/130710_Conley_ArcticEconomics_WEB.pdf.

Cotula, L. (2017). *Responsible investment provisions in international investment treaties: Where next?* Available at https://www.iied.org/responsible-investment-provisions-international-investment-treaties-where-next. Accessed 12 Aug 2020.

ECCHR. (2020). *Definitions of hard and soft law*. https://www.ecchr.eu/en/glossary/hard-law-soft-law/#:~:text=The%20term%20soft%20law%20is,that%20are%20not%20legally%20binding.&text=Hard%20law%20refers%20generally%20to,legally%20enforced%20before%20a%20court. Accessed 11 Aug 2020.

Environmental Impact Assessment and Meaningful Engagement in the Arctic. (2019). https://oaarchive.arctic-council.org/handle/11374/2377. Accessed 11 Aug 2020.

EU-Canada Comprehensive and Economic Trade Agreement (CETA). (2017). https://ec.europa.eu/trade/policy/in-focus/ceta/ceta-chapter-by-chapter/. Accessed 11 Aug 2020.

Federal Law 160-FZ on Foreign Investment in the Russian Federation. (1999). https://www.wto.org/english/thewto_e/acc_e/rus_e/WTACCRUS58_LEG_41.pdf. Accessed 25 June 2020.

Federal Law N57-FZ Procedures for Foreign Investments in the Business Entities of Strategic Importance for Russian National Defence and State Security. (2008). http://en.fas.gov.ru/documents/documentdetails.html?id=13918. Accessed 25 June 2020.

Federal Law N 193 On State Support for Entrepreneurial Activity in the Arctic Zone of the Russian Federation. (2020). http://www.consultant.ru/document/cons_doc_LAW_357078/. Accessed 15 Dec 2020.

Fedorinova, Y., & Tanas, O. (2019). *Russia's thawing permafrost may cost economy $2.3 billion a year*. https://www.bloomberg.com/news/articles/2019-10-18/russia-s-thawing-permafrost-may-cost-economy-2-3-billion-a-year. Accessed 25 June 2020.

Fransen, L. (2012). Multi-stakeholder governance and voluntary programme interactions: Legitimation politics in the institutional design of corporate social responsibility. *Socio-Economic Review, 10*, 163–192. https://doi.org/10.1093/ser/mwr029.

Freeman, R. E. (1984). Strategic management: A stakeholder approach. Pitman.

Garriga, E., & Melé, D. (2004). Corporate social responsibility theories: Mapping the territory. *Journal of Business Ethics, 53*, 51–71. https://doi.org/10.1023/B:BUSI.0000039399.90587.34.

Gordon, K., & Pohl, J. (2011). *Environmental concerns in international investment agreements: A survey* (OECD working papers on international investment, No. 2011/01). OECD Publishing. https://doi.org/10.1787/5kg9mq7scrjh-en.

GRI. (2020). *Stakeholder council*. https://www.globalreporting.org/about-gri/governance/stakeholder-council/. Accessed 25 June 2020.

Guggenheim Partners. (2019). *Financing sustainable development in the Arctic responsible investment solutions for the future*. https://www.guggenheiminvestments.com/GuggenheimInvestments/media/PDF/Financing-Sustainable-Development-in-the-Arctic.pdf. Accessed 25 June 2020.

Guggenheim partners endorses world economic forum's arctic investment protocol. (2016, Jan 21). *NASDAQ OMX's News Release Distribution Channel*. Retrieved from https://search.proquest.com/docview/1758659517?accountid=13031. Accessed 15 June 2020.

Heininen, L., Everett, K., Padrtova, B., & Reissell, A. (2020). *Arctic policies and strategies – Analysis, synthesis, and trends*. IIASA. https://doi.org/10.22022/AFI/11-2019.16175.

Iceland Act on Investment by Non-residents in Business Enterprises. (1991). https://investmentpolicy.unctad.org/investment-laws/laws/90/iceland-investment-act. Accessed 20 June 2020.

Interview with World Economic Forum (WEF). 18.5.2018.

Interview with Arctic Economic Council (AEC). 15.3.2021 and 19.03.2021.

Investment Canada Act. (1985). https://laws-lois.justice.gc.ca/eng/acts/i-21.8/. Accessed 15 June 2020.

Investment Canada Regulations. (1985). https://laws-lois.justice.gc.ca/eng/regulations/SOR-85-611/index.html. Accessed 15 June 2020.

IWGIA. (2020). https://www.iwgia.org/en/russia/2245-russia-denial-of-indigenous-peoples-rights-concern.html. Accessed 5 June 2020.

ISO. (2020). *Additional guidance from the TMB on stakeholder engagement*. https://www.iso.org/files/live/sites/isoorg/files/developing_standards/docs/en/additional_guidance_tmb.pdf. Accessed 5 Dec 2020.
Joint Communication. (2021). *Stronger EU engagement for a peaceful, sustainable and prosperous Arctic*. https://eeas.europa.eu/sites/default/files/2_en_act_part1_v7.pdf.
Kahn, M. (1995). *Concepts, definitions, and key issues in sustainable development: The outlook for the future*. Proceedings of the 1995 international sustainable development research conference, Manchester, England, March 27–28, 1995, 2–13.
Kotuby, C. T., Egerton-Vernon, J., Rooney, M. C. (2014). *Protecting foreign investments in sub-saharan Africa: The southern African development community and its protocol on finance and investment*. https://www.mondaq.com/unitedstates/inward-foreign-investment/285064/protecting-foreign-investments-in-sub-saharan-africa-the-southern-african-development-community-and-its-protocol-on-finance-and-investment. Accessed 15 Aug 2020.
Lehman, B. J., & Crano, W. D. (2002). The pervasive effects of vested interest on attitude–Criterion consistency in political judgment. *Journal of Experimental Social Psychology, 38*, 101–112. https://doi.org/10.1006/jesp.2001.1489.
Lov om krigsmateriel. (2012). https://www.retsinformation.dk/eli/lta/2012/1004.
Malik, M. (2011). Definition of investment in international investment agreements. *The International Institute for Sustainable Development*. https://www.iisd.org/sites/default/files/publications/best_practices_bulletin_1.pdf. Accessed 10 June 2020.
Miller, C. H., Adame, B. J., & Moore, S. D. (2013). Vested interest theory and disaster preparedness. *Disasters, 37*, 1–27. https://doi.org/10.1111/j.1467-7717.2012.01290.x.
Montiel, I. (2008). Corporate social responsibility and corporate sustainability: Separate pasts, common futures. *Organization Environment, 21*(5), 245–269.
Norway National Security Act. (2018). https://investmentpolicy.unctad.org/investment-laws/laws/288/norway-national-security-act.
Peacock, N., Goldberg, S., Cannon, A., & Maxwell, I. (2019). *The changing relationship between environmental protections and BITs*. Herbert Smith Freehills.
Regulation (EU) 2019/452 of the European Parliament and of the Council of 19 March 2019 establishing a framework for the screening of foreign direct investments into the Union. (2019). https://eur-lex.europa.eu/eli/reg/2019/452/oj. Accessed 6 June 2020.
Republic of Benin for the Promotion and Reciprocal Protection of Investments. https://investmentpolicy.unctad.org/international-investment-agreements/treaty-files/438/download. Accessed 5 June 2020.
Rosen, M. E., & Thuringer, C. B. (2017). Unconstrained foreign direct investment: An emerging challenge to arctic security. *CNA Analysis and Solutions*. https://www.cna.org/cna_files/pdf/COP-2017-U-015944-1Rev.pdf. Accessed 5 June 2020.
Scherwin, P. (2020). *The Trillion-Dollar Reason for an Arctic Infrastructure standard*. http://polarconnection.org/arctic-infrastructure-standard/. Accessed 1 Aug 2020.
Section 721 of the Defense Production Act of 1950. https://www.treasury.gov/resource-center/international/foreign-investment/Documents/Section-721-Amend.pdf.
Shaffer, G. C., & Pollack, M. A. (2009). Hard vs. soft law: Alternatives, complements, and antagonists in international governance. *Minnesota Law Review, 94*, 706.
Sherwin, P., & Bishop, T. (2019).The trillion-dollar reason for an arctic infrastructure standard. *The Polar Connection*. http://polarconnection.org/arctic-infrastructure-standard/. Accessed 13 Mar 2020.
Spears, S. A. (2010). The quest for policy space in a new generation of international investment agreements. *Journal of International Economic Law, 13*(4), 1037–1075.
Staalesen, A. (2019). *Chinese money for Northern Sea Route*. https://thebarentsobserver.com/en/arctic/2018/06/chinese-money-northern-sea-route.
Stockholm Convention on Persistent Organic Pollutants. (2004). https://www.wipo.int/edocs/ 671 lexdocs/treaties/en/unep-pop/trt_unep_pop_2.pdf. Accessed 8 Aug 2020.

Summary report of the Arctic stakeholder forum consultation to identify key investment priorities in the Arctic and ways to better streamline future EU funding programmes for the region. https://op.europa.eu/en/publication-detail/-/publication/6a1be3f7-f1ca-11e7-9749-01aa75ed71a1/language-en/format-PDF/source-60752173. Accessed 5 Dec 2019.

Summary of the results of the public consultation on the EU Arctic policy. https://op.europa.eu/en/publication-detail/-/publication/497bfd35-5f8a-11eb-b487-01aa75ed71a1. Accessed 13 Mar 2021.

The Arctic Council signs Memorandum of Understanding with Arctic Economic Council. https://arcticeconomiccouncil.com/the-arctic-council-signs-memorandum-of-understanding-with-arctic-economic-council/#:~:text=The%20Arctic%20Council%20signed%20a,and%20the%20Arctic%20Economic%20Council. Accessed 12 Aug 2020.

The Dutch Model BIT. (2018). https://investmentpolicy.unctad.org/international-investment-agreements/treaty-files/5832/download.

The Dutch Model BIT. (2019). https://www.rijksoverheid.nl/ministeries/ministerie-van-buitenlandse-zaken/documenten/publicaties/2019/03/22/nieuwe-modeltekst-investeringsakkoorden. Accessed 11 Aug 2020.

The Espoo Convention on Environmental Impact Assessment in a Transboundary Context. (1997). https://www.unece.org/fileadmin/DAM//env/eia/eia.htm. Accessed 10 Aug 2020.

The OECD Guidelines for Multinational Enterprises. (2011). http://mneguidelines.oecd.org/guidelines/. Accessed 10 July 2020.

The Paris Agreement. (2015). https://unfccc.int/sites/default/files/english_paris_agreement.pdf. Accessed 9 Aug 2020.

The Rovaniemi Code of Conduct. (1994). http://library.arcticportal.org/1785/.

The Southern African Development Community (SADC) Protocol on Finance and Investments. (2006, 2016). https://www.sadc.int/opportunities/investment/. Accessed 5 Feb 2021.

The Universal Declaration of Human Rights. (1948). https://www.un.org/en/universal-declaration-human-rights/. Accessed 10 July 2020.

The Foreign Investment Risk Review Modernization Act of 2018. https://www.treasury.gov/resource-center/international/Documents/Summary-of-FIRRMA.pdf.

The UN Guiding Principles on Business and Human Rights. https://www.ohchr.org/documents/publications/guidingprinciplesbusinesshr_en.pdf. Accessed 5 Aug 2020.

Titi, C. (2015). International investment law and the European Union: Towards a new generation of international investment agreements. *European Journal of International Law, 26*, 639–661. https://doi.org/10.1093/ejil/chv040.

UN Declaration on the Rights of Indigenous Peoples. (2007). https://www.un.org/development/desa/indigenouspeoples/declaration-on-the-rights-of-indigenous-peoples.html. Accessed 3 Apr 2020.

UNCTAD. (2007). *Foreign direct investment definition*. https://unctad.org/system/files/official-document/wir2007p4_en.pdf. Accessed 10 Dec 2020.

UNCTAD. (2008). *International investment rule-making: Stocktaking, challenges and the way forward*. United Nations. https://unctad.org/system/files/official-document/iteiit20073_en.pdf. Accessed 7 April 2022.

UNCTAD. (2017). *World investment report: Investment and the digital economy*. United Nations. https://unctad.org/system/files/official-document/wir2017_en.pdf. Accessed 7 April 2022.

UNCTAD. (2020). *UNCTAD IIA mapping project*. https://investmentpolicy.unctad.org/international-investment-agreements/iia-mapping. Accessed 7 April 2022.

UNCTAD Investment Policy Hub. https://investmentpolicy.unctad.org/. Accessed 3 Apr 2020.

UNCTAD IIA Mapping Project. https://investmentpolicy.unctad.org/uploaded-files/document/Mapping%20Project%20Description%20and%20Methodology.pdf. Accessed 10 Apr 2020.

UN SDGs. https://www.un.org/ecosoc/en/sustainable-development. Accessed 10 Dec 2020.

Vylegzhanin, A., Korchunov, N., & Tavotrosyan, A. (2020). Arkticheskiy Ekonomicheskiy Sovet: Rol v Mezhdunarodno-Pravovom Mechanizme Prirodoohrannogo Upravleniya Severnyim Ledovitim Okeanom. *Moscow Journal of International Law, 3*. (In Russian).

White & Case LLP. (2018). *CFIUS reform becomes law: What FIRRMA means for industry*. https://www.whitecase.com/publications/alert/cfius-reform-becomes-law-what-firrma-means-industry. Accessed 12 Aug 2020.

Zhang. (2020). *Why a warming Arctic needs an investment review mechanism*. https://www.arctictoday.com/why-a-warming-arctic-needs-an-investment-review-mechanism/.

Zugliani, N. (2019). Human rights in international investment law: The 2016 Morocco-Nigeria Bilateral Investment Treaty. *International & Comparative Law Quarterly, 68*, 761–770. https://doi.org/10.1017/S0020589319000174.

Chapter 8
(Research): The Sustainable Use of Marine Living Resources in the Central Arctic Ocean: The Role of Korea in the Context of International Legal Obligations

Yunjin Kim, Jay-Kwon James Park, and Yeona Son

Abstract Following the conclusion of the Agreement to Prevent Unregulated High Seas Fisheries in the Central Arctic Ocean, the Arctic and non-Arctic States, as well as Arctic Indigenous communities, are facing new challenges in managing the expected increase in human activities in the Central Arctic Ocean and in preserving and protecting the marine environment there. While the Agreement reflects a special responsibility in relation to the sustainable use of marine living resources in the Central Arctic Ocean that will be taken by all States Parties, certain distinctions between the Arctic and non-Arctic States in terms of their legal obligations still exist. Since the Arctic has no single international governance regime, it contains diverse and fragmented legal mechanisms that present questions to those States Parties. What is the spatial scope of the international law applicable to the Central Arctic Ocean? What are the legal obligations that the States Parties are bound to respect for ensuring the long-term conservation of marine living resources beyond national jurisdiction in the Arctic Ocean? Recognizing the role of non-Arctic States in the sustainable management of the Central Arctic Ocean, long-term sustainability would likely require the contribution of key non-Arctic States, such as Korea.

This article reviews key aspects of the evolving international regime relating to the Central Arctic Ocean. It also gives an overview of Korea's international legal obligations and domestic institutional foundations for the pursuing sustainability of the Arctic region.

Y. Kim (✉) · Y. Son
Korea Legislation Research Institute, Sejong, South Korea

J.-K. J. Park
University of Washington, Seattle, Washington, USA

P. A. Berkman et al. (eds.), *Building Common Interests in the Arctic Ocean with Global Inclusion, Volume 2*, Informed Decisionmaking for Sustainability,
https://doi.org/10.1007/978-3-030-89312-5_8

8.1 Introduction

The oceans that cover about three-quarters of the surface of our planet play an integral role in supporting life. Considering that the oceans are fundamental to life on earth, providing natural and energy resources to billions of people who depend on marine areas for their livelihood, increased efforts and interventions to govern human activities are needed for the sustainable use of marine living resources at all levels.[1]

In recent years, human activities such as shipping, commercial fishing and seabed mining have expanded and intensified in marine areas beyond national jurisdiction, which comprise the water column of high seas as well as the sea-bed and ocean floor and subsoil thereof that are not part of the continental shelf of any State. This is true in the Arctic as it is in other parts of the world's ocean. While the receding ice in the Arctic Ocean due to climate change has paradoxically generated more economic opportunities for ocean use, the development of scientific research and governance regimes have struggled to keep pace with these increasing activities. The changing marine environmental conditions will certainly require effective fisheries management, proper assessments of the current status of the Arctic ecosystems and resources, as well as policies and institutional foundations with enforcement mechanisms.

In signing the 2008 Ilulissat Declaration, five Arctic coastal States reaffirmed their commitment to the "extensive international legal framework that applies to the Arctic Ocean." A more comprehensive global legal regime that builds on this framework and also accommodates the perspective of all concerned States is nevertheless likely to be needed for the high seas portion of the Central Arctic Ocean to follow up on the conclusion of a historic agreement to prevent unregulated fishing in that area.

The objective of this paper is to explore the legal obligations that States have in conserving marine living resources in the Central Arctic Ocean area under current international law, and the extent of responsibility that the Republic of Korea (hereinafter 'Korea') has in the international community, recognizing the urgent need for a sustainable future for the Arctic.

8.2 New Challenges Concerning the Central Arctic Ocean

8.2.1 The Central Arctic Ocean as a Common Concern

International environmental law developed from bilateralism to the protection of community interests as a body of law based on common concerns due to a raised

[1]GOAL 14: Life below water, UN Environmental Programme, Conserve and sustainably use the oceans, seas and marine resources for sustainable development.

awareness of the global nature of environmental problems.[2] The 1992 Rio Conference on Environment and Development highlighted the concept of "common concern" in relation to environmental issues; the concept has been incorporated in global regulatory treaties as a "common concern of mankind", for example in the fields of climate change and biological diversity.[3] These global concerns make apparent the need for common action by all States. If successful protection measures are to be taken for the Earth as a whole, they would necessarily require global responsibility towards community interests.

The Arctic Ocean now faces new challenges due to climate change and accelerating human activities, including receding sea ice, increased sea surface temperatures, significantly greater freshwater run-off from melting glaciers, and increasing acidification. These phenomena will lead to a loss of marine biodiversity, destruction of the pristine ecosystem, and potentially unsustainable fishing in this area.[4] Given that the actions of people in all States have contributed to these circumstances in the sense that they are all contributing to climate change, it follows that all States also share the responsibility for addressing the problem, within the framework of international law that recognizes their common values and interests, even in the Central Arctic Ocean.

At the 1992 Rio Conference, States adopted Agenda 21, which sets forth commitments relating to the conservation and sustainable use of marine living resources both within and beyond national jurisdiction.[5] In the case of the Central Arctic Ocean-which includes both a high seas portion and adjacent areas under national jurisdiction-the pursuit of sustainability needs to be addressed in the context of common concerns balancing national and community interests. Although States or a group of States have certain common responsibilities with respect to the conservation and sustainable use of marine living resources in the Central Arctic Ocean, the specific rights and duties of coastal and non-coastal States may differ in some respects.

8.2.2 State Responsibilities in the High Seas

Under the legal zones recognized in UNCLOS, coastal States can claim jurisdiction over fish and seabed resources within 200 nautical miles from the baseline,

[2]Bartenstein. K, 2015, The 'Common Arctic': Legal Analysis of Arctic & non-Arctic Political Discourses, *Arctic Yearbook*, pp.1.

[3]Preamble of the Convention on Climate Change, the Convention on Biological Diversity.

[4]IPCC, Global Warming of 1.5 °C of Global Warming on Natural and Human system.

[5]United Nations Sustainable Development, 1992, Agenda 21, Chapter Protection of the Oceans, All kinds of Seas, including enclosed and semi-enclosed seas, and coastal areas and the protection, rational use and development of their living resources https://sustainabledevelopment.un.org/content/documents/Agenda21.pdf

establishing an Exclusive Economic Zone (EEZ).[6] However, these States are not entitled to the same rights beyond the EEZ that is classified as high seas. The Central Arctic Ocean includes a large high seas area that is entirely surrounded by the EEZs of five Arctic coastal States.[7]

Part VII of UNCLOS ensures the rights of States to exercise freedoms of the high seas: freedom of navigation, fishing, laying submarine cables and pipelines, and conducting scientific research. These rights apply to the high seas portion of the Central Arctic Ocean as they apply elsewhere.[8] Even so, these freedoms shall be exercised by all States with due regard for the community interests and the rights as well as the interests of coastal States under the Convention.[9] In respect of "straddling fish stocks"-fish that occur in both the high seas and adjacent EEZs-Article 63(2) of the Convention lays down the obligation of the coastal States and States fishing on the high seas to seek, either directly or through appropriate subregional or regional organizations, to agree upon the measures necessary for the conservation of such stocks.

Whereas UNCLOS does not specifically prescribe States' participation in such RFMOs, the 1995 UN Agreement on Straddling Stocks and Highly Migratory Stocks (UNFSA)[10] provides that States having a "real interest" in fisheries managed by an RFMO have the right to join that RFMO.[11] The provision does not clearly define the term "real interest," however. The term may imply that flag States can claim to have a real interest in a particular fishery even if they have no history of participating that fishery but that want to fish in the future, or even if they have no intention to fish but want to participate in the RFMO solely for the purpose of safeguarding marine biodiversity.[12,13]

[6]The United Nations Convention on the Law of the Sea of 10 December 1982 Part V. Exclusive Economic Zone, Article 55–75.

[7]Canada, Denmark, Norway, Russia, and the United States.

[8]UNCLOS Part VII. High Seas, Article 87 Freedom of the high seas.

[9]UNCLOS Article 116 Right to fish on the high seas.

[10]The United Nations Agreement for the Implementation of the Provision of the United Nations Convention on the Law of the Sea of 10 December 1982 relating to the Conservation and Management of Straddling Fish Stocks and Highly Migratory Fish Stocks. https://www.un.org/Depts/los/convention_agreements/convention_overview_fish_stocks.htm

[11]UNFSA Article 8(3). para 3. "State having a real interest in the fisheries concerned may become members of such organization or participants in such arrangement. The terms of participation in such organization or arrangement shall not preclude such States from membership or participation; nor shall they be applied in a manner which discriminates against any State or group of States having a real interest in the fisheries concerned."

[12]Molenaar. E, 2000, The Concept of Real Interest and Other Aspects of Co-operation through Regional Fisheries Management Mechanisms, *International Journal of Marine and Coastal Law*, 15(4), pp.496.

[13]As noted above, Article 8(3) of the UNFSA also requires that the terms of participation in an RFMO shall not discriminate against any group of States. The Central Arctic Ocean Fisheries Agreement does not create distinctions between coastal State and non-coastal State Parties in terms of decisionmaking. In the future, however, arguments about such distinctions may arise in

The Central Arctic Ocean, like other parts of the world's ocean in which there is both a high seas portion and EEZs, is of legitimate concern to all States. There is no doubt that efforts by both Arctic coastal States and certain non-coastal States will be needed to ensure that marine living resources in the Central Arctic Ocean are conserved and managed sustainably, taking into account the interests of the international community and the legal obligations in accordance with the balance of rights and responsibilities reflected in UNCLOS and UNFSA.

8.3 The International Legal Regime for Marine Living Resources of the Central Arctic Ocean

8.3.1 Global Legal Framework of Marine Living Resources

The sustainable use of marine living resources and their proper management are essential for the long-term conservation of these resources and biological diversity. The current international legal framework for marine living resources is comprised of bilateral and multilateral regional agreements as well as global conventions. These international law mechanisms apply to Arctic Ocean fisheries resources, including the legal obligations to cooperate to conserve the marine environment and marine natural resources both within and beyond national jurisdiction. The most relevant global legal regime relating to the fisheries management in the Central Arctic Ocean includes, but is not limited to, the 1982 UNCLOS,[14] the 1982 Convention on Future Multilateral Cooperation in Northeast Atlantic Fisheries,[15] the 1992 UNCBD, the 1993 Agreement to Promote Compliance with International Conservation and Management Measures by Fishing Vessels on the High Seas, the 1995 UN UNFSA, and the 2018 Agreement to Prevent Unregulated High Seas Fisheries in the Central Arctic Ocean.[16]

UNCLOS established a fundamental legal framework for the conservation of marine living resources under Articles 61 to 67 and Articles 116 to 119 that are relevant to the Central Arctic Ocean. These provisions recognize the aims of optimum utilization and conservation of marine living resources, including for

considering whether and how to allow commercial fishing to start in the high seas area of the Central Arctic Ocean. Similar arguments may also arise concerning the question of whether CAOFA States Parties should have a privileged role in the development of additional resource management measures for the Central Arctic Ocean. See Balton. D, What will the BBNJ Agreement mean for the Arctic Fisheries Agreement?

[14]UNCLOS Part VII. Section 2. Conservation and Management of the Living Resources of the High Seas, Article 116–119.

[15]The competence area of the NEAFC Convention is limited to a small amount of the Central Arctic Ocean and there is yet no precedent to adopt management measures in this area. See also NEAFC Convention Article 1, a) "*The Convention Area*".

[16]The Agreement entered into force on 25 June 2021.

stocks occurring both within EEZs and the high seas. It also provides the legal basis to take measures necessary for the management of such resources of the high seas. Articles 117–118 further require States to cooperate with other States and to enter into negotiations with a view to taking such measures for their nationals that may be necessary for the conservation of the living resources of the high seas in the form of establishing RFMOs. However these provisions do not offer detailed rules on how to manage such fisheries resources, nor do they give any specific guidance concerning the cooperation of States.[17] Additionally, vague language such as "best available scientific evidence" does not have much practical effect with respect to fisheries resource management.[18,19]

The fact that UNCLOS only provides general legal obligations for States to cooperate in the management of marine living resources may have contributed to the lack of political will among States to take appropriate conservation measures and, therefore, to the failure to achieve the sustainable use of marine living resources. Since the adoption of UNCLOS, however, new approaches in international law and practice based on the "precautionary approach" or "precautionary principle" have arisen.[20] It remains to be seen whether and to what extent these approaches can contribute to resolving the problems under the current international legal framework.

8.3.2 Precautionary Principle

The precautionary principle (or approach) aims to guide the application of international environmental law and the taking of other international legal acts where there is scientific uncertainty.[21] While the precise status and best formulation of this principle have been debated, the international community has mostly embraced it as a general principle of international law. At the most general level, it means that States should take action or adopt decisions based upon careful foresight when their activities may be expected to cause damage to the environment.[22] Implementation of the precautionary principle may nevertheless differ as each State seeks to apply it in accordance its own legal context and culture.

---

[17]Tanaka. Y, 2011, The Changing Approaches to Conservation of Marine Living Resources in International Law, Max Plank Institut für ausländisches öffentliches Recht und Völkerrecht, pp.300.

[18]UNCLOS Article 119 Conservation of the living resources of the high seas; State shall take measures which are designed, on the best scientific evidence available to the State concerned.

[19]Hassan. D, 2009, Climate Change and the Current Regimes of Arctic Fisheries Resources Management: An Evaluation, *Journal of Maritime Law & Commerce*, Vol. 40, No. 4, pp.524.

[20]Tanaka. Y, 2011, *supra* note 17, pp.293.

[21]Sands. P, 2003, Principle of International Environmental Law, Cambridge University Press, pp.267.

[22]*Id.* pp.267–272.

In the case of international fisheries law, the UNFSA enshrined the precautionary approach in Article 6(2). It requires that "States shall be more cautious when information is uncertain, unreliable or inadequate. The absence of adequate scientific information shall not be used as a reason for postponing or failing to take conservation and management measures", obligating States to apply the precautionary principle widely in the conservation and management of straddling and highly migratory fish stocks.[23]

Under Part XII of UNCLOS, the general obligations of States to protect and preserve the marine environment also implies the precautionary approach. In the 1999 *Southern Bluefin Tuna Cases*,[24] the decision of the International Tribunal for the Law of the Sea (ITLOS) called upon the parties to that case to exercise caution in managing the stock of tuna in question in light of the scientific uncertainty concerning the effects of fishing for the stock. In particular, ITLOS justified its grant of provisional measures pending final resolution of the dispute by citing the need not to hinder or postpone the taking of measures necessary for the conservation of the fish stocks. This version of "precaution" does not necessarily require a State to prove that environmental harm is certain or even likely; the evidence that such harm is foreseeable is enough to trigger an obligation for States to act.[25]

Although States have introduced versions of the precautionary principle (or approach) in a variety of international agreements, and although there is now considerable State practice in implementing it, the precise meaning of the precautionary approach is still evolving.[26] International fisheries law is a prime example of an area in which States have introduced the precautionary principle, but its specific formulation and use depends on the individual case as framed in the applicable fisheries agreement.[27]

8.3.3 2018 Central Arctic Ocean Fisheries Agreement

The 2018 Agreement to Prevent Unregulated High Seas Fisheries in the Central Arctic Ocean, signed by nine States[28] and the European Union, can be seen as an

[23] See also Annex II Guidance for the Application of Precautionary Reference Points in Conservation and Management of Straddling Fish Stocks and Highly Migratory Fish Stocks.

[24] ITLOS, 1999, *Southern Bluefin Tuna Cases (provisional measures)* (Australia v. Japan; New Zealand v. Japan).

[25] P. W. Birnie, A. E. Boyle and C. Redgwell, 2009, International Law & Environment, Oxford University Press, pp.163.

[26] Sands. P, 2003, *supra* note 21, pp.279.

[27] Schatz. V, Proelss. A, and Liu. N, 2019, The 2018 Agreement to Prevent Unregulated High Seas Fisheries in the Central Arctic Ocean: A Critical Analysis, *The International Journal of Marine and Costal Law* 34, pp.25.

[28] Canada, China, Demark in respect of the Faroe Island and Greenland, Iceland, Japan, South Korea, Norway, Russia, and the United States.

application of the precautionary approach embedded in a fisheries agreement.[29] Although there is no commercial fishing currently occurring and unlikely to become viable in the high seas portion of the Central Arctic Ocean in the near future, fish species may move northward and become accessible due to the melting sea ice.[30] On this account, the Agreement calls for precautionary conservation and management measures to ensure the sustainable use of fish stocks as part of a long-term strategy that States exercise caution in applying freedom of fishing in areas of the high seas.[31,32]

The Agreement fills a legal lacuna in the fisheries regime in the high seas portion of the Central Arctic Ocean. It imposes a 16-year moratorium on the start of commercial fishing, during which time the States Parties may learn more about the impacts of climate change and thus should be better able to manage any fishing effectively. The Agreement will be extended for additional five-year increments unless any State Party objects to such extension.[33] In this context, the moratorium on high seas fishing can be seen as a highly precautionary measure in support of the long-term sustainable use of marine living resources in the Central Arctic Ocean. In the absence of scientific evidence with which to manage commercial fishing in this area, the States involved agreed not to allow commercial fishing for at least 16 years, during which they will seek to obtain such evidence.

In a nutshell, the Agreement requires the States Parties undertake two basic commitments: to prohibit commercial fishing in the "Agreement Area" and to establish a Joint Program of Scientific Research and Monitoring. Proper implementation of the Agreement will primarily depend on the political will of all States Parties and how well they can constrain national interests and balance those with common interests they share concerning the Central Arctic Ocean.[34] The collective capacity of both Arctic coastal and non-Arctic States Parties will greatly advance the increasing knowledge of such marine ecosystems and management of the Central Arctic Ocean area. In the implementation of the Joint Program envisaged under the Agreement, contribution and commitment from non-Arctic States Parties to promote scientific knowledge will also help in developing a data sharing protocol, which is to include relevant scientific-technical specifications.[35]

---

[29] European Commission, the Agreement to prevent unregulated high seas fisheries in the Central Arctic Ocean https://eur-lex.europa.eu/legal-content/EN/TXT/?uri=COM:2018:453:FIN

[30] Heidar. T, 2017, The Legal Framework for High Seas Fisheries in the Central Arctic Ocean, chap. 6., *International Marine Economy: Law and Policy*, Koninklijke Brill NV, Leiden, pp.179.

[31] Agreement to Prevent Unregulated High Seas Fisheries in the Central Arctic Ocean 2018, preamble, para 11–12, Article 2.

[32] Vylegzhanin. A, Young. O, and Berkman. P, 2020, The Central Arctic Ocean Fisheries Agreement as an Element in the Evolving Arctic Ocean Governance Complex, *Marine Policy*, pp.6.

[33] Balton. D, 2018, The Arctic Fisheries Agreement: Looking to 2030 and Beyond, *The Arctic in World Affairs*, Korea Maritime Institute and East-West Center, pp.88.

[34] Schatz. V, Proelss. A, and Liu. N, *supra* note 27, pp.3.

[35] Chairs' Statement: 5th Meeting of Scientific Experts on Fish Stocks of the Central Arctic Ocean, Ottawa, Canada, October 24–26, 2017.

8.4 The Role of Korea in the Arctic

8.4.1 Legal Obligations of Korea under International Frameworks

As noted above, the Central Arctic Ocean is subject to the same global legal framework as other parts of the world's ocean.[36] UNCLOS provisions concerning fisheries, conservation and management of marine living resources, the outer limits of the continental shelf, navigation rights, the conduct of marine scientific research and ice-covered areas[37] all apply to the Central Arctic Ocean. The 2018 Central Arctic Ocean Fisheries Agreement complements the general provisions of the Convention, particularly in the matter of conservation of the marine environment. Hence, the international legal obligations that Korea undertakes concerning the Central Arctic Ocean include responsibilities under international conventions including UNCLOS and the specific responsibilities it will have as a State Party to the 2018 agreement.

Recognizing that Agenda 21 calls upon States to take actions in accordance with international law and commit themselves to the conservation and sustainable use of marine living resources on the high seas, Korea must comply with the provisions of UNCLOS and CBD regarding the duty to cooperate with other States Parties. Article 5 of the CBD obligates States to work "directly or where appropriate, through competent international organizations, in respect of areas beyond national jurisdiction and on other matters of mutual interest for the conservation and sustainable use of biological diversity." UNCLOS Articles 116–120 requires States to adopt measures for the conservation of the living resources of the high seas with respect to their nationals and vessels and to cooperate with other States in taking such measures. Article 7 of the UNFSA builds on these general obligations in the context of straddling and highly migratory fish stocks, with the aim of achieving compatible measures for those stocks in areas both within and beyond national jurisdiction. Article 6 and Annex II of the UNFSA also require the application of the precautionary approach in adopting conservation and management measures for such stocks.

The 2015 Oslo Declaration, adopted by the five Central Arctic Ocean coastal States, recognized the interests of other States in relation to potential fisheries in the high seas portion of the Central Arctic Ocean. In 2016, Korea, along with China, Japan, Iceland, and the EU, joined the negotiations that produced the Central Arctic Ocean Fisheries Agreement. Following the successful conclusion of those negotiations, the Republic of Korea completed the ratification process for the Agreement

[36] Heidar. T, *supra* note 30, pp.181.

[37] Article 234 Ice-covered areas; coastal States have the right to adopt laws and regulations for the prevention, reduction and control of marine pollution from vessels in ice-covered areas within the limits of the EEZ.

in October 2019.[38] Korea became the sixth Signatory to complete the ratification of the Agreement after Canada, the EU, the U.S, Japan, and Russia. To this end, Korea has agreed not to authorize its vessels to engage in high seas fishing in the Central Arctic Ocean except in accordance with the limited exceptions provided in the Agreement, and to participate in developing and implementing the Joint Program of Scientific Research and Monitoring.

8.4.2 The Engagement of Korea in the Arctic

After serving as an ad hoc observer since 2008, Korea was admitted to the Arctic Council as one of the non-Arctic State observers in 2013 and has been actively working with different countries, stakeholders, and the Permanent Participants (Arctic Indigenous peoples) to contribute to the Arctic Council's goals of promoting sustainable development and environmental protection in the Arctic. Despite its distance from the Arctic, Korea has been seeking to better understand the issues surrounding the Arctic and to become an important player in the Arctic. Such Korean efforts can be found both nationally and internationally.

Korea took its first significant step in Arctic scientific research and projects by establishing the Korea Arctic Science Council (KASCO) in 2001. With KASCO as a cornerstone, Korea began its investment in crucial assets for its Arctic scientific research by opening its first research station on Svalbard, and became one of the few countries to own an ice-breaking research vessel. Korea has been conducting its Arctic ship-based research in the part of the Central Arctic Ocean in the vicinity of the Chukchi and East Siberian Seas on a yearly basis to understand the marine environment in the Central Arctic Ocean, and to predict its future changes.[39] Considering that the Central Arctic Ocean research will need icebreakers to conduct surveys in ice-covered water, the scientific research capacity of Korea should be considered significant in this regard.[40]

Korea also has been undertaking polar scientific research in collaboration with many of the Arctic States and has been involved in formal dialogues on Central Arctic Ocean issues with China, Japan, and non-government experts supported by various international institutions starting from 2015.[41] In parallel with such efforts, Korea will contribute to creating opportunities for securing scientific information through the joint research and monitoring program in the Central Arctic Ocean with

[38] MOFA, ROK completes domestic ratification procedure for Agreement to Prevent Unregulated High Seas Fisheries in the Central Arctic Ocean. http://www.mofa.go.kr/eng/brd/m_5676/view.do

[39] Korea Polar Research Institute (KOPRI), Korea-Arctic Ocean Observing System (K-AOOS, 2016–2020) funded by the Korean Ministry of Oceans and Fisheries.

[40] Kim. J and Kim. J, 2017, Korean Perspectives, *The Arctic in World Affairs*, Korea Maritime Institute and East-West Center, pp. 289.

[41] Preventing Unregulated Commercial Fishing in the Central Arctic Ocean: A compilation of reports from meetings of experts in Shanghai, Incheon & Sapporo, March 2017.

such science leadership and capacity based on domestic policy and legal foundations.

8.4.3 The Domestic Institutional Foundation of Korea Arctic Policy

Korea established its first Arctic Policy Master Plan (2013–2017) with the vision of contributing to the sustainable future of the Arctic soon after it achieved observer status in the Arctic Council. Seven different ministries[42] collaborated to create this Plan, a blueprint for Korea's Arctic vision that includes 31 tasks in international cooperation, scientific investigation, Arctic business, legal and institutional fields.[43] However, there was a knowledge gap between government organizations and a lack of domestic institutional foundation to support scientific activities in the Arctic. With lessons learned from the first period, Korea announced the second Arctic Policy Master Plan (2018–2022). The newly adopted second Plan set the goal of long-term Arctic policy development and strengthening Korea's capacity in scientific research activities that includes building a second ice-breaking research vessel.[44]

To support such activities in the Arctic, both Plans have expanded Korea's domestic institutional foundations. Still, Korea has additional steps to take. For example, there is no Korean legislation for the Arctic comparable to its *Act on Activities in the Antarctic Area and the Protection of Antarctic Environment* to contribute to the protection of the Antarctic environment and the development of science and technology by providing for matters necessary for activities in Antarctica.[45] Moreover, although Korea has passed a *Framework Act on Marine Fisheries Development*[46] that supports plans required for the installation of a marine research station in specific areas including the South and the North pole, and for marine

[42]Ministry of Oceans and Fisheries (MOF), Ministry of Foreign Affairs (MOFA), Ministry of Science, ICT and Future Planning (MSIP), Ministry of Trade, Industry and Energy (MOTIE), Ministry of Environment (MOE), Ministry of Land, Infrastructure and Transport (MOLIT), Korea Meteorological Administration (KMA).

[43]Jin. D, Seo. W & Lee. S, 2017, Arctic Policy of the Republic of Korea, 22 *Ocean & Coastal L.J*, pp.90.

[44]Kwon. S, 2018, Korea's Arctic Policy and Activities, *The Arctic in World Affairs*, Korea Maritime Institute and East-West Center, pp.50.

[45]Korea Legislation Research Institute, ACT ON ACTIVITIES IN THE ANTARCTIC AREA (2004), http://elaw.klri.re.kr/kor_service/lawView.do?hseq=46891&lang=ENG

[46]Korea Legislation Research Institute, FRAMEWORK ACT ON MARINE FISHERY DEVEL OPMENT (2017), http://elaw.klri.re.kr/kor_service/lawView.do?hseq=43304&lang=ENG

science survey and research, the law does not cover the overall Arctic activities.[47] To fill this legal gap, Korea is planning to enact the *Polar Activities Promotion Act* (The Act passed into law, and became effective on 14 October 2021), and related enforcement decrees that will extend to both the Arctic and Antarctic research, development and conservation activities and that would likely apply to the Central Arctic Ocean area.[48]

8.5 Conclusion

Commercial fishing is not currently taking place in the high seas portion of the Central Arctic Ocean. The Arctic coastal and non-coastal States nevertheless have the common interests to pursue the conservation and the sustainable use of marine living resources both within and beyond national jurisdiction, including in the Central Arctic Ocean. The advancement of such common interests needs to be undertaken in the context of international law, which enables all States to take action within a common legal framework.

While the international legal framework for marine living resources, which includes the 1982 UNCLOS and other instruments discussed above, set forth the obligations of States to cooperate in Arctic fisheries resources management, they are not yet sufficient for conservation and for preventing the overfishing of species in the Central Arctic Ocean. The provisions of such instruments give no specific guidance to States in establishing subregional or regional fisheries organizations and in judging breaches of international obligations. Existing organizations such as the North-East Atlantic Fisheries Commission (NEAFC) cover a small part of the high seas portion in the north of Greenland and Svalbard, but neither actively address the issues of proper management of the marine living resource in the Central Arctic Ocean. Those organizations also do not include distant water fishing States.

In filling the legal gap with a precautionary approach, the 2018 Central Arctic Ocean Fisheries Agreement imposed a moratorium on the start of commercial fishing in the Central Arctic Ocean until there is a better understanding of the ecosystem in the Agreement Area and a more comprehensive fisheries management regime in place. However, it still leaves open questions as to the extent to which State Parties can constrain their national interests and behavior in the high seas of the Central Arctic Ocean in terms of international legal obligations. The questions may overlap with questions that are likely to arise under the envisioned BBNJ

---

[47] Article 20 Installation of Marine Research Station, and Survey and Research; The Government shall devise and implement support plans required for the installation of a marine research station in a specific area, such as the South Pole and the North Pole, and for the advancement of marine science survey and research.

[48] Korea National Assembly Agricultural and Fisheries Committee, 2017, The examination report on Act on Promotion of Polar Activities (KOREAN), http://likms.assembly.go.kr/bill/billDetail.do?billId=PRC_K1A6R1H2S0Z1P1G7I2W7D0C9N8D2G9

Agreement. Considering the Central Arctic Ocean as of common concern, it inevitably entails a reaffirmation of both the primary responsibility of flag States with respect to their vessels that fish on the high seas and a responsibility to conserve the marine environment over the Central Arctic Ocean. Consequently, it largely depends on the political will of each State Party to balance its own interests and those of the international community.

Korea began engaging significantly in Arctic affairs other than scientific research less than a decade ago. Despite its short presence in the Arctic, Korea has participated in various working group projects under the auspices of the Arctic Council and has undertaken bilateral science programs with many of the Arctic States. Korea's rights and responsibilities under international law in relation to the Central Arctic are to be respected by the other States involved, in particular as a State Party to the 2018 Central Arctic Ocean Fisheries Agreement. As one of the State Parties to the Agreement, Korea is planning to strengthen its scientific research capacity for the implementation plans in parallel with supporting and expanding domestic institutional foundations. Korea will be committed to addressing challenging transboundary issues of the Arctic Ocean and to promoting science diplomacy along with a national political commitment which is addressed in both of its Arctic Policy Master Plans.

Acknowledgements We would like to express our sincere appreciation to Dr. Hyoung Chul Shin (KOPRI) for his encouragement and insights of Korean scientific leadership. We also like to extend our gratitude to Ambassador David Balton and the reviewers for valuable comments on this paper.

References

Agreement to Prevent Unregulated High Seas Fisheries in the Central Arctic Ocean (2018). https://eur-lex.europa.eu/legal-content/EN/TXT/?uri=celex:22019A0315(01)

Balton D (2018) The Arctic fisheries agreement: looking to 2030 and beyond, The Arctic in World Affairs, Korea Maritime Institute and East-West Center, pp.83–91

Balton, D. (2019). What will the BBNJ agreement mean for the Arctic fisheries agreement? *Marine Policy*. https://doi.org/10.1016/j.marpol.2019.103745

Bartenstein K (2015) The 'common Arctic': legal analysis of Arctic & non-Arctic political discourses, Arctic Yearbook1–17

Birnie, P. W., Boyle, A. E., & Redgwell, C. (2009). *International law & Environment*. Oxford University Press.

Chairs' Statement: 5th Meeting of Scientific Experts on Fish Stocks of the Central Arctic Ocean, Ottawa, Canada, October 24–26, 2017., https://www.afsc.noaa.gov/Arctic_fish_stocks_fifth_meeting/pdfs/5th_FiSCAO_chair_statement_final.pdf

Convention on Fishing and Conservation of the Living Resources of the High Seas, Geneva, 29 April 1958. https://treaties.un.org/Pages/ViewDetails.aspx?src=TREATY&mtdsg_no=XXI-3&chapter=21&clang=_en

Convention on the Conservation of Antarctic Marine Living Resources, 1980. https://www.ccamlr.org/en/organisation/camlr-convention-text#II

Hassan, D. (2009). Climate change and the current regimes of Arctic fisheries resources management: an evaluation. *Journal of Maritime Law & Commerce, 40*(4), 511–536.

Heidar, T. (2017). The Legal Framework for High Seas Fisheries in the Central Arctic Ocean, chap. 6. In *International Marine Economy: Law and Policy* (pp. 179–203). Koninklijke Brill NV.

Agreement to promote compliance with International Conservation and Management measures by fishing vessels on the high seas (n.d.). http://www.fao.org/3/X3130m/X3130E00.HTM

MOFA, Press Releases, ROK completes domestic ratification procedure for Agreement to Prevent Unregulated High Seas Fisheries in Central Arctic Ocean (n.d.) https://www.mofa.go.kr/eng/brd/m_5676/list.do

European Commission, the Agreement to prevent unregulated high seas Fisheries in the Central Arctic Ocean (n.d.). https://eur-lex.europa.eu/legal-content/EN/TXT/?uri=COM:2018:453:FIN

IPCC, Global Warming of 1.5°C of Global Warming on Natural and Human systems, https://www.ipcc.ch/sr15/

Jin, D., Seo, W., & Lee, S. (2017). Arctic Policy of the Republic of Korea. *Ocean & Coastal L.J, 22*, 85–96.

Kim J, Kim J (2017) Korean perspectives, The Arctic in World Affairs, Korea Maritime Institute and East-West Center, pp. 281–290

Korea Law Translation Center, Korea Legislation Research Institute, ACT ON ACTIVITIES IN THE ANTARCTIC AREA (2004), http://elaw.klri.re.kr/kor_service/lawView.do?hseq=46891&lang=ENG, FRAMEWORK ACT ON MARINE FISHERY DEVELOPMENT (2017), http://elaw.klri.re.kr/kor_service/lawView.do?hseq=43304&lang=ENG

Korea National Assembly Agricultural and Fisheries Committee, 2017, The examination report on Act on Promotion of Polar Activities, http://likms.assembly.go.kr/bill/billDetail.do?billId=PRC_K1A6R1H2S0Z1P1G7I2W7D0C9N8D2G9

Kwon S (2018) Korea's Arctic policy and activities, The Arctic in World Affairs, Korea Maritime Institute and East-West Center, pp.49–52

Ministry of Oceans and Fisheries, 2013, Korea Arctic Policy Master Plan http://www.arctic.or.kr/files/pdf/m4/korea.pdf

Molenaar, E. (2000). The concept of real interest and other aspects of co-operation through regional fisheries management mechanisms. *International Journal of Marine and Coastal Law, 15*(4), 475–532.

Preventing Unregulated Commercial Fishing in the Central Arctic Ocean: A compilation of reports from meetings of experts in Shanghai, Incheon & Sapporo, March 2017., https://oceanconservancy.org/wp-content/uploads/2018/09/Preventing-Unregulated-Commercial-Fishing-CAO.pdf

Sands, P. (2003). *Principle of international environmental law*. Press.

Schatz, V., Proelss, A., & Liu, N. (2019). The 2018 agreement to prevent unregulated high seas fisheries in the Central Arctic Ocean: a critical analysis. *The International Journal of Marine and Costal Law, 34*, 1–50.

Tanaka Y (2011) The Changing Approaches to Conservation of Marine Living Resources in International Law, Max Plank Institut für ausländisches öffentliches Recht und Völkerrecht, pp. 291–330

United Nations Sustainable Development, 1992, Agenda 21, Chapter 17 Protection of the Oceans, All kinds of Seas, including enclosed and semi-enclosed seas, and coastal areas and the protection, rational use and development of their living resources https://sustainabledevelopment.un.org/content/documents/Agenda21.pdf

Vylegzhanin, A., Young, O., & Berkman, P. (2020). The Central Arctic Ocean fisheries agreement as an element in the evolving Arctic Ocean governance complex. *Marine Policy*. https://doi.org/10.1016/j.marpol.2020.104001

Chapter 9
(Research): Combining Knowledge for a Sustainable Arctic – AMAP Cases as Knowledge Driven Science-Policy Interactions

Rolf Rødven and Simon Wilson

Abstract While the Arctic is often perceived as a pristine environment, it is exposed to local as well as globally transported contaminants and is undergoing severe changes in environmental conditions. Major oceanic currents and wind systems transport contaminants from distant sources, with the Arctic acting as a ecosystems and ways of life «sink» for harmful substances. Likewise, climate warming in the Arctic is happening more than twice as fast as at lower latitudes, causing changes in ecosystems as well as ways of life for many Indigenous people living in the Arctic.

A prerequisite for managing and mitigating the impacts of both pollution and climate change in the Arctic is the acquisition of knowledge of conditions, with adequate geographical coverage and sufficiently high spatial resolution, as well as mechanisms for communicating such knowledge for policymaking. The Arctic Monitoring and Assessment Programme (AMAP) was initiated to fulfill such a role in 1991, later becoming a working group of the Arctic Council at its establishment in 1996. AMAP focuses its work on the interface between science and policy. Due to the nature of the origins of pollution in the Arctic, such work requires a focus on both contributing with a knowledge base for policy making among the Arctic states, as well as to international bodies outside the Arctic. The contribution made by AMAP to the establishment of the Stockholm and Minamata Conventions are examples of science and policy development in the Arctic successfully feeding into global international processes.

While long-term research facilities in the vast Arctic region are scarce, Indigenous groups represent a source of knowledge which may contribute significantly to understanding the changing environmental conditions in the Arctic. Therefore, from the start, AMAP has included Indigenous groups – Permanent Participants to the Arctic Council – both in its decisionmaking structures as well its expert groups.

R. Rødven (✉) · S. Wilson
Arctic Monitoring and Assessment Programme (AMAP), Tromsø, Norway
e-mail: rolf.rodven@amap.no

P. A. Berkman et al. (eds.), *Building Common Interests in the Arctic Ocean with Global Inclusion, Volume 2*, Informed Decisionmaking for Sustainability,
https://doi.org/10.1007/978-3-030-89312-5_9

Co-development of knowledge has informed understanding of climate change and ensured relevance in efforts addressing adaptation and resilience, as discussed in the Adaptation Actions for a Changing Arctic (AACA) reports.

Still, combining Indigenous, traditional and local knowledge and conventional science remains a challenge, both due to their different origin and nature, the diverse spatial diversity across the Arctic, and also due to the speed of change which challenges the predictive power of all knowledge-based systems. Methods to address these challenges need to be discussed.

9.1 The Misperception of the Arctic as a Pristine Area

While for a long time the Arctic was considered to be a remote and pristine area, relatively undisturbed by human activities, research has shown that major ocean and air currents as well as large river inflows bring long-range transported pollutants such as persistent organic pollutants (POPs) and heavy metals including lead and mercury to the Arctic from industrial source areas at lower latitudes (AMAP, 1998), where they are deposited on sea ice and snow and accumulate in waters, soils and glaciers (AMAP, 2010, 2011, 2017c). Local sources also exists for some contaminants, including chemicals of emerging Arctic concern (AMAP, 2017c).

Physiological characteristics of Arctic biota, such as the significant seasonal storage and mobilization of fat in their tissues make Arctic animals susceptible to fat-soluble pollutants that accumulate and biomagnify in food chains, to levels which may affect their health significantly (AMAP, 1998, 2018b) This in turn leads to exposure and associated health risks to humans, in particular for certain Indigenous groups that consume these animals as part of their traditional diet. Because some contaminants can be passed from mothers to their foetus and infants, they are particularly susceptible (AMAP, 1998, 2015b).

Regarding climate change, the Arctic is warming at three times the rate of more temperate regions (AMAP, 2021), due to northward heat transfer and increased absorption of solar radiation as snow and ice melt exposing bare ground or sea water - contributing to a process known as Arctic Amplification (ACIA, 2005; AMAP, 2017d). Maximum Arctic winter sea ice aereal extent in 2015, 2016, 2017, and 2018 were at record low levels, and the volume of Arctic sea ice present in the month of September declined by 75% from 1979 to 2018 (AMAP, 2019b). Arctic glaciers, with the Greenland Ice Sheet, are the largest land-ice contributors to global sea level rise. Even if the Paris Agreement is successful, they will continue to lose mass over the course of this century. (AMAP, 2017d, 2019b). Hence, while anthropogenic drivers for climate change mainly take place outside the Arctic region, the Arctic warming impacts are profoundly affecting the Arctic region, but also have global consequences through sea level rise and global climate teleconnections (AMAP, 2017d, 2019b).

9.2 The Arctic Environmental Initiative and the Establishment of AMAP

Prior to the 1990s, Arctic environmental threats were addressed primarily through national actions by some Arctic states, combined with some international agreements, such as the Svalbard Treaty. The knowledge gained through scientific research during the 1970s and 1980s revealed the idea of a pristine Arctic to be an illusion, raising an urgent need to assess the circumpolar environmental state of the Arctic. At the same time, the Cold War, which had been a major obstacle to cooperation in the region was ending with the break-up of the Soviet Union. Together, these two factors provided the background for the Arctic Environmental Protection Strategy (AEPS, 1991), an agreement between the eight Arctic States that led to the establishment of the Arctic Monitoring and Assessment Programme (AMAP) in 1991. The AEPS was also ground-breaking in the way that Indigenous Peoples Organizations were given a key role in the process (Stone, 2015).

AMAP was established as a pan-arctic monitoring program with a mandate *"to monitor the levels of, and assess the effects of, anthropogenic pollutants in all components of the Arctic environment."*. The AEPS specified that actions should be undertaken in a step-by-step fashion:

- *"Distinguishing human-induced changes from changes caused by natural phenomena in the Arctic will require estimates and regular reporting by the Arctic countries of contaminant emissions and discharges, including accidental discharges, as well as transport and deposition. In addition, monitoring of deposition and selected key indicators of the Arctic biological environment.*
- *As far as possible build upon existing programs. [...] one of the important tasks [...] will be to review and coordinate existing national programs, establish a data directory, and to develop these programs when appropriate in an international framework.*
- *As an initial priority [..] focus on persistent organic contaminants and on selected heavy metals and radionuclides, and ultimately to monitor ecological indicators to provide a basis for assessments of the status of Arctic ecosystems.*
- *[summarize AMAP results in] regular State of the Arctic Environment Reports."*

And as a result of these actions, AMAP should

> *"provide information for: i) integrated assessment reports on status and trends in the condition of Arctic ecosystem;*
> *ii) identifying possible causes for changing conditions;*
> *iii) detecting emerging problems, their possible causes, and the potential risk to Arctic ecosystems including Indigenous peoples and other Arctic residents; and*
> *iv) recommending actions required to reduce risks to Arctic ecosystem."*

(Rovaniemi declaration, 1991)

In subsequent directions from Ministers, the AMAP mandate was extended in several areas, notably:

> *"... assessment of the effects of [...] climate change on Arctic ecosystems."*

"... human health impacts and the effects of multiple stressors."

(Alta Declaration, 1997)

As a result of the establishment of the Arctic Council in 1996, the AEPS was subsumed into the work of the Arctic Council and AMAP became a working group of the Arctic Council together with five other working groups; Conservation of Arctic Flora and Fauna (CAFF, established 1991), Emergency Prevention, Preparedness and Response (EPPR, established 1991), Protection of Arctic Marine Environment (PAME, established 1991), Sustainable Development Working Group (SDWG, established 1998) and the Arctic Contaminants Action Programme (ACAP, established 2006).

9.3 Organization and Deliverables of the Arctic Monitoring and Assessment Programme

AMAP was organized, with a permanent Secretariat in Oslo, in August 1992, and relocated to Tromsø, Norway in 2018. The decisive strategic level lies with the AMAP Heads of Delegations, consisting of representatives from all the eight arctic states; Canada, Kingdom of Denmark, Finland, Iceland, Norway, the Russian Federation, Saami Council, Sweden and United States of America, as well as representatives from the six Permanent Participants of the Arctic Council, that is Indigenous organizations; Arctic Athabaskan Council (AAC), Aleutian International Association (AIA), Gwich'in Council International (GCI), Inuit Circumpolar Council (ICC), and the Russian Association of Indigenous Peoples of the North (RAIPON) and the Saami Council. Observers, both observer states and observer organizations are invited to participate in AMAP working groups meetings as well as contribute to AMAP work, for example by nominating experts to join AMAP Expert Groups (AMAP, 2019a) and as such contribute as authors to the AMAP assessments.

AMAP's main deliverable are thematic peer reviewed scientific assessments. Since its first report on Arctic Pollution Issues in 1998 (AMAP, 1998), AMAP has published more than 30 such assessment reports, with five new reports being published in 2021. These comprehensive reports are condensed into summaries for policymakers, that include a scientific summary of key findings and recommendations for consideration by policy-makers. Hence, these assessments provide the scientific basis for recommendations on Arctic environmental issues that are addressed to the Arctic Council Ministers and Senior Arctic Officials. In addition, the assessment process is coordinated with international processes, feeding data and information to international bodies such as the IPCC (e.g. the IPCC Special Report on the Ocean and Cryosphere in a Changing Climate (IPCC, 2019)), UN Environment Programme, Stockholm Convention on Persistent Organic Pollutants (2001), Minamata Convention on Mercury (2013) and the Convention on Long-Range Transboundary Air Pollution, (CLRTAP, 1979) (Rottem et al., 2020).

9.4 Does it Work – The Black Carbon Case

While AMAP's mandate focuses on monitoring and assessment, the Black Carbon case may be used to illustrate how different Arctic Council bodies interact to monitor, develop and implement actions, execute mitigation projects, and evaluate effects of these actions. Black carbon, or soot, is a short-lived climate forcer that, through both heating the atmosphere by absorbing solar radiation and decreasing the albedo of snow in the Arctic, causes climate warming. Black carbon is also a constituent of particulate matter and an air pollutant that causes health effects through respiratory illnesses that can affect Arctic communities, including those reliant on diesel generators and wood burning for heat and energy (AMAP, 2015a, 2015c). Although Arctic States are responsible for only about 10% of global black carbon emissions, emissions located within and close to the Arctic have a disproportionately high impact on Arctic climate warming (AMAP, 2015a; Arctic Council, 2019). A major reason for this is the pronounced effect black carbon has a climate driver when deposited on snow (AMAP, 2015a). Within the Arctic, the main sources of black carbon are domestic heating, transportation, and flaring in the petroleum industry (AMAP, 2015a).

AMAP's 2015 report on black carbon informed policy makers that significant reduction in black carbon emissions could be achieved using existing technologies and good practices, including reducing emissions from residential and commercial use of fossil fuels, reducing emissions from wood-burning in residential heating, agricultural burning, and changing flaring practices at oil and gas fields (AMAP, 2015c). The scientific background was translated into the Arctic Council's Framework for Action on Black Carbon and Methane, including, e.g. national implementation plans delivered through the Arctic Council's Expert Group on Black Carbon and Methane (EGBCM), and demonstration projects for black carbon emissions reduction organized through the Arctic Councils ACAP working group. The Framework includes an aspirational goal to collectively reduce black carbon emissions from AC member countries by at least 25–33% below 2013 levels by 2025. In 2019, the EGBCM reported a 16% decrease in black carbon emissions by 2016 relative to 2013 (Arctic Council, 2019). As a follow up to this, new inventory-based estimates of black carbon emissions has been made available in the updated AMAP report for 2021.

9.5 Does it Work – The Mercury Case

Mercury is a highly toxic heavy metal that poses serious risks for detrimental health effects on both wildlife and humans. This was brought to international attention when mecury released from an industrial plant caused severe effects on the nervous systems of inhabitants in the Japanese city of Minamata who were exposed by eating fish. The extractation, use and emissions of mercury are now being regulated internationally through the Minamata Convention. As described by Platjouw et al. (2018), the information on trends and levels of mercury in the Arctic reported through AMAP assessments played an important role in the process of establishing the Minamata Convention. Mercury was one of the priority contaminant addressed in the first AMAP assessments (AMAP, 1998, 2002), reports heavily cited in UNEP's first Global Mercury Assessment (UNEP, 2002). The AMAP reports documented spatio-temporal trends and levels of mercury througout the Arctic, as well as its consequences on ecosystems and human health. Data showed that despite the long distance from major sources, mercury levels in Arctic air can on occasions be five to fifty times higher than levels measured in Europe and North America (UNEP Chemicals Branch, 2008), emphasizing the importance of long-range transport of contaminants in to the Arctic, as well as the need to global mechanims for regulating mercury. The data and information compiled in the 2011 AMAP mercury assessment (AMAP, 2011) fed into the process leading up to the UN Environment 2013 Global Mercury Assessment (UNEP Chemicals Branch, 2013), where the scientific technical background report (AMAP/UNEP, 2013) was prepared as a cooperation between AMAP and UNEP (Platjouw et al., 2018). This collaborative effort was repeated in preparing the 2018 Global Mercury Assessment (UN-Environment, 2019). The assessment work done by AMAP therefore played a key role in both documenting effects as well as facilitating the process leading up to the Minamata Convention that was adopted in 2013 and entered into force in 2017 (Platjouw et al., 2018).

9.6 Why Did It Work?

Both the black carbon and the mercury case are examples of an active policy-oriented approach where AMAP has taken the role as a science-broker. Similar examples are given for AMAPs role in the establishment of the Stockholm Convention (Steindal et al., 2021). In the case of mercury, AMAP's comprehensive assessments increased the awareness of the trends and levels at a circumarctic spatial scale, emphasizing the need for emissions to be treated on a global rather than just national scale (Platjouw et al., 2018). Documenting high levels of mercury in what was perceived as a pristine Arctic environment (AMAP, 2002; UNEP Chemicals Branch, 2008), the report stated the urgent need for action to mitigate on the threat to Arctic ecosystems, as well as human health (Platjouw et al., 2018). According to Platjouw et al., (2018), the timing of the mercury assessment was essential for its successful contribution in the Minamata process. The report prepared the ground for the negotiations, by feeding in data in time for a scientific consensus to be achieved. Hence negotiations could focus on legal aspects of the regulation process, rather on scientific disputes (Platjouw et al., 2018; Selin et al., 2018). Also long-term, sustainable funding gave AMAP the possibility to strategically feed in science-based knowledge to the process over a longer time, and through several steps in the negotiation process. AMAPs work allowed the Arctic Council members to have an active policy-oriented approach as a science broker, which, according to (Platjouw et al., 2018) played an essential role in the development and ratification of the Minamata Convention.

9.7 Future Aspects – Increased Understanding of the Arctic Environmental State by Combining Knowledges

In order to produce comprehensive reports, data are needed that reflect both the large spatial variation in environmental parameters due to abiotic factors (such as climate and weather systems, oceanic and atmospheric currents relative to emission sources, etc.) and biotic factors (such as food webs and species trophic level). In addition, anthropogenic factors including local sources of contaminants need to be considered in supporting both Indigenous Knowledge, Traditional and Local Knowledge that can provide resilience to abrupt changes. Hence an important aspect of AMAPs ability to deliver comprehensive assessments is maintaining its coordinated monitoring program. However, infrastructure is limited in the Arctic, including infrastructure for conducting scientific observations of such factors.

Still, the Arctic is not deserted, but has been inhabited since historical times by Indigenous groups and local people for whom the nature of the Arctic has required awareness of the elements as well as an evolving knowledge transition to allow societies to survive and thrive over time. Indigenous Peoples have, through their long-term presence in the Arctic, adapted to their living conditions, and developed knowledge systems and language to describe the environment they live in. Figure 9.1 shows the diversity of Indigenous languages spoken throughout the Arctic, reflecting the diversity of societies; it also illustrates a diverse source of knowledge that to a certain degree has been neglected in scientific work.

Indigenous Knowledge (IK) and Traditional and Local Knowledge (TLK) may provide an essential additional source of information and environmental knowledge of the Arctic by providing access to otherwise inaccessible data, especially where systematic observation and measurement infrastructure are scarce. While the development of satellite observations and autonomous sampling (for example for meteorological data or air measurements, as well as buoys or gliders providing oceanic data) have increased tremendously, environmental data in the Arctic are limited; the area is too vast for such instruments to able to provide complex data at high spatial resolution. Likewise, understanding of trends is dependent on historical records of environmental data. Often gaps due to lacking data need to be filled by extrapolation or methods which introduce variation and uncertainty into models. Combining research and Indigenous Knowledge has been proposed as an approach to increase the understanding of a changing Arctic environment due to climate change or other stressors (e.g. Eira et al., 2013; Krupnik & Jolly, 2002; Lennert, 2017; Lennert,

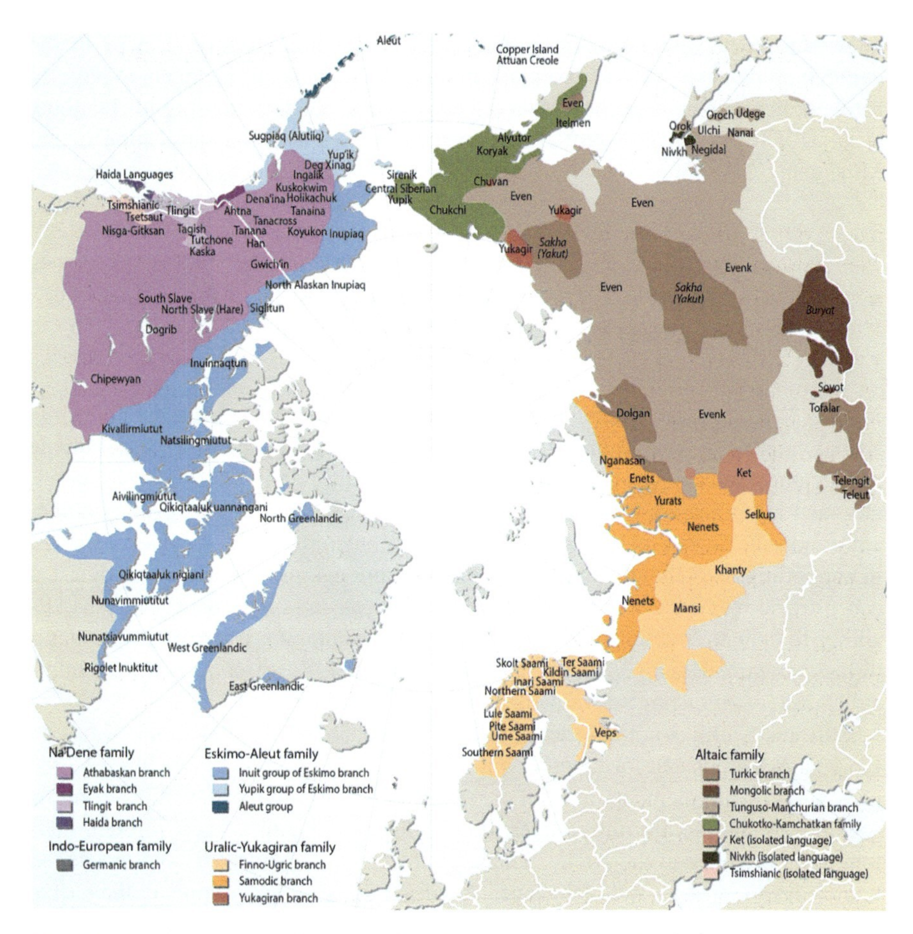

Fig. 9.1 Map of languages of Indigenous Peoples in the Arctic (CAFF, 2013). (Image retrieved at: http://geo.abds.is/geonetwork/srv/eng/catalog.search#/metadata/9c47173b-4774-436f-ae3f-1925f1173ec6)

2016). For instance, it has been advocated that combining Traditional Knowledge and scientific observations may identify important factors acting as additional stressors on marine mammals exposed to climate change and contaminants (Lennert, 2016). In AMAPs assessment on *Biological effects of contaminants on Arctic wildlife and Fish* (AMAP, 2018b), the decreasing trend in persistent organic pollution (POPs) concentration in polar bear (*Ursus maritimus*) tissue in the eastern Canadian Arctic were observed to be levelling off (AMAP, 2018b; Mckinney et al., 2013). This may be due to changes in feeding behavior of the bears, shifting from ice-associated seal species to open-water seal species, where the latter have higher tissue concentrations of POPs. This feeding change corresponds with a climate driven change of reducing sea-ice extent in the area (AMAP, 2018b; Mckinney et al., 2013).

Terminology developed by Indigenous people over many years to describe their living conditions is a further source of Indigenous Knowledge and Traditional and Local Knowledge that could be used to provide better resolution of data or improved understanding of ecosystem impacts in a changing Arctic. One such example is the description of snow and its impacts on Saami reindeer herding. The Saami language contains at least 318 different descriptions of snow and their relation to reindeer feeding conditions and behavior and hence reindeer herding (Eira, 2012; Eira et al., 2013). Such linguistic richness may be an important tool both for understanding the relation between meteorological data, the historical record, and abiotic impacts on snow, dependent on landscape characteristics, as well as increased understanding on how large scale climate variation may have societal impacts for reindeer herders (Eira et al., 2013; Maynard et al., 2011).

In addition, Indigenous Knowledge and Traditional and Local Knowledge may provide information about the societal relevance and importance of data. For instance, the three regional AMAP reports on *Adaptation Actions for a Changing Arctic* (AMAP, 2017a, b, 2018a) initially focused on adaptations to environmental impacts of climate change. However, local inhabitants emphasized that change in societal factors like infrastructure, development, and education was also important to them. While these societal factors may seem less critical than changes in climate per se, the structure and diversity of such factors may influence societal resilience, and hence the ability of local communities to adapt and meet the challenges associated with climate change (Mathis et al., 2015).

9.7.1 Future Perspectives – Common Challenges and Opportunities for Arctic Knowledge

As described above, combining knowledges from different sources and knowledge systems has potential to give a more diverse input to AMAP assessments and thereby make them more relevant as well as robust in meeting new challenges for predicting Arctic change. Co-production of knowledge feeding into co-management processes in the Arctic may also facilitate conditions that allow for adaptation in a rapidly changing environment (Ådnøy et al., 2003; Armitage et al., 2011; Eira et al., 2018; Frainer et al., 2020). While co-production may imply a need for transformative changes in translation between knowledge systems (Norström et al., 2020; Robards et al., 2018; Wheeler et al., 2020), in this case between natural sciences and Indigenous Knowledge and Traditional and Local Knowledge, data from different systems may be combined and compared for instance by using a mixed method framework (Maxwell, 2016; Teddlie & Tashakkori, 2009), or by semi-quantitative methods such as fuzzy cognitive mapping (e.g. Giles et al., 2007), similar to pathway analyses used in ecology (e.g. Focardi & Tinelli, 1996; Johnson et al., 2001).

For instance, mixed methods have been used to investigate community-based management of pastures among reindeer herders in Finnmark, using quantitative analysis of structural variable and qualitative methods for investigating explanatory mechanisms (Hausner et al., 2012).

Co-produced knowledge based on natural sciences and Indigenous Knowledge and Traditional and Local Knowledge, may face challenges in assessing environmental conditions in a changing Arctic. For conclusions to be made that are relevant on a circum-arctic scale, this knowledge needs to be generalized spatially as well as over time, which, if based on interviews, would be very resource demanding. However, community-based monitoring has a potential of capturing large amounts of data if organized in a structural framework (Johnson et al., 2016). Implementing information technology and mobile platforms into the monitoring or dissemination of knowledge may facilitate a better spatial resolution and over time temporal resolution as well. Such platforms have been initiated, with examples including the Inuit Siku Atlas on Inuit sea ice knowledge and use (https://sikuatlas.ca) and the Local Environmental Observer Network (https://www.leonetwork.org), developed by the Alaska Native Tribal Health Consortium (ANTHC) in 2009, now being expanded under the Arctic Council working group ACAP to create a foundation for a Circumpolar Local Environmental Observer (CLEO) Network.

However, both Indigenous Knowledge, Traditional and Local Knowledge, and science are based on empirical evidence, by definition seen in retrospect. As the Arctic is changing to conditions not known in modern times, and as Artic ecosystems may be susceptible to non-linear changes or abrupt tipping points, interpreting ecosystem responses by extrapolation at the margins of normal range of variation may be challenging (e.g. Heinze et al., 2021). Hence, both knowledge systems are facing similar challenges when it comes to using empirical data for predictions and projections of future conditions. Such challenges for weather predictions based upon Indigenous Knowledge has been described for the Canadian Arctic (ACIA, 2005; Krupnik & Jolly, 2002). Similar findings have been experienced by the first author of this article:

Growing up during at the very northern end of Europe, where the continent meets the Arctic Ocean and its seas the first part of the 1900's, my grandfather from he was 9-10 years old were, together with his brothers sent up in the highlands in winter to trap ptarmigan. And every summer they spent fishing salmon, to contribute to the family's income.

The long life in the mountains provided him with experience on weather patterns. Ever since I was a kid, we used to discuss every spring when the ice was leaving the river so we could start the salmon fishing. My grandfather's predictions were fairly accurate, some years we started the fishing early and some late.

However, I remember clearly a day in late spring in the end of the eighties, when I, as every year; asked: "So grandfather, when will the ice leave the river this year, and we can start fishing salmon?". My grandfather sat silent for some minutes. Then he looked at me and said: "I don't know. The signs in nature I have learnt throughout my life do not tell anymore. Something has changed"

The speed of current Arctic change challenges our ability to understand its dynamics. According to the Bayesian framework, however, science progress can be achieved by continuously adjusting prior expectations and models to current data (Chalmers, 1999). Hence, understanding a rapidly changing Arctic may be better facilitated by combining our knowledge on the Arctic Environment, from both scientific, Indigenous, Traditional and Local Knowledge systems.

References

ACIA. (2005). *Arctic climate impact assessment. ACIA overview report*. Cambridge University Press. Retrieved from http://www.amap.no/documents/doc/arctic-arctic-climate-impact-assessment/796

Ådnøy, T., Vegarud, G., Gulbrandsen Devold, T., Nordbø, R., Colbjønsen, I., Brovold, M., ... Lien, S. (2003). *Effects of the 0 - and F-alleles of alpha S1 casein in two farms of northern Norway*. Proceedings of the International Workshop on Major Genes and QTL in Sheep and Goat.

AMAP/UNEP. (2013). *Technical background report to the global atmospheric mercury assessment 2013*. Retrieved from https://www.amap.no/documents/doc/technical-background-report-for-the-global-mercury-assessment-2013/848

AMAP. (1998). AMAP assessment report: Arctic pollution issues. Oslo, Norway. Retrieved from https://www.amap.no/documents/doc/amap-assessment-report-arctic-pollution-issues/68

AMAP. (2002). Arctic Monitoring and Assessment Programme: AMAP Assessment report – Heavy Metals in the Arctic. Assessment 2002. Retrieved from http://www.amap.no/documents/doc/amap-assessment-2002-heavy-metals-in-the-arctic/97

AMAP. (2010). AMAP assessment 2009 - persistent organic pollutants (POPs) in the Arctic. *Science of The Total Environment Special Issue, 408*, 2851–3051.

AMAP. (2011). Arctic monitoring and assessment program 2011: mercury in the Arctic. Assessment. Oslo, Norway. Retrieved from http://www.grida.no/amap

AMAP. (2015a). *AMAP assessment 2015: Black carbon and ozone as Arctic climate forcers. AMAP assessment report*. Oslo

AMAP. (2015b). *AMAP assessment 2015: human health in the Arctic*. Oslo https://doi.org/10.3402/ijch.v75.33949

AMAP. (2015c). *Summary for policy-makers: Arctic climate issues 2015. AMAP summary report*. Oslo, Norway

AMAP. (2017a). Adaptation actions for a changing Arctic: perspectives from the Barents area. . Retrieved from https://www.grida.no/publications/382

AMAP. (2017b). Adaptation actions for a changing Arctic: perspectives from the Bering-Chukchi-Beaufort region.

AMAP. (2017c). AMAP assessment 2016: Chemicals of emerging Arctic concern. . Retrieved from https://www.amap.no/documents/doc/AMAP-Assessment-2016-Chemicals-of-Emerging-Arctic-Concern/1624

AMAP. (2017d). Snow, water, ice and permafrost in the Arctic (SWIPA) 2017. . Retrieved from https://www.amap.no/documents/doc/snow-water-ice-and-permafrost-in-the-arctic-swipa-2017/1610

AMAP. (2018a). *Adaptation actions for a changing Arctic: perspectives from the Baffin Bay/Davis Strait region*. Oslo.

AMAP. (2018b). AMAP assessment 2018: biological effects of contaminants on Arctic wildlife and fish. Oslo, Norway. Retrieved from www.amap.no

AMAP. (2019a). AMAP strategic framework 2019+. Tromsø, Norway https://doi.org/10.4324/9781351047722-1

AMAP. (2019b). Arctic climate change update 2019: an update to key findings of snow, water, ice, and permafrost in the Arctic (SWIPA) 2017. Oslo, Norway. Retrieved from https://www.amap.no/documents/doc/amap-climate-change-update-2019/1761

AMAP. (2021). Arctic climate change update 2021: key trends and impacts. Tromsø, Norway

Arctic Council. (2019). Expert Group on Black Carbon and Methane - Summary of Progress and Recommendations 2019. (2019)

Arctic Environmental Protection Strategy. (1991). Signed 14 June 1991. Rovaniemi, Finland

Armitage, D., Berkes, F., Dale, A., Kocho-Schellenberg, E., & Patton, E. (2011). Co-management and the co-production of knowledge: Learning to adapt in Canada's Arctic. *Global Environmental Change, 21*(3), 995–1004. https://doi.org/10.1016/j.gloenvcha.2011.04.006

CAFF. (2013). Arctic biodiversity assessment. Akureyri, Iceland

Chalmers, A. F. (1999). *What is this thing called science?* (3rd ed.). Open University Press.

Convention on Long-Range Transboundary Air Pollution (1979). Signed 13 November 1979, effective 16 March 1983. Geneva, Switzerland

Eira, Inger Marie G (2012). Muohttaga jávohis giella: Sámi árbevirolaš máhttu muohttaga birra dálkkádatrievdanáiggis/the silent language of snow: Sámi traditional knowledge of snow in times of climate change. UiT - The Arctic University of Norway

Eira, I. M. G., Jaedicke, C., Magga, O. H., Maynard, N. G., Vikhamar-Schuler, D., & Mathiesen, S. D. (2013). Traditional Sámi snow terminology and physical snow classification-two ways of knowing. *Cold Regions Science and Technology, 85*(October), 117–130. https://doi.org/10.1016/j.coldregions.2012.09.004

Eira, I. M. G., Oskal, A., Hanssen-Bauer, I., & Mathiesen, S. D. (2018). Snow cover and the loss of traditional indigenous knowledge. *Nature Climate Change, 8*(11), 928–931. https://doi.org/10.1038/s41558-018-0319-2

Environment, A. (2019). Technical background report to the global mercury assessment 2018. Oslo, Norway & Geneva, Switzerland

Focardi, S., & Tinelli, A. (1996). A structural-equations model for the mating behaviour of bucks in a lek of fallow deer. *Ethology Ecology and Evolution, 8*(4), 413–426.

Frainer, A., Mustonen, T., Hugu, S., Andreeva, T., Arttijeff, E. M., Arttijeff, I. S., ... Pecl, G. (2020). Cultural and linguistic diversities are underappreciated pillars of biodiversity. *Proceedings of the National Academy of Sciences of the United States of America, 117*(43), 26539–26543. https://doi.org/10.1073/pnas.2019469117

Giles, B. G., Findlay, C. S., Haas, G., LaFrance, B., Laughing, W., & Pembleton, S. (2007). Integrating conventional science and aboriginal perspectives on diabetes using fuzzy cognitive maps. *Social Science and Medicine, 64*(3), 562–576. https://doi.org/10.1016/j.socscimed.2006.09.007

Hausner, V. H., Fauchald, P., & Jernsletten, J.-L. (2012). Community-based management: under what conditions do Sámi pastoralists manage pastures sustainably? *PloS One, 7*(12), e51187. https://doi.org/10.1371/journal.pone.0051187

Heinze, C., Blenckner, T., Martins, H., Rusiecka, D., Döscher, R., Gehlen, M., ... Author contributions, N. (2021). The quiet crossing of ocean tipping points and m Arctic Monitoring and Assessment Programme Secretariat, *118*(9). https://doi.org/10.1073/pnas.2008478118/-/DCSupplemental

IPCC. (2019). In H.-O. Pörtner, C. Roberts, V. M.-D. Debra, P. Zhai, M. Tignor, E. Poloczanska, et al. (Eds.), *IPCC special report on the ocean and cryosphere in a changing climate.*

Johnson, C. J., Parker, K. L., & Heard, D. C. (2001). Foraging across a variable landscape: behavioral decisions made by woodland caribou at multiple spatial scales. *Oecologia, 127*(4), 590–602.

Johnson, N., Behe, C., Danielsen, F., Krümmel, E.-M., Nickels, S., & Pulsifer, P. L. (2016). *Community-based monitoring and indigenous knowledge in a changing Arctic: A review for the sustaining Arctic observing networks*. Final report to sustaining Arctic observing networks. Ottawa, ON

Krupnik, I., & Jolly, D. (2002). *The earth is faster now: indigenous observations of Arctic environmental change.* Frontiers in Polar Social Science. Arctic Research Consortium of the United States in cooperation with the Arctic Studies Center, Smithsonian Institution

Lennert, A. E. (2016). What happens when the ice melts? Belugas, contaminants, ecosystems and human communities in the complexity of global change. *Marine Pollution Bulletin, 107*(1), 7–14. https://doi.org/10.1016/j.marpolbul.2016.03.050

Lennert, A. E. (2017). A millennium of changing environments in the Godthåbsfjord. In *West Greenland - Bridging cultures of knowledge*. University of Greenland. https://doi.org/10.13140/RG.2.2.16091.36640

Mathis, J. T., Cooley, S. R., Lucey, N., Colt, S., Ekstrom, J., Hurst, T., ... Feely, R. A. (2015). Ocean acidification risk assessment for Alaska's fishery sector. *Progress in Oceanography, 136*, 71–91. https://doi.org/10.1016/j.pocean.2014.07.001

Maxwell, J. A. (2016). Expanding the history and range of mixed methods research. *Journal of Mixed Methods Research, 10*(1), 12–27. https://doi.org/10.1177/1558689815571132

Maynard, N. G., Oskal, A., Turi, J. M., Mathiesen, S. D., Eira, I. M. G., Yurchak, B., ... Gebelein, J. (2011). Impacts of arctic climate and land use changes on reindeer pastoralism: Indigenous knowledge and remote sensing. *Eurasian Arctic Land Cover and Land Use in a Changing Climate*. https://doi.org/10.1007/978-90-481-9118-5_8

Mckinney, M. A., Iverson, S. J., Fisk, A. T., Sonne, C., Rigét, F. F., Letcher, R. J., ... Dietz, R. (2013). Global change effects on the long-term feeding ecology and contaminant exposures of East Greenland polar bears. *Global Change Biology, 19*(8), 2360–2372. https://doi.org/10.1111/gcb.12241

Minamata Convention on Mercury (2013). Signed 10 October 2013, effective 16 August 2017. Kumamoto, Japan

Norström, A. V., Cvitanovic, C., Löf, M. F., West, S., Wyborn, C., Balvanera, P., ... Österblom, H. (2020). Principles for knowledge co-production in sustainability research. *Nature Sustainability, 3*(3), 182–190. https://doi.org/10.1038/s41893-019-0448-2

Platjouw, F. M., Steindal, E. H., & Borch, T. (2018). From Arctic science to international law: the road towards the Minamata convention and the role of the Arctic council. *Arctic Review on Law and Politics, 9*, 226–243. https://doi.org/10.23865/arctic.v9.1234

Robards, M. D., Huntington, H. P., Druckenmiller, M., Lefevre, J., Moses, S. K., Stevenson, Z., ... Williams, M. (2018). Understanding and adapting to observed changes in the Alaskan Arctic: actionable knowledge co-production with Alaska native communities. *Deep-Sea Research Part II: Topical Studies in Oceanography, 152*, 203–213. https://doi.org/10.1016/j.dsr2.2018.02.008

Rottem, S. V., Prip, C., & Soltvedt, I. F. (2020). Arktisk råd i spennet mellom forskning, forvaltning og politikk. *Internasjonal Politikk, 78*(3), 284. https://doi.org/10.23865/intpol.v78.1504

Selin, H., Keane, S. E., Wang, S., Selin, N. E., Davis, K., & Bally, D. (2018). Linking science and policy to support the implementation of the Minamata convention on mercury. *Ambio, 47*(2), 198–215. https://doi.org/10.1007/s13280-017-1003-x

Steindal, E. H., Karlsson, M., Hermansen, E., Borch, T., & Platjouw, F. M. (2021). From Arctic science to global policy – Addressing multiple stress under the Stockholm convention. *Arctic Review on Law and Politics, 12*, 80–107.

Stockholm Convention on Persistent Organic Pollutants (2001). Signed 22 May 2001, effective 17 May 2004. Stockholm, Sweden

Stone, D. P. (2015). *The changing Arctic environment: The Arctic messenger. The changing Arctic environment: the Arctic messenger.* Cambridge University Press. https://doi.org/10.1017/CBO9781316146705

Teddlie, C., & Tashakkori, A. (2009). Foundations of mixed methods research: Integrating quantitative and qualitative approaches in the social and behavioral sciences. SAGE. Retrieved from https://books.google.no/books?hl=no&lr=&id=c3uojOS7pK0C&oi=fnd&pg=PP1&dq=maxwell+mixed+methods&ots=QbpAWngROG&sig=Vf7cLEiONQbEyW3aFqUn1RMXfFg&redir_esc=y#v=onepage&q=maxwell mixed methods&f=false

UNEP. (2002). Global mercury assessment.

UNEP Chemicals Branch. (2008). The global atmospheric mercury assessment: Sources, emissions and transport. *UNEP-Chemicals, Geneva*. Geneva. Retrieved from http://scholar.google.com/scholar?hl=en&btnG=Search&q=intitle:The+Global+Atmospheric+Mercury+Assessment+:+Sources+,+Emissions+and+Transport#2

UNEP Chemicals Branch. (2013). The global mercury assessment: sources, emissions, releases and environmental transport. Geneva, Switzerland. Retrieved from http://www.unep.org/PDF/PressReleases/GlobalMercuryAssessment2013.pdf

UN-Environment. (2019). Global Mercury Assessment 2018. UN-Environment Programme, Chemicals and Health Branch, Geneva

Wheeler, H. C., Danielsen, F., Fidel, M., Hausner, V., Horstkotte, T., Johnson, N., … Vronski, N. (2020). The need for transformative changes in the use of indigenous knowledge along with science for environmental decision-making in the Arctic. *People and Nature, 2*(3), 544–556. https://doi.org/10.1002/pan3.10131

Chapter 10
(Research): The Value of High-Fidelity Numerical Simulations of Ice-Ship and Ice-Structure Interaction in Arctic Design with Informed Decisionmaking

Marnix van den Berg, Jon Bjørnø, Wenjun Lu, Roger Skjetne, Raed Lubbad, and Sveinung Løset

Abstract Long-term changes in the climate lead to an increase of activities in the Arctic. Ships and structures operating in Arctic waters may encounter sea ice and must be designed to withstand the loads resulting from ice-ship or ice-structure interactions. Traditionally, design rules for ships, structures and operations in ice-covered waters have been based on full-scale measurements and model-scale tests. Full-scale measurements and model-scale tests will remain important sources of data and knowledge in the future. Numerical simulations can give valuable

M. van den Berg (✉) · W. Lu · R. Lubbad · S. Løset
Sustainable Arctic Marine and Coastal Technology (SAMCoT), Norwegian University of Science and Technology (NTNU), Trondheim, Norway

Department of Civil and Environmental Engineering, Norwegian University of Science and Technology (NTNU), Trondheim, Norway

Arctic Integrated Solutions AS (ArcISo), Trondheim, Norway
e-mail: marnix.berg@ntnu.no; wenjun.lu@ntnu.no; raed.lubbad@ntnu.no; sveinung.loset@ntnu.no

J. Bjørnø
Sustainable Arctic Marine and Coastal Technology (SAMCoT), Norwegian University of Science and Technology (NTNU), Trondheim, Norway

Department of Marine Technology, Norwegian University of Science and Technology (NTNU), Trondheim, Norway
e-mail: jon.bjorno@ntnu.no

R. Skjetne
Sustainable Arctic Marine and Coastal Technology (SAMCoT), Norwegian University of Science and Technology (NTNU), Trondheim, Norway

Arctic Integrated Solutions AS (ArcISo), Trondheim, Norway

Department of Marine Technology, Norwegian University of Science and Technology (NTNU), Trondheim, Norway
e-mail: roger.skjetne@ntnu.no

P. A. Berkman et al. (eds.), *Building Common Interests in the Arctic Ocean with Global Inclusion, Volume 2*, Informed Decisionmaking for Sustainability,
https://doi.org/10.1007/978-3-030-89312-5_10

additional insights into the processes and loads resulting from ice-ship and ice-structure interactions. The primary advantage of numerical modelling compared to full-scale measurements and model-scale tests is the marginal cost of testing. This enables the testing of a multitude of design options and ice conditions. Challenges in the application of numerical models are related to finding a balance between model efficiency and accuracy and model validation and calibration.

This chapter briefly introduces a numerical model referred to as the Simulator for Arctic Marine Structures (SAMS), and five examples of its applications to modelling ice-ship and ice-structure interactions for Arctic design with informed descisionmaking. The goal of this chapter is to show how numerical modelling can aid informed decisiomaking by regulators, project owners and contractors in cases involving the safety, efficiency and viability of ships, structures and operations in ice-covered waters. Therefore, the application examples mainly focus on the insights obtained by numerical modelling, and do not provide an in-depth technical description of the modelling process. The reader is referred to the literature referenced within this chapter for more detailed descriptions of the modelling processes and assumptions. The application examples in this chapter cover three application phases: research and development (R&D), design and operational assistance.

10.1 Introduction

The extent and thickness of the Arctic sea ice cover are decreasing as a result of long-term changes in the climate (NSIDC, 2020). The decreasing ice cover creates opportunities for tourism, fisheries, merchant shipping and the exploration of natural resources. As these activities move further North, the risks posed by the presence of ice remain relevant.

Structures and ships operating in areas where floating ice may occur must be designed for ice loads. Currently, design loads for ships and structures operating in ice-covered waters are mainly determined using empirical formulas and model-scale tests. The empirical formulas are based on full-scale measurements of the mechanical properties of the ice and measurements of the ice loads on existing structures. Full-scale measurements of ice loads and model-scale tests will remain important sources of information for the safe design of structures and ships operating in areas where floating ice may occur. However, both full-scale measurements and model-scale tests have their limitations.

Full-scale measurements can only be performed on existing structures or ships. It is not always clear how loads measured on existing structures or ships can be translated to novel structure designs. In addition, it is often challenging to measure all parameters relevant to the ice-structure interaction processes. Furthermore, the conditions in which loads can be measured are governed by environmental conditions and cannot be controlled.

Model-scale tests can be performed on novel structure designs, and the ice conditions can be controlled to a larger extent than in full-scale measurements. However, it is often challenging to appropriately scale the physical parameters of importance to the ice-structure or ice-ship interaction processes, leading to uncertainty in how the model-scale test results can be translated to full-scale values. Finally, both model-scale tests and full-scale measurements are expensive and time consuming to conduct.

Numerical simulations, if properly verified and validated as described and defined in Oberkampf, W., & Roy (2010), can offer a valuable additional tool to further understand the loads and interaction phenomena resulting from ice-ship and ice-structure interactions. The primary advantage of numerical models is the marginal cost of testing a large set of parameter variations. By using numerical simulations, the number of ice conditions and possible ship or structure geometries tested can be greatly expanded compared to model-scale tests. In comparison to empirical design formulas, the main advantage of numerical models is that the structure or ship geometry and the environmental conditions can be captured more completely. For instance, the actual ship or structure geometry can be tested numerically, whereas the ship or structure geometries must be captured by a limited set of parameters and assumptions when applying empirical formulas.

Naturally, there are also challenges related to the development and application of numerical models. We identify two primary challenges in the development of numerical models for ice-structure and ice-ship interaction:

- One must find a balance between model accuracy and numerical efficiency. Sub-processes must be simplified in order to simulate the global ice-structure or ice-ship interaction process with an acceptable computation time.
- Numerical models must be validated using full-scale or model-scale measurements. The limited availability of complete and accurate datasets complicates the model validation and calibration process.

This chapter briefly introduces a numerical model referred to as the Simulator for Arctic Marine Structures (SAMS), and five examples of its applications to modelling the ice-ship and ice-structure interactions for Arctic design and decisionmaking. SAMS was originally developed at the Norwegian University of Science and Technology (NTNU) as a part of the Centre for Research-Based Innovation; *Sustainable Arctic Marine and Coastal Technology* (SAMCoT). SAMS was further developed and commercialized by the NTNU spin-off company Arctic Integrated Solutions AS ("ArcISo," 2020).

The application examples included in this chapter cover potential applications of numerical modelling in research and development (R&D), design, and operational assistance related to ships or structures operating in ice-covered waters. In this context, R&D refers to the development and validation of generalized models, testing procedures and operational procedures for structures and operations in ice covered waters. Design refers to numerical modelling applied in the design phases of a specific structure or operation. Operational assistance refers to the application of

Table 10.1 Application cases discussed in this chapter, covering ships and structures in broken and glacial ice conditions and applications related to R&D, design and operational assistance

Case	Ship/Structure type	Ice condition	Application phases
Validation against full-scale data; ship transit in broken ice	Ship (icebreaker)	Broken ice	R&D
Co-analysis of broken ice loads with model-scale tests and numerical simulations	Jack-up structure (model scale)	Broken ice	R&D, design
Assessment of ice management strategies	Ship (icebreaker), generic protected structure (circular)	Broken ice	R&D, design, operational
Analysis of glacial ice impacts	Semi-submersible	Glacial ice	R&D, design, operational
Estimation of ice actions on the grounded trawler Northguider in the Hinlopen straight	Ship (grounded)	Broken ice	Operational

numerical modelling to support decisionmaking concerning the immediate safety and feasibility of operations.

Across R&D, design and operational assistance applications, there is a change in the specificity of the data and knowledge to be obtained by numerical simulations and there is an increase in the urgency with which the data and knowledge is required. For instance, in an operational assistance setting, the model must be specifically tailored to a specific ship/structure and ice condition, and there is a degree of urgency in obtaining the modelling results, thus requiring a higher computational efficiency compared to R&D related applications. In R&D, the findings must be more generalized, and the computational efficiency is less important than in operational assistance. Table 10.1 summarizes the application cases discussed in this chapter.

10.2 Main Characteristics of SAMS

The Simulator for Arctic Marine Structures (SAMS) is a numerical simulator intended for the simulation of interaction between floating ice and different kinds of structures and ships. This section gives a brief overview of the technical background of SAMS. A more comprehensive descriptions of the different sub-modules can be found in the publications referenced in this section.

SAMS simulates the motions, collisions and failure of ice floes and the motions of ships and structures. In the context of SAMS, ice floe sizes can be in the order of kilometres, where they effectively act as level ice, or as small as ice rubble 'blocks' resulting from the ice-ice or ice-structure interaction processes. The simulations are performed in a three-dimensional (3D) domain, as ice-structure interaction is often a 3D process. Both ice-ice and ice-structure interactions are considered.

SAMS uses the discrete element method (DEM) (see, e.g., Cundall and Strack (1979)) to determine the contact forces between bodies. The discrete element method can be further subdivided in the *smooth* (SDEM) and the *non-smooth* (NDEM) discrete element method. The difference between the two can be seen as the difference between implicit and explicit time integration. SAMS uses the non-smooth discrete element method, and thus applies implicit time integration. An advantage of this method is that the simulations remain stable over a wide range of time step sizes. Because of this feature, the simulation time step size can be chosen by the user based on the desired accuracy and efficiency. For instance, when numerical simulations are to be used in an operational environment, say as an on-board system to aid navigation, simulations must produce results in real-time, or even faster. In such a case a larger time step size can be used, reducing somewhat the simulation accuracy. When the simulator is used for design purposes, a smaller time step size can be chosen, leading to more accurate ice load predictions.

The simulations are performed in the time domain. Time domain simulations are needed because of the nonlinear nature of the ice-structure interaction process. The contact forces between bodies are calculated in each time step. In the calculation of contact forces, the ice material properties, the contribution of non-contact forces, and the influence of forces at other contacts are considered. The magnitude of contact forces is limited by the contact geometry and the ice crushing strength. The maximum frictional force is governed by the normal force following the Coulomb model of friction. The applied discrete element method, including the contact crushing and frictional assumptions, are described in detail by van den Berg et al. (2018).

In addition to the contact forces, non-contact forces are also considered on the ice floes and the structure. These include drag forces from wind, current and propeller flow, gravity, and buoyancy forces. Further details on the inclusion of propeller flow in SAMS can be found in Tsarau et al. (2017). Figure 10.1 gives an overview of the most important model components.

In each time step, once the contact forces are solved, an ice fracture module is activated to assess whether and how the ice fails. The fracture module is based on a series of fracture mechanics–based analytical solutions (Lu, 2014). This strategy of fracture modelling was first introduced by Lubbad and Løset (2011). The major advantage of adopting an analytical fracture treatment is its high computational efficiency. However, several assumptions must be introduced to use the analytical algorithms in characterizing the fracture of an ice floe:

- **Ice can only fail by in-plane splitting, out-of-plane bending and by local crushing**

In the numerical simulations, the ice can only fail in the failure modes for which analytical solutions have been developed. In SAMS, ice can fail by in-plane splitting, out-of-plane bending and by local crushing. The solutions for in-plane splitting failure and out-of-plane bending failure as applied in SAMS are described in Lu et al.

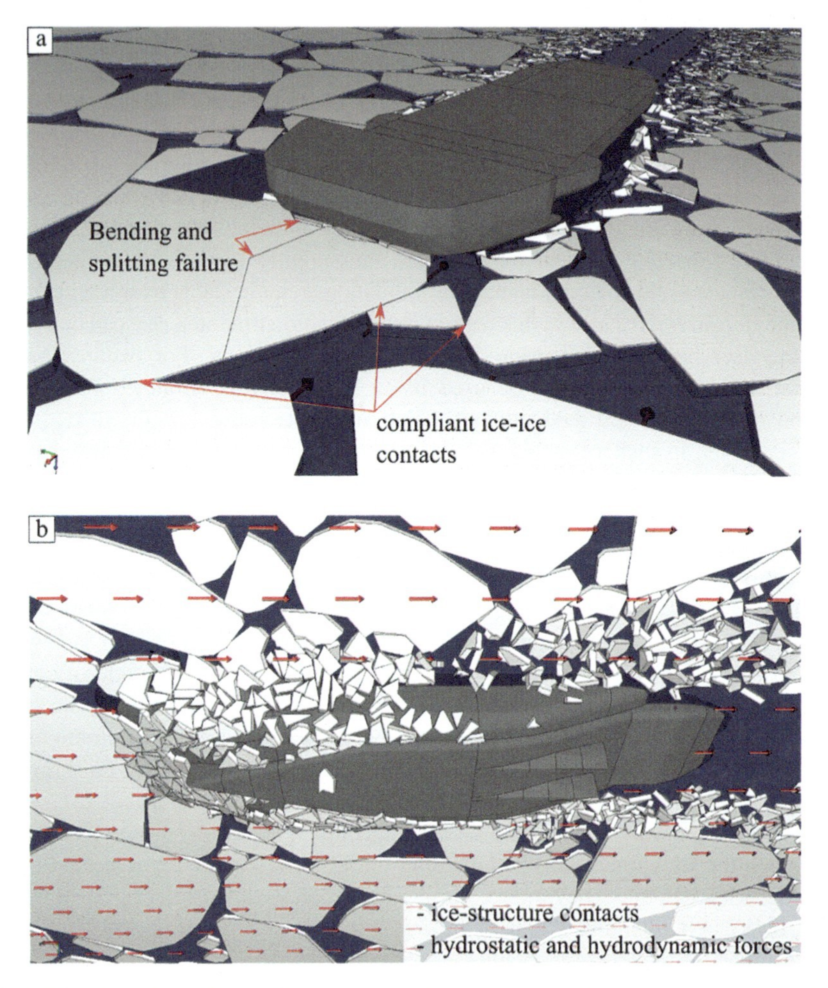

Fig. 10.1 Above (**a**) and below (**b**) water view of the icebreaker Oden sailing in broken ice. The sketch shows the most important model components of SAMS (van den Berg, 2019)

(2015a, b). In-plane splitting and out-of-plane bending failure are the dominant ice failure modes in the interaction between ice and sloping structures.

- **The ice floe failure strengths are derived for simplified floe geometries**

The analytical solutions for ice failure as applied in SAMS consider the size, confinement and geometry of the ice floe. Analytical solutions have been derived for a range of floe geometries and confinement conditions (Lu et al., 2018c, b). The floe geometries and loading conditions that occur in the numerical simulation are mapped to the most representative case for which an analytical solution is available. The ice floe geometry for which the analytical solution was developed is not necessarily the same as the ice floe geometry to which the solution is applied. The effect of this geometry simplification was studied by Lu et al. (2018a) using an eXtended Finite Element Method (XFEM) model. The results of this study show that the error introduced by the geometry simplification is smaller than the variability caused by uncertainty in the ice mechanical properties. The ice failure mode that occurs in the simulation results from the contact forces occurring in the simulation and the loads needed to initiate bending failure, splitting failure or local crushing (Lu et al., 2016a).

The use of NDEM and analytical fracture algorithms leads to an efficient simulator capable of modelling large spatial and temporal domains. This enables SAMS' application in both Arctic offshore structural design (in the scale of hundreds of meters) and Arctic marine operations (in the scale of tens of kilometres).

10.3 Application Examples

The following sections describe selected studies and validation exercises that have been carried out using SAMS. The first section describes a validation study of SAMS against measured full-scale data of the Icebreaker Oden sailing in broken ice conditions. The second section summarizes a study in which SAMS is used to evaluate the variability in model-scale test results. The third section describes an ongoing research project in which SAMS is used in the design of operations. SAMS is used to evaluate the effectiveness of several different ice management strategies. The fourth section describes an evaluation of glacial ice impacts on a semi-submersible structure. Lastly, the fifth section describes how SAMS has been used to evaluate the ice loads on the trawler Northguider, which grounded in the Hinlopen Strait on the 28th of December 2018. The application examples discussed in this section demonstrate how numerical modelling can be used in application related to R&D, design, and operational assistance.

10.3.1 Validation Against Full-Scale Data: Ship Transit in Broken Ice

SAMS provides a powerful tool to quantify the performance of ships transiting in different ice conditions. It allows the calculation of global ice resistance and the maximum achievable transit speed. An advantage of SAMS compared to existing empirical formulas for the calculation of ship resistance is that no idealization of the ship hull geometry or any prior assumptions on the ice accumulation and clearing patterns around the ship are needed. This is especially important for the case of broken ice, where different limiting mechanisms co-exist (i.e., limit stress, limit force, and limit momentum).

Lubbad et al. (2018) used SAMS to simulate the transit of the Swedish icebreaker "Oden" in the Marginal Ice Zone (MIZ) of the Arctic Ocean and validated the results against full-scale data from the Oden Arctic Technology Research Cruise 2015 (OATRC2015). Image processing techniques were used to digitize helicopter images of the MIZ and to create numerical ice fields for the SAMS simulations as shown in Fig. 10.2. The Oden geometry was accurately digitalized. Figure 10.3 shows the digitized Oden hull geometry.

In the simulations, Lubbad et al. (2018) applied measured full-scale propulsion forces to Oden at the centre of gravity. Furthermore, Oden was subjected to hydrodynamic resistance and wind drag forces. A visualization of transit simulations by Lubbad et al. (2018) is given in Fig. 10.4. The times series of full-scale and simulated ice loads on Oden in the surge direction are compared in Fig. 10.5. A comparison between the full-scale and simulated velocity of Oden in the surge direction is given in Fig. 10.6.

The mean simulated ice load has a deviation with the measured full-scale ice load of 4.0%. The mean simulated surge velocity has a deviation with the measured velocity of 3.3%. The differences between the means of the simulation results and the full-scale measurements were within reasonable bounds, considering the uncertainty in model input parameters, such as the ice mechanical properties, and the parameters measured in full-scale (Kjerstad et al., 2018).

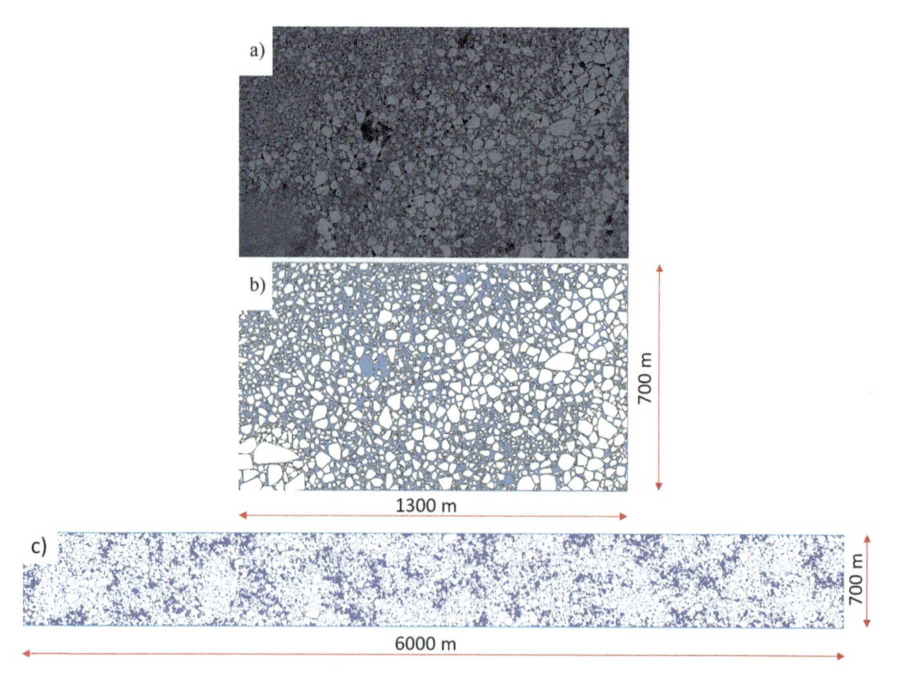

Fig. 10.2 (**a**) Initial helicopter camera image; (**b**) The digitalized ice field (700 m by 1300 m containing 2890 ice floes); and (**c**) the extended ice field (700 m by 6000 m containing 13,518 ice floes). Note that the figure is not drawn to scale (Lubbad et al., 2018)

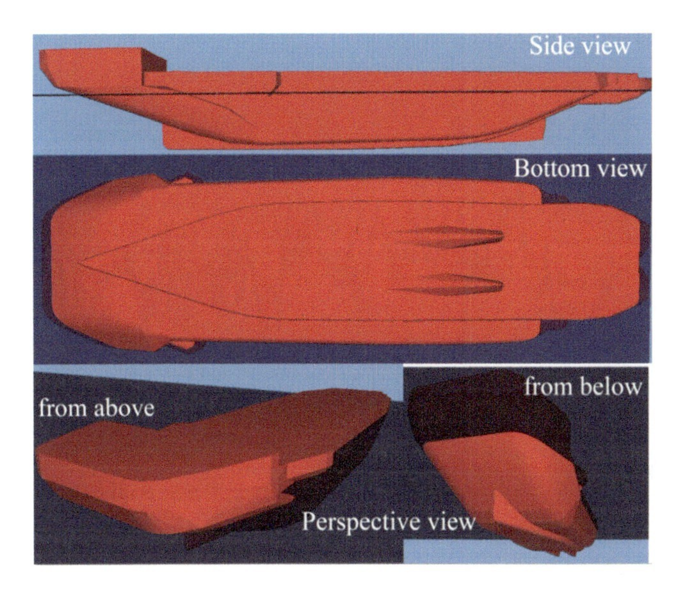

Fig. 10.3 Geometric representation of Oden in SAMS (Lubbad et al., 2018)

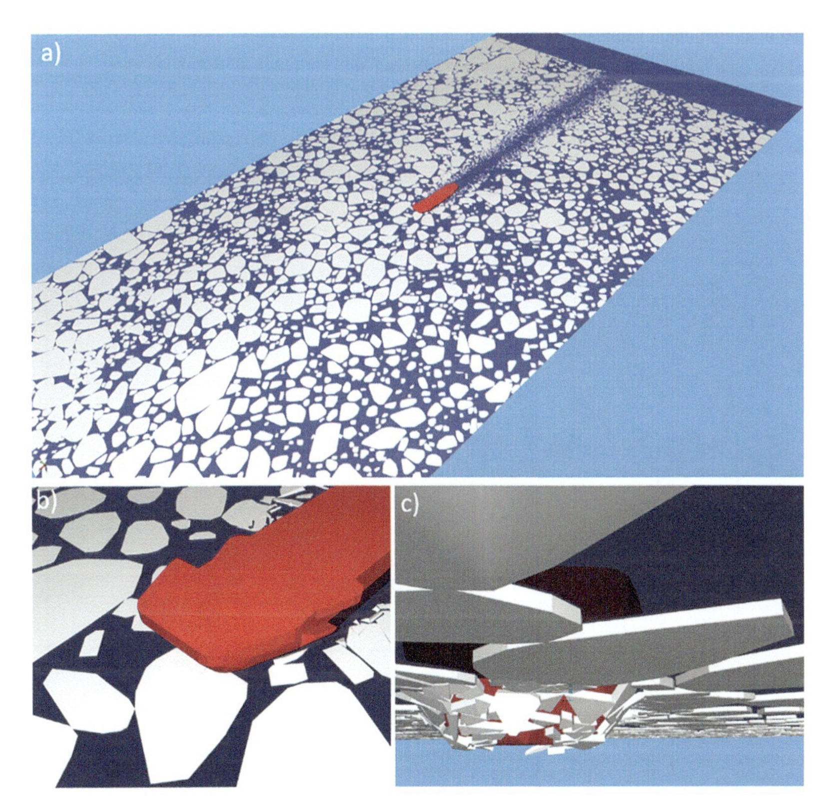

Fig. 10.4 Visualization of the icebreaker Oden transit simulation; (**a**) helicopter view; (**b**) zoom in; and (**c**) underwater view (Lubbad et al., 2018)

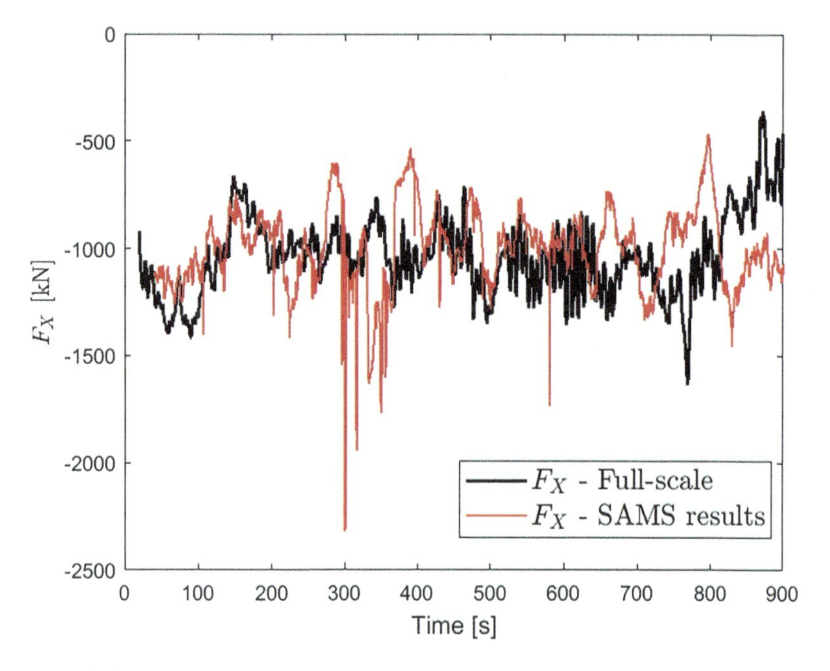

Fig. 10.5 Full-scale (black) and simulated (red) ice loads on Oden in the surge direction (Lubbad et al., 2018)

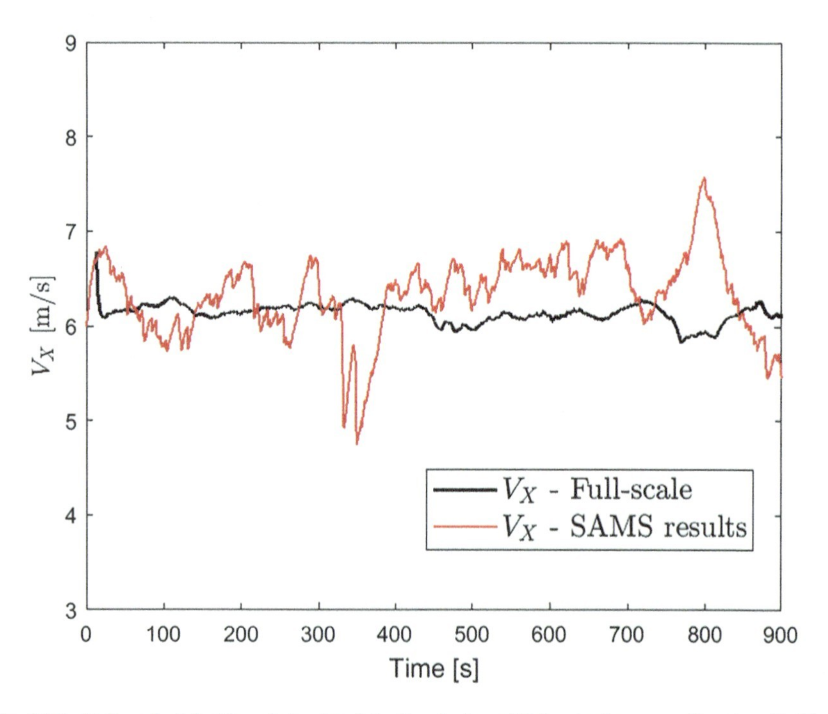

Fig. 10.6 Full-scale (black) and simulated (red) velocity of Oden in the surge direction (Lubbad et al., 2018)

The standard deviation of the simulated ice load has a deviation with the measured full-scale standard deviation of 1.3%. The standard deviation of the simulated surge velocity has a deviation with the measured velocity standard deviation of 377.8%. A possible reason for the lower standard deviation in the measured velocity is that Oden was trying to avoid areas with large ice floes or a higher ice concentration during transit or applied a higher thrust when transiting through these areas. In the numerical simulations, Oden transits with a constant heading and thrust, likely leading to a higher standard deviation in the simulated velocity. The simulated ship's acceleration and velocity directly follows from an integration of the net forces on the ship, considering the ice loads, thrust, and hydrodynamic resistance, and has been verified to be consistent with the simulated forces. More details on this validation exercise can be found in Lubbad et al. (2018).

10.3.2 Co-analysis of Broken Ice Loads with Model-Scale Tests and Numerical Simulations

In this application example, numerical simulations are used to examine the variability in the results of model-scale tests with broken ice. Model-scale tests are often used when designing structures or operations for ice-covered waters. Although model-scale tests are an important part of the design process, tests also have several shortcomings. For instance, it is difficult to scale the material properties of ice, and there is often a significant statistical uncertainty in the test results because of an insufficient test length.

Numerical simulations can be used in conjunction with model-scale tests to complement the test results. As an example, numerical simulations can be used to investigate the influence of boundary and scaling effects. In addition, numerical simulations can be used to investigate the variability and statistical significance of model-scale test results. This section describes a study in which the variability of a model-scale test result is investigated. This study was conducted using publicly available data from model-scale tests performed at the Hamburg Ship Model Basin (HSVA) and was published in van den Berg et al. (2020). Please refer to this publication for more details.

The model-scale tests used in this study were performed as part of the European Community's HYDRALAB IV research program. The tests were performed with a 4-legged, vertical-walled structure in several broken and level ice conditions. The primary research goal of the tests was to identify a relationship between the level ice loads and the broken ice loads for a range of ice thicknesses and ice concentrations. The mean and standard deviation of the global ice load on the structure were assessed. The tests, and a preliminary analysis of the test results, are described by Hoving et al. (2013).

Because of the limited test length, the statistical properties of the test results were influenced by individual random interaction events. It was also observed that the conditions changed during the test runs. As the structure progressed, the ice concentration ahead of the structure increased, and the interaction process appeared to be influenced by the tank walls. Because of these processes, it was difficult to determine the reliability of the obtained test results.

The model-scale tests were partly reproduced using SAMS in order to determine the potential variability in test results due to changes in the initial positions of ice floes. The broken ice fields from the model-scale tests were digitized from top-view images of the ice tank. Figure 10.7 shows the initial ice conditions of one of the tests analysed using SAMS.

In order to assess the possible variability in the test results, the model-scale test was numerically reproduced 20 times using SAMS. Each time, the initial positions of the ice floes was slightly varied by introducing an initialization phase in which each ice floe was initialized with a random linear horizontal velocity between 0 and 0.2 m/s. Otherwise, the conditions were the same. The simulation results represent possible alternative outcomes of the model-scale tests.

The simulation results show that there may be a significant influence of the initial ice floe positions on the mean ice load and load standard deviation measured in the ice tank test. Figure 10.8 shows time series of the ice load on the structure as

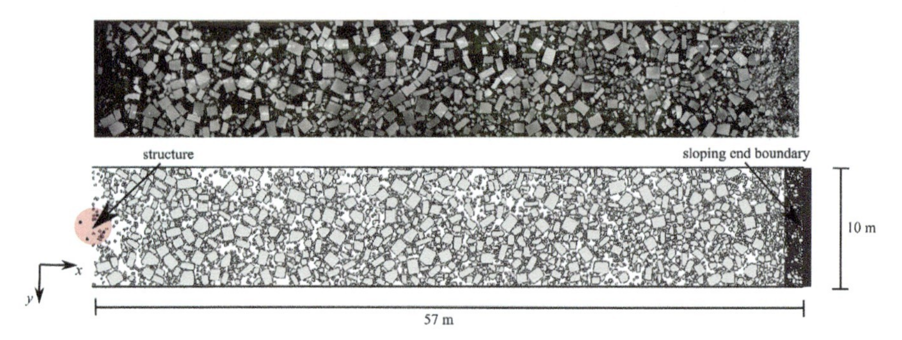

Fig. 10.7 Top-view photo of the ice conditions during the ice tank test (top) and the digitized broken ice field used in one of the numerical simulations (bottom), highlighting the multi-leg structure being tested

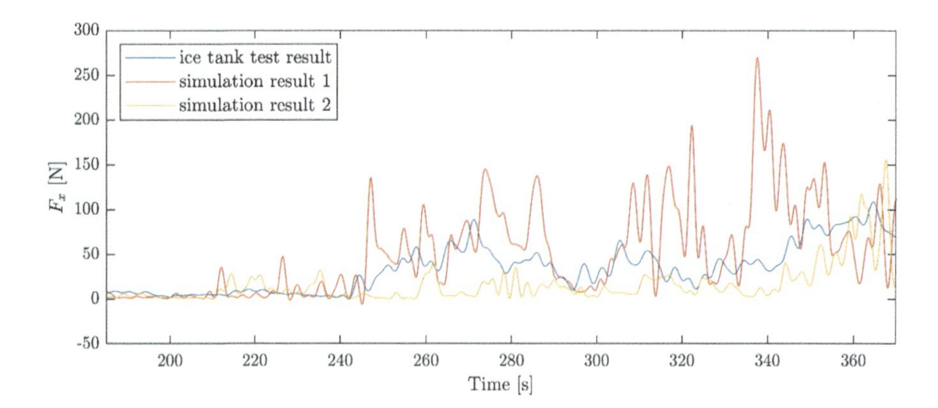

Fig. 10.8 Time series of the load on the structure as measured in the ice tank tests and two simulation results of the same test, representing possible alternative outcomes. Results are filtered for clarity

measured in the ice tank test and the simulated ice loads from two numerical simulations of the ice tank test, representing alternative test outcomes. Figure 10.9 shows the mean and standard deviation of the global ice load time series measured in the ice tank test, in comparison to the mean ice loads and load standard deviations resulting from the 20 numerical simulations of the test.

The maximum mean load from the 20 simulations is more than twice as high as the minimum mean simulated load. The processes responsible for the variability in the simulation results can be identified by studying the simulation visualizations. An example of a process that may influence the mean ice load and load standard deviation is the jamming of ice floes between the structure legs. The numerical simulations show that jamming of ice occurs in some simulations, but not in others, resulting in major differences in the simulation outcome. A similar result variability is expected if the ice tank tests would be repeated several times. Figure 10.10 shows three examples of jamming behaviour.

Based on the results of this study, it is recommended that the possible variability in the results of ice tank tests with broken ice is further assessed by performing multiple repetitions of the same test during ice tank testing campaigns.

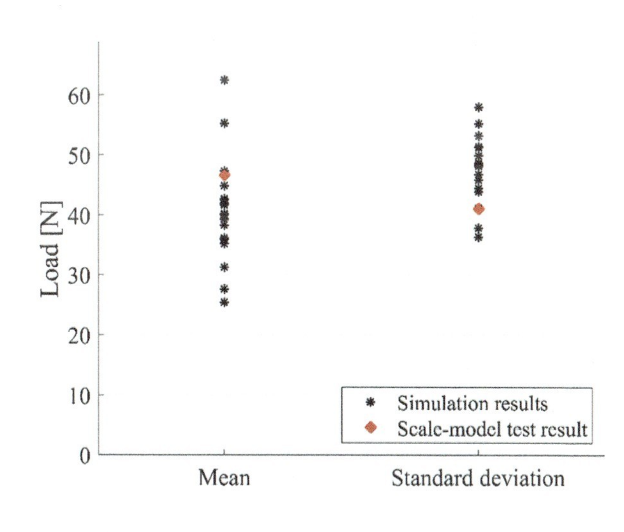

Fig. 10.9 Mean and standard deviation of the global ice load on the structure measured in the model-scale test, and the means and standard deviations of the simulated load-time series

Fig. 10.10 Jamming of ice between the structure legs in the model-scale tests (left) and jamming in two instances of the numerical simulations; significant jamming (middle) and no jamming (right)

10.3.3 Operation Design: Assessment of Ice Management Strategies

This example demonstrates how numerical simulations can be used in the design and performance evaluation of ice management operations. Ice management (IM) is a comprehensive and interconnected system that involves 'detection', 'tracking', 'forecasting', 'decisionmaking', and eventually 'breaking/fracturing' the identified threatening ice features (ISO 19906, 2018). Currently, practical questions such as 'how many icebreakers are needed?' and 'how to deploy the available icebreaker fleet to effectively defend an offshore structure?' are often addressed with kinematic models (Hamilton et al., 2011a, b; Hamilton, 2011). IM operations are currently primarily based on operational experiences with few quantifiable criteria or guidance available. This is largely due to the following two factors: 1) limitations in the ice surveillance system, i.e., the current incapability to accurately retrieve a structure's ambient ice information (floe size distribution and thickness) in real time (Lu et al., 2016b), and 2) lack of physically based models, which can efficiently characterize the ice–structure interaction processes at operational scale (Lu et al., 2018a). This section presents the results of a method for quantifying the effectiveness for a 'racetrack' IM pattern using SAMS by a set of performance indicators. Figure 10.11 shows a snapshot of the simulated IM operation in SAMS. More details on the methods and results of this study can be found in Bjørnø et al. (2020).

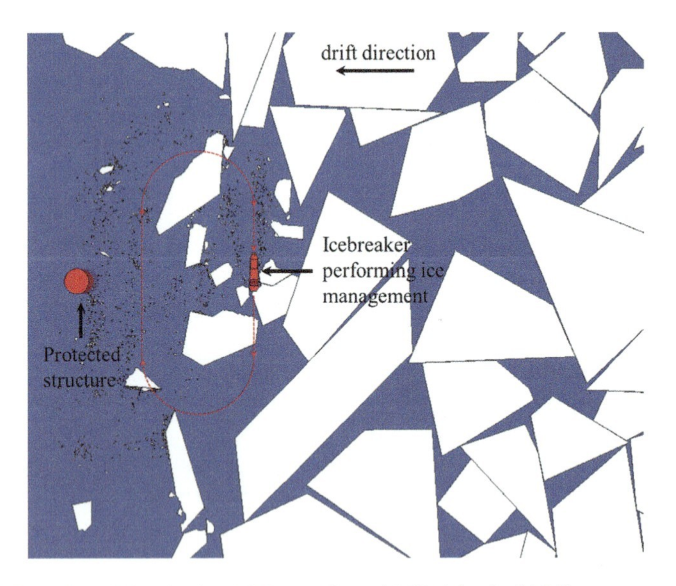

Fig. 10.11 Snapshot of the simulated IM operation with IB Oden in SAMS

10.3.3.1 Performance Indicators

The ice management performance can be considered by several indicators, depending on the specific operation. The floe size reduction is the one measure of performance that is used in many previous studies. Key performance indicators (KPIs) are defined to measure the performance of an ice management operation. The following KPIs are used in this section;

- The mean ice load on the protected structure.
- The number of ice floes that exceeds a given threshold/limit.

These two KPIs are just two out of many that can be used. Other KPIs such as the total work done by the icebreaker, total fuel consumption of the icebreaker, maximum ice floe size after IM, maximum momentum of ice on the protected structure and floe size distribution may also be used. These KPIs (and others) can be combined with different weights to create a final score for different IM strategies, depending on which KPI is valued the most.

Numerical simulations enable the testing of a wide range of ice management strategies and performance indicators. Simulations can be used, in combination with operational trials, in the design and operational phases of ice management operations.

10.3.3.2 Assessment of the Ice Management Results

The performance of the 'racetrack' ice management operation described in Sect. 3.3 is assessed according to the KPI's of mean ice load on the protected structure and the number of ice floes exceeding a given threshold.

Figures 10.12 and 10.13 show the KPI values as a function of ice thickness and ice drift velocity resulting from the simulated ice management operation. It is possible to compare the effect of changes to the ice management operation, like a different ice management pattern, on the defined KPIs. Contrary to the expected result, the KPI value of mean ice loads shows a decreasing trend from an ice thickness of 1.7 m to an ice thickness of 1.8 m. This mean load decrease is attributed to randomness in the numerical modelling results. It is expected that the decreasing trend would disappear if the simulation duration would be increased.

Further details on the quantification of effectiveness of different ice management patterns can be found in Bjørnø et al. (2020).

10.3.4 Design Evaluation: Analysis of Glacial Ice Impacts

The challenges and limitations of full-scale measurements and model-scale tests are especially relevant in the analysis of glacial ice impacts. Impacts are mainly avoided

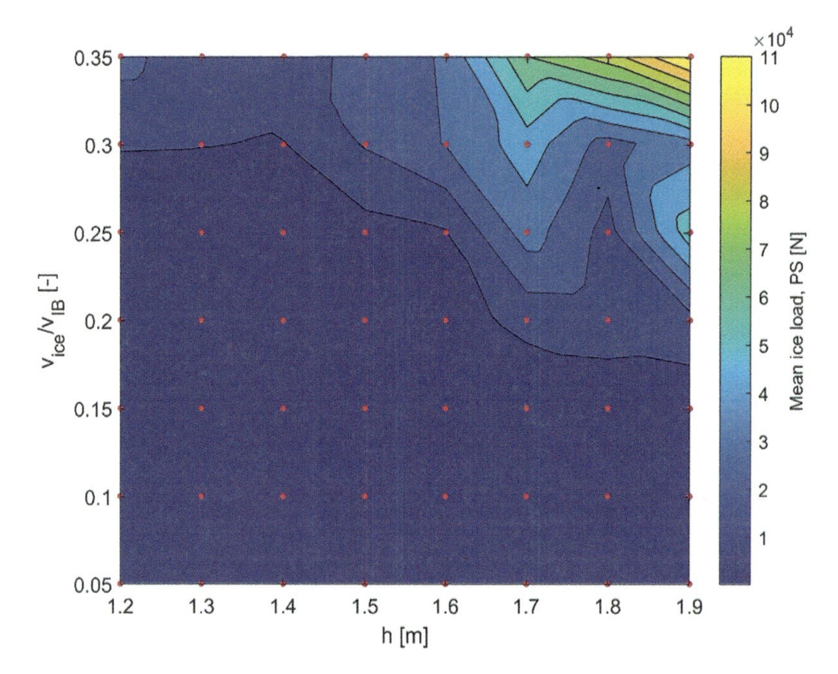

Fig. 10.12 Mean ice load on the protected structure as a function of ice thickness and drift speed, in 80% concentration broken ice

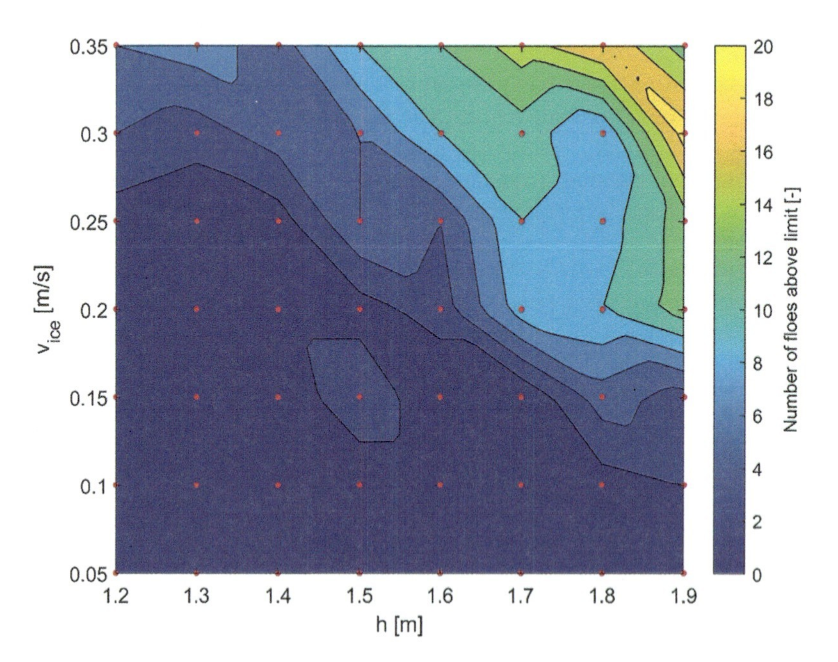

Fig. 10.13 Number of ice floes that exceeds a given limit as a function of ice thickness and drift speed, in 80% concentration broken ice

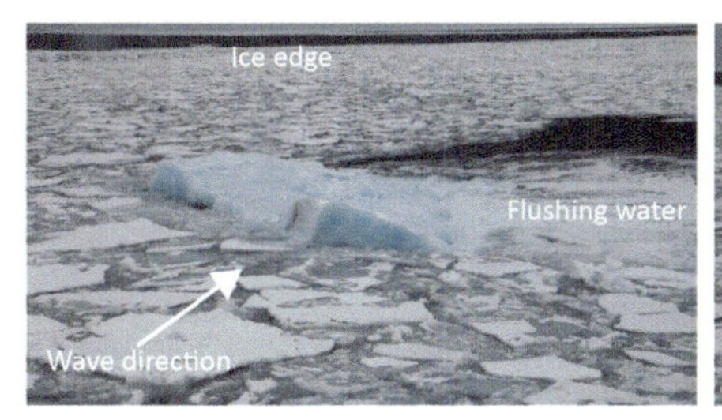

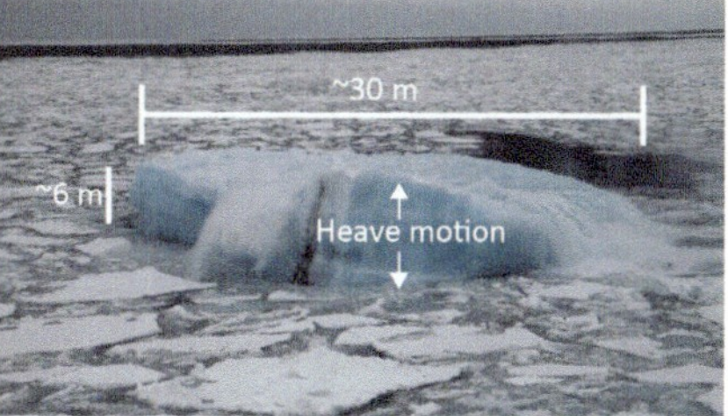

Fig. 10.14 Illustration of a glacial ice feature embedded in a broken ice field and with its motion excited by waves (video by Løset, S)

in full-scale, leading to limited measured data. Glacial ice impacts are difficult to model in model-scale tests because it involves structure deformation as well as crushing of the glacial ice. SAMS has been used in combination with Nonlinear Finite Element Analysis (NLFEA) software to analyse the probabilities and effects of glacial ice impact on a semi-submersible structure.

In the Barents Sea, icebergs with origins from Franz Josef Land, Nordaustfonna, Edgeøya and Novaya Zemlya drift southward mainly due to forcing from current (Løset, 1993). In their drifting course, thermal and wave erosion takes place, leading to morphological changes. Icebergs eventually become smaller glacial ice features with a more rounded shape. Small glacial ice features are more susceptible to wave-driven motions. Figure 10.14 shows an example. The images show a 30 m long glacial ice feature with a strong heave and pitch motion (around 6 m) due to wave action. Compared to the linear drifting velocity (normally smaller than 1 m/s (Yulmetov et al., 2013)), the wave driven velocity is much higher (Lu and Amdahl, 2019). As the impact energy is scaled with velocity squared, the impact from small glacial ice features under wave-driven motion is by no means 'small'. In addition, these small glacial ice features are much more difficult to detect by current ice surveillance systems, e.g., marine radar detection (Lu et al., 2019), and it is more challenging to apply ice management operations. Therefore, it is important to understand the risk and effects of impacts from wave-driven small glacial ice features on structures operating in regions where such features might occur.

The analysis of glacial ice impact probabilities and effects consist of the following phases:

1. Identification of glacial ice probability of occurrence.
2. Identification of critical sea states.
3. Analysis of glacial ice motions under wave forcing.
4. Analysis of impact probability as a function of structure location.
5. Analysis of impact damage.

This section focusses on steps 4 and 5. Further details on the procedures used in the design for glacial ice impacts can be found in (Lu et al., 2019).

10.3.4.1 Impact Probability Assessment

The results of the glacial ice feature's motion analysis (Step 3) are the distribution of impact velocity and impact probability as a function of height. This data is visualized in Fig. 10.15. In addition to the glacial ice feature's motion, the geometry of the structure and the ice feature also influences the impact probability. The influence of geometry on the impact probability and impact energy across the structure was analysed using SAMS. Simulations were performed with 1800 different initial conditions, covering a uniform grid of drift directions and glacial ice positions. Figures 10.16 and 10.17 show the simulated conditions. Combining the data on the glacial ice trajectory from Step 3, and the data on the influence of glacial ice and structure geometry from Step 4, an impact energy map can be constructed. The impact energy map from a glacial ice feature with a mass of 765 metric tons in a sea state with a significant wave height of 13.8 m is shown in Fig. 10.18.

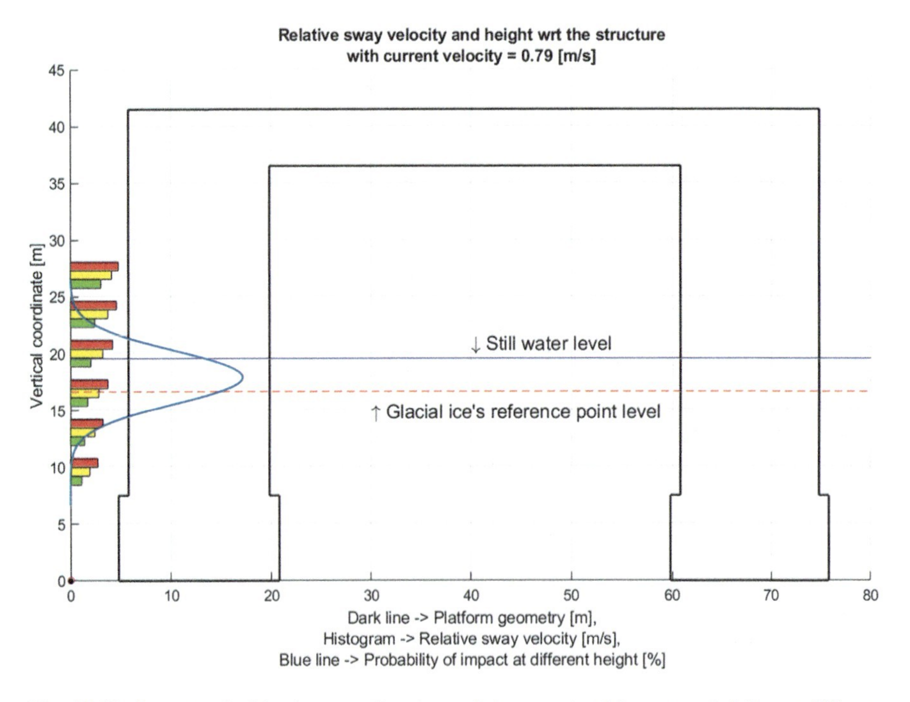

Fig. 10.15 Impact velocities in sway direction and the associated impact probability at different heights with reference to the structure (different colour bars on the left represent the impact velocity with different level of non-exceedance, red: 99%, yellow: 90% and green: 50% non-exceedance level)

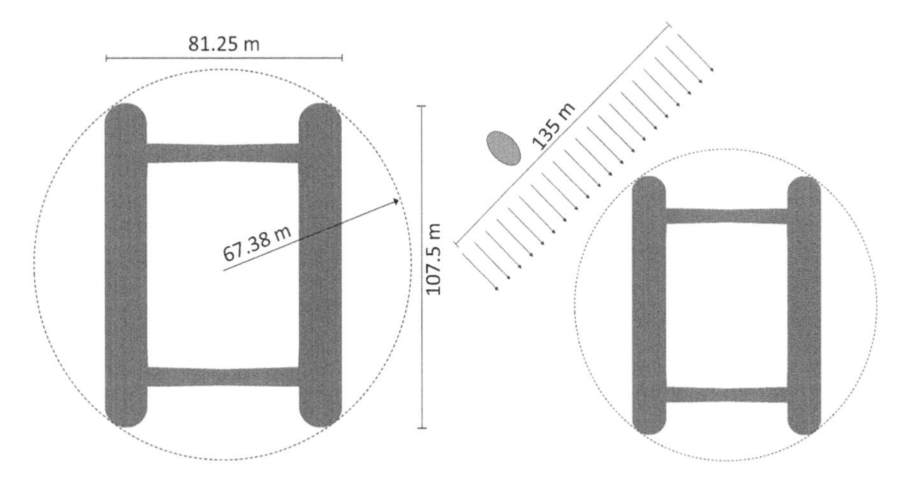

Fig. 10.16 Top view of the semi-submersible structure, showing main structure dimensions and the radius of the minimum bounding circle (left). Ice feature horizontal offset as applied for each impact direction and impact height (right)

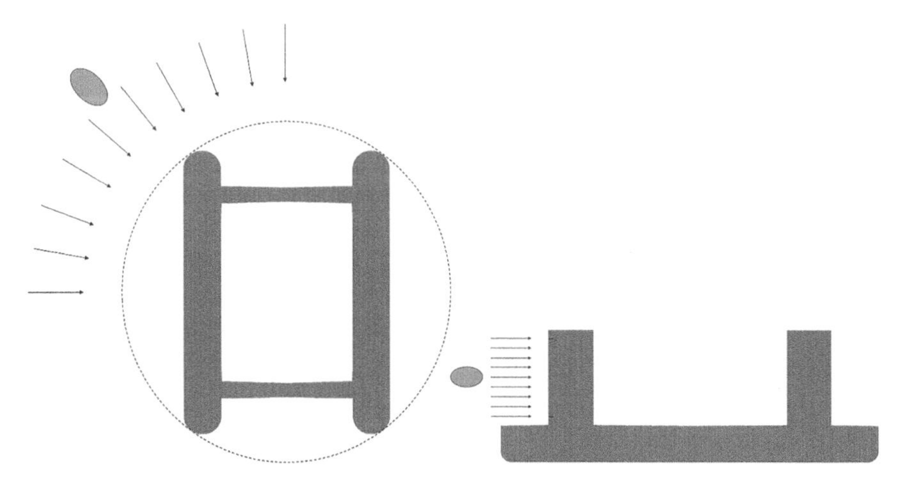

Fig. 10.17 Simulated drift directions (left) and vertical offsets (right)

10.3.4.2 Structural Damage Assessment

The energy map is constructed using analysis results regarding the impact likelihood and the impact velocity. Using the data provided by the energy map, critical impact cases were identified. For these critical cases, non-linear finite element analyses were performed in order to assess the structural damage that might occur. Figure 10.19 shows the result of such an analysis. The results show that the damage to the structure is related to the impact energy and the local geometries of the glacial ice feature and structure. Based on the numerical simulation results, a critical local

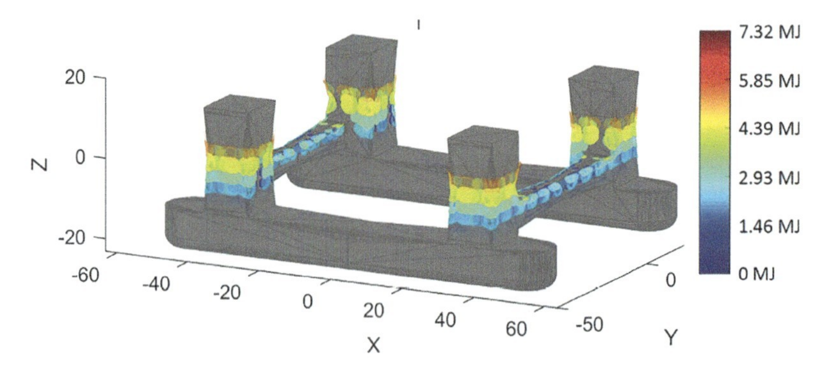

Fig. 10.18 Impact energy map of a glacial ice feature with a mass of 765 metric tons in a sea state with a significant wave height of 13.8 m

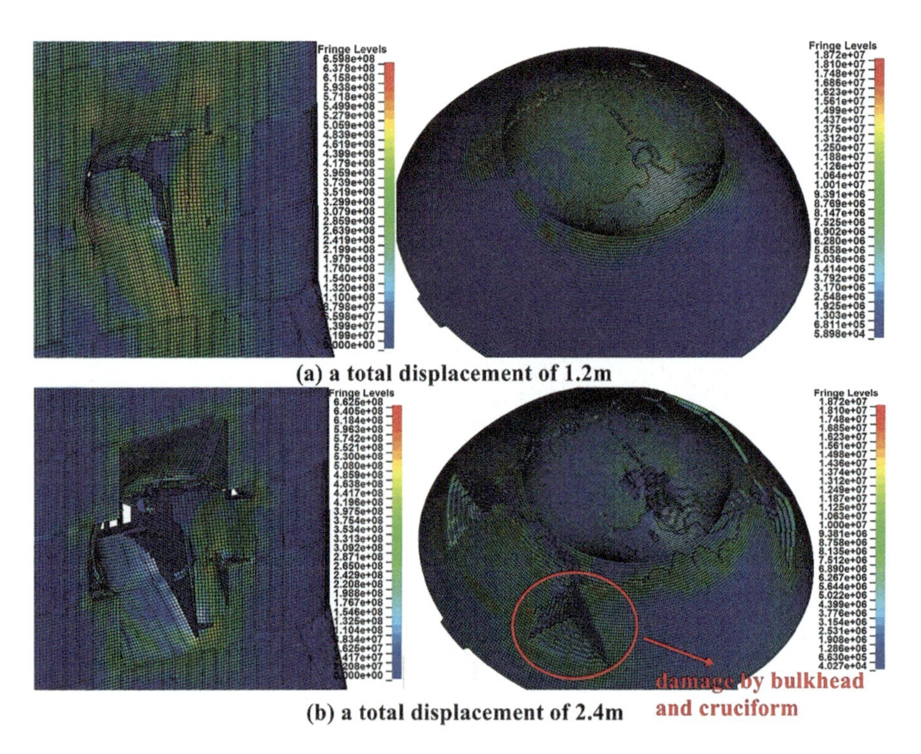

Fig. 10.19 Assessment of structural damage from iceberg impact on a semi-submersible structure using NLFEA

iceberg geometry can be defined. The critical geometry is defined as the geometry that leads to the most structural damage for a given impact energy. Further details on the analysis method can be found in Yu et al. (2020) and Lu et al. (2019).

10.3.5 Estimation of Ice Actions on the Grounded Trawler Northguider in the Hinlopen Straight

The last application example demonstrates the use of numerical simulations in the decisionmaking process related to a salvage operations of a grounded fishing trawler. The trawler Northguider was grounded at Sparreneset in the Hinlopen Strait on December 28th, 2018. Drifting sea ice is usually present in the Hinlopen Strait in the late winter and spring. Possible displacement of the ship due to forces from the drifting sea ice was a concern because of a submarine valley with a water depth of 400–500 m that is in close vicinity of the grounded ship's location. It was feared that the ship might be pushed into the deep water, making salvage difficult. SAMS was used to assess the possible ice loads on the grounded ship. Figure 10.20 shows the location of the grounded ship.

Two ice conditions were simulated: loading of the ship by a field of broken ice floes and loading of the ship by a single large ice floe. Figure 10.21 shows a visualization of the simulations performed for the broken ice condition. Figure 10.22 shows a visualization of the simulations performed for the single large floe condition.

Fig. 10.20 Northguider's grounding location with reference to Svalbard

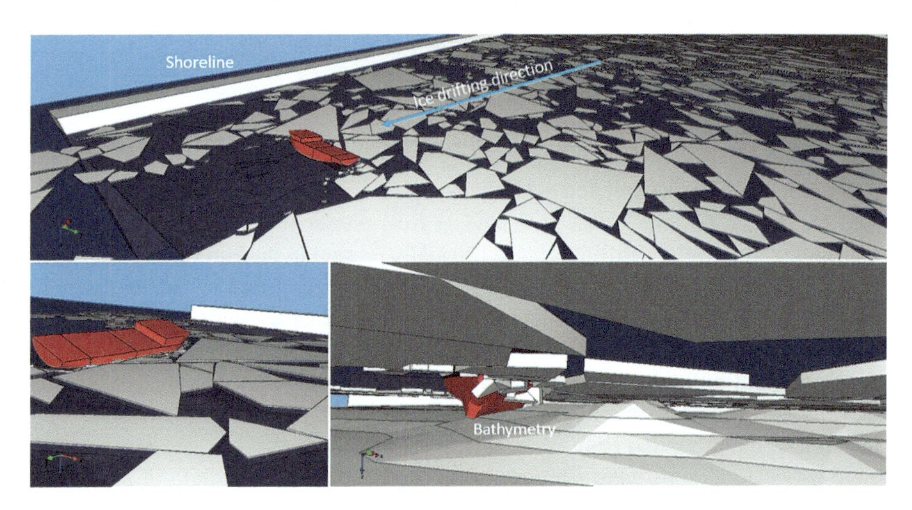

Fig. 10.21 Visualization of the vessel loaded by a broken ice field

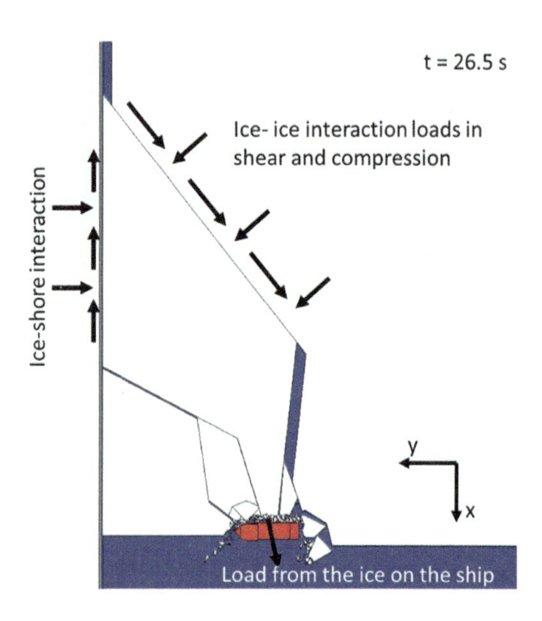

Fig. 10.22 Vessel loaded by a single large ice floe (governing loading condition)

The simulation results show that interaction with a single large ice floe is the governing loading scenario. The maximum load in this scenario is governed by the crushing strength of the ice, in combination with the ice-ship interaction area. The maximum load predicted by SAMS would likely displace the vessel from its stranding location. There was insufficient environmental data to quantify the likelihood of a loading event that would push the grounded trawler into deeper water. However, the simultaneous occurrence of the specific conditions (ice drift direction, ice floe position, shape and size) that could lead to an adverse displacement of the grounded Trawler was estimated to be unlikely.

Considering the load assessment conducted with SAMS and other health, safety and environment (HSE) considerations related to a salvage operation in the winter months, it was decided to postpone the salvage operation until next summer season. By the time the salvage operation was started, the vessel had tilted further onto the rocks on which it was grounded, likely as a result of ice loading. The change in the vessel tilt further complicated the salvage operation.

10.4 Model Validation and Calibration

In the previous sections, we have discussed several cases in which SAMS was applied in R&D, design or operational assistance related to ships, structures or operations in ice-covered waters. These cases demonstrate how numerical simulations can be used as a valuable additional analysis tool. When using numerical simulations, it is important to critically assess the validity and limitations of the simulation tool, and of the obtained simulation results. The reliability and limitations can be more accurately quantified the more validation cases are considered. The validation and calibration of SAMS is an ongoing process. Validation studies have been conducted against both model-scale and full-scale data sets. The published validation studies conducted to far are summarized in Table 10.2.

In addition to the studies listed in Table 10.2, many unpublished validation studies have been part of the software development process, including, but not limited to, checks on (statistical) result convergence with decreasing time step size and mathematical accuracy assessments on the continuous body dynamics (whereas the mathematical accuracy check listed in the table involves a discontinuous contact). As part of the code verification process, a code quality review by an external software company was conducted. The following challenges are encountered in the validation and calibration process:

Table 10.2 Published validation studies conducted with SAMS

Case	References
A full-scale ship transit in a broken ice field	Lubbad et al. (2018)
Comparison of simulation results against measured icebreaker performance in level ice	Raza et al. (2019)
A moored conical structure in model ice	Tsarau et al. (2018)
A jack-up-structure in model ice	van den Berg et al. (2020)
A ship in model ice	van den Berg et al. (2020)
Mathematical accuracy assessment for a single discontinuous contact	van den Berg et al. (2018)
Energy balance assessment	van den Berg et al. (2019a, b)
Statistical convergence of results	van den Berg et al. (2019b)

- Calibration and validation data are not always available in open literature.
- Existing data sets are often incomplete, i.e., not all parameters influencing the studied interaction processes are measured. The values of unknown parameters must be estimated by engineering judgement or must be determined by model calibration.
- There is often a substantial uncertainty and randomness in the physical properties that are measured, resulting from inhomogeneity in the spatial and temporal test conditions.

The uncertainty and randomness in measured data is a limitation to the validation of sub-processes of the global ice-structure interaction process. For all sub-processes, there is a level of detail where a theoretically more accurate approximation of the process will not lead to measurably more accurate simulation results when compared to measured data. For example, as mentioned in Sect. 2, a more accurate approximation of the actual ice floe geometries in the calculation of splitting forces has a limited effect on the main simulation results, which is generally the global load on a ship or structure. The areas where numerical simulations can offer most added benefit are also the hardest to validate; the fewer measured data there is available, the more numerical models can contribute, but the more difficult it is to validate the simulation results.

For the model applications presented in this chapter, the uncertainties resulting from a lack of data are described in more detail in the references included in each section. These uncertainties are considered in the interpretations and recommendations following from the numerical simulation results. Because of the validation and calibration challenges mentioned in this section, numerical simulations should be seen, at this stage, as an analysis tool that can be used in combination with model-scale tests and empirical formulas, and not as a replacement of model-scale tests and/or empirical formulas.

10.5 Conclusions

This chapter describes, by discussing various application examples, how numerical simulations can be used for the better understanding and more accurate assessment of interaction processes and resulting ice loads on structures, ships and operations in ice-covered waters. The application examples cover applications in the R&D, design, and operational assistance phases. Each phase requires a different level of specificity and computational efficiency.

A numerical simulator, if properly verified and validated as described and defined in Oberkampf, W., & Roy (2010) can be a valuable analysis tool that can improve the insight on the ice-ship and ice-structure interaction processes, thereby assisting in the development of more robust procedures, safer and more economic designs, and better-informed operational decisions. Current design practice relies mainly on model-scale tests and empirical formulas. Compared to model-scale tests, numerical

Table 10.3 Synthesis of main results obtained by numerical simulations for the application cases discussed

Case	Main results
3.1: Validation against full-scale data; ship transit in broken ice	Validated capability of the numerical simulator SAMS.
3.2: Co-analysis of broken ice loads with model-scale tests and numerical simulations	Better understanding of the variability in the results of ice tank tests with broken ice.
3.3: Assessment of ice management strategies	Improved quantification tools of ice management effectiveness.
3.4: Analysis of glacial ice impacts	Better understanding of impact probabilities and structural damage from glacial ice impacts.
3.5: Estimation of ice actions on the grounded trawler Northguider in the Hinlopen straight	Guidance on the operational planning of the salvage operation.

simulations generally have a shorter preparation time and lower test costs per tested condition. Therefore, numerical simulations can be used to test a wider range of conditions and interaction scenarios than what would be possible with model-scale tests alone. However, numerical models are not yet at the stage where model-scale tests can be replaced. Rather, simulations should be performed to complement and refine model-scale tests.

The application examples discussed in this chapter were performed with the numerical simulator for ice-structure and ice-ship interactions SAMS. SAMS is a 3D multi-body simulator which uses the non-smooth discrete element method. Both ice and structure are fully dynamic. Environmental forces from wind, current and propeller flow are considered. Failure of ice is implemented using analytical solutions. The combination of using the non-smooth discrete element method with an analytical approach to ice failure leads to efficient simulations. The efficiency of the simulations enables the simulation of large spatial and temporal domains.

Five application examples have been discussed: A comparison of simulation results to full-scale ship transit data, a co-analysis of ice tank tests with numerical simulations, the use of SAMS in evaluating an ice management process, the use of SAMS to model the interaction between a semi-submersible structure and a glacial ice feature, and the assessment of maximum ice loads on a grounded trawler. These application examples demonstrate how numerical modelling can be used in R&D, design and operational assistance related to ships, structures and operations in ice-covered waters. The main results for each case are summarized in Table 10.3.

References

ArcISo [WWW Document]. (2020). URL arciso.com (accessed 9.29.20).

Bjørnø, J., van den Berg, M., Lu, W., Skjetne, R., Lubbad, R., & Løset, S., 2020. Quantifying icebreaker performance in ice management operations by high-fidelity numerical simulations. Proc. Int. Offshore Polar Eng. Conf. 2020-Octob, 824–830.

Cundall, P. a., & Strack, O. D. L. (1979). A discrete numerical model for granular assemblies. *Géotechnique, 29*, 47–65. https://doi.org/10.1680/geot.1979.29.1.47

Hamilton, J.M. (2011). The challenges of deep-water Arctic development, in: Proceedings of the twenty-first international offshore and polar engineering conference. Maui, Hawaii.

Hamilton, J., Holub, C., Blunt, J., Mitchell, D., & Kokkinis, T. (2011a). Ice Management for Support of Arctic Floating Operations, in: OTC Arctic Technology Conference.

Hamilton, J., Holub, C.J., & Blunt, J., (2011b). Simulation of ice management fleet operations using two decades of Beaufort Sea ice drift and thickness time histories. Proc. Int. Soc. Offshore Polar Eng. (ISOPE).

Hoving, J.S., Vermeulen, R., Mesu, A.W., & Cammaert, G. (2013). Experiment-based relations between level ice loads and managed ice loads on an Arctic jack-up structure, in: Proceedings of the 22nd International Conference on Port and Ocean Engineering under Arctic Conditions. Espoo, Finland.

ISO 19906. (2018). *Petroleum and natural gas industries: Arctic offshore structures*. Int. Organ. Stand.

Kjerstad, Ø. K., Lu, W., Skjetne, R., & Løset, S. (2018). A method for real-time estimation of full-scale global ice loads on floating structures. *Cold Regions Science and Technology, 156*, 44–66.

Løset, S. (1993). Thermal energy conservation in icebergs and tracking by temperature. *Journal of Geophysical Research, 98*, 10001–10012.

Lu, W. (2014). *Floe ice – sloping structure interactions*. Dr. thesis. Norwegian University of Science and Technology, Trondheim.

Lu, W., & Amdahl, J. (2019). Glacial ice and offshore structure impacts under wave and current excitation, in: Proceedings of the 25th International Conference on Port and Ocean Engineering under Arctic Conditions. Delft, The Netherlands.

Lu, W., Lubbad, R., & Løset, S. (2015a). In-plane fracture of an ice floe: A theoretical study on the splitting failure mode. *Cold Regions Science and Technology, 110*, 77–101.

Lu, W., Lubbad, R., & Løset, S. (2015b). Out-of-plane failure of an ice floe: Radial-crack-initiation-controlled fracture. *Cold Regions Science and Technology, 119*, 183–203.

Lu, W., Lubbad, R., Løset, S., & Kashafutdinov, M. (2016a). Fracture of an ice floe: Local out-of-plane flexural failures versus global in-plane splitting failure. *Cold Regions Science and Technology, 123*, 1–13.

Lu, W., Zhang, Q., Lubbad, R., Løset, S., & Skjetne, R. (2016b). A shipborne measurement system to Acquire Sea ice thickness and concentration at engineering scale. *Arctic Technology Conference, 2016*.

Lu, W., Heyn, H.-M., Lubbad, R., & Løset, S. (2018a). A large scale simulation of floe-ice fractures and validation against full-scale scenario. *International Journal of Naval Architecture and Ocean Engineering, 10*, 393–402.

Lu, W., Lubbad, R., & Løset, S. (2018b). Parallel channels' fracturing mechanism during ice management operations. Part II: Experiment. *Cold Regions Science and Technology, 156*, 117–133.

Lu, W., Lubbad, R., Shestov, A., & Løset, S. (2018c). Parallel channels' fracturing mechanism during ice management operations. *Cold Regions Science and Technology, 156*, 102–116.

Lu, W., Yu, Z., van den Berg, M., Monteban, D., Lubbad, R., Hornnes, V., Amdahl, J., Løset, S., & Kim, E. (2019). Loads, design and operations of floaters in the North. Stavanger, Norway: Petroleumstilsynet (Petroleum Safety Authority Norway) report. URL https://www.ptil.no/contentassets/d16ac78b023f4ad5b06225856db6950e/rapport%2D%2D-laster-design-og-operasjon-av-flytere-i-nordarmadene.pdf

Lubbad, R., & Løset, S. (2011). A numerical model for real-time simulation of ship-ice interaction. *Cold Regions Science and Technology, 65*, 111–127.

Lubbad, R., Løset, S., Lu, W., Tsarau, A., & van den Berg, M. (2018). An overview of the Oden Arctic Technology Research Cruise 2015 (OATRC2015) and numerical simulations performed with SAMS driven by data collected during the cruise. *Cold Regions Science and Technology, 156*, 1–22.

NSIDC – National Snow and Ice Data Center [WWW Document], 2020. URL https://nsidc.org/data/seaice_index (accessed 9.3.20).

Oberkampf, W., & Roy, C. (2010). *Verification and validation in scientific computing*. Cambridge University Press. https://doi.org/10.1017/CBO9780511760396

Raza, N., van den Berg, M., Lu, W., & Lubbad, R. (2019). Analysis of oden icebreaker performance in level ice using Simulator for Arctic Marine Structures (SAMS), in: Proceedings of the 25th International Conference on Port and Ocean Engineering under Arctic Conditions. Delft, The Netherlands.

Tsarau, A., Lubbad, R., & Løset, S. (2017). A numerical model for simulating the effect of propeller flow in ice management. *Cold Regions Science and Technology, 142*, 139–152.

Tsarau, A., van den Berg, M., Lu, W., Lubbad, R., & Løset, S. (2018). Modelling results with a new Simulator for Arctic Marine Structures – SAMS, in: ASME 2018 37th International Conference on Ocean, Offshore and Arctic Engineering. ASME.

van den Berg, M. (2019). Discrete numerical modelling of the interaction between broken ice fields and structures. Doctoral thesis, Norwegian University of Science and Technology. URL https://ntnuopen.ntnu.no/ntnu-xmlui/handle/11250/2630787

van den Berg, M., Lubbad, R., & Løset, S. (2018). An implicit time-stepping scheme and an improved contact model for ice-structure interaction simulations. *Cold Regions Science and Technology, 155*, 193–213.

van den Berg, M., Lubbad, R., & Løset, S. (2019a). The effect of floe shape on the interaction of vertical-sided structures with broken ice, in: Proceedings of the 25th international conference on port and ocean engineering under Arctic conditions. Delft, The Netherlands.

van den Berg, M., Lubbad, R., & Løset, S. (2019b). The effect of ice floe shape on the load experienced by vertical- sided structures interacting with a broken ice field. *Marine Structures, 65*, 229–248. https://doi.org/10.1016/j.marstruc.2019.01.011

van den Berg, M., Lubbad, R., & Løset, S. (2020). Repeatability of ice-tank tests with broken ice. *Marine Structures, 74*, 102827. https://doi.org/10.1016/j.marstruc.2020.102827

Yu, Z., Lu, W., van den Berg, M., Amdahl, J., & Løset, S. (2020). Glacial ice impacts, part II: Damage assessment and ice-structure interactions in accidental limit states (ALS). *Marine Structures, 75*, 102889.

Yulmetov, R., Løset, S., & Eik, K.J. (2013). Analysis of drift of sea ice and icebergs in the Greenland Sea. Proc. 22nd Int. Conf. Port Ocean Eng. under Arct. Cond.

Chapter 11
(Action): Sustainable Arctic Ocean

Manuel Barange

Minister, secretary, dignitaries, ladies and gentlemen, let me just start by asking you to reflect on two important realities in the world today. First is that for the first time in history we are moving towards a future that in many ways we can predict. Of course I am referring to climate change. We don't know the magnitude of it in the future, but we know it will be warmer with changes in precipitation, in storms, in glacier and ice cover. Never before in history have we had the privilege and responsibility of this level of knowledge. Second, is that we have never been more aware of the enormity of the impact of the human species. I am not just referring to population size, but also that social media and information overload have amplified an overwhelming feeling of crowdedness and hopelessness. One of the consequences of this is that we amplify our differences rather than our commonalities, and that we easily challenge the validity of your solution if it is not my solution. You may be wondering what this has to do with arctic fisheries, but just bear with me. Last week an academic paper was published in the Proceedings of the National Academy of Sciences of the United States of America (PNAS) showing that those fish stocks around the world subject to scientific assessments have on average been growing in biomass and on average are in good health. You might be surprised by this because it was not reported in the media. This result shows that fisheries management works, giving credibility to the fisheries management and governments around the world that are willing to take strong action. This is important for the Arctic because Arctic countries have by and large a history of exploiting their resources sustainably and cooperatively. It is not the case in many parts of the world, specially in places subject to political instability, poverty and hunger, where I have spent most of my professional time.

M. Barange (✉)
FAO Fisheries and Aquaculture Division (NFI), Food and Agriculture Organization of the United Nations (FAO), Rome, Italy
e-mail: manuel.barange@fao.org

P. A. Berkman et al. (eds.), *Building Common Interests in the Arctic Ocean with Global Inclusion, Volume 2*, Informed Decisionmaking for Sustainability,
https://doi.org/10.1007/978-3-030-89312-5_11

Now, there are two main things that will happen in the Arctic and subartic fisheries in the era of climate change. First, resources will shift distributions. This means new species will arrive, like hake and tuna, in Norway already. While some will move out. And second, species productivity will change, with some benefiting from enhanced primary production and ocean warming - for example, cod in the Barents Sea -, while others will suffer. Overall, science tells us that climate change is expected to improve fish yields in arctic and subartic regions, while the opposite will be experienced in the tropics. But these potential positive impacts among a myriad of negative ones are not assured unless we do three things. First, adapt existing and create new institutions. Many resources are shared among countries and some regional arrangements are already in place. But as species shift, they will bring conflict between stakeholders if current institutions do not have the geographical coverage and political mandate to cope and evolve. Second, adapt management. This is the easiest adaptation in places where management is already up to a good standard. But we will need adaptations as the target species will change, and thus so will management reference points, fishing permits, etcetera. And third, we must educate consumers. This is the most difficult of all. We should all eat the fish of the day. I mean the fish available on the day. We capture over two thousand species of fish, molluscs and crustaceans. We culture over six hundred of them. We are going to need profound market and consumer adaptation programs around the world if we are to change our relationship with our natural environment to a more dynamically adaptive one, including our taste and our dietary preferences.

The above comments apply mostly to coastal and continental shelf fisheries. The central Arctic, as mentioned yesterday, is subject to a legally binding moratorium to prevent commercial fishing. It is not yet known if there are exploitable fish resources in the central Arctic, or if there will be some in the future. But the moratorium says that scientific monitoring and research must be conducted before it can be revisited.

My time is almost up. But let me finish by commenting on yesterday's conversation on the appropriateness of current institutions to cope with the changes that are coming. And let me link it to my initial contention that we are moving towards a future with much knowledge but much uneasiness over how to work out collective solutions. Yes, it may well be that our institutions are not as fit for purpose as we would like, but note that the current political climate is not one in favor of multilateralism. I see this tendency every week in my job. Given the chance there will be voices that would claim unfitness of purpose in order to eliminate multilateral processes that currently provide significant checks and balances. Not all global problems demand global solutions, but all global problems require multilateral mechanisms and institutions to discuss options and trade offs. Without them we will continue to hope we can solve problems by pointing fingers. And I have not seen any problem being truly solved in that way, no matter how much knowledge we have at our disposal.

https://www.youtube.com/watch?v=X81xsMQbm8o&list=PLpwWVxYVoO1muxaLTCD5uVyUXaXLlDw1i&index=27

Chapter 12
(Action): Sustainable Arctic Ocean

Hide Sakaguchi

Excellencies, ladies and gentlemen,

First of all, thank you very much for inviting me as a keynote speaker despite Japan is not Arctic country.

As Steven introduced, I would like to ask you how much you know about Japan. Japan has an area of 0.37million square kilometers while Norway is 0.38 million square kilometers, which is quite similar but Japan is a little bit smaller than Norway. But in terms of population, we have more than 100 million people in such small area. That means Japan is a very noisy or very busy country. But in reality as some of you may know, Japanese is very quiet and quite shy especially at this kind of meeting. And since they are too quiet, we are sometimes sort of being mysterious or sometimes thought to be stupid because we say nothing.

There is a reason why Japanese are really quiet. For a long time, being quiet has been thought to be the beauty in life. In addition to this Japanese attitude, Japan once has been in the exclusion time for more than 200 years from 17th century to 19th century, completely apart from any other country, language, culture, etc. It is 21st century today but it still has an effect for a lot of Japanese.

But before that time of exclusion, we had a very deep exchange with other countries because all of our language, culture, etc. come from China and Korea.

Also, we have some very good exchange from Arctic region. For example, my wife is from Hokkaido Island, which is in the northern part of Japan. There are still many Indigenous people in Hokkaido. They have quite a similarity with Indigenous people in Arctic region. That means, before the exclusion time Japan was open, far

H. Sakaguchi (✉)
Japan Agency for Marine-Earth Science and Technology: (JAMSTEC), Yokosuka, Kanagawa, Japan

Present Address: Ocean Policy Research Institute, Sasakawa Peace Foundation, Tokyo, Japan
e-mail: h-sakaguchi@spf.or.jp

P. A. Berkman et al. (eds.), *Building Common Interests in the Arctic Ocean with Global Inclusion, Volume 2*, Informed Decisionmaking for Sustainability,
https://doi.org/10.1007/978-3-030-89312-5_12

open to all over the world and not only human beings but fish, plankton and any other marine organisms floating in the ocean current or in the ocean circulation without our human being's permission or without us being aware.

In that sense, being quiet is a bit of nonsense. Especially, when we think about Arctic region science. In Arctic region, quietness has been already broken due to climate change, global warming and pollution. We must make our eyes wider and wider to take a look and take an action to support each other.

From now on, we Japanese, especially from my side or a scientific community, would like to declare that we take a big action to contribute to the Arctic frontier community and also Arctic region problem and also to support each other. That is our mission as a science research institute. Science has no border and science is a good tool to communicate with all over the world and also to enhance mutual understanding which eventually brings peace and friendship among us. That is JAMSTEC's mission.

Lastly, additional information is that in this year's November, Japan is going to have Arctic science ministerial meeting in Tokyo. You are all welcome to attend and to come to see Japan. Another information is that last December, JAMSTEC was asked to build a new ship for research with an icebreaker function. That means, we JAMSTEC will make a full effort to enhance the science in Arctic region.

Thank you very much

Chapter 13
(Action): Sustainable Arctic Ocean – Ocean Wealth Is Ocean Health

Jens Frølich Holte

A sustainable ocean economy is a key priority for the Norwegian Government.

A sustainable Arctic ocean is vital if we are to achieve this goal, given that 80% of Norway's sea areas are north of the Arctic Circle. What happens in the Arctic doesn't stay in the Arctic. And what happens outside the Arctic has a big impact on the region.

Global solutions are required to solve global problems. The challenges we face illustrate this fact perfectly.

The Arctic ice cap is melting at a record pace.

The permafrost is thawing.

The Arctic ocean is getting warmer, more acidic and losing oxygen.

This will impact the four million people living in the Arctic, but also the millions of people worldwide who rely on the ocean as a source of food, energy, employment, welfare and recreation.

That is why the sustainability of the Arctic ocean concerns us all.

In 2018, Prime Minister Erna Solberg took the initiative to establish a High-level panel on a Sustainable Ocean Economy.

The panel consists of 14 Heads of State and Government. They are presently hard at work developing a set of concrete recommendations. We hope the result will be a "to-do"-list for the ocean, showing how the right policies can enable us to protect, produce and prosper.

The good news is that, although there are serious challenges, the ocean can be a major solution to climate change, as well as a key to many of the sustainable development goals.

J. F. Holte (✉)
State Secretary MFA, Oslo, Norway
e-mail: Jens.frolich.holte@mfa.no

P. A. Berkman et al. (eds.), *Building Common Interests in the Arctic Ocean with Global Inclusion, Volume 2*, Informed Decisionmaking for Sustainability,
https://doi.org/10.1007/978-3-030-89312-5_13

A report commissioned by the High-level panel shows that ocean-based industries can contribute up to 21 per cent of the emission reductions needed to meet the Paris climate target.

Another report shows that the ocean as a nutritious food basket, can be pivotal to combat hunger and malnutrition.

Given the right conditions, the ocean could provide over six times more food than it does today.

It is imperative that we create these conditions, so that we can continue to harvest from the ocean's riches, as Arctic people have done for millennia.

The key word here is "integrated ocean management". It might sound technocratic, but it should actually be the hottest sustainability buzzword! In short, it is a method for thorough planning to ensure value creation while maintaining the structure, functioning, productivity and diversity of the ecosystems. In order to achieve ocean wealth, we need ocean health.

Through the new aid programme Oceans for Development, Norway will share our experiences with partner countries so that they can achieve a sustainable and inclusive ocean economy. This experience include Norway's successful use of integrated ocean management.

People in the Arctic have always lived off the ocean. Through fishing, aquaculture, shipping and oil and gas production we have created wealth for our own communities and for the world at large.

In the future, the potential for sustainable development and job creation within ocean industries in the Arctic are immense, such as as offshore wind, green shipping and pharmaceuticals.

But in order to avail of the solutions represented by the ocean, we need to confront the challenges. In that effort, we need the combined knowledge of governments, international organisations, civil society, research environments, business and coastal communities. Thank you to Arctic Frontiers for providing an arena for such cross-fertilization. That's how we ensure a sustainable Arctic ocean.

Chapter 14
(Action): Sustainable Arctic Ocean

Sam Tan

Question: When you come to an event like this, what is the message you are bringing? MOS Tan: I think the most important message that I want to bring to the audience is that what happens in the Arctic, does not stay in the Arctic. The converse is also true – what happens outside the Arctic, does not stay outside the Arctic. Singapore is a small island state, and we are surrounded by seas and oceans. On the eastern side, we have the South China Sea; on the western side, we have the Indian Ocean. We have a very small land area of only 720 square kilometres, and one third of this is actually very low-lying. We have 5.7 million people living on this tiny island, and we have literally built our city to the brink of the land. Our people live next to the water. Thus, we are very, very careful with our water management. If we are not careful and we pollute our water bodies, seas and oceans, the consequences will be disastrous for us.

With climate change, a melting Arctic and rising sea levels, Singapore is also in trouble. We have always heard that mean sea levels will rise by up to 1 metre within this century. If this happens, our future Prime Minister will have to conduct Cabinet meetings in a scuba diver suit, for we will be submerged underwater! For our survival, we have embarked on an ambitious programme to deal with rising sea levels, and we expect to invest more than S$100 billion over the next 50 to 100 years for all kinds of infrastructure projects to protect our coastlines.

We are doing a lot to make sure that we do not contaminate the oceans. We have also been doing a lot to control and reduce our carbon emissions. Even though we are going to have an election either this year or next year first quarter, we introduced a carbon tax that covers all industries with no exemptions. Come 2030, we want to

S. Tan (✉)
Minister of State, Ministry of Foreign Affairs and Ministry of Social and Family Development, Singapore, Singapore
e-mail: TAN_Tah_Jiun@mfa.gov.sg

P. A. Berkman et al. (eds.), *Building Common Interests in the Arctic Ocean with Global Inclusion, Volume 2*, Informed Decisionmaking for Sustainability,
https://doi.org/10.1007/978-3-030-89312-5_14

reduce our emissions intensity by 36% from 2005 levels. This is a very painful process. But compared to the pain of Singapore being overrun by seawater, it is a price we think we should pay.

Post-Script

Singapore submitted an enhanced Nationally Determined Contribution (NDC) to the UNFCCC in end-March 2020, with the headline target reflecting our commitment to an absolute peak emission level of 65 million tonnes of carbon dioxide equivalent (MtCO2e) around 2030. These enhancements – an absolute emissions limitation target (in place of our previous emissions intensity target) with a clear peaking level (i.e. 65 MtCO2e) – will provide greater clarity and transparency of the level and when Singapore's emissions are expected to peak, and facilitate the tracking of progress.

Part III
The Broader Arctic Setting

Framing Questions

1. **What are the major drivers of change, including but not limited to climate change, in the Arctic?**
2. **Who are the key players whose actions will affect the future of the Arctic?**
3. **What is the role of business in introducing innovations to secure sustainable development in the Arctic?**

Chapter 15
(Research): Evolution of Arctic Exploration from National Interest to Multinational Investment

Eda Ayaydin

Abstract The history of exploring the Arctic, long considered a remote and obscure region, has profoundly shaped the regime for governing the Arctic that exists today. The impacts of climate change and the increasing pressure of globalization that are affecting the Arctic are also changing the nature of exploration. The interconnected phenomena of exploration and climate change have led to heightened awareness of the Arctic worldwide and have sparked geopolitical interest and competition. This chapter mainly examines the central aspects of changing Arctic exploration from nation-state interests to the endeavors of oil and gas companies, among others. It will also consider the ways in which exploration is increasing the geopolitical and economic importance of the region for states. The chapter thus will contribute to the debate over the impact of exploration on political and economic interests in the Arctic.

15.1 Introduction

Exploration might be an act of discovering an unexplored place or discovering an unfamiliar subject; in both cases, an "unknown" is required. John Franklin sought to sail through an unknown passage; Robert Peary and Frederick Cook tried to reach the mythic North Pole. Explorers made the imaginary Thule real.

Depending on the era of exploration, the principal actors in the Arctic have changed over the centuries. Until the start of the twenty-first century, traditional "explorers" acted in the Arctic. Compared to the explorations in the previous centuries, contemporary efforts to explore the Arctic originate from oil and gas companies and scientists, who employ different methods and techniques and have different motives for exploration. Consequently, as the "unknown" elements of the Arctic have evolved and changed, so has the "subject" of exploration transformed as well.

E. Ayaydin (✉)
Sciences Po Bordeaux, Bordeaux, France
e-mail: eda.ayaydin@scpobx.fr

P. A. Berkman et al. (eds.), *Building Common Interests in the Arctic Ocean with Global Inclusion, Volume 2*, Informed Decisionmaking for Sustainability,
https://doi.org/10.1007/978-3-030-89312-5_15

Previous explorations leading up to the twenty-first century certainly increased the geopolitical importance of the Arctic, as nations developed new fishing grounds and new shipping routes. Subsequent colonization of parts of the Arctic (Russian rule in Alaska, the Hudson Bay Company in Canada, and the Bergen Greenland Company of Norway and Denmark) paved the way for additional economic activities, such as fur trading (Norman, 2018: 2).

Early economic and scientific explorations were financed by states primarily to promote their interests in expanding their reach in the areas of the Arctic considered *terra nullius*. Since then, the increasing pressure of globalization on the nation-state system in the twenty-first century promoted new actors in Arctic exploration, such as IGOs (intergovernmental organizations), NGOs (non-governmental organizations), private companies and multinational investors. Indeed, the term "explorer" today is largely used as a metaphor, since the type of exploration has evolved from discovering unknown territory to developing new economic activities, such as the exploration of oil and gas deposits by companies. In such cases, the new explorers are companies.

Climate change has become one of the most important motives for scientific exploration, as we seek to understand the causes and impacts of the warming Arctic. These explorations are not only financed by states but also by NGOs, private companies or international organizations. Scientists, Indigenous peoples and other Arctic residents also take part in these projects.

Science has always been a form of soft power, which is the ability to pursue interests by attraction such as culture, technology, political ideals and policies rather than coercion (Nye, 2004: 5). Nations often pursue scientific endeavors, including at the international level, for reasons not wholly related to science, including with motives concerning the promotion of their national interests. Sometimes they ended up with national gains, as in the case of International Polar Years. Karl Weyprecht, one of the organizers of the First International Polar Year in 1882–1883, intended the initiative to be solely for scientific aims (Weyprecht, 1875:33). However, not all the participating countries and financial sponsors (states) agreed to limit the IPY only to scientific purposes. Today, non-Arctic powers such as China conduct scientific research on Arctic climate change as a way to make their presence felt in the Arctic. According to Aki Tonami, China and other Asian states see their scientific efforts in the Arctic as means of attaining the long-term political and economic goals in the region (Tonami, 2019: 28).

In the twenty-first century, it is no longer possible to explore a territory and to occupy it in the Arctic region. States can, however, gain access to Arctic resources by building and maintaining trusted international partnerships. Yet, the current political atmosphere pervading the Arctic—flowing from the international sanctions imposed on Russia, the hostile attitude of the United States toward China articulated in conjunction with the Arctic Council meeting in 2019, the efforts of the Arctic 5 in 2008 to claim leadership of the region, and the growing militarization of the region—does not ensure the kind of stability that will allow such partnerships to flourish. In such circumstances, states have difficulty using soft power as a means to advance their strategic, economic and political aims. (Nye, 1990: 160) The international institutions that exist today in the Arctic foster cooperation amongst states in the region. But despite strong regional governance in terms of institutions in the region, states remain the main actors who use science to promote their strategic, economic and political interests.

In order to trace the evolution in the nature of Arctic exploration, this chapter, based primarily on first hand in-depth interviews and media archives, will compare the twenty-first century to the previous centuries, analyze the liberalization and internationalization of the Arctic, and scrutinize the role of climate change and growing economic activity in the region. It will examine how and why Arctic exploration evolved from efforts based on national interest to multinational investment.

15.2 Nineteenth Century: Nation States Are the Principal Actors of the Exploration

During previous centuries, explorations were primarily sponsored by states, whose main purpose was to find new territories and new routes for reasons based on their national interests that were as much political (sovereignty) and economic as they were scientific. During 1818, Captain John Ross commanded the *Isabella* with the intention to transit the Northwest Passage. At one point, Ross thought that the passage was a dead end and decided to go back to England. Although he came back with remarkable scientific results, the reaction of the Lord John Barrow shows that the true expectations for the voyage were of political prestige and economical gain from the Arctic: "his career in the Royal Navy is shattered. He will return to the Arctic, however, on its own, and prove its value unjustly questioned" (Le Brun, 2018: 124). Moreover, John Ross, ridiculed for his premature U-turn in 1818, wanted to make up for it with a brilliant success on a subsequent expedition, financed not by England but by the whiskey distiller Felix Booth; it was the first commercially sponsored trip of its kind in history (Le Brun, 2018: 127).

Arctic exploration accelerated by the nineteenth century as did Antarctic exploration, especially as John Franklin's ill-fated expedition to explore the Northwest Passage wound up paving the way for others. Following the disappearance of Franklin's vessels in 1845, many expeditions were arranged both to understand the fate of those vessels and their crews and to continue exploring the Northwest Passage and new other straits and passages. The loss of the Franklin expedition caused a renewed interest in the Arctic areas. Napoleon III, for example, prepared an expedition to Greenland with the *Queen Hortense Corvette* (Le Brun, 2018: 626). Moreover, during the nineteenth century, the exploration team of Austria-Hungary Empire under the guidance of Karl Weyprecht (who, as noted earlier, would later organize the First International Polar Year) discovered Franz Josef Land in 1873 (Luedecke, 2004: 56). Thus, if the nineteenth century was the golden age of nation-building processes, it was also the period of nation state driven explorations in this remote region. During this period, we may say that science was national and served national preoccupations.

15.3 From the 19th to the Twentieth Century: Scientific Cooperation Increases

With the twentieth century and the consolidation of the concept sovereignty on the one hand and of rationalism and scientism on the other hand, exploration purposes in the Arctic evolved.

Karl Weyprecht, mentioned above, realized the need for comprehensive meteorological measurements to get better results. He developed a motto during 1870s: "Forschungswarten statt Forschungsfahrten", (research observatories instead of research voyages) (Lüdecke et al., 2010: 10). This idea was accepted during the First International Meteorological Congress in Vienna in 1873 (Lüdecke et al., 2010: 10). The International Polar Year was the first global scientific initiative in the history of Arctic exploration.

Science was also required for economic and political gain in the Arctic; it was needed to understand geography, physical structure, sea ice and meteorology. Weather forecasting became particularly important once economic trade involving the Arctic region began. Therefore, the idea of conducting substantial joint explorations for meteorological measurements was useful both for states and scientists.

Although the polar explorers often employed international and multinational crews, they were typically conducted in a spirit of international rivalry; therefore, expedition leaders vacillated between scientific and national goals (Bulkeley, 2010: 2). That is why Weyprecht, while planning the IPY, stated, "nationally-focused and often short-lived expeditions were an ineffective way to study" (Bulkeley, 2010: 2). He also thought that discovery and topography had been the main objectives of the Arctic expeditions, but now it was time for science; the states need to put aside the rivalry between them and look for common good for mankind (Weyprecht, 1875: 33).

The First International Polar Year was organized for better comprehension of the atmospheric and meteorological events, but the scientists also met Arctic Indigenous peoples and observed and photographed their way of living.[1] Of course, these lands were not vacant; they were inhabited by Indigenous peoples whose ancestors had been living there for thousands of years. The IPY-1 was important in terms of opening the way for international cooperation among scientists from all over the world, and of course the cooperation of states. The First IPY played an important role in liberalizing the Arctic region as well by making science more collaborative

[1] Archives of the First International Polar Year, National Oceanic and Atmospheric Administration of US Department of Commerce – University of Washington (https://www.pmel.noaa.gov/arctic-zone/ipy-1/US-Barrow-P1.htm) Access in August 2020.

and less subject to national pressures. The Second International Polar Year was realized during the interwar era in 1932–1933 with the participation of 44 countries and 27 research observatories (Lajus, Lüdecke, 2010: 165). These research observatories were established for atmospheric and meteorological researches. However, the states could not establish stations in the Antarctic because of financial restrictions of the time. The Third International Polar Year, which is known as International Geophysical Year, took place in 1957–1958. Even though this scientific collaboration faced some interruptions because of the Cold War, the Committee of IPY-3 assigned high significance to the storage and access to the results. The Fourth International Polar Year took place in 2007–2008 with a grand support of science based non-governmental organizations such as the Scientific Committee on Antarctic Research (SCAR) and the International Arctic Science Committee (IASC), who organized joint conferences and brought together hundreds of researchers from the Arctic and Antarctic fields within the perspective of "opening the Arctic for science". The fourth international polar year became so successful at the international and national political level that scientists who worked hard to organize this year received political support as well.[2]

There is no doubt that international scientific cooperation had existed before the First IPY. However, states most likely engaged in such cooperation in order to better prepare their navy personnel who would venture to the Arctic, including on foreign ships. For example, Peter the Great, who was genuinely enthusiastic about Western European culture, hired German-speaking scholars and naval officers to teach at the Academy of St. Petersburg (Cracraft, 2003: 34). Vitus Bering, a Danish national, led the Great Northern Expeditions of Russia in the eighteenth century and, among other accomplishments, explored the Bering Sea and the Bering Strait and laid the foundation for Russian settlement of modern-day Alaska. Moreover, the cooperation between states was also the case in Western Europe. For example, French navy personnel worked aboard English naval vessels engaged in exploration (Le Brun, 2018: 136).

During the nineteenth century, Arctic explorations were largely the result of state-sponsored expeditions. The famous Franklin expedition of 1845 and trips to the North Pole and to the Northwest Passage are good examples. Yet, there were also very active companies in the Arctic trade and Arctic exploration from the seventeenth to the nineteenth century. These explorations were in a political and economic framework prior to the advent of nation states where the economy of large business companies was centered on mercantilism. The main actors were big state or private charter companies.

Once Russians settled in Alaska, the Russian-American Company sponsored further exploration of the region. Similarly, the Hudson Bay Company (which still exists) and the North West Company, two private charter entities, explored the Canadian Arctic. By 1870, HBC exercised *de facto* corporate sovereignty over a

[2]IASC book, p:56. (https://view.joomag.com/iasc-25-years/0102946001421148178?short&) Access in August 2020.

great deal of territory in the Canadian Arctic. Indeed, the company's acronym—HBC—came to be identified as "Here Before Christ". HBC was viewed as the first bearer of civilization in that area (Kaufman, Macpherson, 2005:444). Although such companies had considerable freedom of action in overseeing the territories in question, they still acted under the banner of the King of England or the Russian Czar. Ultimately, HBC transferred the *de facto* sovereignty it exercised over land in the Canadian Arctic to the Dominion of Canada.

All this cooperation from the end of the nineteenth century and during the twentieth century, especially through the IPY pushing to scientific collaboration,[3] changed shape starting by the end of the bipolar world and the rise of liberalism.

15.4 Twenty-First Century: Multinationalization of Exploration

The nature of exploration in the Arctic has changed in the twenty-first century because of the increasing effects of climate change, which has created a new political and economic Arctic agenda in terms of physical changes of the ocean, easier access to oil and gas deposits, new shipping routes, new fishing areas and growing global interest in the Arctic Council. Climate change has also magnified the effect of globalization in the Arctic as a new global phenomenon. Thus, the Arctic is attracting the attention of non-Arctic world.

During this new era in the Arctic, scientific and economic exploration go hand in hand for some Arctic states such as Russia, where including scientists in economic exploration efforts is compulsory. Scientists from Russia interviewed for this chapter indicate that they joined several explorations with Rosneft and conducted their own scientific research on sea ice.[4] In Norway, Equinor has always cooperated with universities and other research institutions. Ørjan Birkeland[5] says that it is expected that the industry creates sustainable ripple effects in the societies in which they work, and this also implies engagement of universities. Equinor has agreements and sponsorship programs with several universities and supports research programs in their areas of interest, such as geology and energy distribution; sometimes the company sponsors independent research projects of universities (e.g., from Norway, United Kingdom, Germany and other countries). In other situations, Equinor funds research projects to examine a specific issue that it wants to be analyzed. In these

[3]International Polar years are held in 1882–1883, 1932–1933, 1957–1958 (also known as International Geophysical Year) and finally in 2007–2008.

[4]For example, the interview conducted in 26 May 2019 with a scientist from Arctic Institute of St Petersburg, the name of the interlocutor is confidential.

[5]Interview, Ørjan Birkeland, Project manager for the Northern Area Project at Equinor. (22 May 2020)

ways, scientific involvement on the part of private companies is implemented on different levels.[6]

On the other hand, oil and gas exploration involves very different methods and means from the type of exploration undertaken in previous times. Modern efforts related to oil and gas development require expertise in geology and seismology and in related technological fields. Oil and gas companies must plan their explorations years in advance, and sometimes face failure. For example, Norway conducted an important oil exploration in Barents Sea in 2017 with disappointing results. However, the chief executive of Statoil (by 2018 Equinor) stated that the government-controlled company was "very patient" when it came to exploring new areas and that it would return to the Norwegian and Barents Seas the following year with "optimism but realism" (Milne, 2017). Indeed, Statoil resumed exploration between 2017 and 2019 in the Barents Sea, developing exploration wells in the areas of Jøkåsen, Gjøkåsen Deep, Korpfjell Deep, Sputnik and Mist.[7] The "exploration" page of the Norwegian state-owned company Equinor's website displays its purpose as: "Inspiration from a history of explorers. Building on a heritage of adventurers, Equinor has become a world-leading explorer for oil and gas."[8] It is obvious that scientific exploration and economic exploration methods differ greatly, but the notion of "exploration" is fundamental in this new era.

In the field of exploration, state-owned enterprises (SOE) are in some cases playing roles as important as those of states. The most effective SOEs in oil and gas sector in the Arctic are in Norway and Russia. The economies of both countries are largely based on natural resources. 80% of Russian exportation is based on oil and gas and the half of it is produced in the Arctic. 45% of Norwegian exportation is also oil and gas (Finger, Heininen, 2019: 60).

Gazprom and Rosneft are Russian oil and gas enterprises with overlapping activities. Gazprom now owns an oil company (Sibneft) and Rosneft has produced gas as well since 2007. Even though Russia would benefit economically from foreign investment in its oil and gas sector, Moscow gives exploration licenses only to Russian state-owned enterprises on its Arctic shelf. According to Russia's 2008 federal law on the subsoil, "licenses for Arctic offshore may only be granted to legal entities that are more than 50% controlled by the Russian state" (Finger & Heininen, 2019: 50).

But because Russia's offshore technology is limited, it needs partners. The Russian Federation accordingly bought production equipment from Norway, and Rosneft became partners with Exxon, Eni and Statoil in the Arctic. The partnership between Exxon and Rosneft in the Russian Arctic was significant in terms of exploring new oil fields in Kara Sea until 2018, when it ended because of the US

[6] *Ibidem.*

[7] Equinor website, (https://www.equinor.com/en/what-we-do/responsible-drilling-in-the-barents-sea.html) Access in February 2020.

[8] *Ibid.*

sanctions imposed following Russia's 2014 invasion of Crimea.[9] The EU and the US sanctions both targeted Russia's trade and financial sectors, and also included a ban on support for Arctic deep-water exploration and oil production in Russia.[10] The sanctions affected Russia's cooperation with Norway as well, as Rosneft-Equinor activities in Russia declined. Ørjan Birkeland stated that, while the sanctions certainly had an impact on the cooperation in terms of limiting joint oil and gas activities, the situation does not require the termination of the partnership.[11] He added that Norwegian Equinor still has some investments and activity in the Russian shelf but that sanctions prohibit Norwegians from using western technology and from undertaking deep water exploration. Therefore, this situation limits the tools for cooperation. On the other hand, Pierre Giboin stated that Total did not withdraw from the project in Yamal, because they had already invested significant amounts for this project.[12]

Because of the Western sanctions, Russia needs to cooperate with Asian states more than before. There are significant Chinese and some Japanese investments in Russian oil and gas projects in the Arctic, particularly in Yamal. The mega project of Gazprom with grand Chinese investment in Yamal and LNG projects of Novatek are very important steps for development of the region, but also for enhancing Russia's cooperation with Asian states. Japan is one of the Asian states cooperating with Russia in the LNG 2 project in Yamal. India is another new partner in the Russian Arctic that will invest in the new Vostok project of Rosneft which will take place in the remote Taymyr Peninsula. According to Igor Sechin, CEO of Rosneft, this project is going to be the biggest oil project in modern day oil industry.[13]

According to the international relations theory of "complex interdependency," as transnational and transregional relations increase among states, as well as among non-state and private actors, the need for each other grows in the absence of military use and coercive power (Keohane & Nye, 2012: 12). It is obvious that this concept is very applicable in the new global and multinational Arctic. Conducting explorations together by Multinational Companies (MNCs) and SOEs is advantageous for all sides, since the SOEs know the region better than a foreign investor and because such arrangements can simplify the bureaucratic processes. MNCs in the Arctic are

[9]Scheyder Ernest, Soldatkin Vladimir, « Exxon quits some Russian joint ventures citing sanctions », *Reuters*, 28 February 2018. (https://www.reuters.com/article/us-exxon-mobil-russia-rosneft-oil/exxon-quits-some-russian-joint-ventures-citing-sanctions-idUSKCN1GC39B) Access in May 2020.

[10]"How far do EU-US sanctions on Russia go?", *BBC News*, 15 September 2014. Access in August 2020. (https://www.bbc.com/news/world-europe-28400218).

[11]Interview, Ørjan Birkeland, Project manager for the Northern Area Project at Equinor. (22 May 2020).

[12]Interview, Pierre Giboin, Responsable of Yamal LNG Shipping Department at Total. (13 June 2019).

[13]Staalesen Atle, « Rosneft tells Putin its new Arctic project will be biggest in global oil", *The Barents Observer*, 11 February 2020. Access in May 2020. (https://thebarentsobserver.com/en/2020/02/rosneft-tells-putin-its-new-arctic-project-will-be-biggest-global-oil).

worldwide companies that increase confidence in exploration, and they are also important actors as financial investors and often technology providers.

Today, a primary challenge that the MNCs and SOEs are facing is climate change, which has strengthened the opposition to oil and gas activities. This has created significantly more pressure on these entities because they operate in an environmentally sensitive region that has attracted public attention. This pressure exists, directly or indirectly, on the states as well. Even if companies adopt more environmentally sensitive policies, growing environmental opposition will not recognize the legitimacy of their explorations. As in the case of potential commercial fishing in the Central Arctic Ocean, the economic exploration and production of oil and gas will remain on the agenda of the environmental NGOs and the public.

These same environmental concerns lead to other types of scientific exploration, both to understand the patterns and impacts of climate change and also prevent environmental damage from commercial activities. For example, in Norway, "zero-discharge requirements" and oil spill contingency planning are prepared to meet expectations of environmental lobbies that the oil and gas exploration might be realized safely in environmentally sensitive areas. (Hasle; Kjellén, et all, 2009; 833). Moreover, scientists are already present at oil and gas exploration sites since this type of exploration requires science, but also because involving scientists in the exploration is important to address environmental concerns as well. Engaging representatives of NGOs in such exploration would be another option to convince these lobbies that their concerns are being taken into account.

On a broader level, however, participants involved in Arctic Council working groups whom we interviewed mentioned the inability of the Council to negotiate restrictions on fossil fuel development and use in the Arctic. According to them, the international community would need to strengthen the Paris Agreement regime significantly in order to force changes in the production of oil and gas. On the other hand, Rasmus Bertelsen[14] thinks that NGOs in some western States may have the political influence to affect exploration, for example in Scandinavia or North America, but that they have absolutely no influence in Russia. NGOs in Norway drove public opinion in a manner that led to the cancellation of an exploration project of Equinor in the southern part of the Lofoten Islands. Ørjan Birkeland says that Equinor has the capacity to conduct exploration in a sustainable manner wherever they operate, but the public became more concerned with the offshore areas around Lofoten and with the impact of their activity may have on the marine environment. He also adds that future oil and gas activity in general are becoming more challenged in the public domain.[15] Nevertheless, Alexander Sergunin[16] says that NGO expertise

[14]Interview, Rasmus Bertelsen, Professor of Political Science at the Arctic University of Norway. (12 May 2020).

[15]Interview, Ørjan Birkeland, Project manager in the Northern Area Unit at Equinor. (22 May 2020).

[16]Interview, Alexander Sergunin, Professor of Political Science at the St. Petersbourg State University. (08 June 2020).

is required for each oil and gas activity in the Russian Arctic for environmental concerns. He gives the example of an oil rig that was constructed for use in the Kara Sea in 2011. Russia postponed postponed drilling operations until a dialogue between Gazprom Neft and NGOs ended in a consensus in 2013. However, when we asked a former Total executive to describe the environmental policies of Total in the Arctic and the impact of NGOs on the environmental policies, the answer was "your question - while clear- involves confidential aspects"[17]!

The image of the Arctic in people's minds is a clean and resource-rich region. In part for this reason, the public seems to view oil and gas development in the Arctic as more environmentally dangerous, even if such activities contribute less to global climate change than similar activities elsewhere in the world. The comments from oil and gas company decisionmakers supports this assessment. Moreover, as long as this public perception remains, the symbolic significance of the Arctic as a region particularly vulnerable to climate change will also remain.

15.5 Recent Era: Current Governance and the Exploration in the Region

The creation of the Arctic Council in 1996, which coincided with the rise of the region's geo-economic importance, reinforced the distinction between Arctic states and non-Arctic states. Only Arctic States are members of the Council; non-Arctic States are only eligible to be Observers. The Arctic Council now has among its Observers 13 non-Arctic States, 13 IGOs, and 12 NGOs, which reflects the growing interest in the Arctic among a wide range of entities.

As the predecessor of the Arctic Council, in 1991, officials of the Arctic nations came together in Rovaniemi as part of an initiative launched by Finland to adopt the Arctic Environmental Protection Strategy (AEPS) for the preservation of the Arctic environment. Following the AEPS, when the Ottawa Declaration established the Arctic Council, the Council initially focused on environmental protection and sustainable development in the region. However, it started to be difficult for the Council to focus only on these two issues, since the region has become an important political and economic subject. As the region's significance and visibility advanced, the region has also internationalized. This brought new actors as well as an emerging international legal regime (Scopelliti et al., 2019: 7).

The Arctic Council is an inter-governmental forum with a non-binding structure. However, in the past decade, the Council has served as the venue for the development of three binding agreements: the Agreement on Cooperation on Aeronautical and Maritime Search and Rescue in the Arctic (2011), the Agreement on Cooperation on Marine Oil Pollution Preparedness and Response in the Arctic (signed 2013)

[17] Attempt of interview with an executive at Total. (15 May 2020).

and the Agreement on Enhancing International Arctic Scientific Cooperation (signed 2017).[18] In this sense, the Arctic Council has become a law-making mechanism.

Moreover, the five Central Arctic Ocean (CAO) coastal states and other actors—China, the European Union, Iceland, Japan and the Republic of Korea—have signed the Agreement to Prevent Unregulated High Seas Fishing in the Central Arctic Ocean in 2018. When all 10 ratify, the Agreement will come into force.[19] (European Commission, 2018).

Despite these agreements, the remoteness and climatic conditions of the Arctic region still present serious challenges to the realization of economic and scientific projects. Taking these various factors into consideration, the importance of international cooperation remains at the forefront. In recent years, however, considerable scientific inquiry, both at the global level and in the Arctic, has focused on climate change. Nevertheless, the fact that the administration of one of the world's superpowers denies the reality of climate change affects the quantity, quality and the funding of such endeavors.

In 2019, for the first time in its history, the Arctic Council could not reach agreement on a Ministerial Declaration, reportedly because the U.S. Administration refused to accept language about climate change that the other members had insisted on. More generally, the United States is no longer supporting the work on climate change undertaken by the Arctic Council working groups. According to Timo Koivurova, the United States is challenging the nature of the Arctic Council, pushing other Arctic states to concentrate on other intergovernmental forums in the Arctic where the United States is absent, such as the Nordic Council or the Barents Euro-Arctic Council (Koivurova, 2019a).

Conducting science-based explorations in order to understand the how and whys of climate change is necessary to mitigate its effects. However, if the United States, as a superpower, remains unsupportive of these kinds of scientific endeavors, it would have significant economic and symbolic impacts. First, the conduct of long-term or short-term scientific exploration in the Arctic is very expensive because of the remoteness of the region. Therefore, U.S. financial aid is crucial. Second, the contribution and presence of one of the great powers of the Arctic symbolically enhances the significance of scientific efforts. The Third International Polar Year was organized in 1957 in the middle of the Cold War and suffered from political tensions. The bipolar world of those days undermined both eastern and western scientific effort. Ironically, in today's multipolar era, science suffers again because of power politics.

Despite these recent tensions, the Arctic region is still a relatively cooperative part of the world. It is not a new Middle East, even if it also has significant hydrocarbon resources. Unlike the Middle East, the Arctic region has, since the end of the Cold War, experienced low tension and has benefitted from multiple regimes that have

[18] https://arctic-council.org/index.php/en/our-work/agreements Access in January 2020.

[19] European Commission, 2018 https://ec.europa.eu/fisheries/eu-and-arctic-partners-enter-historic-agreement-prevent-unregulated-fishing-high-seas---frequently_en Access in January 2020.

promoted international cooperation (Young, 2019: 7). Some recent facts are nevertheless alarming. British and U.S. warships sailing into the Barents Sea in 2020 and increasing Russian military activities at its Arctic coast might threaten the strategic stability of the region. The U.S. attitude at the Arctic Council meeting in 2019 threatened the political stability. And sanctions towards Russia, even if justified in the aftermath of the invasion of Crimea, might threaten the economic stability.

As these examples demonstrate, global issues and processes have a direct impact on the Arctic region and therefore affect the governance of Arctic itself. As Koivurova shows, in examining global and semi-global perspectives, much of the governance of the region in effect involves all the states in the world (Koivurova, 2019b: 25). Oran Young thinks that the Arctic region was peripheric. However, it is now one of the centers of vital global issues as well as a stage of cooperation whose actors now include a growing number of non-Arctic states, as evidenced by the Central Arctic Ocean Fisheries Agreement negotiated among the Arctic 5, China, Iceland, Japan, South Korea and European Union. Therefore, one may state that the Arctic is a well internationalized region and no longer a peripheric region (Young, 2019: 2). That's why it is possible to see proliferation of regimes dealing with needs for governance issues in the Arctic. (Young, 2019: 7).

15.6 Conclusion

The principal actors in Arctic exploration have changed in the twenty-first century due to the changing physical, political and economic conditions of the world. Previously, the main actors were explorers and states as sponsors; in the twenty-first century, scientists and oil and gas companies replaced explorers, as the needs of the world changed. But one thing has remained constant: exploration is still undertaken to a great extent for national gains. They are conducted by companies but mostly by state owned ones, for the benefit of state economies.

Even the "multinational" companies involved in modern Arctic exploration remain largely identified with the state in which they first incorporated, as in the case of Total. Pierre Giboin from Total states that when Total operates in Yamal, it is seen as a French company and represents the French culture, even if Total is multinational. If it is true that multinational companies are perceived in this way, the very same companies act in some sense as instruments of the nation state, projecting a national image.

Augmentation of cooperation has been important in the Arctic shelf in order to increase the international investment in the region. Oil and gas exploration in the region is very difficult and requires significant cooperation in economic and technological terms. When the exploration is bilateral or multilateral, the home country utilizes the technology of the investor and the MNCs gain experience operating in the region.

The evolution of different kinds of exploration also increased the visibility of the region in general. Economic explorations brought non-Arctic states and companies to the region, and with them, global attention. While the region was not well-known until the twenty-first century, climate change now requires explorers to acquire knowledge of both the environmental and economic aspects of their activities. Being the most affected by climate change, the region has come to the forefront of the scientific agenda and hosted several explorations for the sake of climate change.

Before the twenty-first century, Arctic exploration did not require more than the needs of explorers. Modern Arctic exploration involving oil and gas exploration development is complex and sophisticated, requiring advanced technology and collaboration, therefore, partnership with other states is often needed. On the other hand, new seaports, airports or railroads are required for this kind of an exploration, the construction of which requires employment. Thus, we can say that the impacts of exploration have also changed and grown.

In sum, we might conclude that in order to maintain national interests, states compete to increase the attraction of the activities in the Arctic. Competitiveness increases the charm of the region which brings international companies to the region and this paves the way for exploration and exploitation of new resources. Therefore, there is not a direct transition from national to multinational era, it is a gradual change. However, despite the impact of globalization, either exploration or governance of the region still serves national interests in the twenty-first century. Nevertheless, it is an undeniable fact that science constitutes a great bridge in order to link the states to reach national targets.

References

Bulkeley, R. (2010). The first three polar years – a general overview. In C. Lüdecke & S. Barr (Eds.), *The history of the international polar years (IPYs)*. Springer.

Cracraft, J. (2003). *The revolution of Peter the great*. Harvard University Press.

Finger, M., & Heininen, L. (2019). *The global Arctic handbook*. Springer.

Kaufman, W., & Macpherson, S. H. (2005). *Britain and the Americas: culture, politics and history*. ABC-CLIO.

Keohane, R. O., & Nye, J. S. (2012). *Power and interdependence*. Longman.

Koivurova, T. (2019a). "Is This The End Of The Arctic Council And Arctic Governance As We Know It?", *The Polar Connection*, 11 December 2019. (http://polarconnection.org/arctic-council-governance-timo-koivurova/) Access in February 2020)

Koivurova, T. (2019b). In S. Marzia, S. Nikolas, S. Akiho, & Z. Leilei (Eds.), *Emerging legal orders in the Arctic: the role of non-Arctic actors*. Routledge Research in Polar Law.

Le Brun Dominique. (2018). *Arctique l'Histoire Secrète: De Pythéas à Poutine, Un Combat de 2500 Ans*. Omnibus.

Lüdecke, C. (2004). The first international polar year (1882–1883): a big science experiment with small science equipment. *Proceedings of the International Commission on History of Meteorology, 1*(1), 56.

Lüdecke, C., Sukhova, G. N.'y., & Tammiksaar, E. (2010). The international polar year 1882–1883. In C. Lüdecke & S. Barr (Eds.), *The history of the international polar years (IPYs)*. Springer.

Milne, R. (2017). "*Statoil will not give up on exploration in Arctic*", *Financial Times*, 30 October 2017. (https://www.ft.com/content/ba437158-bb25-11e7-8c12-5661783e5589) Access in February 2020

Nye, J. (1990). Soft power. *Foreign Policy, 80*, 153–171.
Saul, N. (2018). California-Alaska trade, 1851–1867: the American Russian commercial company and the Russian America company and the Sale/purchase of Alaska. *Journal of Russian American Studies, 2*(1) May 2018, 1–14.
Scopelliti, M., Sellheim, N., Shibata, A., & Zou, L. (2019). *Emerging legal orders in the Arctic: the role of non-Arctic actors*. Routledge Research in Polar Law.
Tonami Aki (2019), Emerging legal orders in the Arctic: the role of non-Arctic actors, Scopelliti Marzia, Sellheim Nikolas, Shibata Akiho, Zou Leilei, Routledge Research in Polar Law,
Weyprecht, K. (1875). Scientific work of the second Austro-Hungarian polar expedition, 1872–4. *The Journal of the Royal Geographical Society of London, 45*, 19–33.
Young, O. (2019). Is it time for a reset in Arctic governance? *Sustainability, 4497*, 1–12.

Webography

European Commission. (2018). https://ec.europa.eu/fisheries/eu-and-arctic-partners-enter-historic-agreement-prevent-unregulated-fishing-high-seas-—-frequently_en Access in January 2020

Chapter 16
(Research): Indigenous Community-Based Food Security: A Learning Experience from Cree and Dene First Nation Communities

Colleen J. Charles and Ranjan Datta

Abstract This chapter is responding to food security in Indigenous communities in Canada. Using an autoethnography research framework, Indigenous meaning was explored in view of community-based food security and why it became a challenging issue for many northern Indigenous communities. The ways of Indigenous knowledge have much to offer in support of resilience against food insecurity, considering intercultural reconceptualization of research methodologies with environmental sustainability and educational programs that support Indigenous communities. The goal of this contribution is to enhance the capacities of Indigenous communities to make informed decisions about their food security short-to-long term by developing new ways of food sovereignty.

16.1 Introduction

Indigenous communities in Canada[1] have rich histories of sustainable **food sources** (Lambden et al., 2007; Schuster et al., 2012), which have been utilized with resilience in the face of climate changes historically over geologic time scales. With current climate changes (Guyot et al., 2006), the Inuit and other First Nation peoples (descending from original inhabitants thousands of years ago) are once again facing a crisis with **food security** (Ehrlich et al., 1993; United Nations, 2003; Chan et al., 2006; Douglas et al., 2014; Islam & Berkes, 2016; Deaton et al., 2019), impacting their community health and well-being. This crisis is compounded by

[1] Refers to the collective group of First Nations, Metis and Inuit peoples with an understanding that each is unique and diverse in their culture, traditions, language and worldviews.

C. J. Charles
Woodland Cree, Lac La Ronge Indian Band, La Ronge, Northern Saskatchewan, Canada
e-mail: colleen.charles@usask.ca

R. Datta (✉)
Indigenous Studies, Department of Humanities, Mount Royal University, Calgary, AB, Canada
e-mail: rdatta@mtroyal.ca

P. A. Berkman et al. (eds.), *Building Common Interests in the Arctic Ocean with Global Inclusion, Volume 2*, Informed Decisionmaking for Sustainability,
https://doi.org/10.1007/978-3-030-89312-5_16

settler colonization, which has become another severe threat to the traditional food security of remote and northern Indigenous communities (Wolfe, 2006; Council of Canadian Academies, 2014).

Recent studies show that many Indigenous people now live with **food insecurity** in Canada (St-Germain et al., 2019; Inuit Tapiriit Kanatami, 2014; Nutrition North Canada, 2015; Subnath, 2017; Arriagada, 2017). Food insecurity has the highest documented rate among First Nations and Inuit populations residing in remote and northern Canada. The women and children are particularly vulnerable, noting in Nunavut that nearly 90% of children experience hunger regularly, with 46% of households experiencing food insecurity in 2016, increasing from 33.1% in 2011 (Arriagada, 2017; St-Germain et al., 2019).

While many studies discuss food security and insecurity in Indigenous communities in Canada, few focus on **food sovereignty** (Rudolph & McLachlan, 2013; Coté, 2015) aligned with Indigenous perspectives and ways (Elliott et al., 2012; Skinner et al., 2013; Sorobey, 2013; Council of Canadian Academies, 2014). With autoethnographic storytelling (Wastasecoot, 2017) as part of formal as well as informal research and education, like planting a garden filled with questions (Datta, 2016), the goal of this chapter is to contribute to sustained Indigenous community resilience in Canada, around the Arctic and globally in view of relationships between food security, insecurity and sovereignty.

16.2 Situating Researchers with Autoethnography

Discussing who we are, where we came from and why we are writing this chapter are critical to contributing as Indigenous researchers with relational responsibility (Wilson, 2008; Denzin et al., 2008). As a lifelong unlearning and relearning process, decolonizing autoethnography (Smith, 1999, 2012; Battiste, 2013; Datta, 2018) can be applied to transform ideas into practice through education, commitment, and collaborative engagement. This methodology is applied herein with: Indigenous identity; cross-cultural knowledge; and experiences with Inuit, First Nations and Métis peoples of Northern Canada to address relationships between food security, insecurity and sovereignty in Indigenous communities facing disruptive change in the Arctic. Options for the sustainability of Indigenous communities (Datta et al., 2015; Datta 2019) are considered in view of building community-based food security. In the sense of science diplomacy and informed decisionmaking (Berkman et al., 2020), these options can be used or ignored explicitly, respecting the decision-makers inclusively. Following an Indigenous worldview, the authors developed a trusted and respectful working relationship with each other during the past decade to situate their research and education contributions from distinct Indigenous backgrounds.

Situating Author: Colleen J. Charles (CJC) I am from the Lac La Ronge Indian Band of the Woodland Cree in Saskatchewan, Canada, where there are many First Nations: Cree, Dene (Chipewyan), Saulteaux, Dakota, Lakota, Nakota and Homeland of the Metis. I was born and raised in La Ronge until the age of 16, then moved down south to the city of Prince Albert to complete my education. My mother married my biological father at age 18. They had three children together and were only married for 5 years. Both my mother and father had gone to residential school. They were physically abused for speaking their Cree language. My mother did not pass on this knowledge to her children because she did not want the same fate for her children. However, my mother continues to speak Cree fluently. My mother's parents were Cree from Montreal Lake Cree Nation and Dene from Fish River, an old fur-trading post northwest of La Ronge. My mother's father was from Fish River and his mother, my chapan (great grandmother), was from Hatchet Lake (Wollaston Lake) in the far north of Saskatchewan.

Although my father had gone to residential school, he did not lose his Cree language and is a skilled hunter and trapper today. He has a trapline by the Bow River on Highway 165, going east towards Creighton, Saskatchewan. After the separation and later divorce, I lost touch with my father and his family relations. I grew up not knowing my father's family until later in my adult years. My father's parents were Cree from the La Ronge and Stanley Mission area. My grandmother had relations with the Scottish fur traders. Currently, my father works as a Community/Cultural Support Worker for the Lac La Ronge Indian Band. He utilizes his trapline for cultural events for all age groups to teach the traditional ways of the Woodland Cree.

Situating Author: Ranjan Datta I was born and raised in minority[2] communities in Bangladesh. Like many minority communities in Bangladesh, we are displaced and killed and our women are raped. Moreover, the identity and land rights of Indigenous peoples are not recognized by the Bangladesh Government's Constitution, fomenting ongoing racism from the Bangladeshi Muslim majority (Adnan, 2004; Datta, 2020). Through graduate and post-doctoral research into professional activities, a strong relationship with Saskatchewan's Indigenous and non-Indigenous communities was developed to situate me in this autoethnographic research.

Although there are significant differences between our minority backgrounds, we focused on the similarities to learn from each other in view of community-based food security and resilience. In this study, we start from the position that Cree First Nation Elders, Knowledge-Keepers, and Leaders are scientists and educators (Fig. 16.1), whose stories offer authentic and vital metrics of change without need to be validated by Western research (Bastien, 2004).

[2]The term *minority* refers as religious and ethnic minorities, including Indigenous people in Bangladesh (Datta, 2020).

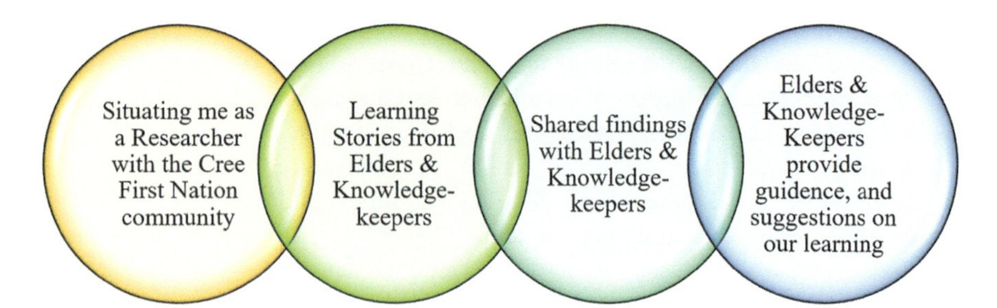

Fig. 16.1 Steps connecting our learning process with stories from Cree First Nation Elders, Knowledge-Keepers and Leaders

16.3 Indigenous Perspectives on Food Security

In terms of Indigenous food sovereignty, it is important to grasp the notion of native spirituality or cosmology and myths, as illustrated in *The Orders of the Dreamed*, piecing together the journals of George Nelson, an early fur trader in the early eighteenth century (Brown & Brightman, 1988). Nelson's manuscript contains two lengthy myths from Cree informants at Lac La Ronge: the birth and exploits of the subarctic Algonquian trickster-transformer or *Wisahkecahk* ('Wee-suck-a-jock') and the successive conflicts of transformer's son *Nehanimis* ('Nay-han-nee-mis') or the North Wind. As a young child, I (CJC) remember visiting my father's trapline and recall the stories that his father and grandmother told about *Wisahkecahk* and *Nehanimis*. The stories were scary for children, so they do not wander off in the bush and get lost. When I got older, the stories contained truths about shapeshifting.

In the *Treaty Elders of Saskatchewan*, Cardinal and Hildebrandt (2000) compare definitions and ceremonial contexts for First Nations people such as 'peoples' or 'sovereign peoples.' The meanings originate in the spiritual traditions of each of the First Nations. They form foundational components for "witaskewin" – a term that refers, in the context, to peoples establishing relationships that are to be governed by the laws of "wahkohtowin" and which are reflected in the kinds of land-sharing arrangements created between the parties. One of the Elders noted: "*And the relationships between witaskewin and wahkohtowin, they are all the same. They all have the same connotation concerning the relations of the land*." The most important values that I (CJC) was taught as a child were to 'share' and 'show respect' as basic features of the Woodland Cree culture to pass on to our children.

One of the ceremonies that was conducted is the Shaking Lodge Ceremony, where a person would go into a small barrel-shaped lodge. The shaking from side to side attributed it to the spirits entering the lodge. A primary purpose of 'conjuring' was to obtain information about persons or events distant in time and space or otherwise inaccessible. This information most often pertained to the future, but could concern the past or present. Typical questions addressed to the spirits in the lodge related the diagnosis and treatment of sickness, the location of game animals or lost articles, when the game would next be killed, the welfare of absent relatives, and the whereabouts and arrival time of visitors (Brown & Brightman, 1988).

Fast forward to the nineteenth century, the Truth and Reconciliation Commission (2015) wrote a report, and in their executive summary, they argued:

> For over a century, the central goals of Canada's Aboriginal policy were to eliminate Aboriginal governments, ignore Aboriginal rights, terminate the Treaties, and through a process of assimilation, cause Aboriginal peoples to cease to exist as distinct legal, social, cultural, religious, and racial entities in Canada. The establishment and operation of residential schools were a central element of this policy, which can be best described as "cultural genocide."

Residential schools took children from their parents for five generations. I (CGJ) attended a residential school for a brief time, but it still scarred me for my life. Including my grandparents, both my parents attended residential schools. For a long time, people did not talk about what happened in the schools, and it is only in the past decade that First Nations people started sharing their stories and began to heal from their trauma. However, there is a lot more work to be done in terms of educating the public. Although residential schools took their toll on Indigenous communities, the traditional cultural practices are making a comeback.

'The Trapline' (Fig. 16.2) represents:

Fig. 16.2 This artwork ('The Trapline') was painted by Colleen J. Charles

> As long as the river flows, there is life for the four-legged, winged, and two-legged. On the trapline, there is an abundance of chatter surrounding the trees and amongst the wind. As Woodland Cree people, that life was taken away, but it waited for us, to come back. To nourish our spirits and bodies. To give back as resilient Indigenous people that we are. For building Indigenous community-based food security, we need to instill the traditional cultural teachings to our youth. To continue for generations to come. In return, we must protect the waters, plants, animals, and the land.

Dene Ancestry The Europeans became aware of the Dene through the fur-trade system stemming from the Cree middlemen. The Europeans called them the Northern Indians or Caribou Eaters. The Cree word was "cipewayanawuk" where 'ci' is pronounced 'ch' and the English word became "Chipewyan" that means pointed skins, referring to the clothing they wore. The Chipewyans lived across the northern regions of Alberta, Saskatchewan and Manitoba. They belong to the Athapaskan language family, a group that includes the Navajo and Apache in the United States. As the legend goes, two Chipewyan brothers fought and went their separate ways. One traveled south and the other to the north. This is why they have relations with the Navajo and Apache tribes. They are referred to as the Dine. Also, the Dogrib, Gwich'in and Slavey of the Northwest Territories are a part of this family.

In Saskatchewan, the word "Chipewyan" has a derogatory term amongst the Cree due to the historical conflicts that these tribes had, including raids, stealing women and children, etc. The Cree drove the Chipewyans further north where they reside now in Treaty 10 Territory and northwest in Treaty 8 Territory. Today, they called themselves "Dene" which means "The People" or Dene Suhne, "The Real or Genuine People."

Dene Spirituality The Dene utilizes the drums made from caribou skins. They play the drums in ceremonies to pay homage to the Creator and give thanks for the food (caribou and fish), water and the land. They sing songs in their traditional language, sharing stories. In *They Will Have Our Words: The Dene Elder Project, Volume 2,* Holland and Kkailther (2003) wrote: *"The shoulder blade of the caribou – ehgala – is taken out and prepared for something, not to eat, but for be k'e?izj (marking on something). This is used to locate people."* They go on to mention, *"Another thing is about dzagor tsoe (kneecap bone). It is used to locate caribou."* Sometimes the caribou would not come in the area for a long time. When the Dene were able to hunt caribou, they made dry meat and pemmican so it will last a long time.

Colonization in the Far North Like the Cree First Nations across the Canadian provinces, the Dene and their Northwest Territory relations suffered the same impacts of colonization. Following the *Pan-Territorial On-The-Land Summit*, in a *Special Issue of Northern Public Affairs Magazine,* it was noted (Trout et al., 2018):

> The land is a source of life for all Northern People. It provides the basis in the physical, emotional, mental and spiritual wellness. Colonization, residential schools, and interference with people's ability to make decisions about their own lives have disconnected many Northern and Indigenous Peoples from their land. This disconnection had in turn led to breakdowns in connection to language, culture and identity.

The last residential school closed in 1996. In retrospect, it was not that long ago. During the last decade, Indigenous people have reclaimed their languages, ceremonies and gone back to living off the land like their ancestors did for many years. They are teaching the young people land-based education in order to pass down the traditional skills for generations to come.

Similarities Between the First Nation and the Inuit Food security between these two groups of Indigenous people include the caribou which is a staple in their diet. However, there has been a ban to hunt caribou because of the decline in numbers in the last decade, corresponding to "*spikes in mineral staking in the 1990s and 2000s*" even though "*the scapegoat is the Indigenous person who depends on caribou for subsistence*" (Weber, 2018).

Communities feasted on caribou more than two times per week. When the government banned caribou hunting for a period of time, the Indigenous people felt a huge impact when it came to store-bought food because it was very expensive. Moreover, as noted by Vowel (2016):

> going out and hunting is an essential part of a hunting culture, but the act itself is not everything. The focus on hunting informs the language, the traditions, the stories, the music, and the art.

In a Woodland Cree perspective, my (CJC) Cree father continues to hunt moose, deer, bears, beaver, fish, rabbits, ducks, as well as, picking berries and mushrooms in northern Saskatchewan. The chanterelles (wild mushrooms) are a delicacy in some countries around the world.

Cree and Dene First Nation community's focus is on the youth to show and teach them the ways of traditional life to live off the land. The concept of sharing and showing respect for all living things and Elders, especially, is to be grounded and connected to families, First Nation communities, the land and the Creator. Indigenous food sovereignty encompasses all aspects of First Nations traditional ways of knowledge. One cannot live without the other.

Residential schools broke spirits, but not the souls. Some are stronger than others. For example, I (CJC) am a single mother of three children with a Master's degree in Education from the University of Saskatchewan. My mother told me education is essential and I completed my Master's degree by driving every week for 2 years from La Ronge to Saskatoon and back. It is about determination and resilience. Currently, I live on a trapline (Fig. 16.2) near Two Forks River, south of La Ronge. It is about getting back to nature and learning the old ways.

As an Indigenous Studies instructor from Northlands College, I (CJC) gave a university class assignment to interview an Elder in the community. One of my former students interviewed an Indigenous judge who was born and raised in La Ronge and was a student at Timber Bay Boarding School. When he was in grade 11, he questioned his education and noted that he did not need degree 12 to be a hunter, trapper or fisherman. Throughout his education and employment, he maintained his practice of traditional lifestyle in the north. In 2006, he was appointed a Judge with the Queen's Bench. Now that he is retired, he spends his free time on his trapline hunting and lake fishing (Otter, 2019). This is an example of how we can continue our traditional food practice.

16.4 Why Is Food Insecure in Indigenous Communities?

Many interconnected factors contribute to food insecurity in the Cree communities in Canada. In sharing their stories, the Elders and Knowledge-Keepers discuss many challenges. These challenges to the Cree First Nation's everyday food security are interconnected with their land rights, as well as with their values and priorities (Fig. 16.3).

Lack of Indigenous Land Rights with Ongoing Colonization Many Elders and Knowledge-Keepers suggest that the lack of Indigenous rights to their land and water is one of the significant challenges for building food security in Indigenous communities. For instance, if we have our land (land refers to both land and water) rights, we can build our food security. Without our land rights, we will not be able to build our food security.

Cree Elders and Knowledge-Keepers suggest that without understanding "*our past and ongoing colonization, there would not be any food security for the community.*" One of the Cumberland House Cree Nation Elders gave examples of how colonization created food insecurity in the community: "*We did not know that we were poor until colonizers told us. Colonizers killed or displaced our caribou and gave us fake foods (i.e., store canned foods).*" A knowledge-keeper from the same community suggests further that the term food security is a colonial term, noting: "*why should we learn the western concept of food security when we (i.e., Cree people)*

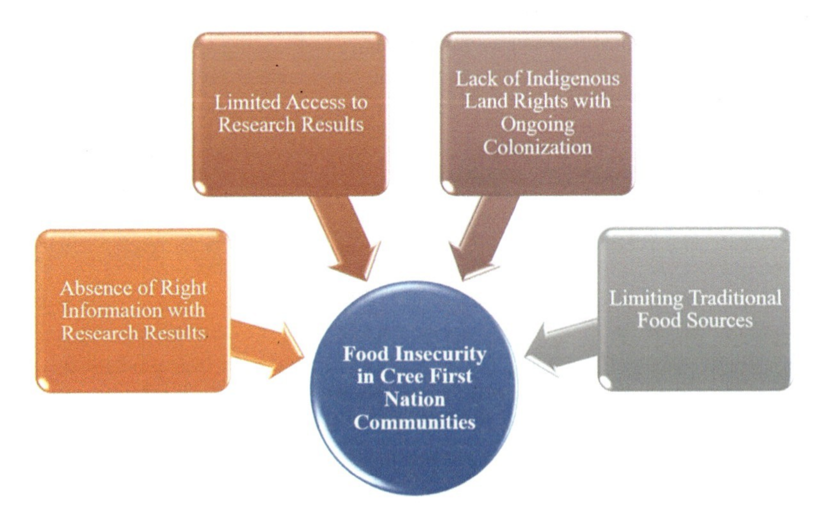

Fig. 16.3 Some factors that create food insecurity in Cree First Nation communities

do know how to build our food sustainability". Elders and Knowledge-Keepers additionally listed some of the significant measures of ongoing colonization in the Cree communities, including forced relocation, politics of assimilations, degradation of cultural practice and beliefs, increased self-reliance on dependency, decreased individual and community capacity, changed food habits, and created food myths.

Limiting Traditional Food Sources Traditional food sources have been decreasing significantly because of settlers' industrial projects (such as pipeline, diamond mining and industrial projects), according to Knowledge-keepers. Many Elders indicated that populations of traditional animals (such as moose, caribou, grizzly bears, muskrats, geese, char, and various fishes) are listed as decreasing, some of them dangerously so. Similarly, many plant-based food sources are also declining, including blackberries, cranberries, akpiks (fish and berries), and blueberries. Another James Smith Cree Nation Community Elder said: "*we should not consider all food as foods, such as sugar pops*".

Limited Access to Research Results Without current research, the communities are unable to access updated information on food deficiency. Many Elders explained that although academic, governmental, and non-governmental bodies are doing significant research, little of it involves Cree First Nation people in their study. For instance, an Elder from James Smith Cree Nation explained:

> Current scientific research demonstrates a limited understanding of Cree holistic systems. Scientific researchers only take a scientific approach. Such research is commonly focused on the identification of unique attributes based on specific hypotheses and vulnerabilities and is centered on cause and effect correlation. They do not understand that a single quality or theory cannot explain our knowledge system.

Absence of Right Information with Research Results Another Knowledge-Keeper from James Smith Cree Nation further noted:

> There is little documentation of indicators of health and wellness throughout an entire ecosystem as defined by Cree First Nation Elders and Knowledge-Keepers. There is a lack of Cree First Nation-initiated and -defined food research.

For the community Elders, and Knowledge-Keepers, many factors are causing Cree First Nation food insecurity and the issues outlined above (Fig. 16.3) need to be taken seriously so that Cree First Nation communities can build food security.

16.5 Community-Based Food Security with Food Sovereignty

While there is no single, widely accepted definition of community-based food security, it is connected with Indigenous ways of knowing and doing, according to many Elders and Knowledge-Keepers. Cree First Nation Elders and Knowledge-Keepers explained that community-based food security is part of their food sovereignty, referring to Indigenous ways of life, their sacred meanings, collaborative work, self-determination and self-governance, land rights, responsibilities, and collective decisionmaking. Cree First Nation Elders and Knowledge-Keepers suggest that the end goal of community-based food security is long-term food sovereignty.

Land, Water, Culture, and Language Protection Rights According to Cree First Nation Elders and Knowledge-keepers, community-based food security is interconnected with the protection of Indigenous land, water, culture, and language. As Elder from Little Pine First Nation explained:

> without achieving First Nation Treaty rights [i.e., land rights, language protection], the First Nation will not get food security. Our treaty rights on our land, water, culture, and language are our food security.

Another Knowledge-Keeper from James Smith Cree Nation clarifies how Cree First Nation rights on their land, water, culture, and language will create community food security:

> Our language and culture help us to understand animals' behaviors when to hunt when to protect. Through our ancestors' knowledge, we learned how to hunt successfully. Our animals and we are relationally connected. We need to protect each other. Therefore, we need our land rights so that we can protect our animals.

In a similar point, another Knowledge-Keeper from Cumberland House Cree Nation suggests:

> Indigenous people know the best way to their food security as they were living in their land for thousands of years. Indigenous people should have land rights to developing their food security. Indigenous should have their full rights on their hunting and gathering for their food security.

First Nation land, water, culture, and language rights are an inseparable part of First Nation food security. Without these rights, food security is impossible.

Holistic Perspective We apply the term 'holistic perspective' in view of informed decisionmaking with contributions from diverse participants (Berkman, 2019), referring to an "*international, interdisciplinary and inclusive (holistic) process*" with science diplomacy (Berkman et al., 2020). Holistic perspective is a continuous informed decisionmaking process for creating balance among different stakeholders. For instance, Cree First Nation food security refers to a comprehensive understanding of their ecosystem, including environmental, physical, cultural, mental and spiritual worldviews. Learning is always holistic, a lifelong process, connected with local ways of knowing and doing, interconnected with experimental, formal and informal knowledge.

The Cree First Nation food-related knowledge, values, and wisdom that guide the present-day food sovereignty movement in First Nation communities was built up over thousands of years. For instance, a Knowledge-Keeper from James Smith Cree Nation offered that "*food sovereignty is when we have access to land and access to seeds to grow foods*". Another Elder explains perspectives further by saying, "*if you're producing enough corn on your land to last through the entire year, you have both equity and food security.*"

Indigenous Ways of Life Food security is connected with the First Nation community's ways of life, ways of knowing, and ways of doing. The community-based food security model is connected with all of the means of life, and it is foundational to Cree First Nation culture. A Knowledge-Keeper explained why Cree First Nation ways of knowing are essential for their community:

> Our ancestors have been sustainability living in this land for hundreds of years, they knew how to build sustainable food source for their family and community, their knowledge is very significant for our future generation.

Another Knowledge-Keeper explained further, why Indigenous knowledge is essential for food security: "*It is not just about hunting, fishing, it is our way of life. Our food defines who we are. We need to commit collectively to fight for food security*".

Sacred The Indigenous meanings of community-based food security are sacred and divine, highlighted by a Knowledge-Keeper with the observation that food is a "*gift from the Creator; in this respect, the right to food is sacred and cannot be constrained or recalled by colonial laws, policies, and institutions.*" An Elder further noted:

> Every food that we eat, we pray to them from the time we wake up to sleep from when we born to when we die. Our way of living with our food relative is sacred.

Consequently, food sovereignty is a sacred responsibility to maintain a healthy relationship with our land, plants, and animals that provide us with our food, as observed by another Elder: "*everything is alive; the animals, fish, and plants are just like you and me.*"

Collaborative Decisionmaking The Cree First Nation community refers to foods as collaborative ways of knowing and doing. Collaborative decisionmaking is fundamentally based on collaborative action or the day-to-day practice of maintaining cultural harvesting strategies. One of the Elders from Cumberland House Cree Nation suggested that:

> Recognizing the need for such crosscutting systemic action, the Cree food security strategy proposes a collective vision and common agenda for impact rooted in Cree First Nation values and knowledge.

This collaborative decisionmaking includes accessible health and wellness decisionmaking power and management of the tools that are needed to support those components, including knowledge sources. Collaborative decisionmaking is about "*making sure that everyone is fed and enjoying their time.*" Thus, to maintain Cree First Nation food sovereignty as a living reality for both present and future generations continued participation in cultural harvesting strategies at all levels, individual, family, community and regional is vital.

16.6 How to Develop Indigenous Community-Based Food Sovereignty?

Self-Determination and Self-Governance The self-determination in Indigenous community-based food security refers to the ability of the Cree First Nation to respond to their own needs for healthy, culturally adapted Indigenous foods. An Elder from Little Pine explains, "the ability to make decisions over the amount and quality of food we hunt, fish, gather, grow, and eat. Freedom from dependence on grocery stores or corporately controlled food production, distribution, and consumption in industrialized economies." Similarly, a Knowledge-Keeper explained that self-determination in community-based food security is "where they promote leadership and establish communication and support networks within each community". This Knowledge-Keeper again suggested that self-determination is vital for achieving Cree First Nation community's food security as it allows them to:

> generate their ideas of their vision for their production systems and how they want them to be, and then go through the communication skills to build a proposal for support for that and the elaboration of the designs.

Cree First Nation community-based food security relates to their self-decisionmaking power. The self-decisionmaking power in the Indigenous family and community refers to self-governance, appreciating the collaboration is involved (von der Porten et al., 2016; Delaney, 2017). According to Elders and Knowledge-keepers, through self-governance, the Cree First Nation will gain several essential "abilities" to:

- Use and value Cree First Nation knowledge to manage daily activities;
- Build and rely on self-governance across space and time;
- Use their knowledge system in synergy with other knowledge systems, such as Western science;
- Manage human activities within the Arctic environment and to better understand changes occurring;
- Apply holistic knowledge to understanding the Indigenous environment through Indigenous knowledge philosophies and methodologies;
- Manage activities within the Indigenous in a way that ensures younger generations will have healthy and nutritious foods to harvest;
- Control over their fate;
- Use their cultural value systems.

Natural Law Food security is the natural right of all Cree First Nation to be part of the ecosystem, to access food, and to care for, protect, and respect all of life's land, water, and air. The essence of food sovereignty is the right as Cree First Nation to define its hunting-gathering, fishing, plant, and water policies. The right to define what is sustainable socially, economically, and culturally appropriate to maintain ecological health. The right to determine environmental and culturally sustainable practices for the distribution of food and the right to define and manage their food systems.

Indigenous Lead Research Almost all Elders and Knowledge-Keepers argue for Cree First Nation lead research. Cree First Nation people should manage their research according to the traditions and culture of their communities. The Elders and Knowledge-Keepers should guide all forms of research so that they have full access to the results and science (Fig. 16.3). During my (CJC) conversation with the Elders and Knowledge-Keepers, they suggested research should follow the 5Rs—Relationship, Responsibility, Relevance, Respect, and Relearn—in community-based food sovereignty (Fig. 16.4). They explain that developing the 5Rs begins by asking questions about your role as a researcher. Questions also are the foundational level with informed decisionmaking (Berkman et al., 2020).

- *Relationships*: What relationships are formed?
- *Relevance:* What do Indigenous communities need or want?
- *Respect*: Do I respect, acknowledge, and honour community knowledge and practice?
- *Responsibility*: What must I do to achieve community-centered success?
- *Relearn*: Am I willing to learn from the community?

An Elder further explains that responsibility for researchers in Cree First Nation food security is "*making sure that the project is for community capacity building. The community needs to lead the projects for their community food security*".

Intergenerational Knowledge Indigenous community-based food security includes the responsibility and ability to pass on knowledge to the younger generation, the taste of traditional foods rooted in place. It is also concerned with how to

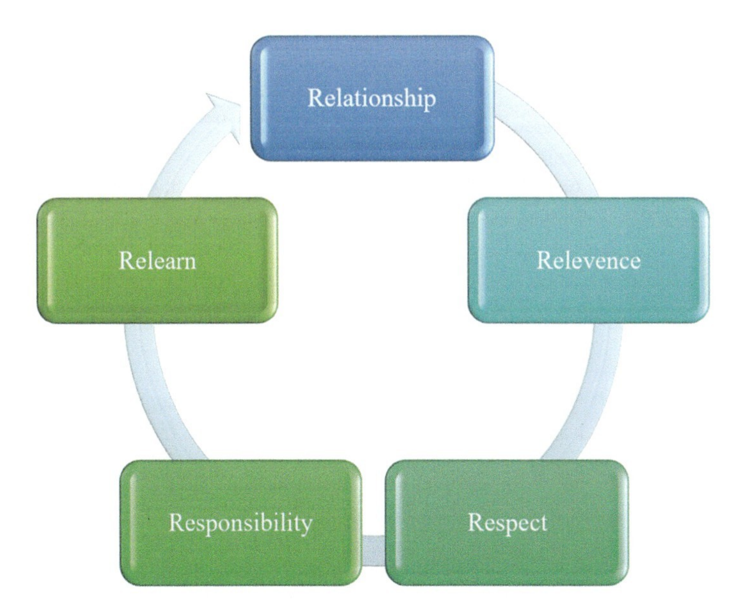

Fig. 16.4 This relational circle shows the '5Rs' that are interconnected with Cree First Nation research

safely obtain and prepare traditional foods for medicinal use, clothing, housing, nutrients. Overall, understanding food security more broadly includes the recognition that food is a lifeline and a connection between the past, present, and future, and between the self and cultural identity.

Elders and Knowledge-Keepers Teachings The Elders and Knowledge-Keepers are considered teachers in the many Cree First Nation communities. They are respected persons within the communities. A Little Pine Chief explained:

> We will not have self-sufficiency without our Elders and Knowledge-Keepers. Our Elders and Knowledge-Keepers are our scientists for our food security.

Similarly, thinking across generations, a Knowledge-Keeper from James Smith community said:

> The kids are in the dark if you don't teach them. Sit down, talk with our kids, and explain things to our kids. Our Elders and Knowledge-Keepers can play a significant role in this.

Another Elder from Cumberland House Cree Nation suggested they could build traditional food security through "*revitalizing knowledge on learning how to relearn the traditional food system.*"

Elders' and Knowledge-Keepers' teachings are the pathways to accomplish food security, learning how animals behave and then using that knowledge to learn how to hunt successfully.

> When you live in an area, you become part of the environment; we are part of the environment. We have been sustaining this environment for thousands of years without degrading it.

As another Elder noted: "*We need to hunt as we were taught. Our knowledge tells us how to be within this world.*"

The Bridge Between Academics and Communities Building a bridge between academics and communities is fundamental to build common interests in Indigenous perspectives in areas beyond national jurisdictions, as in the high seas of the Central Arctic Ocean. For instance, a Little Pine Chief suggested that the bridging between scholars and community is essential to:

> build meaningful networks from the ground up, and academic researchers do need to respect participants and count community members as equals and bearers of knowledge.

Besides, academics need to prioritize the Cree First Nation's needs in deciding where to start and encourage people to use Indigenous languages and hire as many people as possible who speak Indigenous languages and are from the communities.

Food Storage The need for food storage is an integral part of community-based food security. A Knowledge-Keeper explained how limited traditional food storage is in their community. He further observed the limited use of traditional food storage could have many negative impacts in the community, including reducing income-earning opportunities, disrupting sharing networks all the while limiting opportunities for youth to acquire harvesting knowledge and skills.

Building community-based food security in Indigenous communities requires community informed policy changes that strengthen community-based harvester support programs (e.g., funding for hunter and trapper organizations), it requires investment in infrastructure and skills development and support for community wellness programs. Always it must be part of a broader effort toward poverty reduction, community development, reconciliation and healing.

16.7 Conclusion

We learned from my conversations with Cree First Nation Elders, Knowledge-Keepers, and Leaders that the consequences of food insecurity are far-reaching with adverse effects on physical and mental health for both children and adults. Food insecurity leads to increasing reliance on store foods and disrupted sharing networks as well as limited opportunities for youth to acquire harvesting knowledge and skills. In turn, store foods require income-earning, which underlies many other crises, including poverty.

The Cree First Nation community-based food security is about the communities surviving and protecting their culture and heritage. We learned that food insecurity remains a severe public health and human rights issue for Indigenous communities. According to Elders and Knowledge-Keepers, First Nation food security is all about life, it is not a project, it is not a program, and it is not a service and not an event. It is an ongoing lifelong process. We learned that Cree First Nation Elders and Knowledge-Keepers stressed the importance of community-based food security, which relies on relational ways of understanding and doing. According to Elders and Knowledge-Keepers, the Cree First Nation community-based food security framework is:

- *Community-driven*: the community establishes direction and goals;
- *Locally connected*: built on the particular cultural strengths, traditions, and values of the communities they serve;
- *Capacity building*: create lasting value and build capacity for the community;
- *Deep collaboration*: aspire to build authentic partnerships that support the long-term development goals of Indigenous communities.

Inuit Circumpolar Council-Alaska (2015) also suggests a similar food security framework for Indigenous food sovereignty in Alaska. This study suggests that food sovereignty is the right of Indigenous peoples to define their hunting, gathering, fishing, land and water policies; the right to define what is sustainable, socially, economically and culturally appropriate for the distribution of food and to maintain ecological health; the right to obtain and maintain practices that ensure access to tools needed to obtain, process, store and consume traditional foods.

The Cree First Nation Elders and Knowledge-keepers suggested a similar understanding of community-based food security where everyone brings their strengths and resources, The Elders, for example, bring their knowledge about food, and the hunters contribute the meat.

The Elders, Knowledge-Keepers and other community members suggested several options to build community-based food security in Indigenous communities (Figs. 16.3 and 16.4). Valuing respectful relationships with Indigenous communities is significant, as is:

- Respecting Indigenous laws and values;
- Respecting Indigenous environmental stewardship in the community;
- Knowing the importance of Indigenous food collection and sharing;
- Caring for a collective decisionmaking process;
- Becoming collective, valuing and practicing Indigenous skills; and
- Seeking resourceful solutions.

The Cree First Nation community-based food security is a community-led model of growing their food, being in control of what they eat and how they produce their food, and their right to food sovereignty. One of the Elders said it best:

> When I think about the idea that you do not have to go hungry, the idea that you can feed yourself. That you do not have to worry about where your next meal is coming from. It is part of our food sovereignty. I think of Cree First Nation Nations of people having the ability to feed the entire family, the entire region. They can feed themselves either next year, even generations to generations.

References

Adnan, S. (2004). *Migration land alienation and ethnic conflict: Causes of poverty in the Chittagong Hill Tracts of Bangladesh*. Research & Advisory Services.

Arriagada, P. (2017). *Food insecurity among Inuit living in Inuit Nunangat*. Statistics Canada. https://www150.statcan.gc.ca/n1/en/pub/75-006-x/2017001/article/14774-eng.pdf?st=biMKT0Ld

Bastien, B. (2004). *Blackfoot ways of knowing*. University of Calgary Press.

Battiste, M. (2013). *Decolonizing education: nourishing the learning spirit*. Purich Press.

Berkman, P. A. (2019). Evolution of science diplomacy and its local-global applications. *European Foreign Affairs Review., 24*(2/1), 63.

Berkman, P. A., Young, O. R., & Vylegzhanin, A. N. (2020). Book series preface: Informed decisionmaking for sustainability. In O. R. Young, P. A. Berkman, & A. N. Vylegzhanin (Eds.), *Informed decisionmaking for sustainability. Volume 1. Governing Arctic seas: Regional lessons from the Bering Strait and Barents Sea* (pp. v–xxv). Springer. https://link.springer.com/content/pdf/bfm%3A978-3-030-25674-6%2F1.pdf

Brown, J. H. S., & Brightman, R. (1988). *The orders of the dreamed: George Nelson on Cree and Northern Ojibwa, religion and myth, 1823*. University of Manitoba Calgary Press.

Cardinal, H., & Hildebrandt, W. (2000). *Treaty elders of Saskatchewan: Our dream is that our peoples will one day be clearly recognized as nations*. University of Calgary Press.

Chan, H. M., Fediuk, K., Hamilton, S., Rostas, L., Caughey, A., Kuhnlein, H., Egeland, G., & Loring, E. (2006). Food security in Nunavut, Canada: Barriers and recommendations. *International Journal of Circumpolar Health, 65*(5), 416–431. https://www.tandfonline.com/doi/abs/10.3402/ijch.v65i5.18132

Coté, C. (2015). "Indigenizing" food sovereignty. Revitalizing indigenous food practices and ecological knowledges in Canada and the United States. *Humanities, 5*(57), 1–14. https://www.mdpi.com/2076-0787/5/3/57

Council of Canadian Academies. (2014). *Aboriginal food security in northern Canada: An assessment of the state of knowledge* (p. 193). Expert Panel on the State of Knowledge of Food Security in Northern Canada, Council of Canadian Academies.

Datta, R. (2016). Community garden: A bridging program between formal and informal learning. *Cogent Education, 3*(1(1177154)), 1–15. https://www.tandfonline.com/doi/full/10.1080/2331186X.2016.1177154

Datta, R. (2018). Decolonizing methodologies: A transformation from science-oriented sociology researcher to relational/participant-oriented researcher. *American Indian Culture and Research Journal, 42*(1) http://uclajournals.org/doi/pdf/10.17953/aicrj.42.1.datta

Datta, R. (2019). Clarifying the process of land-based research, and the role of researcher(s) and participants. *Ethics in Science and Environmental Politics*. https://www.int-res.com/articles/esep2019/19/e019p001.pdf

Datta, R. (2020). *Land-water management and sustainability in Bangladesh: Indigenous practices in the Chittagong Hill Tracts*. Routledge.

Datta, R., Khyang, U. N., Khyang, H. K. P., Kheyang, H. A. P., Khyang, M. C., & Chapola, J. (2015). Understanding Indigenous sustainability: A community-based participatory experience. *Revista Brasileira de Pesquisa em Educação em Ciências, 14*(2), 99–108.

Deaton, B. J., Scholz, A., & Lipka, B. (2019). An empirical assessment of food security on First Nations in Canada. *Canadian Journal of Agriculture Economics, 68*(1), 5–19. https://onlinelibrary.wiley.com/doi/abs/10.1111/cjag.12208

Delaney, D. (2017). The Master's tools: Tribal sovereignty and tribal self-governance contracting/compacting. *American Indian Law Journal, 5*(2), Article 1.

Denzin, N., Lincoln, Y., & Tuhiwai Smith, L. (2008). *Handbook of critical and indigenous methodologies*. Sage.

Douglas, V., Chan, H. M., Wesche, S., Dickson, C., Kassi, N., Netro, L., & Williams, M. (2014). Reconciling traditional knowledge, food security, and climate change: Experience from Old Crow, YT, Canada. *Progress in Community Health Partnerships: Research, Education, and Action, 8*(1), 21–27. https://muse.jhu.edu/article/545087

Ehrlich, P. R., Erlich, A. H., & Daily, G. C. (1993). Food security, population, and environment. *Population and Development Review, 19*(1), 1–32. https://doi.org/10.2307/2938383

Elliott, B., Jayatilaka, D., Brown, C., Varley, L., & Corbett, K. K. (2012). "We are not being heard": Aboriginal perspectives on traditional foods access and food security. *Journal of Environmental and Public Health*, Article ID 130945. 9p. https://doi.org/10.1155/2012/130945.

Guyot, M., Dickson, C., Paci, C., Furgal, C., & Chan, H. M. (2006). Local observations of climate change and impacts on traditional food security in two northern Aboriginal communities. *International Journal of Circumpolar Health, 65*(5), 403–415. https://www.tandfonline.com/doi/abs/10.3402/ijch.v65i5.18135

Holland, L., & Kkailther, M. (2003). *They will have our words: The Dene Elders Project* (Vol. 2). Holland-Dalby Educational Consulting.

Inuit Circumpolar Council-Alaska. (2015). *Alaskan Inuit Food Security Conceptual Framework: How to assess the Arctic from an Inuit perspective*. Technical Report. Anchorage, AK. https://iccalaska.org/wp-icc/wp-content/uploads/2016/05/Food-Security-Full-Technical-Report.pdf

Inuit Tapiriit Kanatami. (2014). *Social determinants of Inuit health in Canada*. Inuit Tapiriit Kanatami.

Islam, D., & Berkes, F. (2016). Indigenous peoples' fisheries and food security: A case from northern Canada. *Food Security, 8*, 815–826. https://doi.org/10.1007/s12571-016-0594-6

Lambden, J., Receveur, O., & Kuhnlein, H. V. (2007). Traditional food attributes must be included in studies of food security in the Canadian Arctic. *International Journal of Circumpolar Health, 66*(4), 308–319. https://www.tandfonline.com/doi/abs/10.3402/ijch.v66i4.18272

Nutrition North Canada. (2015). *Nutrition North Canada*. Indigenous and Northern Affairs Canada.

Otter, M. (2019). *Traditional lifestyle in northern Saskatchewan: Interview with Donald Bird.* (Personal Communication with author-2). Indigenous studies 262. La Ronge: SK Northlands College

Rudolph, K. R., & McLachlan, S. M. (2013). Seeking Indigenous food sovereignty: Origins of and responses to the food crisis in northern Manitoba, Canada. *Local Environment, 18*(9), 1079–1098. https://www.tandfonline.com/doi/abs/10.1080/13549839.2012.754741

Schuster, R. C., Wein, E. E., Dickson, C., & Chan, H. M. (2012). Importance of traditional foods for the food security of two First Nations communities in the Yukon, Canada. *International Journal of Circumpolar Health, 70*(3), 286–300. https://www.tandfonline.com/doi/abs/10.3402/ijch.v70i3.17833

Skinner, K., Hanning, R. M., Desjardins, E., & Tsuji, L. J. S. (2013). Giving voice to food insecurity in a remote indigenous community in subarctic Ontario, Canada: Traditional ways, ways to cope, ways forward. *BMC Public Health, 13*(427), 1–13. https://bmcpublichealth.biomedcentral.com/articles/10.1186/1471-2458-13-427

Smith, L. (1999). *Decolonizing methodologies: Research and Indigenous peoples*. Zed Books; University of Otago Press.

Smith, L. T. (2012). *Decolonizing methodologies: Research and Indigenous peoples* (2nd ed.). Zed Books.

Sorobey, M. (2013). *Northwest Company*. Paper presented at northern exposure 2 conference: Realities of remote logistics, Winnipeg, MB.

St-Germain, A. F., Galloway, T., & Tarasuk, V. (2019). Food insecurity in Nunavut following the introduction of Nutrition North Canada. *Canadian Medical Association Journal = journal de l'Association medicale canadienne, 191*(20), E552–E558. https://doi.org/10.1503/cmaj.181617

Subnath, M. (2017). Indigenous food insecurity in Canada: An analysis using the 2012 aboriginal peoples survey. *Electronic Thesis and Dissertation Repository., 4459*. https://ir.lib.uwo.ca/etd/4459

Trout, L., Kirk, T., Erickson, M., & Kleinman, A. (2018). Place-based continuing medical education in the rural north. *Northern Public Affairs Magazine*. http://www.northernpublicaffairs.ca/index/volume-6-special-issue-2-connectivity-in-northern-indigenous-communities/place-based-continuing-medical-education-in-the-rural-north/

Truth and Reconciliation Commission of Canada. (2015). *Honoring the truth, reconciling for the future*. Summary of the Final Report for the Truth and Reconciliation of Canada.

United Nations. (2003). *Trade reforms and food security: Conceptualizing the linkages*. United Nations Food and Agriculture Organization. http://www.fao.org/3/y4671e/y4671e00.htm#Contents

von der Porten, S., de Loë, R. C., & McGregor, D. (2016). Incorporating indigenous knowledge systems into collaborative governance for water: Challenges and opportunities. *Journal of Canadian Studies, 50*(1), 214–243.

Vowel, C. (2016). *Indigenous writes: A guide to first nations, Metis & Inuit issues in Canada*. Canadian Council for the Arts, Highwater Press.

Weber, B. (2018). *Caribou hunting quotas make scapegoats out of northern First Nations*. The Canadian Press. https://www.ctvnews.ca/canada/caribou-hunting-quotas-make-scapegoats-out-of-northern-first-nations-study-1.3822892?cache=yesclipId104062%3FclipId%3D89950

Wastasecoot, B. I. (2017). *Showing and telling the story of Nikis (my little house): An arts-based autoethnographic journey of a Cree adult educator*. Doctoral dissertation, University of Toronto. https://tspace.library.utoronto.ca/handle/1807/82394.

Wilson, S. (2008). *Research is ceremony: Indigenous research methods*. Fernwood Publishers.

Wolfe, P. (2006). Settler colonialism and the elimination of the native. *Journal of Genocide Research, 8*(4), 387–409.

Chapter 17
(Research): From Global to Local Climate Change Governance: Arctic Cities' Perceptions of the Uses of Expert Knowledge

Nadezhda Filimonova

Abstract This chapter explores the uses of expert knowledge in local climate change policymaking based on publicly available official documents. The views of Russian city administrators in Norilsk and Yakutsk regarding engagement with expert knowledge are used as case studies. The chapter concludes that Norilsk and Yakutsk authorities apply expert knowledge to create a comfortable and safe urban environment and foster the cities as centres for scientific research, innovation, and education. The data analysis reveals three major purposes for expert knowledge in (1) developing and implementing policies; (2) becoming centres for science and innovation, and (3) training and retraining of specialists. The research on Arctic cities' engagement with expert knowledge is important in connection to their future sustainable and climate-resilient development at a time of ongoing rapid transformations in the region.

17.1 Introduction

In face of the projected continuing climate change (AMAP, 2017), scholars emphasize the present and future key roles of cities in climate change governance (Hirte et al., 2018; Woodruff, 2018; Vogel & Henstra, 2015). Different studies have examined factors accounting for encouraging and discouraging cities' engagement with climate change issues (Broto, 2017; Bulkeley & Betsill, 2013; Patterson & Van der Grijp, 2019). Several authors have identified the importance of scientific information and expertise for municipal climate change actions (Dodman & Carmin, 2011; Hughes & Romero-Lankao, 2014).

N. Filimonova (✉)
University of Massachusetts Boston, Boston, MA, USA
e-mail: n.filimonova001@umb.edu

P. A. Berkman et al. (eds.), *Building Common Interests in the Arctic Ocean with Global Inclusion, Volume 2*, Informed Decisionmaking for Sustainability,
https://doi.org/10.1007/978-3-030-89312-5_17

However, most current studies are focused on national and global levels (Hughes & Romero-Lankao, 2014), whereas the engagement of cities with scientific expertise represents an emerging area of academic research (Lundin & Öberg, 2014; Van Stigt et al., 2015). Notably, grasping cities' views on uses of expert knowledge helps to unpack the role of expertise in local decisionmaking (Lundin & Öberg, 2014). Consequently, the study of how local governments engage with expert knowledge allows one to examine why urban climate change policies exist in their present form and what actors and factors shape urban climate change policymaking.

The present study aims to contribute to this body of work by focusing on how cities perceive the uses of expert knowledge to address climate change impacts. In the context of this chapter, expert knowledge of climate change impacts refers to an understanding of and acquaintance with a body of information and facts produced through research-based methods by scientists in academic and research institutions. It is also generated through constant use in advising and consultancy by practitioners to enhance local climate change decisionmaking. Experts thus are represented by both scientists and practitioners in climate change impacts. However, as opposed to formal scientific knowledge, "much expert knowledge is informal and undocumented, remaining hidden until it is expressed for a specific application" (Drescher et al., 2013, p. 2). Therefore, there is a need for more research to shed light on the role of expert knowledge in urban climate change governance.

Informed by the science-policy interface and urban climate change governance literature, this chapter presents preliminary research on Arctic urban authorities' perceptions of the uses of expert knowledge, particularly in the Russian cities of Norilsk and Yakutsk. The topic of Arctic cities' engagement with expert knowledge is important in connection with their future sustainable and climate-resilient development at a time of the ongoing rapid transformations in the region.

These cities were selected based on several criteria: as being the largest cities located on permafrost, having educational and research institutions, and the status of their climate change policies. Norilsk and Yakutsk have long-standing histories of urban infrastructure development on permafrost. In the 1940s, the first buildings resting on piles were constructed in Yakutsk; in the 1950s, this technique of building construction was applied by engineers in Norilsk. Additionally, the cities envision their development as scientific and innovative centers by interacting with a scholarly community. Finally, Norilsk and Yakutsk authorities have not yet adopted any climate change strategy.

Despite the absence of direct access to the Arctic Ocean, Norilsk and Yakutsk capacities and development are associated with the Northern Sea Route (NSR). Specifically, the port of Dudinka links Norilsk to Europe and Asia since its materials (copper, nickel, cobalt) are shipped from this port via the NSR to global markets. Additionally, several scholars discuss the prospects for cargo transportation from Yakutsk via the port of Tiksi and the NSR to Russian and European markets (Tarasov et al., 2018).

Furthermore, the impacts of climate change on urban infrastructure might result in soil and river pollution that is linked to the marginal seas of the Arctic Ocean. A prominent example is the spill of 20,000 tons of diesel oil that occurred in Norilsk in

May 2020. This spill may have been the result of damage from thawing permafrost (Fedorinova, 2020). Such accidents prove the necessity for increased attention from scientists and politicians to the impacts of climate change in the Arctic urban context. Generally, the selection of Russian Arctic cities as case studies is dictated by the fact that the topic of urban climate change governance in Russia has not been addressed systematically in the scholarly literature (Martus, 2019; Van der Heijden et al., 2019).

The chapter is structured as follows: (1) the first section presents an overview of theoretical and empirical research on the uses of expert knowledge in urban governance; (2) the second section presents the case studies of Norilsk and Yakutsk and outlines the applied methodology; and (3) the following sections discuss the cities' framing of climate change impacts and their perceptions of the uses of expert knowledge. The chapter concludes that the city administrations of Norilsk and Yakutsk frame climate change as one of the factors impacting permafrost degradation and negatively affecting the construction and maintenance of roads and buildings. The research reveals that Norilsk and Yakutsk authorities apply expert knowledge to create comfortable and safe urban environments and develop the cities into centers for scientific research, innovation, and education. Data analysis reveals three major purposes for using expert knowledge: (1) developing and implementing policies; (2) becoming centres for science and innovation, and (3) training and retraining of specialists.

17.2 Uses of Expert Knowledge by Local Governments

To understand how local governments apply expert knowledge to address climate change impacts, this section discusses existing academic approaches regarding the role of expert knowledge in urban governance. Baker et al. (2006) and Hardoš (2018) conclude that major topics emerging from academic approaches to the conceptualization of the notion of the expert include possession of knowledge, information, and expertise on a topic. In more specific terms, Brand and Karvonen (2007, pp. 27–29) identify four types of experts in relation to sustainable development such as an outreach expert who communicates knowledge and information to the general public; an interdisciplinary expert who possesses knowledge from different disciplines; a meta-expert who identifies linkages between various disciplines and translates the relevance of each discipline for a problem solution; and a civic expert who enhances communication with experts and non-experts.

Expert knowledge might be acquired from personal experience, education, and training, as well as being a result of the obtained personal skills and academic research (Burgman et al. 2011; Larrick & Feiler, 2016; Hardoš, 2018). Reay (2007) claims that an expert's authority rests upon their professional connection to a body of scientific knowledge and their recognition in various organizational settings as possessors of important skills. One of the discussed topics in the academic literature is the importance of expert knowledge for policymaking (Hordijk & Baud, 2006, Spruijt et al., 2014; Van Stigt et al., 2015). Hordijk and Baud (2006) state that, like national governments, local governments need expert knowledge and information to elaborate, monitor, legitimize and evaluate their policies, as well as to incorporate target groups and their needs while developing their local actions. Furthermore, Van Stigt et al. (2015) add that expert knowledge helps decisionmakers to better comprehend a political issue. Dodman and Carmin (2011) note that cities might use knowledge for understanding their climate vulnerabilities, planning, and setting climate priorities and responses, as well as to minimize uncertainties regarding climate change predictions.

Several scholars take further steps by examining how local administrators view and make use of expert knowledge (Lundin & Öberg, 2014; Van Stigt et al., 2013; Hordijk & Baud, 2006). Lundin and Öberg (2014) explore the extent to which local administrators in Swedish municipalities apply expert knowledge when they prepare political advice and recommendations for politicians. They find that local administrators are more inclined to use these data in situations involving political disagreements. The authors also discover that politicians are less reluctant to use the information provided by administrators in such an environment (Lundin & Öberg, 2014). Based on these outcomes, Lundin and Öberg (2014) conclude that the political context impacts the role of experts in the decisionmaking process. Additionally, Hordijk and Baud (2006) emphasize the role of local governments as coordinators and mediators of different actors' interests. In this regard, cities require scientific information and knowledge on "an interplay of interests and power structures within the city and on linkages with institutions and regions outside the city" (Hordijk & Baud, 2006, p. 675). For example, Van Stigt et al. (2015) conclude that some municipalities in the Netherlands use expert knowledge to deliver a political decision that will be supported by stakeholders and public opinion. Dodman and Carmin (2011) claim that cities might apply expert knowledge to mobilize political support for climate adaptation actions. They provide an example of politicians using scientific data and risk maps for visualizing climate risks in the city of Copenhagen (Denmark) to obtain political support among dwellers for their climate adaptation policies (Dodman & Carmin 2011).

To better understand the role of expert knowledge, scholars unpack forms of the science-policy interface (Harris & Moore, 2014; Hughes & Romero-Lankao, 2014; Craft & Howlett, 2012). Several works apply an insider-outsider approach that is based on whether advisors are located internally or externally in relation to policymakers (Craft & Howlett, 2012; Fraussen & Halpin, 2017). Halligan (1995) and Craft and Howlett (2013) assert that internal advisors have more power than outside advisors in general. As an example, Hughes and Romero-Lankao (2014) compare science-policy interactions in two cities: Delhi and Mexico City. They conclude that Mexico City has relied on an informal policy network for its climate policy initiation, whereas Delhi has applied internal expertise for funding and policy priority setting for its climate change actions (Hughes & Romero-Lankao, 2014). Besides, Craft and Howlett (2012) define forms and content of advice and expertise by subdividing them along two dimensions: procedural (technical) vs substantive (policy formation and implementation) and short-term reactive vs long-term anticipatory. Furthermore, various scholars subdivide scientific knowledge into several forms, including process, normative, empirical, and predictive knowledge (Tennøy et al., 2016; Rydin, 2007).

Additionally, a number of studies examine the forms of expert involvement in urban environmental policies: the legitimization of best practices (Harris & Moore, 2014); creation of independent scientific data; personal investment in a city's future development through education and training of younger generations; engagement with wider society through hosting of special events and establishing scientific centers (Mabon et al., 2019). Spruijt et al. (2014, p. 23) argue that the role of experts depends on several factors such as "the complexity of the issue at stake, the type of knowledge and values that the experts have and contextual factors, such as the types of organizations in which the experts are employed or the broader societal context".

To summarize, discussions of forms of local governments' engagement with expert knowledge include (1) reliance on internal and external expertise for procedural and substantive knowledge; (2) usage of scientific data for climate policy formation, implementation, and minimization of climate uncertainties; (3) sharing of knowledge for education and training on climate change and (4) application of scientific data for the legitimization of climate change actions. In a later section, this chapter applies these forms for identifying the application of expert knowledge in the two cases.

17.3 Methodology

Through a review of publicly available data, the case studies of Norilsk and Yakutsk demonstrate expert knowledge use in urban decisionmaking, with an emphasis on the impacts of climate change and associated permafrost degradation on urban infrastructure.

17.3.1 Case Study Selection

This study applies a comparative case study analysis of Norilsk and Yakutsk authorities' perceptions of uses of expert knowledge to address climate change impacts. Pickvance (1986, p. 166) identifies two reasons for conducting comparative analysis: "to discover whether a theoretically-derived model holds in empirical case studies and to see whether an empirically-based relationship derived from the one study holds in another." This study, therefore, examines whether certain theoretically derived characteristics of expert knowledge use are present in the cases of Norilsk and Yakutsk and whether the two cities share similarities in their views on the role of expert knowledge. The study also identifies specific characteristics of uses of expert knowledge in the two cities as a larger contribution to the academic literature. The analysis examines the importance of climate change impacts on the cities' political agenda. Finally, the research identifies the role of expert knowledge in the development and implementation of local climate change actions.

The choice of the two cities is based on several factors. Norilsk is located above the Arctic Circle (69°N). Although Yakutsk is situated below the Arctic Circle (62°N) and is not included in the Arctic Russian land territory based on the 2014 Presidential Decree № 296, its geographical position and climatic conditions make Yakutsk a northern city (Volosnikova, 2012).

Source: Climates to Travel

The population of Norilsk is 180,239 (The City of Norilsk, 2018b), while the size of Yakutsk's population is 311,760 (Cities of Russia, 2018). Both cities are located on continuous permafrost defined as "ground which remains at temperatures below 0 °C for at least two consecutive years" (Streletskiy et al., 2019, p. 1). Currently, in Norilsk, "about 60 percent of buildings have been damaged by permafrost thaw, and 10 percent of the houses in the city have been abandoned" (Schreiber, 2018). By 2023, it is estimated that 66 buildings will sustain more than 70% structural damage (The City of Norilsk, 2018a). Concerning Yakutsk, scientific estimations suggest that the increase in annual mean temperature by 1.5 °C could damage nearly all foundations in the city (Shiklomanov et al., 2016). In general, "the projected climate-induced decrease in bearing capacity will exceed 55% around the 2040s" (Shiklomanov et al. 2016, p. 137).

At the same time, the cities differ in their administrative status and the structure of their economies. Yakutsk is the capital of the Republic of Sakha (Yakutia). It hosts the republican government bodies and territorial authorities of the federal government bodies, as well as representatives of foreign states. Among Yakutsk's administrative functions as the republican capital are the development of local socio-economic programs and republican target programs; the implementation of federal and republican target development programs (The President of the Republic Sakha, 2011). Norilsk is not one of the administrative centers of the Krasnoyarsk Kray. According to Laruelle (2020), the Norilsk administration has less autonomy since it has to answer directly to the regional authorities in Krasnoyarsk and to state officials in Moscow depending on the relevant issue. In general, based on the Constitution of the Russian Federation (1993), republics in Russia possess more autonomy in comparison to other federal subjects (i.e. Kray and oblast), including a right to establish and promote their own languages. The government of the Republic of Sakha (Yakutia) has been proactive in developing and promoting legislation on permafrost protection at the regional and federal levels (Gazyeva, 2020).

Furthermore, Norilsk is a single-industry city whose economic activities are centered on Nornickel's factories, which produce more than 80% of Russian nickel, copper, and platinum ore (Veselov, 2005). In 2014, "the company employed 56,000 persons, or 45% of the city's active population" (Laruelle & Hohmann, 2017, p. 316). On the contrary, Yakutsk's economy is more diverse. Around 76% of the working population is involved in the following economic sectors: education, public health, public administration, transport, and finance (The City of Yakutsk, 2015).

Finally, the administrations of both cities stress the importance of fostering innovation and education for future development (The City of Norilsk, 2018b; The City of Yakutsk, 2015). Norilsk and Yakutsk possess long-standing experience in research on various problems associated with construction on permafrost. Pilyasov (2013) states that Arctic cities are following several scenarios while restructuring their economies. Specifically, for an administrative center, like Yakutsk, it is important to become an innovation hub for diffusing knowledge to its neighboring territories. For a single-industry city, like Norilsk, it is critical to diversify its economy for its future sustainable development (Pilyasov, 2013) (Table 17.1).

Considering everything mentioned above, Norilsk and Yakutsk present ideal case studies to reflect the theoretically defined forms of science-policy interface regarding local climate change decisionmaking.

17.3.2 Data Collection, Analysis, and Limitations

This study is based on preliminary research. The primary data source consists of socio-economic strategies, general plans, reports, and programs issued by the city administrations of Norilsk and Yakutsk from 2015 through 2018. To identify relevant documents, I consulted official web pages of Norilsk and Yakutsk administrations (i.e. news and documents sections). I also searched for information on the published documents in local mass media and social media. These policy documents are cited individually where appropriate, although they are referred to collectively as "the documents" for simplification purposes. The decision to focus on these documents is based on the fact that these materials represent a primary source of information about the local governments' official stances on the research topic.

Table 17.1 Major differences and similarities between Norilsk and Yakutsk

Characteristics	Norilsk	Yakutsk
Population size	180, 239	311, 760
Location on permafrost	X	X
Port city	a railway connects Norilsk to the port of Dudinka	X
Administrative status	An industrial city in the Krasnoyarsk Kray	Capital of the Republic of Sakha (Yakutia)
Economy structure	Single – industry city	Diverse economy
Status of climate change policies	No official climate change strategy. It has a long-standing experience of construction on permafrost.	No official climate change strategy. It has a long-standing experience of construction on permafrost. The government of the Republic of Sakha (Yakutia) has been proactive in developing legislation on permafrost protection.
Fostering education and innovation for development	To become a center for Arctic competency and engineering.	To develop scientific research and university education with its rebranding as a hub for innovations, education, and science.

Source: The City of Yakutsk (2018), The City of Norilsk (2018a b)

Also, these documents are publicly accessible on the Russian official websites. Data limitations will be addressed in future research. Specifically, official documents are declarative and contain general information about the reasons for the uses of expert knowledge. Additionally, these data lack information on the reasons for the non-use of expert knowledge.

Critical discourse analysis was implemented to obtain information from the official documents on how the city administrations of Norilsk and Yakutsk have framed the notion of climate change impacts and their views on uses of expert knowledge. The coding of data was conducted in the following steps: (1) identification of sentences that contained the relevant topics/categories; (2) breaking down of sentences into separate words or groups of words; (3) identification of categories and connections between them, and (4) reexamination of categories. Data collection, analysis, and reexamination were done by applications of the identified characteristics of the forms of uses of expert knowledge in local climate change policies.

17.4 Norilsk's and Yakutsk's Framing of Climate Change Impacts

The notions of climate change and its impacts are not explicitly expressed in Norilsk's and Yakutsk's official documents. They mainly relate to one of the cities' socio-economic problems. In particular, the words "global climate change" or "global warming" are used in connection with ground thawing and waterlogging, while referring to housing and road problems and permafrost thawing (The City of Yakutsk, 2018; The City of Norilsk, 2018b). Climate change impacts are pictured by the documents with some level of uncertainty since they are mainly mentioned as future negative impacts.

Notably, climate change impacts are not identified by the documents as a separate challenge for infrastructure, and instead are listed as one of the factors (see Table 17.2). For instance, Norilsk's Strategy for Socio-Economic Development highlights that the combination of different factors will reduce by half the period of pre-repair operation of buildings by 2020 (The City of Norilsk, 2018b). In this regard, the Strategy's approach correlates with the scientists' positions identifying that the problem of collapsing housing has emerged as a result of "the poor quality of construction, improper operation of the city infrastructure, difficulties associated with socio-economic transitions, and unanticipated climatic changes" (Shiklomanov et al. 2017, p. 285).

Although the cities share similar viewpoints on climate change impacts, Norilsk and Yakutsk authorities have different views on their importance. For instance, Yakutsk's Strategy for Socio-Economic Development emphasizes that the current inefficient surface water drainage system negatively impacts the quality of roads and permafrost thawing. The document underscores that this may become a major problem for Yakutsk in the future as a result of climate change.

Norilsk's Strategy identifies a challenging situation with permafrost thawing as an important risk for the city that directly impacts the housing situation. The document has a strong emphasis on the proposition that the lack of housing leads to a shortage of skilled workers as one of the threats for the city's development (The City of Norilsk, 2018b). Generally, the functioning of the city is ultimately connected to the Russian state's national security policies since Norilsk officials position the city as a center for production of non-ferrous metals in Russia (The City of Norilsk, 2018a; The City of Norilsk, 2018b).

Table 17.2 List of factors that impact social infrastructure in Norilsk and Yakutsk

Factors	Norilsk	Yakutsk
Old housing	X	X
Poor quality of construction planning	X	X
Climatic conditions	X	X
Anthropogenic impacts	X	
Lack of surface water drainage system		X
Demand for new technologies	X	
Permafrost thawing and degradation	X	X
Financial issues	X	X
Climate change's impacts	X	X
Lack of local construction legislation		X
Lack of permafrost monitoring	X	X
Waterlogging		X

Source: The City of Yakutsk (2018), The City of Norilsk (2018a b)

To summarize, the city administrations of Norilsk and Yakutsk frame climate change as one factor that negatively impacts their infrastructure such as roads and buildings. This finding is consistent with the academic literature stating that climate change is not the primary factor for climate change actions (Araos et al., 2016) and that these actions are driven by several factors (Ford et al., 2011). Furthermore, their framing of climate change as a future challenge reflects academic research that highlights the necessity for a city to undertake innovative ways for scientific data production due to the uncertainties associated with climate change (Dodman & Carmin, 2011).

17.5 Norilsk's and Yakutsk's Perceptions of Uses of Expert Knowledge

My data analysis identified three main themes concerning Norilsk's and Yakutsk's perceptions of uses of expert knowledge to manage permafrost degradation: (1) development and implementation of policies; (2) becoming centers for science and innovation, and (3) training and retraining of specialists. These themes reflect visions of Norilsk and Yakutsk administrations regarding their role in creating a comfortable and safe urban environment for residents, as well as becoming centers for scientific research, innovation, and education (The City of Norilsk, 2018b; The City of Yakutsk, 2018).

The use of expert knowledge in the development and implementation of policies related to managing thawing permafrost is broadly defined in the documents as monitoring and controlling of permafrost-geocryological conditions; analysis, maintenance, and provision of data on permafrost, and preparation of recommendations and initiatives (The City of Norilsk, 2018b; The City of Yakutsk, 2012; The City of Yakutsk, 2018). For instance, in 2019 the mayor of Yakutsk Sardana Avksentieva announced plans to restore permafrost monitoring with the participation of experts to discuss construction and landscaping projects (Yakutia.info, 2019).

In this regard, Norilsk's and Yakutsk's approaches correlate with the existing academic approaches to the use of expert knowledge for development, monitoring, and evaluation of urban policies, analyzing political issues and minimizing uncertainties relating to climate change (Hordijk & Baud, 2006; Van Stigt et al., 2015; Dodman & Carmin, 2011). Expert knowledge also provides technical advice, knowledge, and expertise with short-term and long-term policy actions (Craft & Howlett 2012).

Generally, Norilsk and Yakutsk authorities express positive perceptions of the science-policy interface by referring to the necessity of establishing partnerships between business, academia, and society (The City of Norilsk, 2018b), and implementing scientific knowledge in finding solutions for socio-economic issues (The City of Yakutsk, 2018). My data analysis reveals that Norilsk and Yakutsk officials apply a diverse array of policy-science interactions such as a city administration expert council, scientific projects, agreements, conferences, and workshops (The City of Norilsk, 2018b; The City of Yakutsk, 2018).

Both cities have expert councils within the structure of their administrations consisting of experts, businesses, and city officials to advise the mayor on various issues. In 2014, for instance, an Advisory Council was established to inform the mayor of Norilsk on issues related to permafrost and sustainability in building construction (Zapolarnaya Pravda, 2014). Although the document analysis identifies several forms of the science-policy interface, they are limited in detail. As an example, local media analysis reveals that the meetings of the Scientific and Technical Council within the city administration of Yakutsk encompass expert's presentations, discussions, and recommendations (Yakutia.info, 2020).

However, the presence of these expert councils indicates direct communication between experts and local officials on permafrost degradation. The significance of such science-policy interactions is that they may help to avoid potential impediments to the incorporation of expert knowledge in local policy development and implementation. Hordijk and Baud (2006) point out that research outcomes are sometimes kept at universities, NGOs, and other agencies as the result of the ignorance of local officials about the relevance of the knowledge. Furthermore, they also highlight a potential gap between scientific research and political needs caused by researchers' lack of comprehension of political requirements (Hordijk & Baud, 2006).

In the case of Norilsk and Yakutsk, a challenge is to bridge the gap between expert knowledge, the city administration's responsibilities for creating sustainable living conditions for its residents, and the existing federal legislation that regulates construction on permafrost. For instance, the city administration of Norilsk called for the creation of a unified permafrost monitoring system (The City of Norilsk, 2018b). In May 2018, the Republic of Sakha (Yakutia) adopted a regional level law on permafrost protection, and it plans to promote its adoption at the federal level (Vasilieva, 2018).

At the state level, the Russian government has advanced legislation regulating construction on permafrost. In December 2018, the Russian government introduced amendments to the Comprehensive Plan for Climate Doctrine Realization by assigning to the Ministry of Construction, Housing, and Utilities of the Russian Federation tasks for the development of adaptation scenarios for construction in permafrost zones. The next year, the Ministry adopted a set of rules for projects involving railway construction in permafrost zones.

Norilsk and Yakutsk authorities identify high construction costs and financial constraints of local budgets as the major challenges for their policies (The City of Norilsk, 2018b; The City of Yakutsk, 2018). For instance, the Norilsk administration notes that the intensive development of housing construction is only possible with the allocation of funding from the federal and regional budgets (The City of Norilsk, 2018b). In this regard, there is the presence of a strong dependence of cities on federal and regional financial means regarding their socio-economic policies.

Several scholars suggest that while being disadvantaged financially, medium-sized and small cities might use the subject of climate change as a means to attract funding for the realization of their socio-economic policies (Major & Juhola, 2016;

Jonas et al., 2017). Wurzel et al. (2019) suggest that small cities might decide to embrace cognitive leadership for rebranding their external image to attract skilled people and investment. In this regard, the topic of climate change might be used for an urban settlement's rebranding from a climate-vulnerable city to a leader in climate change policies (Wurzel et al., 2019).

Both administrations declare in the documents their intentions to develop expert knowledge to become centers for research and innovation (The City of Norilsk, 2018b; The City of Yakutsk, 2018). The Norilsk administration envisions the city becoming a center for Arctic competency and engineering. To realize this vision, it plans to establish a center for monitoring of permafrost-geocryological conditions in the construction area to support the application of new building materials and technologies. Norilsk authorities foresee the creation of a scientific hub that will attract experts and contribute to the launching of new housing policies (The City of Norilsk, 2018b). The Yakutsk administration links the development of scientific research and University education to its rebranding as a hub for innovations, education, and science for North-East Russia (The City of Yakutsk, 2018).

The visions of Yakutsk and Norilsk authorities for becoming innovative centers for knowledge and expertise reflect the discussions in the academic literature about the importance of large cities becoming platforms for innovation and experimentation, particularly in climate change policies (Bulkeley & Betsill, 2005; Broto & Bulkeley, 2013). In this regard, the documents identify two purposes of expert knowledge: (1) conducting research to develop innovations and new technologies, and (2) training and retraining specialists (The City of Norilsk, 2018b; The City of Norilsk, 2018a; The City of Yakutsk, 2018). Despite the similarity of their interests, Norilsk and Yakutsk follow different approaches. The Norilsk administration focuses on training and retraining of specialists primarily for local employment opportunities, especially in Nornickel's factories (The city of Norilsk, 2018b). The city administration in Yakutsk positions M. K. Ammosov North-Eastern Federal University (NEFU) as the scientific and education platform of North-East Russia (The City of Yakutsk, 2018).

For the conduct of training and research, both cities underscore their reliance on local expertise as well as on cooperation with national institutions and international experts (The City of Norilsk, 2018b; The City of Yakutsk, 2018). Analysis of the documents reveals that in comparison to Yakutsk, Norilsk's authorities have specific cooperation strategies. For instance, there is an agreement between the Norilsk administration and the Siberian State University for the development of the town-planning strategy and the training of relevant specialists (The City of Norilsk, 2018b). This is due to Norilsk's limited innovation and resource capacities in comparison to Yakutsk, resulting from the differences in their administrative status, economic structures, and population size. In addition to being a single-industry city, Norilsk identifies its inefficient level of innovations and lack of highly-qualified specialists as additional challenges (The City of Norilsk, 2018b). Previous research also found that medium-sized and small cities are often at a disadvantage because they do not have the economic, technological, and scientific capacities available to large cities (Major et al., 2018).

In both Norilsk and Yakutsk, authorities' perceptions of the uses of expert knowledge reflect the academic discussions on the role of expert knowledge in policymaking (Hordijk & Baud, 2006, Spruijt et al., 2014; Van Stigt et al., 2015). In both cases, uncertainty exists regarding power relations between the cities and experts, especially considering the urban settlements' financial constraints for development.

17.6 Conclusion

This chapter explores the uses of expert knowledge in local policymaking based on publicly available official documents. It examines how authorities in Norilsk and Yakutsk think about the uses of expert knowledge to address climate change impacts on the local built environment. Specifically, the chapter focuses on forms of expert knowledge used by the two cities administrations to address problems arising from permafrost degradation. Although the topic of the role of experts and epistemic communities in decisionmaking has been discussed in the academic literature, few works have explored local administrators' views regarding its application (see Van Stigt et al., 2015; Hughes & Romero-Lankao, 2014). Three major observations about the application of expert knowledge by Norilsk and Yakutsk authorities have emerged from this analysis.

Norilsk and Yakutsk authorities' framing of climate change impacts supports the literature on urban climate change governance, which argues that climate change represents one of the factors for cities to undertake actions. The city administrations of Norilsk and Yakutsk frame climate change as one of the factors impacting permafrost degradation that negatively affects the construction and maintenance of roads and buildings. In this regard, challenges associated with climate change are viewed by Norilsk and Yakutsk officials as a part of larger socio-economic problems. Concurrently, the city administration of Norilsk links the challenging housing situation with the capacity of fulfilling its functions as the centre for non-ferrous metals production in Russia.

The approaches of the Norilsk and Yakutsk administrations in addressing challenges associated with thawing permafrost reveal that the use of expert knowledge is predominantly applied for the development and implementation of policies, turning the cities into centers for science and innovation, and training and retraining of specialists. This finding shows a correlation between Norilsk's and Yakutsk's approaches and academic literature on the roles of experts in sharing knowledge for education and using scientific data for local decisionmaking. The city administrations apply expert knowledge to fulfill their functions for the creation of comfortable and safe living conditions, along with establishing centers for scientific research and education.

Despite the prominence of references to the use of expert knowledge in the official documents, it remains unclear how effectively Norilsk and Yakutsk authorities actually implement available expertise. Further research is needed to document the extent to which the application of expert knowledge produces impacts on environmental and living conditions in the Arctic.

References

AMAP. (2017*). Snow, water, ice and permafrost in the Arctic (SWIPA)*. Arctic Minitoring and Assessment Program (AMAP). Oslo, Norway, xiv + 269 pp.

Araos, M., Berrang-Ford, L., Ford, G., Austin, S., Biesbroek, R., & Lesnikowski, A. (2016). Climate change adaptation planning in large cities: A systematic global assessment. *Environmental Science and Policy, 66*, 375–382. https://doi.org/10.1016/j.envsci.2016.06.009

Baker, J., Lovell, K., & Harris, N. (2006). How expert are the experts? An exploration of the concept of the 'expert' within Delphi panel techniques. *Nurse Researcher, 14*(1), 59–70. https://doi.org/10.7748/nr2006.10.14.1.59.c6010

Brand, R., & Karvonen, A. (2007). The ecosystem of expertise: Complementary knowledges for sustainable development. *Sustainability: Science, Practice and Policy, 3*(1), 21–31. https://doi.org/10.1080/15487733.2007.11907989

Broto, V. C. (2017). Urban governance and the politics of climate change. *World Development, 93*, 1–15. https://doi.org/10.1016/j.worlddev.2016.12.031

Broto, V. C., & Bulkeley, H. (2013). Maintaining climate change experiments: Urban political ecology and the everyday reconfiguration of urban infrastructure. *International Journal of Urban and Regional Research, 37*(6), 1934–1948. https://doi.org/10.1111/1468-2427.12050

Bulkeley, H., & Betsill, M. (2005). Rethinking sustainable cities: Multilevel governance and the 'urban' politics of climate change. *Environmental Politics, 14*(1), 42–63. https://doi.org/10.1080/0964401042000310178

Bulkeley, H., & Betsill, M. (2013). Revisiting the urban politics of climate change. *Environmental Politics, 22*(1), 136–154. https://doi.org/10.1080/09644016.2013.755797

Burgman, M., Carr, A., Godden, L., Gregory, R., Mcbride, M., Flander, L., & Maguire, L. (2011). Redefining expertise and improving ecological judgment. *Conversation Letters, 4*(2), 81–87. https://doi.org/10.1111/j.1755-263X.2011.00165.x

Cities of Russia. (2018). Общие сведения и описание Якутска (*General information about Yakutsk)*. http://города-россия.рф/sity_id.php?id=65

Craft, J., & Howlett, M. (2012). Policy formulation, governance shifts, and policy influence: location and content in policy advisory systems. *Journal of Public Policy, 32*(2), 79–98. https://doi.org/10.1017/S0143814X12000049

Craft, J., & Howlett, M. (2013). The dual dynamics of policy advisory systems: The impact of externalization and politicization on policy advice. *Policy and Society, 32*(3), 187–197. https://doi.org/10.1016/j.polsoc.2013.07.001

Dodman, D., & Carmin, J. (2011). *Urban adaptation planning: the use and limits of climate science*. IIED Briefing Papers, July 10, 2012.

Drescher, M., Perera, A. H., Johnson, C. J., Buse, L. J., Drew, C. A., & Burgman, M. A. (2013). Toward rigorous use of expert knowledge in ecological research. *Ecosphere, 4*(7), 1–26. https://doi.org/10.1890/ES12-00415.1

Fedorinova, Y. (2020). *Huge spill stains arctic and climate change could be the cause*. Bloomberg. https://www.bloomberg.com/news/articles/2020-06-04/russia-declares-state-of-emergency-over-arctic-city-fuel-spill

Ford, J. D., Berrang-Ford, L., & Paterson, J. (2011). A systematic review of observed climate change adaptation in developed nations. *Climatic Change, 106*(2), 327–336. https://doi.org/10.1007/s10584-011-0045-5

Fraussen, B., & Halpin, D. (2017). Think tanks and strategic policy-making: The contribution of think tanks to policy advisory systems. *Policy Sciences, 50*(1), 105–124. https://doi.org/10.1007/s11077-016-9246-0

Gazyeva, E. (2020). Госсобрание Якутии предложило узаконить охрану вечной мерзлоты (*State Assembley of Yakutia proposes to legalize permafrost protection*). Kommersant. https://www.kommersant.ru/doc/4378882

Halligan, J. (1995). Policy advice and the public sector. In B. G. Peters & D. J. Savoie (Eds.), *Governance in a changing environment* (pp. 138–172). McGill-Queen's University Press, Canadian Centre for Management and Development.

Hardoš, P. (2018). Who exactly is an expert? On the problem of defining and recognizing expertise. *Sociológia, 50*(3), 268–288.

Harris, A., & Moore, S. (2014). Planning histories and practices of circulating urban knowledge. *International Journal of Urban and Regional Research, 37*(5), 1499–1509. https://doi.org/10.1111/1468-2427.12043

Hirte, G., Nitzsche, E., & Tscharaktschiew, S. (2018). Optimal adaptation in cities. *Land Use Policy, 73*, 147–169. https://doi.org/10.1016/j.landusepol.2018.01.031

Hordijk, M., & Baud, I. (2006). The role of research and knowledge generation in collective action and urban governance: How can researchers act as catalysts? *Habitat International, 30*(3), 668–689. https://doi.org/10.1016/j.habitatint.2005.04.002

Hughes, S., & Romero-Lankao, P. (2014). Science and institution building in urban climate-change policymaking. *Environmental Politics, 23*(6), 1023–1042. https://doi.org/10.1080/09644016.2014.921459

Jonas, A. E. G., Wurzel, R. K. W., Monaghan, E., & Osthorst, W. (2017). Climate change, the green economy, and reimagining the city: The case of structurally disadvantaged European maritime port cities. *Journal of the Geographical Society of Berlin, 148*(4), 197–211. https://doi.org/10.12854/erde-148-49

Larrick, R. P., & Feiler, D. C. (2016). Expertise in decision making. In G. Keren & G. Wu (Eds.), *Wiley-Blackwell handbook of judgment and decision making* (1st ed., pp. 696–721). Malden, MA: Blackwell.

Laruelle, M. (2020). Urban regimes in Russia's northern cities: Testing a concept in a new environment. *Arctic, 73*(1), 53–66. https://doi.org/10.14430/arctic69933

Laruelle, M., & Hohmann, S. (2017). Biography of a polar city: Population flows and urban identity in Norilsk. *Polar Geography, 40*(4), 306–323. https://doi.org/10.1080/1088937X.2017.1387822

Lundin, M., & Öberg, P. (2014). Expert knowledge use and deliberation in local policy making. *Policy Sciences, 47*, 25–49. https://doi.org/10.1007/s11077-013-9182-1

Mabon, L., Shih, W., Kondo, K., Kanekiyo, H., & Hayabuchi, Y. (2019). What is the role of epistemic communities in shaping local environmental policy? Managing environmental change through planning and greenspace in Fukuoka City, Japan. *Geoforum, 104*, 158–169.

Major, D., & Juhola, S. (2016). Guidance for climate change adaptation in small coastal towns and cities: A new challenge. *Journal of Urban Planning and Development, 142*(4), 1–4. https://doi.org/10.1061/(ASCE)UP.1943-5444.0000356

Major, D. C., Lehmann, M., & Fitton, J. (2018). Linking the management of climate change adaptation in small coastal towns and cities to the sustainable development goals. *Ocean and Coastal Management, 163*, 205–208. https://doi.org/10.1016/j.ocecoaman.2018.06.010

Martus, E. (2019). Russian industry responses to climate change: The case of the metals and mining sector. *Climate Policy, 19*(1), 17–29. https://doi.org/10.1080/14693062.2018.1448254

Patterson, J. J., & Van der Grijp, N. (2019). Empowerment and disempowerment of urban climate governance initiatives: an explanatory typology of mechanisms. In J. van der Heijden, H. Bulkeley, & C. Certomà (Eds.), *Urban climate politics: Agency and empowerment* (pp. 39–58). Cambridge: Cambridge University Press. https://doi.org/10.1017/9781108632157.003

Pickvance, C. G. (1986). Comparative urban analysis and assumptions about causality. *International Journal of Urban and Regional Research, 10*(2), 162–184. https://doi.org/10.1111/j.1468-2427.1986.tb00010.x

Pilyasov, A. (2013). Russia's policies for Arctic cities: problems and prospects. In M. Laruelle & R. Orttung (Eds.), *Urban sustainability in the Arctic: Visions, contexts, and challenges* (pp. 213–217). Institute for European, Russian and Eurasian Studies/The George Washington University.

Reay, M. (2007). Academic knowledge and expert authority in American economics. *Sociological Perspectives, 50*(1), 101–129. https://doi.org/10.1525/sop.2007.50.1.101

Rydin, Y. (2007). Re-examining the role of knowledge within planning theory. *Planning Theory, 6*(1), 52–68. https://doi.org/10.1177/1473095207075161

Schreiber, M. (2018). *The race to save Arctic cities as permafrost melts*. CityLab. https://www.citylab.com/environment/2018/05/the-race-to-save-arctic-cities-as-permafrost-melts/559307/

Shiklomanov, N., Streletskiy, D., Swales, T., & Kokorev, V. (2016). Climate change and stability of urban infrastructure in Russian permafrost regions: Prognostic assessment based on GCM climate projections. *Geographical Review, 107*(1), 125–142. https://doi.org/10.1111/gere.12214

Shiklomanov, N., Streletskiy, D. A., Grebents, V. I., & Suter, L. (2017). Conquering the permafrost: Urban infrastructure development in Norilsk, Russia. *Polar Geography, 40*(4), 273–290. https://doi.org/10.1080/1088937X.2017.1329237

Spruijt, P., Knol, A. B., Vasileiadou, E., Devilee, J., Lebret, E., & Petersen, A. C. (2014). Roles of scientists as policy advisers on complex issues: A literature review. *Environmental Science and Policy, 40*, 16–25.

Streletskiy, D., Suter, L. J., Shiklomanov, N. I., Porfiriev, B. N., & Eliseev, D. O. (2019). Assessment of climate change impacts on buildings, structures and infrastructure in the Russian regions on permafrost. *Environmental Research Letters, 14*, 025003.

Tarasov, P. I., Zyryanov, I. V., & Khazin, M. L. (2018). Транспортный коридор через западную якутию (Transport passage through West Yakutia). *Gorniy Informatsionno-Analiticheskiy Bulletin, 6*, 170–184. https://doi.org/10.25018/0236-1493-2018-6-0-170-184

Tennøy, A., Hansson, L., Lissandrello, E., & Næss, P. (2016). How planners' use and non-use of expert knowledge affect the goal achievement potential of plans: Experiences from strategic land-use and transport planning processes in three Scandinavian cities. *Progress in Planning, 109*, 1–32. https://doi.org/10.1016/j.progress.2015.05.002

The City of Norilsk. (2018a). Приложение к Стратегии социально-экономического развития муниципального образования город Норильск до 2030 года (*Addendum to the Strategy for socio-economic devlopment of the municipality of Norilsk until 2030*). Norilsk: Administration of the City of Norilsk.

The City of Norilsk. (2018b). Стратегия социально-экономического развития муниципального образования город Норильск до 2030 года (*The Strategy for socio-economic development of the municipality of Norilsk until 2030*). Norilsk: Administration of the City of Norilsk.

The City of Yakutsk. (2012). Положение о создании Научно-технического совета при Главе городского округа «город Якутск» (*Regulation on the establishment of the Scientific and Technical Council under the Head of the city district "City of Yakutsk"*). Yakutsk: The city of Yakutsk.

The City of Yakutsk. (2015). Решение «О Стратегии социально-экономического развития городского «округа город» Якутск на период до 2032 года» (*The decision on the "Strategy for socio-economic development of the City District of Yakutsk City for the period until 2032"*). Yakutsk: The Administration of the city of Yakutsk.

The City of Yakutsk. (2018). Решение о внесении изменений в решение Якутской городской Думы от 25 ноября 2015 года РЯГД-21-3 «О Стратегии социально-экономического развития городского округа «город Якутск» на период до 2032 года» (*The Decision on introducing changes to the decision of the Yakutsk City Duma of November 25, 2015, Ryagd-21-*

3 "On the Strategy for Socio-Economic Development of the City District of Yakutsk City for the Period until 2032"). Yakutsk: The Administration of the city of Yakutsk.

The President of the Republic of Sakha. (2011). Закон "О статусе столицы Республики Саха (Якутия)" 924-З N 753-IV (Law "On the status of the capital of the Republic of Sakha (Yakutia)" 924-3 N 753-IV). Yakutia: President of the Republic of Sakha Yakutia.

Van der Heijden, J., Luckmann, O., & Cherkasheva, A. (2019). Urban climate governance in Russia: Insights from Moscow and St. Petersburg. *Journal of Urban Affairs, 42*(7), 1047-1062. https://doi.org/10.1080/07352166.2019.1617036.

Van Stigt, R., Driessen, P. P. J., & Spit, T. J. M. (2013). Compact City development and the challenge of environmental policy integration: A multi-level governance perspective. *Environmental Policy and Governance, 23* (4), 221–233. https://doi.org/10.1002/eet.1615

Van Stigt, R., Driessen, P. P. J., & Spit, T. J. M. (2015). A user perspective on the gap between science and decision-making. Local administrators' views on expert knowledge in urban planning. *Environmental Science and Policy, 47*, 167–176. https://doi.org/10.1016/j.envsci.2014.12.002

Vasilieva, T. (2018). В Якутии принят региональный закон об охране вечной мерзлоты (*Yakutia adopts a new law on permafrost protection*). State Assembly of the Republic of Sakha. http://old.iltumen.ru/content/v-yakutii-prinyat-regionalnyi-zakon-ob-okhrane-vechnoi-merzloty

Veselov, A. (2005). Комбинат, который построил всех (Factory, which gave life to everybody). *Expert Sibir, 24*(76) http://expert.ru/siberia/2005/24/24si-30-01_67186/

Vogel, B., & Henstra, D. (2015). Studying local climate adaptation: A heuristic research framework for comparative policy analysis. *Global Environmental Change, 31*, 110–120. https://doi.org/10.1016/j.gloenvcha.2015.01.001

Volosnikova, E. A. (2012). Северный город: понятие и типология (Northern city: concept and typology). *The Surgut State Pedagogical University Bulletin, 2*, 98–103.

Woodruff, S. C. (2018). City membership in climate change adaptation networks. *Environmental Science and Policy, 84*, 60–68. https://doi.org/10.1016/j.envsci.2018.03.002

Wurzel, R. K. W., Moulton, J. F. G., Osthorst, W., Mederake, L., Deutz, P., & Jonas, A. E. (2019). Climate pioneership and leadership in structurally disadvantaged maritime port cities. *Environmental Politics, 28*(1), 146–166. https://doi.org/10.1080/09644016.2019.1522039

Yakutia.info. (2019). Сардана Авксентьева: в Якутске появится собственный мерзлотный надзор *(Sardana Avksentieva: in Yakutsk will appear its own permafrost control)*. Yakutioa. Info. https://yakutia.info/article/190428

Yakutia.info. (2020). Проблему изменения состояния Вечной мерзлоты в Якутске предлагают обсуждать в рамках научно - практической конференции *(The problem of the changing state of permafrost in Yakutsk is proposed to discuss at a scientific conference)*. Yakutioa.Info. https://yakutia.info/article/193174

Zapolarnaya Pravda. (2014). Глава города Олег Курилов подписал постановление о создании консультативного совета по сохранению устойчивости зданий (The head of the city Oleg Kurilov signed a decree on the creation of an advisory council on the preservation of buildings). Zapolarnaya Pravda. http://gazetazp.ru/news/gorod/27446n-glava-goroda-oleg-kurilov-podpisal-postanovlenie-o-sozdanii-konsultativnogo-soveta-po-sohraneniyu-ustoychivosti-zdaniy-norilska.html

Chapter 18
(Research): Separate Arrangements of the People's Republic of China, Japan and South Korea on the Arctic: Correlation with the Arctic Council's Policy

Elena V. Kienko

Abstract The chapter examines how the Arctic policies of China, Japan and South Korea, as well as their recent activities in the international fora on Arctic issues, correlate with their commitments taken within the Arctic Council. The main characteristics of the Arctic policies of the People's Republic of China, Japan and South Korea are examined through an analysis of their strategic national documents and international agreements with Arctic States. It is noted that these legal documents are focused on environmental protection, issues of navigation and scientific research in the Arctic Ocean.

Introduction Special attention is paid to the analysis of the strategic documents on the Arctic policy of the People's Republic of China aimed at expansion of their ambitious claims to mineral resources in the Arctic. Statements of Chinese officials also cited in this chapter and some provisions of China's Arctic policy are not harmonized with the common will of the Arctic coastal States relating to the contemporary legal regime of the Arctic Ocean as reflected in the Ilulissat Declaration, 2008.

The alarming reaction to such documents among the Arctic coastal States, especially the USA and Canada, is also considered in the chapter. The chapter concludes that prevailing interpretations of such documents is negative only in relation to China. Possible explanations lie, primarily, in the fact that China's economic clout is greater than either Japan's or South Korea's as well as the USA's and Canada's view of China as a strategic competitor and a superpower. The research provides evidence that up till now there are no grounds for such a negative view of China's role in the Arctic. This is possibly more true in the light of the understanding that Japan and

E. V. Kienko (✉)
Candidate of Law, Primakov National Research Institute of World Economy and International Relations, Russian Academy of Sciences, Moscow, Russia

P. A. Berkman et al. (eds.), *Building Common Interests in the Arctic Ocean with Global Inclusion, Volume 2*, Informed Decisionmaking for Sustainability,
https://doi.org/10.1007/978-3-030-89312-5_18

South Korea are close allies with Canada and the United States on a range of other issues which can exclude the Arctic.

In this context separate legal documents relating to the regime of the Arctic region and its natural resources (which are agreed upon by China, Japan and South Korea as a result of the Trilateral High-Level Dialogues on the Arctic organized by their governments) are scrutinized in the chapter. Overall, the development by China, Japan and South Korea of separate trilateral legal documents relating to the regime of the Arctic Ocean seems to cause unpredictable consequences.

1. ***Forming the China-Japan-South Korea forum dealing with Arctic issues***

In 2013 China, Japan and South Korea were granted observer status in the Arctic Council at the 8th session of this "high-level forum" held in Kiruna, Sweden (along with Italy, India, and Singapore).[1]

The Ottawa Declaration ("Declaration on the Establishment of the Arctic Council", 1996) and *the Rules of Procedure of the Arctic Council* do not give observers rights that are equal to those of Arctic States, even if the observers are economically and militarily powerful. According to new criteria for the admitting observers stated in *the Senior Arctic Officials Report to Ministers* at the Nuuk Ministerial Meeting 2011, the Arctic States shall take into account the extent to which observers accept and support the objectives of the Arctic Council defined in the Ottawa declaration; recognize Arctic States' sovereignty, sovereign rights and jurisdiction in the Arctic; recognize that an extensive legal framework applies to the Arctic Ocean including, notably, the Law of the Sea, and that this framework provides a solid foundation for responsible management of this ocean; respect the values, interests, culture and traditions of Arctic Indigenous peoples and other Arctic inhabitants; have demonstrated a political willingness as well as financial ability to contribute to the work of the Permanent Participants and other Arctic Indigenous peoples; have demonstrated their Arctic interests and expertise relevant to the work of the Arctic Council; and have demonstrated a concrete interest and ability to support the work of the Arctic Council, including through partnerships with member states and Permanent Participants bringing Arctic concerns to global decisionmaking bodies.[2]

Non-Arctic States that obtain observer status, nevertheless, consider the Arctic Council as an institutional and legal springboard for strengthening mutually beneficial cooperation with the Arctic States on Arctic issues and for concluding bilateral and multilateral agreements with them. The involvement of China, Japan, and South Korea in specific economic projects in the Arctic depends, primarily, on building effective bilateral partnerships with the Arctic States.

However, the engagement of non-Arctic States in Arctic issues and an active promotion of their national interests in the region raise crucial questions. The first

[1] See about "particular role of the Arctic Council" (Fife, 2013).

[2] Senior Arctic Officials (SAO) Report to Ministers, Nuuk, Greenland, May 2011. URL: http://library.arcticportal.org/1251/1/SAO_Report_to_Ministers_-_Nuuk_Ministerial_Meeting_May_2011.pdf

and foremost is: can non-Arctic States such as China, Japan and South Korea, by concluding separate arrangements on Arctic matters among themselves, bring instability to the longstanding Arctic legal regime,[3] which is based on the Law of the Sea and numerous other sources of the international law on global, regional or bilateral levels (Lackenbauer & Manicom, 2013)?

To answer this question, we should take into account that the formalization of the Arctic dialogue among China, Japan and South Korea is developing impressively, even in comparison with the dialogue among the Arctic States. For example, there has never been a high-level summit of the Arctic States (neither in the format of the "five" Arctic coastal States, nor in the format of the "eight" – the Arctic Council's member States).

By contrast, the highest governmental level of officials of China, Japan and South Korea discussed Arctic issues in a meeting on November 2015.[4] According to *Joint Declaration for Peace and Cooperation in Northeast Asia, 2015*, trilateral cooperation between the People's Republic of China, Japan and South Korea has been developing since 1999 through the regular holding of the Trilateral Summits in the three States since 2008. This trilateral cooperation has been further institutionalized through the establishment in 2011 on the basis of equal participation of three Asian States of the *Trilateral Cooperation Secretariat*. The initial goal of this international organization is "to promote peace and common prosperity among the People's Republic of China, Japan and South Korea".[5] Such format of international cooperation provides additional opportunities for their governments to discuss a wide range of issues, including the Arctic. As stated in the Joint Declaration for Peace and Cooperation in Northeast Asia, 2015, "acknowledging the global importance of Arctic issues" three States decided to launch a trilateral high-level dialogue on the Arctic "to share Arctic policies, explore cooperative projects and seek ways to deepen cooperation over the Arctic". The 2015 Summit (not only devoted to the Arctic, but also to "peace and cooperation issues in Northeast Asia") was followed by a special meeting on Arctic issues in April 2016 in Seoul at the level of Foreign Ministers.

As stated in the Joint Press Release of the First Trilateral High-Level Dialogue on the Arctic among the People's Republic of China, Japan and South Korea the parties "expected that the Dialogue would serve as a platform to seek ways to strengthen the three countries' cooperation over the Arctic"; that "the three countries would develop their Arctic cooperation in various areas including science and research"; also discussed "the guiding principles of the trilateral Arctic cooperation and shared the view that the three countries should continue their commitments of contribution

[3]Berkman et al., (2019). XL, 734 p.

[4]Joint Declaration for Peace and Cooperation in Northeast Asia, 1 November, 2015. The Sixth Trilateral Summit. URL: https://www.mofa.go.jp/a_o/rp/page1e_000058.html

[5]Upon the agreement signed and ratified by each of the three governments, the Trilateral Cooperation Secretariat was officially inaugurated in Seoul in September, 2011. URL: https://tcs-asia.org/en/about/overview.php

to the Arctic Council and enhance their cooperation within various international fora"; shared the view that "with regard to the specific Arctic cooperation, *scientific research* is among the most promising areas for their joint activities and trilateral cooperative activities in this area need to be encouraged".[6]

It is noteworthy that the parties reaffirmed that the Trilateral High-Level Dialogue on the Arctic is an important arrangement for the three countries to keep the momentum of their continued cooperation over the Arctic. Three representatives of this States also decided to report the discussions of this Dialogue to the Arctic Council through appropriate channels.

The First Dialogue was followed by the series of annual trilateral meetings. In 2017, on the occasion of *the Second Trilateral High-Level Dialogue on the Arctic* the representatives of the People's Republic of China, South Korea and Japan "underscored the importance of further strengthening their respective contributions to the work of the Arctic Council through engagement at the Working Groups, Task Forces and Expert Groups, and in particular of enhancing engagement with other various international fora based on the shared recognition that the challenges and opportunities over the Arctic have global ramifications beyond the region".[7] Moreover, according to *the Joint Statement on the Second Trilateral High-Level Dialogue on the Arctic* the three States reconfirmed that "scientific research presents the most promising area for their joint activities and trilateral cooperative activities"; "requested their experts to identify specific cooperative projects on scientific research, such as cooperative research for environmental changes in the Pacific side of the Arctic Ocean as a major contribution to the Pacific Arctic Group, and Pan-Arctic Ocean observation project in the international coordinated cruises in summer 2020 under Synoptic Arctic Survey". Basically, the parties confirmed the importance of following up on these activities on a regular basis with opportunity of "identifying cooperative projects in the other areas".

In 2018, three leaders of China, Japan and South Korea met on the occasion of *the Seventh Trilateral Summit*. Among other crucial topics of this Summit parties highlighted their success in elaborating of the Trilateral High-Level Dialogue and expressed their willingness to continue such cooperation: "we reconfirm the importance of the trilateral cooperation on the Arctic, especially in the area of scientific research".[8]

In June 2018, the representatives of China, Japan and South Korea met again in Shanghai on *the Third Trilateral High-Level Dialogue on the Arctic*. According to the Joint Statement China, Japan and South Korea "recognized the positive role of the Trilateral High-Level Dialogue on the Arctic as one of the outcomes of the Sixth

---

[6]Ministry of Foreign Affairs of South Korea. URL: http://www.mofa.go.kr/eng/brd/m_5676/view.do?seq=316483&srchFr=&srchTo=&srchWord=Outcome&srchTp=&multi_itm_seq=0&itm_seq_1=0&itm_seq_2=0&company_cd=&company_nm=&page=161&titleNm

[7]Ministry of Foreign Affairs of Japan. URL: https://www.mofa.go.jp/files/000263104.pdf

[8]Joint Declaration of the Seventh Japan-China-South Korea Trilateral Summit. URL: https://tcs-asia.org/en/data/documents.php?s_topics=T1553563153&s_gubun=&s_year=&s_txt=

Trilateral Summit in 2015"; and also "reaffirmed the importance to promote trilateral cooperation on the Arctic, especially in the area of scientific research, as endorsed in the Joint Declaration of the Seventh Trilateral Summit on May 9, 2018". It is supposed that through this Dialogue, the People's Republic of China, Japan and South Korea "addressed common challenges over the Arctic, from the perspective of *East Asian countries*, and reiterated their intention to make contributions to promoting peace, stability and sustainable development in the Arctic, continued to promote scientific research as priority for cooperation among the three countries". As an aside, it should be noted that three States here declared themselves as a "group" of East Asian countries. In the document the parties also supported the main direction of three States Arctic policies to develop scientific research in the Arctic through "the exchange of information on Arctic expeditions", and "the sharing of scientific data and further development of collaborative surveys" with the possibility of exploring other areas for cooperation in the Arctic.

In the view of the main idea of this chapter more attention should be given to the perspective of these States to the Arctic Council. As stated in the Joint Statement on the Third Trilateral High-Level Dialogue on the Arctic three States "valued the positive role of the Arctic Council, especially in environmental protection and sustainable development in the Arctic"; "the three countries, all became accredited observers to the Council in 2013"; and all three States "are willing to further strengthen their respective contributions to the work of the Council, by cooperating with the Arctic States, Permanent Participants and other observers, through engagement including in the Working Groups, Task Forces and Expert Groups". Also, they shared the intention "to communicate the discussions of this Dialogue to the Council in line with previous practice".[9]

The Fourth Trilateral High-Level Dialogue on the Arctic was held in June 2019. In comparison with previous Dialogues dedicated to the Arctic issues here it was stated meaningful discussions and potential areas of cooperation between the Trilateral Arctic Expert Group and the Trilateral High-level Dialogue.[10]

Although these three States have produced a variety of significant documents concerning the Arctic region—including the documents adopted as a result of their trilateral meetings, their respective regulations and policy pronouncements relating to the Arctic, as well as bilateral individual agreements of each of them with some Arctic States—their actual policy about the Arctic remains ambiguous. This is particularly true of China, whose strategic interests in the region are the most ambitious.

[9]Ministry of Foreign Affairs of the People's Republic of China. URL: https://www.fmprc.gov.cn/mfa_eng/wjdt_665385/2649_665393/t1567103.shtml

[10]Ministry of Foreign Affairs of South Korea. URL: http://www.mofa.go.kr/eng/brd/m_5676/view.do?seq=320574

2. *China's legal perception of the Arctic*

China is the world's largest State in terms of population (about one and a half billion people) and is expected to maintain this position throughout the twenty-first century. By most indicators, China has the world's second largest economy, behind only the United States. China is a permanent member of the UN Security Council and a strong nuclear power with growing military strength. China does not have a coastline in the Arctic Ocean; accordingly, China does not have its own territorial sea, exclusive economic zone, or continental shelf in the Arctic. Because China also has no land territory above the Arctic Circle, China is not a member of the Arctic Council and, accordingly, does not have the same rights in this "high-level" intergovernmental forum as the States that are members of the Arctic Council.

China started to participate in Arctic Council activities in 2007 as an *ad hoc* observer. According to the statement of Chinese ambassador to Sweden, Lan Lijun, "the participation of more non-Arctic States as observers would have a "positive significance to the work of the Council"; "he recognized that much of the region fell under national jurisdiction of the Arctic states"; and that "the participation of observers in the work of the Council is based on the recognition of Arctic states' sovereignty, sovereign rights and jurisdiction in the Arctic as well as their decisionmaking power in the Council" (Lackenbauer et al., 2018, 140).

After China gained permanent observer status in the Arctic Council, the Chinese presence in the Arctic and activities in the Arctic Council increased dramatically. According to China's report as a permanent observer to the Arctic Council, 2016, China has attended all the governmental meetings opened to observers under the umbrella of the Arctic Council; as well as the Working Groups, Task Forces and Expert Groups of the Arctic Council, including the meetings of the Protection of the Arctic Marine Environment working group, the Conservation of Arctic Flora and Fauna working group, the Arctic Monitoring and Assessment Program working group and the Scientific Cooperation Task Force.[11] As stated in the document, China continues the existing bilateral and multilateral dialogue and cooperation with the Arctic States and non-Arctic States, while welcoming more inclusive, comprehensive and diversified cooperation with all relevant stakeholders regarding the Arctic affairs. In this regard, a seminar on Arctic issues with the Institute of Arctic Research of Finland and the fifth China-Nordic Arctic Cooperation Symposium in 2017 are mentioned.

In 2018, China submitted its second observer's report to the Arctic Council and expressed its willingness to continue to contribute to the Arctic Council work.[12] According to this document, China presented practical suggestions, which were incorporated in the Arctic Migratory Birds Initiative Work Plan of the Conservation of Arctic Flora and Fauna working group. Other spheres of China's interest in the

[11] Arctic Council. URL: https://oaarchive.arctic-council.org/bitstream/handle/11374/1860/EDOCS-4018-v1-2016-11-26_China_Observer_activity_report.pdf?sequence=1&isAllowed=y

[12] Arctic Council. URL: https://oaarchive.arctic-council.org/bitstream/handle/11374/2251/CHINA_2018-05_Review-Report.pdf?sequence=1&isAllowed=y

framework of the Arctic Council, as stated, include black carbon and methane related projects and research reports (Arctic Contaminants Action Program); meteorology; a joint program of building the Aurora Observatory in Iceland; welfare of Indigenous peoples, etc.

Against the background of strengthening of bilateral and multilateral cooperation with Arctic States, especially with the Nordic States, China confirmed that it would continue its efforts to develop cooperation with non-Arctic States. In this regard, the China's second observer report stated that China, Japan and South Korea launched High-Level Trilateral Dialogues on Arctic affairs to promote exchanges on practices and experiences regarding Arctic scientific research and commercial cooperation. On the multilateral level, China attends various meetings relating to the Arctic, including the remarkable "Arctic Circle" and "Arctic Frontiers".

In January 2018, China issued its first official "Arctic Policy" paper,[13] which raised serious concern in the USA and Canada.[14] Analysts are considering what this paper—and similar policy papers issued by Japan and South Korea and the materials of the Trilateral meetings of officials from these three Asian States—signify for the current legal regime of the Arctic Ocean and its adjacent seas (Vylegzhanin et al., 2018). Is it true that China has organized a forum of non-Arctic States, as opposed to the established bilateral and regional legal mechanisms of the Arctic States, in order to introduce, in the interests of the latter, changes in the established legal order in the Arctic? Or do China, Japan, and South Korea, although they have established their own trilateral mechanism of meetings on Arctic issues, still respect the current legal status of the Arctic? In comparison with the pronouncements of the European Union on the Arctic, one might see nothing wrong with the three-party arrangements between China, Japan and South Korea on the Arctic. However, it is noteworthy that the European Union, unlike China, is not a nuclear power.

The reason why some analysts claim that China's current stance on the status of the Arctic is still evolving and should be of grave concern to all Arctic States, based on the following:

1. In March 2010 Chinese Admiral Y. Zhou stated that "the Arctic belongs to all the peoples of the world, and no state has sovereignty" over it; that "China should play an integral role in Arctic exploration because we are a fifth of the world's population"; and that the "current scramble over Arctic sovereignty" among the Arctic states "affects the interests of many other countries".
2. China, acting "in its own interest, is playing the role of advocate for the common heritage of mankind", thereby establishing a legal basis for access to the natural resources of the Arctic's marine subsoil, albeit without having a coastline in the Arctic (Chircop, 2011). However, as discussed below, the five Arctic coastal States have not yet completed the process for determining the outer limits of their

[13]The State Council Information Office of the People's Republic of China. URL: http://www.xinhuanet.com/english/2018-01/26/c_136926498.htm

[14]Thomson J. *What does China's new Arctic policy mean for Canada?* 2018. URL: http://www.cbc.ca/news/canada/north/what-does-china-s-new-arctic-policy-mean-for-canada-1.4506754

respective continental shelves in the Arctic Ocean.[15] So, it is premature for China to assert that any of the seafloor in the Central Arctic Ocean will be part of the Area and that any of the seabed resources are part of the common heritage of mankind.

China's increasing activity in the Arctic is of particular concern to the USA. The official US position on this issue is reflected in strategic documents of the Defense Department and the Coast Guard. Thus, the 2019 Strategic Forecast of the US Coast Guard in the Arctic region identified China as one of the main "US competitors in the Arctic" and the main "threat to US national interests" in the Arctic region. The document highlights the sharp increase in China's economic and political presence in the Arctic; it recalls that China has declared the Arctic as its strategic priority and has already conducted six scientific expeditions in the region; that in addition to conducting scientific research, China is interested in developing its transport potential; and that "in 2017, about forty percent of ships passing through the Northern Sea Route are associated with China".[16]

According to the US National Defense Strategy 2018, there is growing concern about "China's attempts to extend its influence in the Arctic"[17]; in accordance with the US document, this could hamper "freedom of navigation in the Arctic" as a whole, especially since "this has already happened in the South China Sea". Earlier in 2015, "Chinese warships were seen in the Bering Sea", which is bordered only by the United States and Russia. Despite the fact that international law permits the presence of warships of non-coastal States, the United States is concerned about "China's growing naval activities" in the Arctic and Pacific oceans.[18]

Some Russian scholars also give an ambiguous assessment of China's policy in the Arctic, for instance: "now China's economic presence in the economy of circumpolar States has a significant impact on politics in the region. Some Nordic States have fallen into *partial dependence* on China, such as Norway, which was forced to make political concessions to China during the 2010 conflict in order to resume the necessary level of interaction" (Konyshev & Kobzeva, 2017); other Russian scholars call into question the genuine nature of the China's arguments related to "its legitimacy of interest towards the Arctic" (Zagorskii, 2019).

---

[15] Either by delimiting the continental shelf according to customary international law and Article 83 of UNCLOS or by delineation according to the new rules provided by Article 76 of UNCLOS.

[16] URL: https://climateandsecurity.files.wordpress.com/2019/04/uscg-arctic-strategic-outlook_22-apr-2019-release-date.pdf

[17] It might be suggested that the competition for Arctic dominance is more likely to be in economic and political terms and not military ones. But the reality is that the total number of weapons in the Arctic is constantly growing. For more detail see: Lanteigne M. The changing shape of Arctic security. NATO Review. 2019. URL: https://www.nato.int/docu/review/articles/2019/06/28/the-changing-shape-of-arctic-security/index.html

[18] Thad W. Allen, Christine Todd Whitman. Independent Task Force Report No. 75. *Arctic Imperatives Reinforcing U.S. Strategy on America's Fourth Coast.* https://www.cfr.org/sites/default/files/pdf/2017/02/TFR75_Arctic.pdf

The legitimacy of China's interest reflected in its "Arctic policy", 2018. In addition to the statement that China considers itself geographically as "a one of the continental States that are closest to the Arctic Circle", China's rights in the Arctic are based on the 1982 UN Convention on the Law of the Sea; the 1920 Spitsbergen Treaty (to which China is a party); and on "general international law".[19] Also, in the document, China acknowledges that "States from outside the Arctic region do not have territorial sovereignty in the Arctic, but they do have rights in respect of scientific research, navigation, overflight, fishing, laying of submarine cables and pipelines in the high seas and other relevant sea areas in the Arctic Ocean". It is indicative that China and other non-Arctic States have "rights to resource exploration and exploitation in the Area". In other words, China has already assumed that there is an "Area" in the Arctic Ocean, meaning a portion of the seafloor beyond the continental shelves of the Arctic coastal States. It should be highlighted here that the designation "Area" in the Arctic Ocean is possible if all five Arctic coastal States delimit in the Arctic their continental shelf in accordance with Article 76 of the 1982 Convention. One of the five States (the USA) is not a party to this Convention and does not consider its provisions on the "Area" as customary norms of international law (Vylegzhanin, 2019).

Thus, contrary to China's document, at present there is no "Area" designated in the Arctic Ocean within the meaning of the 1982 Convention, nor do China or other non-Arctic States have rights to any resources in any part of the Arctic Ocean subsoil. At present, at least, only the five Arctic States have such rights with respect to their continental shelves in the Arctic.

The Chinese document emphasizes that China's position in relation to the Arctic region is special; that "China is an important stakeholder in Arctic affairs"; that "natural conditions of the Arctic and their changes have a direct impact on China's climate system and ecological environment, and, in turn, on its economic interests in agriculture, forestry, fishery, marine industry and other sectors".

China states that its goals in the Arctic are to "understand, protect, develop, and participate in the governance of the Arctic so as to safeguard the common interests of all countries and the international community in the Arctic, and promote sustainable development of the Arctic", based on the basic principles of "respect, cooperation, win-win result and sustainability".[20]

According to China's "Arctic Policy", China believes that "the Arctic shipping routes are likely to become important transport routes for international trade" and "the use of shipping routes and the development of resources in the Arctic can have a huge impact on China's energy strategy and economic development as a major trading power and global energy consumer". Therefore, China intends to play a "leading role" in expanding shipping routes which comprise the Northeast Passage,

[19]The State Council Information Office of the People's Republic of China. URL: http://www.xinhuanet.com/english/2018-01/26/c136926498.htm

[20]The State Council Information Office of the People's Republic of China. URL: http://www.xinhuanet.com/english/2018-01/26/c136926498.htm

Northwest Passage, and the Central Passage, pursuant to the "Arctic Policy". The paper proposes the term "Polar Silk Road" to refer to both the "North-East Passage" and the "North-West Passage".[21] But this term is not in the legislation of Russia, Canada or of any other Arctic State. No Arctic Council document uses the term "Polar Silk Road", either.[22] It seems contradictory to the statement that "China respects the legislative, enforcement and adjudicatory powers of the Arctic States in the waters subject to their jurisdiction".

China's "Arctic Policy" also states that "China respects the sovereign rights of Arctic States over oil, gas and mineral resources in the areas subject to their jurisdiction in accordance with international law". To be more precise, China respects the rights of the Arctic States over oil, gas and other mineral resources not only on their continental shelf (which is under the jurisdiction of the Arctic State), but also natural resources of the subsoil of their internal sea waters, territorial sea, as well as land in the Arctic (which are under the sovereignty of the corresponding Arctic State). This position is complemented by China's cooperation agreements with several Arctic States.

3. *China – Iceland Arctic cooperation*

Bilateral cooperation between China and Iceland in the Arctic are numerous and multifaceted, although the military, economic and demographic potential of these two States is obviously different. Unlike China, which has a population of 1.5 billion, Iceland's population is just over 300 thousand people.[23] Iceland is a member of the Arctic Council and a member of NATO. Iceland is not a member of the European Union but is closely associated with the Schengen Agreement. Despite these differences, China and Iceland signed *six cooperation agreements* (including one on Arctic issues) in 2012 during an official visit of the Head of the Chinese State Council to Iceland.[24]

In the *Joint Statement on Comprehensive Deepening of Bilateral Relations between the Governments of Iceland and the People's Republic of China, 2013*, the parties agreed to strengthen cooperation in the Arctic in the fields of environmental protection, climate change, and geothermal and marine research.[25] This bilateral instrument is legally linked, primarily, to the *Framework Agreement on*

[21] The State Council Information Office of the People's Republic of China. URL: http://www.xinhuanet.com/english/2018-01/26/c136926498.htm

[22] China as a member of the International Maritime Organization China supports *the International Code for Ships Operating in Polar Waters (Polar Code)*. Albeit the Polar Code has an impact on marine operations of all States, it is applicable not only to the Arctic waters but also to the Antarctic.

[23] United Nations Department of Economic and Social Affairs. https://esa.un.org/unpd/wpp/Publications/Files/WPP2017KeyFindings.pdf

[24] Government of Iceland. URL: http://eng.forsaetisraduneyti.is/news-and-articles/nr/7144

[25] Government of Iceland. URL: https://www.government.is/media/forsaetisraduneyti-media/media/frettir1/Joint-statement-of-PMs-Iceland-China-2013.pdf

Arctic Cooperation.[26] According to the China's "Arctic Policy", this Agreement "is the first interstate agreement on Arctic issues between China and an Arctic State". In 2012, China and Iceland signed a *Memorandum of Understanding on Cooperation in Marine and Polar Research and Technology.*[27] In the same year a bilateral *Memorandum of Understanding on Geothermal and Geosciences Cooperation* was signed by the Minister for Foreign Affairs of Iceland and the Minister of Land and Resources of China.[28]

In 2013 China and Iceland signed a *Free Trade Agreement,*[29] which provides for the rights and obligations of the two States in order to remove obstacles to relations concerning trade and to promote economic convergence. China's icebreaker diplomacy also contributed to the rapprochement with this Arctic State: in 2012, after crossing the North Pole, China's icebreaker (the Xue Long) arrived in Reykjavik.[30]

During the same period, Iceland provided to China quotas for fishing in the maritime areas under Iceland's jurisdiction. The Chinese company China National Offshore Oil Corporation International Ltd. also obtained a license for joint exploration and production of hydrocarbons in 2014.[31] However, in January 2018 it was announced that the Chinese company withdrew from the project; at present, it is difficult to predict whether this energy project will be implemented.[32]

China and Iceland intend to coordinate their policy on Arctic shipping. Iceland believes that navigation across the North Pole, which will soon be ice-free due to climate change and the shrinking ice cover in the Arctic Ocean, will increase Iceland's importance as an Arctic State and accelerate the growth of its economy. In this regard, Iceland has similar views to China on maritime navigation in the Arctic.[33]

---

[26] Government of Iceland. URL: https://www.government.is/news/article/2012/04/20/Agreements-and-declarations-signed-following-a-meeting-between-Prime-Minister-Johanna-Sigurdardottir-and-Premier-Wen-Jiabao-in-Reykjavik-today/

[27] Government of Iceland. URL: https://www.government.is/news/article/2012/04/20/Agreements-and-declarations-signed-following-a-meeting-between-Prime-Minister-Johanna-Sigurdardottir-and-Premier-Wen-Jiabao-in-Reykjavik-today/

[28] Government of Iceland. URL: https://www.government.is/news/article/2012/04/20/Agreements-and-declarations-signed-following-a-meeting-between-Prime-Minister-Johanna-Sigurdardottir-and-Premier-Wen-Jiabao-in-Reykjavik-today/

[29] Ministry of Foreign Affairs of Iceland. URL: https://www.mfa.is/foreign-policy/trade/free-trade-agreement-between-iceland-and-china/

[30] Embassy of the People's Republic of China in Iceland. URL: http://is.china-embassy.org/eng/xwdt/t971781.htm

[31] Tang I. *China's CNOOC starts preparation work to explore for oil and gas offshore Iceland.* 2014. URL: https://www.platts.com/latest-news/oil/singapore/chinas-cnooc-starts-preparation-work-to-explore-26748948

[32] Lanteigne, Marc. *Stumbling Block: China-Iceland Oil Exploration Reaches an Impasse.* 2018. URL: https://overthecircle.com/2018/01/24/stumbling-block-china-iceland-oil-exploration-reaches-an-impasse/

[33] Government of Iceland. URL: https://www.mfa.is/media/Raedur/framsoguraeda-OS-14-feb-2013.pdf

However, it is not clear how such high-latitude shipping will be implemented, taking into account the rights of the five Arctic coastal States to adopt special measures to prevent pollution of the marine environment in their 200-mile exclusive economic zones in the Arctic according to Article 234 of the 1982 UN Convention on the Law of the Sea. Certainly, without crossing at least one of these five exclusive economic zones, an Icelandic or a Chinese ship cannot pass across the North Pole.

4. ***China – Norway Arctic cooperation***

With regard to bilateral relations between China and Norway, analysts highlight several factors that contribute to the proximity of these States' legal positions on the regime of the Arctic Ocean: firstly, China and Norway are the largest exporters of fish, including fish caught in the Arctic Ocean and adjacent seas[34]; secondly, both China and Norway are interested in expanding long-term scientific cooperation in the Arctic; thirdly, these "nations are united by a desire to better understand the intentions of their common great-power neighbor Russia".[35] It is noteworthy that while Norway and China each have Russia as a neighbor, only Norway has Russia as a neighbor in the Arctic (China and Russia are neighboring States, but not in the Arctic).

Joint statement on normalization of bilateral relations between China and Norway, 2016 noted that "Norway is one of the first Western countries to recognize the People's Republic of China and establish diplomatic relations with it"; that "China and Norway had previously strong interstate relations".[36] Unlike the United States, which provides political and military support to Taiwan, "the Norwegian government stands in solidarity with "the policy of one China, fully respects China's sovereignty and territorial integrity", does not support actions against the People's Republic of China, and will "make every effort in the future to prevent a deterioration of bilateral relations". Pursuant to this Sino-Norwegian document, the two States have agreed to develop mutually beneficial cooperation, particularly in trade, culture, science, education, and "on polar issues". A "normalization" of relations between China and Norway was necessary. The two States had suffered a break in relations after the incident occurred with Liu Xiaobo, a Chinese dissident who won the Nobel Peace Prize in 2010.

Within the framework of the "China-Norway Dialogue on The Changing Arctic and International Cooperation" in Shanghai in 2017, the Prime Minister of Norway Erna Solberg stressed that "Norway and China have cooperated on polar research for more than 15 years"; noted China's increased activity in the high northern latitudes; expressed Norway's intention to "remain at the forefront of developments in Arctic

[34]Top Fish Exporters 2016. URL: http://www.worldsrichestcountries.com/top-fish-exporters.html

[35]Shanghai Institutes for International Studies. URL: http://en.siis.org.cn/Research/EnConferencesInfo/88

[36]Government of the Kingdom of Norway. URL: https://www.regjeringen.no/globalassets/departementene/ud/vedlegg/statement_kina.pdf

science".[37] Norway has been very skilful in using China's investment potential to develop its oil sector, but has not granted to China any legal rights over its territory in the Arctic region. In 2009, a *Memorandum of Understanding* was concluded *between the Ministry of Oil and Energy of the Kingdom of Norway and the National Energy Administration of China.*[38] In accordance with Article 2 of the Memorandum, the parties intend to take relations in this industry to a new level. For this purpose, the parties agreed to facilitate consultations, to exchange of data on oil production activities, technologies and projects. The parties also agreed to create favorable conditions for the development of cooperation between oil enterprises of the two countries: exchange of information on planning issues in the oil and gas sector; exchange of views on the development of global and regional energy markets and environmental measures.

In addition to intergovernmental cooperation, China and Norway have strengthened academic and scientific ties. The Chinese Association for Scientific Expeditions and Svalbard University of Norway signed an *Agreement on Academic Cooperation in the Arctic.*[39] In accordance with the Agreement, the parties shall carry out joint scientific research aimed at the study of geophysics, geology, and biology of the Arctic, as well as the development of technology, and increase the exchange of scientific information, scientists and students. The Agreement between the Chinese Polar Research Institute and the Norwegian Polar Institute on cooperation in polar research defines the exchange of scientific knowledge on climate change as a goal.

5. ***China – Russia Arctic cooperation***

Bilateral cooperation between China and Russia in the Arctic is focused on scientific research, exploration and exploitation of hydrocarbon deposits of the Russian Arctic shelf, navigation along the Northern Sea Route and the creation of infrastructure along this route. For instance, in 1994, two States agreed to promote cooperation in order to meet the needs of international maritime transportation; to ensure maritime safety, including the safety of ships, crews, passengers and cargo; to develop cooperation in the field of freight activities; to expand economic, scientific and technical ties and exchange experience; to exchange views on economic activities in the Arctic region, etc.[40] Furthermore, Russia and China have agreed not to impede the ships of two States from carrying out maritime transportation not only

[37] Shanghai Institutes for International Studies. URL: http://en.siis.org.cn/Research/EnConferencesInfo/88

[38] Government of the Kingdom of Norway. URL: https://www.regjeringen.no/globalassets/upload/oed/pdf_filer/mou/mou-petroleum-kina.pdf

[39] Government of the Kingdom of Norway. URL: https://www.regjeringen.no/globalassets/upload/KD/Vedlegg/Forskning/PRIC_agreement.pdf

[40] Government of the Russian Federation and Government of the People's Republic of China concluded the Agreement on cooperation in the field of maritime navigation. URL: http://www.conventions.ru/view_base.php?id=16911

between Russian and Chinese ports, but also between the ports of third States; in all matters of navigation two States provide most favored nation treatment.

In 2012, Russia and China concluded *the Agreement on Cooperation to Prevent, Deter and Eliminate Illegal, Unreported and Unregulated Fishing of Living Marine Resources*. The purpose of the 2012 Agreement is to prevent illegal fishing. This problem is relevant, including for the areas of the Chukchi Sea, in the Arctic zone of the Russian Federation. In the text of the 2012 Agreement the term "maritime areas" means "inland waters, territorial sea and other maritime spaces in the North-Western Pacific where States exercise sovereign rights and jurisdiction in accordance with international law.[41] This Agreement provides for consultations, which are held at least once a year at a meeting of the Russian-Chinese Commission, established in accordance with the Agreement between the USSR and China on Cooperation in the field of fisheries of October 4, 1988. Such consultations are aimed at solving emerging issues in the field of prevention, deterrence and elimination of illegal, unreported and unregulated fishing of living marine resources.

A new stage of cooperation between China and Russia began in 2017, with an official visit of the President of China, Xi Jinping, to Russia. As a result, the *Joint Statement of the Russian Federation and the People's Republic of China on further deepening of comprehensive partnership and strategic cooperation relations* and *the Joint Statement of the Russian Federation and the People's Republic of China on the current world situation and important international issues* were signed. Moreover, the parties approved the Action Plan for the implementation of *the Treaty on Good Neighborliness, Friendship and Cooperation between the Russian Federation and the People's Republic of China for 2017–2020.*

Thus, the parties agreed to strengthen Russian-Chinese cooperation in the Arctic region, cooperation between the competent authorities of the States, research organizations and enterprises in areas, such as the development and use of the Northern Sea Route, joint scientific expeditions, exploration and exploitation of mineral resources, Arctic tourism, environmental protection. The parties also expressed willingness to continue work on "conjugation of the Eurasian Economic Union and "One Belt, One Road" Initiative in order to "promote the conclusion of trade agreements and agreements on economic cooperation between the Eurasian Economic Union and its members, on the one hand, and the People's Republic of China, on the other hand".[42] This Agreement was concluded in 2018 with a view to promoting economic integration in the Asia-Pacific and Eurasia regions, as well as to join the Eurasian Economic Union and the "One Belt, One Road" Initiative as a tool of strong and stable trade relations in the region. In the future the parties declared the establishment of the Eurasian Economic Partnership.[43]

In the context of developing international law and increasing its role in solving international issues, in June *2016 the Declaration of the Russian Federation and the*

---

[41] URL: http://docs.cntd.ru/document/902395104

[42] URL: http://www.kremlin.ru/supplement/5218

[43] Ibid.

People's Republic of China on enhancing the role of international law was published. The Declaration highlights that Russia and China declare their commitment to the principles of international law as reflected in the United Nations Charter and the 1970 Declaration on Principles of International Law concerning friendly relations and cooperation among States in accordance with the United Nations Charter. Both parties reaffirmed their willingness to strengthen bilateral cooperation in order to establish a just and reasonable international order based on rule of international law. Most significantly, both States emphasize the crucial role of the 1982 United Nations Convention on the Law of the Sea in upholding the rule of law with respect to maritime activities. The provisions of this universal international convention have been applied in such a way as not to prejudice the rights and legitimate interests of States parties to the 1982 Convention, or to compromise the integrity of the legal regime established by the Convention.[44]

International cooperation with Russia is one of the highest priorities for China. This is confirmed by the active development of joint gas project "Yamal LNG",[45] and cooperation on construction of the White Sea – Komi – Ural railway ("Belkomur"),[46] which will run from Arkhangelsk to the Komi Republic and to the Urals. China is also showing constructive activity in joint investment projects with Russian companies. In 2014, China National Petroleum Corporation was granted a stake in RN-Vankor LLC (a subsidiary of "Rosneft"), which is engaged in the development of the Vankor field, one of the largest fields in Siberia.[47]

In 2018, according to the official website of "Rosneft" it was reported on the expansion of cooperation between "Rosneft" and the China National Petroleum Corporation. As part of the IV Eastern Economic Forum, the parties signed the Agreement on Cooperation in Exploration and Production in the Russian Federation. According to the Agreement, the China National Petroleum Corporation will have an opportunity to acquire the minority shares in the major oil and gas projects of "Rosneft", in particular, in Eastern and Western Siberia. Parties agreed to review the China National Petroleum Corporation proposal for provision on market principles of services for these fields in the area of exploration, development and production of hydrocarbons.[48]

In 2019, the China National Oil Offshore Corporation acquired a 10% equity interest in the Arctic LNG 2 LLC in Russia and the acquisition was completed. Through this acquisition, the China National Oil Offshore Corporation can increase

[44]URL: http://thailand.mid.ru/key-issues/1545-deklaratsiya-rossijskoj-federatsii-i-kitajskoj-narodnoj-respubliki-o-povyshenii-roli-mezhdunarodnogoprava

[45]URL: http://yamallng.ru/project/about/

[46]URL: http://www.belkomur.com/belkomur/2.php

[47]URL: https://vankorneft.rosneft.ru/about/Glance/OperationalStructure/Dobicha_i_razrabotka/Vostochnaja_Sibir/vankorneft/

[48]URL: https://www.rosneft.com/press/releases/item/192215/

the proportion of natural gas production, which is in line with its low-carbon and environmental-friendly development model.[49]

Bilateral relations between China and Russia are also developing in the area of joint scientific research. In October 2018, one periodical published an article on the second Sino-Russian expedition in the Arctic (Xie, 2018). The purpose of this scientific expedition was collection of the data necessary to ensure the development of the Polar Silk Road. The expedition was organized in cooperation with the Pacific Oceanological Institute of the Russian Academy of Sciences and the Pilot National Laboratory for Marine Science and Technology of China.

6. ***China – USA cooperation in the Arctic***

International cooperation in the exploration and exploitation of Arctic mineral resources between the USA and China is primarily focused on the development of fossil fuel production from State of Alaska deposits, the budget of which is funded mainly by tax revenues from the oil sector. China sees this cooperation as beneficial in terms of access to the US liquefied natural gas. Thus, in 2017, after the Summit at the residence of President Donald Trump in Mar-a-Lago (USA), President Xi Jinping met with Alaska Governor Bill Walker to discuss the terms of supply of liquefied natural gas to China (Hong, 2020). The parties also signed Memorandums of Understanding between American and Chinese oil and gas corporations.

In November 2017, the USA and China signed the *Joint Development Agreement for Alaska LNG* with the participation of the State of Alaska, Alaska Gasline Development Corporation, China Petrochemical Corporation (Sinopec), China Investment Corporation (CIC Capital Corporation), Bank of China Limited.[50] According to the Agreement, the government of State of Alaska intended to develop the infrastructure of the North Slope natural gas field to export liquefied natural gas to the global market. For this purpose, it was planned to establish an Alaska liquefied natural gas production and supply system, the sole owner of which is currently the State of Alaska. China was entitled to 75% of liquefied natural gas at a reduced cost, and in return provided an equal share of funding. In addition, the Alaska-LNG project provided for joint efforts by the parties to construct additional engineering structures, upstream and downstream facilities; to determine the system capacity; to determine the parties' involvement; to deliver liquefied natural gas; and to complete the necessary documentation by the end of 2018. This US-Chinese project has not been yet implemented due to the "trade war" between the USA and China. In 2019, there were reports that the Agreement had expired and had not been extended by the Parties.[51]

[49] URL: https://www.cnoocltd.com/col/col7321/index.html

[50] Joint Development Agreement for Alaska LNG. URL: https://s3.amazonaws.com/arc-wordpress-client-uploads/adn/wp-content/uploads/2017/11/22061553/01.-Signed-JDA-SOA-AGDC-Sinopec-CIC-BOC.pdf

[51] AGDC president outlines path forward; China deal is dead. URL: https://www.alaskajournal.com/2019-07-24/agdc-president-outlines-path-forward-china-deal-dead

In its main strategic documents related to the Arctic China also highlighted the importance of bilateral cooperation with the USA. According to the China's "Arctic Policy", "China proposes to form cooperative partnerships between Arctic and non-Arctic States, and has carried out bilateral consultations on Arctic affairs with all Arctic States", particularly with the USA. In 2010, "China and the United States set up an annual dialogue mechanism for bilateral dialogues on the Law of the Sea and polar issues".[52] In its observer report to the Arctic Council, it is also mentioned that Chinese delegates participated in the first White House Arctic Science Ministerial held in the United States in September 2016, and signed the Joint Statement of Ministerial.

The analysis of the agreements and other legal documents signed between China and some Arctic States, has shown that China, in parallel with its increasing participation in the Arctic Council (as a permanent observer), is achieving significant results in strengthening its economic and scientific presence in the Arctic.

7. *Japan as a participant in the trilateral dialogue on Arctic issues*

Japan clearly indicated that the Arctic would be a new focus of its maritime policy in 2015. The Headquarters for Ocean Policy approved the first-ever comprehensive and strategic Japanese "Arctic Policy" with further additions to the main concept formulated in the Third Basic Plan on Ocean Policy of 2018.

Japan's "Arctic Policy" is intended to define policy in details. Thus, the "Arctic Policy" implies (1) full use of Japan's strength in science and technology; (2) full consideration to the Arctic environment and ecosystem; (3) ensuring the rule of law, and promoting of international cooperation in a peaceful and orderly manner; (4) respect the right of Indigenous peoples to continuity in their traditional economic and social foundations; (5) full attention to security developments in the Arctic; (6) economic and social compatibility with climate and environmental changes; (7) possible economic chances for the use of the Arctic Sea Route and for the development of resources.[53]

Among these points this document also provides that it is important for Japan to put its scientific knowledge and advanced technology to use in order to make further contributions to the activities of the Arctic Council. In the frame of the Arctic Council's work Japan will participate actively in discussions of expanding the role of observers. Japan expresses its willingness to participate actively in international forums other than the Arctic Council, and to initiate constructive discussions based on its scientific knowledge. According to the "Arctic Policy" in parallel with multilateral initiatives, it is also important to develop bilateral discussions and cooperative relations with Arctic States and other States concerned.

This "Basic Plan" subsequently formed the basis for Japan's key position paper on the status of the Arctic Ocean and its adjacent seas. According to the Basic Plan

---

[52] China's Arctic Policy. First Edition. January 2018 // Xinhua News Agency. URL: http://www.xinhuanet.com/english/2018-01/26/c_136926498.htm

[53] URL: https://www8.cao.go.jp/ocean/english/arctic/pdf/japans_ap_e.pdf

on Ocean Policy, which was enacted in 2007, "the Government shall review the Basic Plan on Ocean Policy at least every five years, and shall make necessary changes." Japan adopted the Third Basic Plan on Ocean Policy in May 2018.[54] This plan addresses the Arctic as one of the most important ocean-related issues.

The Third Basic Plan includes the following main directions of Japan's Arctic policy: (1) strengthening research initiatives in the Arctic region, based on *the Arctic Challenge for Sustainability*[55] where "the government and researchers work together"[56]; (2) strengthening the observational and research system pertaining to the Arctic region, including the construction of an Arctic region research ship with icebreaker capacity; (3) promoting international cooperation in science and technology for the Arctic region with the Arctic States and other relevant countries on the basis of bilateral agreements on cooperation in science and technology; (4) developing human resources to contribute to solutions for the Arctic region; (5) use of multilateral fora including the Arctic Council and bilateral dialogues with the Arctic States so that principles of international law including freedom of navigation is respected in the Arctic Ocean based on the 1982 United Nations Convention on the Law of the Sea; (6) proactive participation in the formulation of international rules based on the rule of law (including issues of the conservation and management of fishery resources); (7) utilizing the Arctic Sea Route; (8) conduct experimental tests to create sea ice flash charts for safe navigation along the Arctic Sea Route, by using sea ice observation data collected by satellites (the Water Circulation Change Observation Satellite and Advanced Land Observing Satellite); (9) Securing Protection of the Marine Environment in the Arctic Sea, including in the frame of the Arctic Council; and (10) proactively participate in international fora such as the Arctic Economic Council and the Arctic Circle so that they can promote the economic activities in the Arctic region.

In terms of bilateral and multilateral cooperation with the Arctic States and others, the document notes that Japan will "make the best use of international frameworks on the Arctic such as the Arctic Science Ministerial, the Arctic Circle, the Arctic Frontiers and the Trilateral High-Level Dialogue on the Arctic among Japan, China and South Korea". As for the contribution to the activities of the Arctic Council, it is noteworthy that Japan "will promote policy dialogues with stakeholders including the Arctic Council Chair and the States, and strengthen contributions to the Arctic issues as an important player"; and also "proactively participate in the discussions on how the Arctic Council should be, including expansion of the role of observers".

---

[54]URL: https://www8.cao.go.jp/ocean/english/plan/pdf/plan03_e.pdf

[55]The biggest Arctic research project of Japan.

[56]Funded by the Ministry of Education, Culture, Sports, Science and Technology, the key players are the National Institute of Polar Research, the Japan Agency for Marine-Earth Science and Technology and Hokkaido University. Moreover, the Government of Japan continues its effort to develop the Arctic Challenge for Sustainability successor project (the ArCS II) as well as the new Arctic research vessel concept. Japan will host the next Arctic Science Ministerial Meeting in Tokyo, on May 2021 with co-organizer, Iceland, the Chair of the Arctic Council. URL: https://arctic-council.org/ru/news/interview-with-arctic-council-observer-japan/

As mentioned in Japan's *observer report* to the Arctic Council, science[57] and technology remain Japan's strong point for its Arctic policy and are also indispensable for the solution of Arctic challenges.[58] Japan's document also includes the goal of harmonizing social and economic factors in the region. Japan's "Natural Resource Development" section of the same document notes that "in view of the harsh climatic conditions in the Arctic", natural resource development should be conducted "in the light of technical progress in the development of natural resources in the ice-covered sea areas and in conjunction with States whose coasts face the Arctic Ocean".[59] The words "jointly" are not, however, supplemented by an indication of the need to reach agreement with the Arctic State concerned in order to develop natural resources in the Arctic. Such clarification would be logical, nevertheless: with respect to natural resources in the Arctic seas, the respective Arctic coastal State exercises sovereign rights and jurisdiction both in its exclusive economic zone and on its Arctic shelf, including in areas of the shelf extending beyond 200 nautical miles from the baselines along the Arctic coast (Vylegzhanin & Dudykina, 2018).

The legal framework for Japan's participation in the development of oil and gas resources in the Arctic is not significant. Japanese companies are involved in Arctic projects only at the design stage, equipment procurement, and less frequently in construction of facilities and transportation of extracted energy resources (e.g. Japan's participation in the Yamal LNG project is very modest). According to Japanese Special Envoy for the Arctic, the Yamal LNG project is a key element of Russian–Japanese relations.[60] These relations, however, are not governed by an intergovernmental agreement, but by private legal contracts. In the northeastern part of the Yamal Peninsula (Russia), an LNG plant is being built with a production capacity of 16.5 million tons of LNG per year with the necessary infrastructure, including energy facility, a seaport and an international airport.[61] Japan's "Arctic

[57] Japan has contributed to the Arctic Council's activities in scientific research, which is Japan's strength. Among Japanese national institutions which are effective instruments of Arctic policy of Japan we can mention the Hokkaido University, the Arctic Environment Research Center, the National Institute of Polar Research, the Japan Agency for Marine-Earth Science and Technology. These leading Arctic research institutions of Japan have been providing scientific data to the Arctic Council. As a representative of Japan said Aerospace Exploration Agency also contributes to the Arctic Council by providing valuable data obtained by its satellites. URL: https://arctic-council.org/ru/news/interview-with-arctic-council-observer-japan/

[58] Observer report of Japan 2018. URL: https://oaarchive.arctic-council.org/bitstream/handle/11374/2259/JAPAN_2018-05_Review-Report.pdf?sequence=1&isAllowed=y

[59] Arctic Council. URL: https://oaarchive.arctic-council.org/bitstream/handle/11374/1868/EDOCS-4031-v1-2016-12-16_Japan_Annex1_to_Observer_activity_report.PDF?sequence=2&isAllowed=y

[60] Pollmann M. 2016. *"How Japan and Russia Cooperate in the Arctic"*. URL: https://thediplomat.com/2016/03/how-japan-and-russia-cooperate-in-the-arctic/

[61] The Yamal LNG Project is operated by JSC Yamal LNG and implemented by a joint-venture of NOVATEK (50.1%), TOTAL (20%), CNPC (20%) and Silk Road Fund (9.9%), the Project is based on the Yamal Peninsula, above the Arctic Circle, and utilizes the resources of the South Tambey Field. URL: http://yamallng.ru/project/about/

Policy" does not mention the Yamal LNG Project; however, it does indicate an intention to continue financial support for Greenland Petroleum Exploration Co., Ltd. which is participating in an exploration project in an ocean area northeast of Greenland, via the Japan Oil, Gas, and Metals National Corporation.

Although Japan does not have an intergovernmental agreement with Denmark on access to natural resources on the Arctic shelf north of Greenland, Japan also has commercial interests in this area. In 2013, Japan Oil, Gas and Metals National Corporation received rights to develop an oil field on the Greenland shelf (together with Chevron and Shell). The Japanese company "Japan Petroleum Exploration Co., Ltd" (JAPEX) has received two licenses for the development of resources of the Greenland continental shelf in the Kanumas area (north-eastern part of the Greenland shelf). Then, the joint venture (Shell, Chevron, JAPEX) entered into a licensing agreement with the Ministry of Industry and Mineral Resources of Greenland.[62]

Japan is also cooperating with Iceland on energy issues. *The Joint Statement on Strengthening Relations between Japan and Iceland in 2014*[63] provides that the parties confirm their intention to cooperate in the field of geothermal energy on the basis of *the Memorandum of Cooperation between the Ministry for Foreign Affairs of Iceland and the Japanese Bipartisan Coalition of Legislators for Promoting Geothermal Power Generation in the area of geothermal energy.*[64] The Joint Statement stipulates the intention of the parties to continue cooperation within the framework of the Nordic-Baltic Eight, including on the Arctic issues. Moreover, the Ministers stressed that environmental changes in the Arctic provide new opportunities and pose challenges for the international community, including Iceland and Japan, and that any action to be taken in the Arctic needs to be based on the rule of law. The Ministers also mentioned the importance of environmental protection, sustainable development and use of natural resources, including energy and fishery, and human rights of Indigenous people. The Ministers expressed their commitment to mobilize their political will, business opportunities and academic resources to promote these elements. The Ministers also shared the recognition that the seas and oceans need to be open, free and secure. They reaffirmed that maritime order based on the rule of law must be maintained and that common principles, such as the freedom and safety of navigation and overflight over the high seas, should be fulfilled. The Ministers decided on the paramount importance of refraining from the use or threat of force and of resolving disputes through peaceful means in accordance with international law, including the 1982 UN Convention on the Law of the Sea.

In *the Joint Statement on Strategic Partnership between Japan and Finland*, the parties call for further strengthening of cooperation in the Arctic, stressing the common interests of the two States, due to the geopolitical position of Japan and

---

[62] Licensing Round 2012/2013 in Greenland Sea Area, December 24, 2013 https://www.japex.co.jp/english/newsrelease/pdfdocs/20131224_Greenland_E.pdf

[63] Ministry of Foreign Affairs of Japan. URL: http://www.mofa.go.jp/files/000059031.pdf

[64] National Energy Authority of Iceland. URL: https://orkustofnun.is/media/mou/MOU-Japan.pdf

Finland and "mutual respect".[65] The Parties share the view that all activities in the Arctic should be carried out by the States "in strict accordance with the norms and principles of international law"; they affirm the importance of environmental protection, sustainable use of natural resources (energy, minerals, forestry, fisheries) and protection of the rights of Arctic Indigenous peoples; and they also affirm their commitment to promote dialogue and cooperation in the Arctic, bearing in mind that Japan and Finland are "on different sides" of the Northern Sea Route. Both States, with advanced technologies applicable to the Arctic, have expressed their intention to strengthen cooperation in the Arctic between relevant government agencies, research institutes, and the business community.

On the academic level, the Finnish Environment Institute and the National Institute for Environmental Studies of Japan signed a *Memorandum of Cooperation* in 2017 aimed at promoting joint environmental research projects, such as researches on climate change in the Arctic. Both institutes are consulting about possible cooperation areas.[66]

Japan is also collaborating with other research institutes of the Arctic States. For instance, Japan is preparing the use of new Japanese research station at Ny-Ålesund in Svalbard, which is constructed by the government of Norway; Japan held the Japan-Norway Symposium "Past, Present, and Future of the Arctic and Antarctic" in Norway in June 2017; Japan held the two workshops on Arctic research with Russian researchers and institutes in 2017 and 2018; and Japanese National Institute of Polar Research concluded a Memorandum of understanding with Russian Arctic and Antarctic Research Institute and started to collaborate the observation at the Research Station Ice Base Cape Baranova. Japan will also start research and observation of the ecosystem at the Canadian High Arctic Research Station in Cambridge Bay, Canada.[67]

8. ***South Korea as a participant in the trilateral dialogue on Arctic issues***

In May 2013, South Korea joined the Arctic Council as an observer, and since that time, as stated in its *observer report 2018*, "South Korea has participated actively in the programs and activities of the Arctic Council's working groups, task forces, and expert groups".[68] The main role in such activities plays the Korea Arctic Experts Network which recruits and dispatches qualified experts to work in subsidiary bodies of the Arctic Council.

According to its observer report, South Korea's engagements in the Arctic Council's work include (1) "organizing seminars of the Protection of the Arctic

[65] URL: http://www.presidentti.fi/public/default.aspx?contentid=342941&culture=en-US

[66] Observer review reports 2018. URL: https://oaarchive.arctic-council.org/bitstream/handle/11374/2419/SAOXFI204_2018_ ROVANIEMI _05_Observer-Review-Reports-Combined.pdf?sequence=1&isAllowed=y

[67] Ibid.

[68] URL: https://oaarchive.arctic-council.org/bitstream/handle/11374/2262/REPUBLIC-OF-KOREA_2018-06_Review-Report.pdf?sequence=1&isAllowed=y

Marine Environment Shipping Experts Group" with participation of the Korean Maritime Institute; (2) participating in working-level projects, namely the Arctic Indigenous Marine Use Mapping; (3) working together with the Conservation of Arctic Flora and Fauna working group on the Arctic Migratory Birds Initiative; (4) participating of the Korea Polar Research Institute in a range of the Arctic Monitoring and Assessment Program's meetings; (5) involving of the Korea Research Institute of Ships and Ocean Engineering in various projects in the frame of the Emergency Prevention, Preparedness and Response working group; (6) exchange program in cooperation with the UArctic, etc.[69] It is demonstrated that the main purpose of the South Korea's Arctic policy is scientific research and international cooperation on the academic level.

South Korea pays much attention to the issues of Indigenous peoples, and supports efforts to help them to mitigate and adapt to the impacts of climate change. In the framework of the Arctic Council, South Korea supported financially the Arctic Indigenous Marine Use Mapping project led by Aleut International Association. The goal of the project is to come up with an information tool by which coastal Indigenous communities can produce scientifically justifiable maps of local marine use.[70]

In 2013, the Government of South Korea adopted with possible prolongation every 4 years the *Arctic Master Plan* for "implementing a comprehensive Arctic policy and follow-up measures".[71] For the specified duration the document set forth the following strategic goals: (a) strengthening international cooperation with the Arctic region; (b) encouraging scientific and technological research capacity; (c) pursuing sustainable Arctic businesses; and (d) securing institutional foundation. The document is focused, mainly, on the development of shipbuilding technologies for the Arctic and for different vessel types (containers, LNG carriers) and materials technologies (that are suitable for operations at very low temperatures); and on the development of port infrastructure along Arctic Shipping Routes. In this regard, it is also stated that South Korea pursues "joint research with Arctic States in the fields of resources development, cargo shipping infrastructure, transshipment ports, and the commercial use of the Northern Sea Route".[72]

In 2018, South Korea published "Second Arctic Master Plan for 2018-2022", aimed at becoming 'leading observer state' in the Arctic. The vision of the "Policy Framework for the Promotion of Arctic Activities of the Republic of Korea 2018-2022" lies in becoming a pioneer and partner in shaping the Arctic future. Thereby, the policy goals are set to (a) promote participation in Arctic economies, (b) increase

[69] Ibid.

[70] URL: https://oaarchive.arctic-council.org/bitstream/handle/11374/2262/REPUBLIC-OF-KOREA_2018-06_Review-Report.pdf?sequence=1&isAllowed=y

[71] The Arctic Policy of the Republic of Korea. URL: http://library.arcticportal.org/1902/1/Arctic_Policy_of_the_Republic_of_Korea.pdf

[72] For more details also see Kim 2014.

participation in Arctic governance, and (c) contribute to the international community and build capacity for addressing challenges in the Arctic.

From 2018 to 2022, South Korea will pursue 13 implementing actions under four major strategic directions, which are (a) mutually reinforcing economic cooperation, (b) responsible partner in Arctic cooperation, (c) research contribution towards addressing common challenges,[73] and (d) capacity building.[74] This Plan also includes South Korea's aim to build a second ice-breaking research vessel.

According to some estimations South Korea spends more on Arctic research than the United States does.[75] The Korea Research Institute for Oceanography is involved in marine oil spill prevention and response projects, given the fact that South Korea has advanced technologies, including the ability to predict the trajectory of oil spills. The demand for such technologies in the Arctic is predicted to grow.[76] South Korea's interest in the Arctic is driven by huge investments in the construction of ice class ships, including LNG tankers, drilling ships, cruise ships. As an example, in 2012, South Korea and Norway signed two Memorandums of Understanding on Arctic shipping and shipbuilding.

The Canada – South Korea strategic partnership since 2014, according to *Joint Declaration*, lays out a strategic direction for stronger relations in key areas of common interest including energy and natural resources, science, technology and innovation, and Arctic research and development.[77]

In October 2010, Russia and South Korea signed the *Agreement on Maritime Transport*. In July 2013 at the regular session of the Russian-Korean joint commission on economy, science and technology the parties signed the *Memorandum of Understanding on construction of port infrastructure*. South Korea has developed even greater cooperation with Denmark (Greenland): *four Memorandums of Understanding on Energy and Natural Resources of Greenland* were signed between two States.

South Korea is also a party to the 1920 Spitsbergen Treaty.

Conclusion

Today, we observe unprecedent activity of the non-Arctic States in the Arctic region. This growing activity demonstrated by a range of legal documents signed by the People's Republic of China, Japan and South Korea, whether between them or with

[73] This goal is imposed mostly on the Korea Polar Research Institute. URL: https://www.kopri.re.kr/eng/html/rsch/030101.html

[74] Policy Framework for the Promotion of Arctic Activities of the Republic of Korea 2018–2022. URL: http://www.koreapolarportal.or.kr/data/Policy_Framework_for_the_Promotion_of_Arctic_Activities_of_the_Republic_of_Korea-2018-2022.pdf

[75] URL: https://www.economist.com/node/21561891

[76] Arctic Council. URL: https://oaarchive.arctic-council.org/bitstream/handle/11374/1862/EDOCS-4020-v1-2016-11-29_Republic_of_Korea_Observer_activity_report.PDF?sequence=1&isAllowed=y

[77] Joint Declaration between Canada and South Korea 2014. URL: https://www.canada.ca/en/news/archive/2014/09/joint-declaration.html

the Arctic States, can already been experienced. These documents are focused on the environmental protection, issues of navigation and, above all, scientific research. Some documents aim at the exploration and exploitation of natural resources, nevertheless.

The individual legal positions of China, Japan and South Korea to the regime of the Arctic Ocean are also reflected in their Arctic policies which have similar provisions. Even though the interpretations of such documents by different scholars from the Arctic States are alarmist only in relation to China. The reason of such attitude might lie in its official document which regards the Arctic as a "common heritage of mankind" and also China's perception of itself as a "near-Arctic State". It is China's "Arctic policy" that is considered by most analysts as a threat to the international legal order established in the Arctic, and reflected in the Ilulissat Declaration of 2008, i.e. the sovereignty of each Arctic State within its national territory, the sovereign rights of Russia, the United States, Canada, Denmark and Norway to the relevant areas of their Arctic shelf and exclusive economic zone in the Arctic Ocean. China is said to be dissatisfied with this international legal order, including how the Arctic States (through the Arctic Council) ensure the proper environmental regime and regulation of shipping in the Arctic waters. In summary, many analysts write that "China seeks dominance" in the Arctic; that "the Arctic Council is flawed"; and that China is "provoking" other non-Arctic States to create separate legal instruments on the Arctic Ocean regime (Lackenbauer & Manicom, 2013).

The analysis of the documents discussed in this chapter has led to another conclusion. It is true that certain statements of some Chinese officials, cited above, as well as some provisions set forth in the China's "Arctic Policy", take positions different from those of the Arctic States regarding the current legal regime of the Arctic Ocean. But the emphasis of China's policy documents shows a commitment to cooperate with the Arctic States on matters of mutual interest and a willingness to pursue such cooperation within the framework of the international legal regime applicable to the Arctic.

In a short term, China is supposed to be seen by many scholars as a threat to a stable order in the Arctic. In the long-term, however, ongoing cooperation between the Arctic States and non-Arctic States will be seen from different angle and bring positive results for both groups of States. In this case two important conditions should be met: (1) the role of Arctic States should remain determinative in the Arctic issues; (2) States should do their best to maintain balance between common interests (associated with sustainable activities) and national interests of Arctic coastal States (environmental safety issues when economic operations are involved in the Arctic).[78]

---

[78]For more information see Berkman and Vylegzhanin (2013).

As noted above, the non-Arctic States do not enjoy the same rights in the region as the Arctic States do. That is why some of them are willing to participate actively in discussions of expanding the role of observers. However, the rights of the Arctic States will not be eroded by activities of China, Japan or South Korea in the Arctic. For this reason, the Arctic States have little reason for concern over Arctic policy statements and bilateral engagements with certain Arctic States of non-Arctic States.

The main point is to regard China, Japan and South Korea as a formal alliance of three powerful and economically developed States that have similar national interests in the Arctic. Besides the geographical proximity and the similarity of their Arctic policies, these three States tend to participate in the same international agreements related to the Arctic, for instance, the Spitsbergen Treaty, 1920 or the Agreement to Prevent Unregulated High Seas Fisheries in the Central Arctic Ocean, 2018.

Furthermore, China, Japan and South Korea have already formed the alliance of three States. It has been institutionalized as the Trilateral Cooperation Secretariat to coordinate the trilateral cooperation in various fields, including the Arctic issues, and for the efficient promotion and management of the trilateral cooperative projects. In the frame of the Trilateral Cooperation Secretariat these States have openly set forth their common interests and purposes and have expressed their common intensions through the established Trilateral Dialogue on Arctic issues.

The Joint Statements which three States publish as a result of the Trilateral Dialogues on the Arctic demonstrate that China, Japan and South Korea emphasized increasingly their contributions to scientific investigations in the Arctic and expressed their willingness to develop other spheres related to the Arctic. All three States underlined that they fully respect the sovereign rights of Arctic coastal States and none of them has claims or disputes in the Arctic. Besides, China, Japan and South Korea highly appreciate the role of the Arctic Council in solving Arctic issues. All three States still might fulfill their broader aspirations of playing visible roles in the Arctic issues only through international cooperation with the Arctic States. Overall, China, Japan and South Korea have indicated a serious commitment to work with the Arctic States on matters concerning the Arctic region.

The Arctic States should ensure that the mechanisms of Pan-Arctic cooperation in the spheres of environmental protection and sustainable development facilitate the involvement of China, Japan and South Korea in constructively maintaining the established legal regime of the Arctic. As a first step, this would be facilitated by the participation of scientists and specialists from the Arctic States in trilateral consultations on Arctic issues between China, South Korea and Japan. This would ensure the openness of the Trilateral Dialogue and prevent these non-Arctic States from creating new trilateral instruments on the legal regime of the Arctic that might differ from the current legal regime and have unpredictable future international legal consequences.

References

Berkman, P. A., & Vylegzhanin, A. N. (2013). *Environmental security in the Arctic Ocean*. Springer. 459 p.

Berkman, P. A., Vylegzhanin, A. N., & Young, O. R. (Eds.). (2019). *Baseline of Russian Arctic laws*. Springer.

Chircop, A. (2011). The emergence of China as a polar-capable state. *Canadian Naval Review*, (7), 3.

Fife, R. (2013). Cooperation across boundaries in the Arctic Ocean: The legal framework and the development of policies. In P. Berkman & A. Vylegzhanin (Eds.), *Environmental security of the Arctic Ocean* (pp. 355–356). Springer.

Hong, N. (2020). *China's role in the Arctic: Observing and being observed*. Routledge. 218 p.

Kim, J. D. (2014). Korea's Arctic policy. In O. R. Young, J. D. Kim, & Y. H. Kim (Eds.), *The Arctic in World affairs*. Korea Maritime Institute and East-West Center. 424 p.

Konyshev, V., & Kobzeva, M. (2017). China's policy in the Arctic: Tradition and modernity. *Sravnitel'naya politika i geopolitika [Comparative Politics and Geopolitics], 8*(1), 78. [In Russian].

Lackenbauer, P. W., Lajeunesse, A., Manicom, J., & Lasserre, F. (2018). *China's Arctic ambitions and what they mean for Canada* (Beyond boundaries: Canadian defence and strategic studies series, No. 8). University of Calgary Press.

Lackenbauer, W., & Manicom, J. (2013). Canada's Northern strategy and East Asian interests in the Arctic. In K. Hara & K. S. Coates (Eds.), *East Asia-Arctic relations: Boundary, security and international politics* (pp. 77–116). McGill Queen's University Press.

Vylegzhanin, A. (2019). Legal regime of the Arctic region as reflected in the documents. In P. Berkman, A. Vylegzhanin, & O. Young (Eds.), *Baseline of Russian Arctic laws* (p. XXVII). Springer.

Vylegzhanin, A., & Dudykina, I. (2018). *Baselines in the Arctic: applicable international law*. MGIMO University. 174 p. [In Russian].

Vylegzhanin, A., Saligin, V., Dudikina, I., & Kienko, E. (2018). Positions of the non-Arctic States regarding the legal regime of the Arctic Ocean. *The Journal "State and Law", 10*, 124–135. [In Russian].

Xie, C. (2018). Sino-Russian expedition provides Arctic data. *China Daily*. 31, October, 2018. p. 4.

Zagorskii, A. (2019). China accepts rules in the Arctic. *Mirovaja jekonomika i mezhdunarodnye otnoshenija [World Economy and International Relations], 63*(7), 77. [In Russian].

Chapter 19
(Research): Innovations in the Arctic: Special Nature, Factors, and Mechanisms

Nadezhda Zamyatina and Alexander Pilyasov

Abstract Arctic innovations are considered in a broad context – as a way of life in northern communities with omnipresent technological, economic and social implications. Consideration of innovations reveals a gap in modern research in the social sciences between the numerous works on innovation in large urban agglomerations and the almost complete absence of efforts to study innovation in the world periphery, including the Arctic. Major features of the human dimension of the innovation process in the Arctic are: (a) prominent position of the individual Schumpeterian-type entrepreneur-innovator, the creative destroyer, whose role and meaning is visible, tangible and concrete; (b) unprecedented role of local knowledge and competencies, which are based on the extremely specific natural and economic conditions of the Arctic; and (c) extreme unevenness in the concentration of talents in space and time that are explained by resource development cycles. As an outcome, six types of innovation systems (IS) are revealed in the global Arctic: (1) IS of multifunctional urban centers; (2) Network IS in the old-developed resource and coastal regions; (3) IS of base city-islands in old-developed resource regions; (4) IS of areas of modern pioneer development (frontier IS); (5) "Privileged" IS of island capitals; (6) West Siberian ISs as a network of resource urban centers. The fundamental specificity of the Arctic innovations stems from differences across developed regions in actors, networks and institutions.

19.1 Introduction

Traditionally, innovations in the Arctic were considered very narrowly as a technological phenomenon that provides the saving of expensive labor costs in the interests of production efficiency (Matveev, 2011). The scientific literature discussed innovative solutions in life support systems (heat and energy supply, food security, etc.) in the Arctic (Pilyasov & Yadryshnikov, 1997). However, innovation has never been

N. Zamyatina (✉) · A. Pilyasov
Lomonosov Moscow state University, Moscow, Russia

P. A. Berkman et al. (eds.), *Building Common Interests in the Arctic Ocean with Global Inclusion, Volume 2*, Informed Decisionmaking for Sustainability,
https://doi.org/10.1007/978-3-030-89312-5_19

considered in a broad context – as a way of life in northern communities, as an «omnipresent» technological, economic and social phenomenon. It seems that the time has come to take a look at Arctic innovations in such a broad, and not technological, but social way of changing the internal foundations of human life.

Unlike many other polar territories, which are closer in terms of socio-economic development to the "mainland" parts of their countries, the Russian Arctic is more specific and more different in terms of the course of the innovation process from the zone of main settlement. In addition, in the Russian Arctic there are many regions, the natural and economic conditions of which are also internally very different from each other, producing a continuum of situations in the deployment of the innovation process in the Arctic periphery. All this makes the study of Arctic innovations and the innovation system here interesting not only for Russia, but for the entire Arctic world.

Arctic innovations are not just an extention to the Arctic of those innovations that were previously spread in densely developed regions of the country, adjusted for the natural extremity and transport remoteness of these polar territories. No, this is an absolutely special holistic phenomenon that needs to be separated from the rest, and not understood as just an extreme, ultimate form of well-studied and well-known phenomena of the more southern regions of the country's main settlement zone.

Currently, innovation processes in the Russian Arctic are multidirectional in nature. On the one hand, the accelerated development of the Arctic means the intensification of contradictions between new technologies and established social institutions and spatial structures of socio-economic systems, such as systems of resettlement and distribution of productive forces, territorial structures of the economy. On the other hand, it is the Russian Arctic that is often ahead of other regions of the country in the development of innovations that facilitate the solution of the most acute problems of socio-economic development of the Arctic (such as remoteness and a rare transport network and a sparse network of settlements). Here, the population and entrepreneurs are more active users of e-commerce, Internet search of business partners, communication capabilities of social media (Pilyasov, 2018).

The task of studying Arctic innovations as an absolutely separate, specific phenomenon has determined the organization of this chapter. In the first section, the authors state a gap in modern research in the social sciences between the numerous works on innovation in large urban agglomerations and the almost complete absence of efforts to study innovation in the world periphery, including the Arctic. The next section attempts to answer the question: what is the phenomenon of Arctic innovation itself? In the third section, specific examples are used to describe the most common mechanisms of innovative development in the Russian Arctic. In the fourth section, an attempt is made to take a holistic view of the phenomenon of Arctic innovations from the perspective of the concept of a peripheral innovation system and its major types. In contemporary conditions of a dynamic and turbulent Arctic, this system is an important mechanism to guarantee resilience for these peculiar and specific territories of the world through informed decisionmaking processes, science diplomacy, and harmonization of multi-actor interests. Finally, the last section provides an answer to the question: how can Arctic innovations be of interest to the rest of the world?

19.2 The Concept of the Geography of Innovations and the Russian Arctic: The Current Gap and Problem Statement

Research on the geography of innovation started in the early 1990s, and it is necessary to note the breakthrough of Maryann Feldman, who introduced the term. In 1994, her pioneer monograph on this subject appeared (Feldman, 1994). Gradually, through the efforts of a large army of researchers, innovations themselves began to be understood much more broadly than traditional technological, production innovations, which were recognized in the industrial era. The interpretation of innovations as a social phenomenon, which depends on the personality of their creator (even, one might say, his biography), on the type of his communication (how wide?), on the institutional environment, and on the historical and cultural context in which it develops, has gradually begun to take hold. And this social phenomenon depends upon all kinds of proximity (spatial, social, organizational, institutional, and cognitive) identified by Boschma (2005).

The talented works of M. Feldman, R. Boschma, R. Florida (Florida, 2008), B. Asheim (Asheim & Gertler, 2005), D. Audretsch (Feldman & Audretsch, 1999) and others were concentrated mainly on large-scale urban areas of high density communications, with excellent infrastructural equipment, with the strong development of knowledge-intensive business services and creative class. In Russia, interesting work in this direction in recent years has been carried out by our colleagues, economic geographers and regional economists V. Baburin (Baburin & Zemtsov, 2017), S. Zemtsov (Zemtsov et al., 2016), E. Kutsenko, and others.

The breakthrough in the development of the topic of the geography of innovations did not affect the sparsely populated and low-density spaces of the world, including the Arctic zone. Powerful and broad research studies of the anatomy of the innovation process, dominating in the developed regions, have stopped at the Arctic's borders.

At the same time, within the Arctic itself there were very interesting studies, but modest and narrow in their design: for example, on the influence of a snowmobile technological revolution on the traditional way of life of small Indigenous peoples of the North (Pelto, 1987; Stammler, 2009), on the topic of "smart specialization" in the Arctic (Healy, 2017), on the patent activity in the State of Alaska (Zbeed & Petrov, 2017), and on the metrics of creative capital in the cities and towns of the Canadian North and Alaska's regions (Petrov, 2008, 2011).

A gap is evident between, on the one hand, the accumulated potential for studying the geography of innovations as a collective social process in densely developed and large urban areas of the European Union, the USA, and Russia and, on the other hand, limited studies in the Arctic, either too narrow or, on the contrary, too general, not reflecting the fundamental features of the Arctic zone. There is a need to link the local, micro-analytical and the national levels in understanding the innovation process in the Arctic: to use the achievements of the school of geography of innovations and apply them creatively to the realities of the Arctic.

19.3 The Special Nature of Arctic Innovation

Summarizing the numerous works of our foreign colleagues in the geography of innovations and our own 35 years of experience in researching economic and social processes in the circumpolar North and in the Arctic of Russia, let us formulate ideas about the phenomenon of Arctic innovations by comparing the Arctic and the "mainland". Significant differences of the Arctic in the innovation process from the territories of the temperate zone are clearly grasped through three slices: key actors; features of the urban settlement system; the nature of knowledge, information exchanges and learning (Table 19.1).

The Arctic as a whole is more "corporate" territory in Russia in the sense of a stronger presence in its economy of large resource corporations of global or national scale. Arctic corporations are the most important generator of production innovations, which include new technologies for the development of mining projects, new growth "poles"/greenfield projects (resource extraction facilities, new elements of the settlement system like shift camps), as well as brownfield projects of technological modernization of old mining enterprises. Given the production nature of a typical Arctic economy, these production innovations often set the context and lead the other (ICT, life-supporting) innovations. For example, corporate winter roads can serve as communication and life-supporting innovations which can be used for transportation and life support for the population of the entire village closest to the mining field.

Do actors change at different stages of the innovation process? Initially, at the search stage, its key actors are individual innovative entrepreneurs, completely independent loners, or part of a small venture firm, or integrated into large state or corporate super-organizations (Pilyasov, 1993). Very quickly, at the stage of pioneering development, they are replaced by subdivisions of global or national resource corporations, public or private. There is no other way to solve the costly tasks of developing a new production project or a new resource territory in the Arctic.

Big corporations also dominate at the next stage of rapid production growth, which provides companies with economies of scale without which they simply cannot exist. But the same economies of scale will kill incentives for innovation in prospecting and production further.

The subsequent inevitable decline in production again strengthens the interest of companies in innovation, but at the same time the innovation process itself is significantly diversifying, and small and medium-sized businesses in exploration,

Table 19.1 Differences between Arctic innovations and «mainland» innovation

Features	Arctic	«Mainland»
Key actors of innovation process	Resource corporations, entrepreneurs, including Indigenous	SME, corporations, state, NGOs etc.
Agglomeration effect	Weak or absent	Strong
Externalities	Narrow specialization	Urban diversity
Type of knowledge	DUI synthetic	STI, DUI[a] analytical, synthetic, symbol
Circulation of knowledge+	Temporary geographical proximity	Constant geographic proximity
Circulation of knowledge -	Closed corporate loop	Fragmentation, distrust of actors
Barriers for absorptive capacity	Overspecialization lock-in	Cognitive lock-in from path-dependency
The main way to "acquire" knowledge	Exploration and search	R&D
The main sources of new knowledge	External networks, tacit knowledge	Internal networks, formal knowledge
The flow of knowledge: Forms	Employee mobility, Internet publications, electronic forums	Spin-offs, cooperation with other actors (suppliers, consumers, competitors)
Learning process	Learning by doing, by experiencing	Retraining courses, formal training
Research subsystem	Interdisciplinary expeditions, experimentation	Universities, research institutes, academic laboratories, etc.
Operational subsystem (dominant local production system)	Mining industry	Manufacturing industry
Key industrial contracts	Vertical (mining, processing, marketing)	Horizontal (subcontracting, etc.)
Source of innovation	Combination of activities, interdisciplinarity, interchange and integration of competences	Division of labor and competencies, micro-specialization

[a]*STI* Science, Technology and Innovation, *DUI* Doing, Using and Interacting mode (Asheim et al., 2019)

production and production services become its participants (along with the R&D divisions of companies). Later on, under the pressure of depletion, the innovation process becomes even more radical - gradual production innovations are replaced by revolutionary ones and the local innovation system itself is reborn from a purely sectoral, corporate one into a territorial one, with simultaneous diversification through the active development of social, life-supporting, service innovations which existed before, but were strictly subordinated to the interests of the main resource-extracting industry.

An intense innovation search at the stage of depletion, in which the structures of small and medium-sized businesses are actively involved, can give rise to a new cycle of economic development of new natural resources or new regions, with the repetition of the indicated patterns of the innovation process.

The type of natural resource significantly concretizes the described scheme (Kryukov, 1998), determines the dynamics of the innovation process and the involvement of various actors, such as large companies and small and medium-sized businesses, both in the mining industry itself and in the structure of knowledge-intensive business services. The rule applies: the more specific is the natural asset and material assets that are geared towards its extraction and transportation, the greater the load on the innovation system in ensuring the effective deployment of the entire resource chain from extraction to the sale of the final product.

Several features of the human dimension of the innovation process in the Arctic can be noted. First is the prominent position of the individual personality of the Schumpeterian entrepreneur-innovator, the creative destroyer, whose role and meaning is visible, tangible and concrete, as rarely happens in densely populated regions of the world.

It is much easier for such original people who are absolute crushers of indisputable truths to find support for their ideas and reach their implementation in the Arctic than anywhere else. The fact is that the conditions for competitive selection of ideas do not work here, so the chance that an adventurous idea will survive and become legitimate is much higher than anywhere else. Tolerance for innovative adventurism in the Arctic is greater than in densely developed areas. All this creates an excellent environment for the most daring and even adventurous experiments. One can call it «the open horizons for crazy ideas» effect.

Second is an unprecedented role of local knowledge and competencies, which are based on the extremely specific natural and economic conditions of the Arctic. Meanwhile, the ability to understand them sharply differs even among highly qualified personnel. Those of them who have the talent for quickly absorbing local tacit knowledge, are capable of making breakthroughs in the economic development of areas for new resource development in the Arctic.

Third is the extreme unevenness in the concentration of talents in space and time, which is explained by resource development cycles: at the exploration and pioneer stage, a unique concentration of talents arises in a new resource project, which then dissipates at the subsequent and more routine stages of rapid growth and stabilization of production and is rarely repeated at the stage of exhaustion and decline.

A researcher who compares the internal anatomy of the innovation process in the "mainland" and in the Arctic, associated with the nature of knowledge, its flows, is faced with a paradox. In the Arctic, new knowledge is generated not in laboratories, not owing to classical achievements of fundamental academic science, but during field expeditions, observations of the production process, and training in the process of field or stationary work. The role of concrete experience in Arctic knowledge generation and innovation is unprecedented.

And this Arctic knowledge often is not analytical, narrowly sharpened, professional knowledge of egg-headed cabinet scientists, but synthetic, engineering knowledge of Arctic practitioners and experts. In this knowledge, the tacit component that is not fully formalized in books and textbooks is very strong, tied to an expert, a carrier of unique competencies and local "field" knowledge.

In full accordance with modern ideas about the innovation process (Asheim et al., 2019), the fundamental specificity of the Arctic stems from its differences from the developed regions in *actors, networks and institutions*: dominant corporate actors, the increased role of external networks, gatekeepers and institutions of temporal proximity in the circulation of knowledge, the dominant institutions of the mining and not manufacturing industries, which all have a multifaceted effect on the nature of knowledge and knowledge spillovers.

If for the "mainland" the research laboratory is the classical birthplace of innovation, the field geological expedition can serve as such a standard image for the Arctic. In such expeditions, all the Arctic specific features of actors, networks and institutions of the innovation process are fully reflected. And the "customer", which drives the demand for geological discovery, is a resource corporation.

19.4 Specific Arctic Mechanisms of Innovative Development

In order to come to terms with the special Arctic mechanisms of innovative development that are not like the mainland, the researcher is reminded of "Alice through the Looking Glass": "You don't know how to manage Looking-glass cakes," the Unicorn remarked. "Hand it round first, and cut it afterwards" (Carroll, 1973).

A powerful mechanism for innovative development in the Arctic is the process of developing a new frontier itself. The frontier is a well-known phenomenon from the history of the United States. Among other features, the frontier went down in history as a generator of political and social, technological and technical innovation. It is believed that it was on the frontier that many innovations were born that eventually determined the national character of Americans (Burstin, 1958).

The innovative potential of the frontier was determined by a rare combination of two factors. On the one hand, the development of new territories required solving many problems arising from the specifics of the new territory: new soils, new social composition, etc. On the other hand, the rapid involvement of large amounts of resources in the economic turnover made it possible to achieve the effect of increasing returns and high profits. Profits delivered sufficient financial resources for the pilot implementation of innovations. The frontier was a true innovation laboratory, where new solutions were not only invented, but immediately tested, and if successful, achieved mass distribution.

In modern conditions, the front-line mechanism of innovative development, tied to the pioneering development of Arctic resources, became manifest when a new Yamal-LNG project was deployed in the shift camp of Sabetta in the north of the Yamal-Nenets autonomous okrug. The pilot project receives the special status of an experimental initiative (as earlier in Soviet times, the status of the all-national –

"vsesoyusnaya" or "vserossiyskaya"- construction) and special tax regimes for its deployment, which subsequent projects of a similar nature do not have.

The economy of developed territories is the economy of large numbers, large quantities, sometimes even overpopulation with ultrahigh density. Therefore, the innovative mechanism here is more reminiscent of the laws of evolution according to Darwin: competition, selection of the most viable option and its consolidation in the course of subsequent development.

On the other hand, the economy of the Arctic is an economy of small numbers, insufficient density and frequent interruptions, developmental delays and even "extinctions" and then "re-development". In conditions of small quantities, an innovative mechanism is formed from a creative reassembly of a few familiar elements in a new unexpected way. And the realities of the catastrophe economy in the form of frequent abandonment of former economic sites lead to the increased importance of pioneer development from scratch, the high role of radical, rather than gradual, innovations. This is not continuous evolution, but discrete catastrophism, which is the "fuel" for Arctic innovation.

Often, innovations in the Arctic are launched in the course of force majeure temporary abandonment of the principle of division of labor and, conversely, the combination of functions caused by a shortage of workers and crisis. This frequently happens suddenly, but it is during these periods of forced combination of occupations, which were previously considered absolutely impossible and unacceptable, that many Arctic innovations arise (rather than simple local adoption of new innovations from outside).

19.5 Peripheral Innovation System

The deepening theoretical ideas on the specifics of the innovation process in remote and peripheral territories is critical to ensure that industrial and innovation policy is based on real knowledge of these territories, and does not routinely repeat theories that reflect the experience of the metropolitan regions but do not work on the periphery.

Summarizing the few works that have appeared in the last 10 years on innovations in the periphery (Ferrucci & Porcheddu, 2006; Virkkala, 2007; Petrov, 2011; Karlsen et al., 2011; Dawley, 2014; Isaksen & Karlsson, 2016; Asheim et al., 2019), allows us to identify their key features.

In these peripheral regions, as a rule, there are no opportunity for the full-blooded manifestation of economy on urbanization, the agglomeration effect, although these factors are the core of modern economic-geographical and regional-economic studies after the work of P. Krugman (Krugman, 1991), R. Florida (Florida, 2008) M. Fujita (Fujita & Krugman, 1999). But what arises here in place of this powerful effect?

Instead of permanent, stationary urban and economic agglomerations, in the remote territories of the Arctic and the North, there are temporary agglomerations and mobile economic associations. We can call them temporary "poles of growth/development" in the terms of F. Perroux (Perroux, 1950). And these temporary concentrations of business entities are based on effects of temporal proximity, a concept that has been developed in recent years by the French school of proximity theory, headed by A. Torre (2008).

Another striking feature of peripheral innovation system is the small number of knowledge organizations, for example, structures of higher and secondary professional education, academic institutions. This defines a "thin" layer of local knowledge. Under these conditions, the knowledge potential of the global resource corporation, the local branches of TNCs with which local small businesses contract (Iammarino & McCann 2013), is of great importance. Resource corporations become agents of new technologies in remote areas, industrial innovations determine the technological path of the territories where they are located (Dosi, 1982).

The projects they implement for new resource development through subcontracting procedures and tight interaction with local small businesses, can have a profound effect on the formation of a local innovation system. One can compare this role with the role of universities and other higher and vocational education institutions in the central regions. Therefore, the creation of a modern theory of the peripheral innovation system without strong integration with the modern theory of TNCs is impossible.

The small number of organizations carrying new knowledge is combined in remote areas with the enormous importance of state support for institutions in the innovation process. The role of such support is significantly higher than it is in the central regions. The state acts here as the main force capable of reducing the information costs of uncertainty for all actors. The dependence of the innovation system on state support measures, on political initiatives, on budget investments (in conditions of weak market forces) is unprecedentedly great here.

But this support itself should be specific. The fact is that modern researchers distinguish between industries and firms with different innovative "modes." Some give rise to innovations according to the "science-technology-innovation" algorithm, others according to the "doing-utilizing-interacting" algorithm. The first relies on the institutions of fundamental science and the implementation of their advanced achievements. It is clear that this is the reality of central, but not peripheral regions.

On the other hand, the second mode is more typical for the remote mining regions of the Arctic and the North. Here many competencies are acquired right in the process. Researchers note that in remote areas a compromise is also possible when the company integrates knowledge from various sources in its innovative projects, on the one hand, based on the achievements of fundamental science, and on the other hand, on its practical experience. Sectors and firms that are subject to different innovation regimes need different types of support in the form of institutions, knowledge and other infrastructure of the regional innovation system.

Inside peripheral innovation systems, interfirm and spatial flows of knowledge are usually weak for the simple reason of the lack of diverse knowledge here. Those types of knowledge that are usually readily accessible "on the side" to firms in large urban centers are not available to neighbors here. There are no knowledge spill-overs nearby.

That is why firms on the periphery are often forced to "internalize" various types of knowledge (e.g. Surgutneftegas does this). The desire to reach a high level of self-sufficiency in technical and engineering, geological and other knowledge among TNCs in the peripheral regions is connected precisely with the fact that it is not possible to find these competencies in local labor markets nearby. This causes the desire of the company to ensure the stability of its qualified and competent personnel.

Another strategy for acquiring knowledge is the entry of firms from peripheral regions into geographically wide networks with external partners. Weak links of peripheral innovation systems with their own sources of new knowledge makes it natural to turn to external sources. Using the expression from a popular scientific article, we can say that in the peripheral regions there is little *buzz*, but a lot of knowledge from *global pipelines* (Bathelt et al., 2004).

Even for those firms that habitually rely on their internal knowledge, it is critical to have networks of external partners, suppliers of new knowledge.

Studies show that ceteris paribus, large and small peripheral firms are indeed more likely to enter into contractual relations with distant (global) partners and are generally more inclined to cooperate than firms in central regions. It is as if they themselves are aware of their information, knowledge vulnerability.

You could even say that each peripheral company should have its own strategy of "sucking in" external knowledge and forming for these purposes temporary and permanent partner networks for familiarization with global knowledge flows. The effectiveness of familiarization with external channels of knowledge depends on the "absorptive capacity" of the company on the periphery, which, in turn, depends on the hiring of educated and competent people. Their presence strengthens the firm's ability to extract external knowledge, mix it with its own and commercialize it.

The features of a particular periphery form specific conditions for attracting new knowledge in some case through labor migration, in others through internships and business trips of its full-time employees, and in others through master classes by world-class professionals. The work of our foreign colleagues describes how on the periphery local "islands" of innovation can arise due to the migration of prominent ("star") scientists (Trippl, 2013).

But how can we identify the model of a peripheral innovation system in practice? The realities of the mono-resource Arctic regions of Russia give us such an opportunity. According to the canonical representations of this concept, this system consists of two subsystems: research and operational production. New knowledge is generated in the first, and it is commercialized in the second in the interests of the local economy and economic development. For the regions of the Arctic, this means that the first subsystem generates new geological knowledge about the mineral resources, fuel and energy resources of the territory (and this can happen in a variety of structures, for example, in the contour of a resource corporation, but, of course, not in the classic system of developed areas in universities or research laboratories).

The second subsystem uses this knowledge in the process of developing new deposits of natural resources discovered by the first subsystem.

So that the process is not interrupted, both systems must be in balance: the decrease in reserves of the first subsystem must not lag behind the repayment of reserves in the process of production by the second subsystem. The practice of the Soviet era shows that it was incredibly difficult to maintain this balance for a long time due to the natural laws of decreasing returns on natural assets from previously and long-discovered mineral deposits. The difficult and dynamic dialectics of the development of these two subsystems determine the overall effectiveness of the entire regional innovation system of a specific resource region of the Arctic.

Is it possible to identify different types of innovation systems in the global Arctic? Features of innovative development are always largely determined by the specifics of the space in which communication and knowledge flows between the actors of the innovation system take place. The properties of a particular space are determined by the characteristics of the settlement system, transport and communication connectivity of the territory.

The global Arctic with its exceptional variety of local options for transport and information accessibility, the presence and absence of the agglomeration effect and sharp changes in population density is a real encyclopedia of options for the development of the innovation process and the corresponding local innovation systems. An idea of this diversity can be obtained from Fig. 19.1 and Table 19.2.

This classification of innovation systems is based on the idea of the leading role of spatial factors in the development of Arctic innovations. As in the rest of the world, an important condition for the innovation process is the concentration of the population in urban agglomerations, but there are some peculiarities here: the largest urban agglomerations of the Arctic concentrate private and/or state structures that control economic processes over many thousand kilometers of the Arctic zone and therefore have similar functions as the global cities.

The presence of a city network simplifies the flow of knowledge between individual actors in the innovation process. But in some areas of the Arctic, cities are isolated from large national and interregional centers by thousands of kilometers. Under these conditions, remote small cities often assume functions that in the larger zones of settlement would be characteristic for much larger urban centers.

Given the remoteness and daily challenges of the harsh Arctic environment, many urban centers are forced to innovate. At the same time, in areas where there are no cities at all, the innovation process is concentrated in the activities of large resource corporations, and here it acquires a complex character, integrating logistics, technological and organizational innovations.

There are also unique cases that have no analogues in other parts of the world. These are relatively isolated administrative capitals in terms of transport (with a small adjacent territory), concentrating - due to their capital position - financial and information resources as well as a significant pool of creative and ambitious people who arrived from different regions. The insular position usually promotes peripheralization. But here, on the contrary, innovation processes are intensified, and isolation acts as a challenge that enhances the innovative search. The opposite

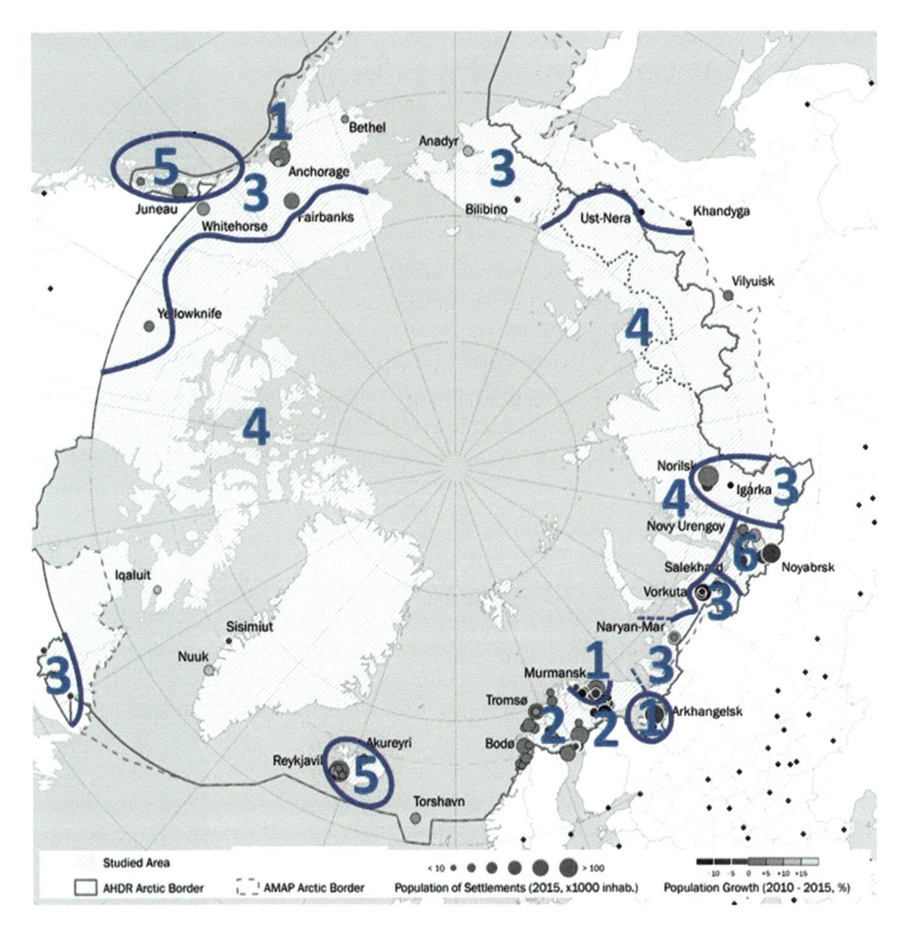

Fig. 19.1 Types of innovation systems in the global Arctic (numbers see in text and Table 19.2) Source of the base map: Zamyatina and Goncharov (2018). Arctic zone of Russia borders as of 2018.

example is the Russian network of cities in the north of Western Siberia, created in the 1970s and 1980s. These cities are characterized by reduced innovation activity.

Let us consider these types of local innovation systems (IS) in the global Arctic in more detail.

1. **IS of multifunctional urban centers.** This type includes the agglomerations of the cities of Anchorage, Arkhangelsk and Murmansk. Here, the innovation processes are the closest to those in densely populated areas of the world, in large urban agglomerations, but there are many specific features. These cities are relatively small by world standards (about 300 thousand residents, with agglomerations up to 500 thousand). The headquarters and administrations of large Arctic corporations and megaprojects are located here (for example, Arkhangelsk

Table 19.2 The relationship between space properties and the type of local innovation system

Type of innovative system (IS)	Spatial features			Features of innovative systems			
	Territory	Features of the settlement system	Transport accessibility (connectivity)	Major actors	Type and characteristics of key innovations	The scope of innovation	Type of IS
Regular types, differentiated by the level of development of the territory							
1. IS of multifunctional urban centers	Cities of Anchorage, Archangelsk, Murmansk	Urban agglomerations	Best in the Arctic: International multimodal transport hubs	Divisions of large corporations, universities, local governments, international organizations	Logistic, managerial, technological and organizational innovations (own and borrowed)	Interregional	Central: Inter-regional level
2.Networks IS in the old-developed resource and coastal regions	Arctic regions of Norway, Sweden, Finland; Murmansk region (outside the Murmansk agglomeration), the republic of Karelia in Russia	A network of cities connected by year-round transport routes, combined with old and young mines.	High: Year-round land transport network	Universities, mining and manufacturing corporations, small businesses, civil society structures	Non-specific (borrowed) technological, social	Local	Peripheral IS
3.IS of base city-islands in old-developed resource regions	Central Alaska, southern Yukon and Northwest Territories, Labrador City in Canada, Chukotka, lower Yenisei, Nenets autonomous okrug and Vorkuta	"Island" base cities in combination with old and young (including rotational) settlements at the deposits (no further than 250 km from cities) and traditional settlements of	Medium: Single highways, partially off-road	Specialized Arctic R&D centers (their presence is a distinctive feature of the type), resource corporations, small businesses, local governments	Highly specific Arctic innovations in life support, manufacturing, Creative crafts; high level of integration of traditional knowledge and borrowed advanced technologies	Interregional	Central (regional level, for regions of the new and old frontier)

(continued)

Table 19.2 (continued)

Type of innovative system (IS)	Spatial features			Features of innovative systems			
	Territory	Features of the settlement system	Transport accessibility (connectivity)	Major actors	Type and characteristics of key innovations	The scope of innovation	Type of IS
		Indigenous peoples					
4.IS of areas of modern pioneer development (frontier IS)	Tundra regions of North America (including Greenland) and eastern Eurasia (from Yamal to Yakutia)	A mix of resource corporations' camps and Indigenous villages; small number of sparsely populated isolated logistics and/or administrative centers	Low: Off -road	Corporations	Integrated development (interdisciplinary) innovations (transport, environmental, social, technological, etc.)	Interregional (limited to the territories of the resource corporation)	Frontier IS
Specific types associated with natural and / or historical unique features of territories							
5. "Privileged" IS of island capitals	Iceland and southern Alaska (with Juneau and adjacent islands), Faroe Islands	Network of cities on the islands	Medium: The island position is offset by the presence of international airports and a good level of development of maritime transport in the absence of restrictions on ice conditions	Universities, governments, international organizations (Iceland), civil society structures, small businesses	Innovations in the field of green economy and environmental management, tourism; social	From local to international	Central (regional level, outside the resource frontier)

6. West Siberian IS- a network of resource urban centers	Southern Yamal-Nenets autonomous okrug	A network of young single-industry cities with hydrocarbon deposits	High: All-season road network	Corporations and city governments, small businesses	Corporate industrial innovations (more often brought from cities - centers of corporate R&D), urban space innovations (borrowed)	Local, in some cases - interregional (limited to the territories of presence of the corresponding corporation)	Peripheral IS

is the center of the entire hydrometeorological service of the western sector of the Russian Arctic). A significant part of personnel training for work in the Arctic is also concentrated here, and large-scale scientific research is being conducted.

The main driving force of the innovation process here is the search for management solutions focused on global and national problems of the development of the Arctic, the synthesis of knowledge about Arctic regional diversity. Relatively small in terms of population, these cities concentrate information from across the Arctic.

2. **Network IS in the old-developed resource and coastal regions.** This type includes Arctic Scandinavia, Finland, enclosing the territories of the Murmansk region, Karelia, Arkhangelsk region (except for large urban agglomerations). In terms of the nature of the space, these territories resemble the peripheral regions of the more developed territories of Europe and North America with the difference that the network of cities is more sparse, and the role of the extractive industries in the economy is relatively increased. The presence of a relatively dense network of small towns determines the main features of this region. Even in the case of new mining operations (for example, powerful gold deposits like Kittilä in Finland), companies can use nearby settlements and a largely ready-made transport infrastructure as a base.

The natural environment is not so harsh as to require special technological solutions. Numerous local universities are focused mainly on solving local problems (including in the resource industries). Due to the relatively good (for the Arctic) accessibility, these regions are ready for the development of mass tourism through the efforts of local entrepreneurs.

The presence of the characteristic problems of the development of single-industry cities (combined with an orientation towards high standards of quality of life) stimulates the development of social and organizational innovations - for example, the complex process of transferring a part of the city of Kiruna to a new location. In Russia, Kostomuksha stands out in this type as one of the first cities in the country to develop its own brand in the interests of local small businesses.

The most important factor and a favorable prerequisite for the deployment of an innovation system here is the best infrastructural arrangement of space in the global Arctic. At the same time, the innovation process is focused on solving local problems and differs little from the innovation process in other old industrial regions of Europe. It is not surprising that the very concept of a peripheral innovation system was born here (Asheim, Isaksen, Trippl, 2019). It is the least "Arctic" of the Arctic innovation systems. An exception is Tromsø, which is close to the next type in its developmental characteristics.

3. **IS of base city-islands in old-developed resource regions.** This type is distinguished by the presence of remote cities and is perhaps the most specific innovation system in the Arctic. An important factor in its development is the functional diversity of the local environment. This category includes both cities and villages of Indigenous peoples, old (often abandoned) and new (developed on

a rotational basis) resource projects located in relative proximity to each other (the distance from cities to deposits usually does not exceed 250 km, which makes it possible to use local cities as reference points for development).

However, a key feature of the development of local innovation systems is their high orientation toward innovative search in the field of life support in the Arctic, and the general high level and rich traditions of this search. The main settlement network was formed in such areas, as a rule, 75–100 years ago, at a time when it was technically impossible to develop on a rotational basis and, accordingly, the development of natural resources was accompanied by the creation of *forced multi-functional* support cities and settlements usually with attempts to develop local agriculture, production of building materials, etc., as well as scientific research.

This type is similar to the previous one, but differs in terms of a sparser network of settlements (especially urban ones), a poorer level of transport accessibility, more severe natural conditions, and, as a consequence, a higher innovative activity aimed at life support. Local cities serve as bases for the development of the surrounding area and often have specialized R&D institutions aimed at developing solutions in the field of Arctic life support in general as well as adapting the experience of Indigenous peoples in the modern economy. Typical examples of such R&D organizations are "markers" such as the Cold Climate Housing Research Center[1] and the Alaska Center for Energy and Power[2] in Fairbanks, scientific research in the field of construction on permafrost and the Research Institute of Agriculture and Arctic Ecology[3] in Norilsk, and the Uelen bone carving workshop (serving as an example of the integration of traditional crafts into the world commodity market) in Chukotka.

4. **IS of areas of modern pioneer development (frontier IS).** Such ISs are developing in areas with an extremely low population density, an almost complete absence of cities (with the exception of small logistics and administrative centers such as Nuuk and Tiksi), extremely difficult climate conditions, low transport accessibility, and often lack of Internet connection. This zone is characterized by the strongest contrast between the traditional subsistence and the powerful processes of Arctic industrialization associated with the activities of large resource corporations. In the absence of large research centers in this zone, the bulk of innovation is brought in from outside, from the locations of R&D units of large TNCs. At the same time, however, successful solutions found in a specific place are often replicated on a global scale. For example, with the arrival of American investors in oil production in the Nenets autonomous Okrug in the 1990s, for the first time in Russia, the method of drilling from frozen ice pads[4] (Ardalinskoye field) was used.

[1] http://www.cchrc.org/

[2] http://acep.uaf.edu/

[3] http://norilsk-niisharctic.ru/

[4] http://www.oilru.com/nr/79/774/

5. **"Privileged" IS of the island capital regions** (e.g. Iceland and Juneau, Alaska and adjacent islands). ***The development of these IS is determined, on the one hand, by the capital status of the largest cities in these territories and on the other, by their relatively high transport isolation. The first challenge is to attract creative, ambitious people to them, including potential innovators, by the concentration of information and administrative resources. The second factor involves an increased need for developing solutions in the field of reducing the cost of life support.*** It is not surprising that the combination of both factors makes the Alaska metropolitan area around Juneau and Iceland attractive to a high level of green energy development. In addition, both districts are characterized by a high level of tourism development and traditional specialization in fishing, both areas of application of local innovation.
6. **West Siberian type: networks of single-industry urban centers.** Despite the concentration of relatively large (from 25 to 100 thousand people) cities, which is unique for the Arctic, the IS developing here is characterized by "stagnation" due to the single-industry resource nature of these cities. The most important factor in the development of innovations here is external relations with the more southern regions of Russia (the zone of main settlement with the main centers of production of innovations) as well as the search for innovative technological solutions (in the fields, relying on the infrastructure of basic cities).

The most important factor and a favorable prerequisite for the deployment of the innovation system here is the highest level of per capita wealth and the highest "density" of resource wealth per unit area. The key actors are large resource corporations as well as local administrations seeking to increase the attractiveness of the urban environment for the population.

19.6 The Global Importance of Arctic Innovation

The Arctic today is a gigantic laboratory, developing solutions for rather specific conditions, including, for example, a very cold climate, strong winds, high migration mobility of the population and the associated socio-cultural challenges, and a sparse network of settlements. Among others, the last point deserves special attention in the context of its potential significance for the global system.

The modern global system is characterized by high mobility of the population and goods. Sociologist John Urry speaks of the mobility paradigm as the basis of modern civilization (Urry, 2007). Mobility is the very paradigm that Urry has put forward as determining for a modern society, which is urban, dependent on oil and on intelligent systems that provide mobility. Incidentally, he considered the rapid spread of infections diseases (through animals) as a consequence of this mobility.

In the event of any cataclysms that would entail a sharp rise in the price of hydrocarbons (or, on the contrary, a voluntary rejection of the excessive mobility that environmental alarmists are calling for today), mobility can plummet. This will inevitably require a complete restructuring of economic, technological, and economic processes.

Urry also gave a gloomy forecast for the end of the era of mobility: "We definitely should not expect that the mobile world of the 20th century will remain an organizational principle in this century. Some even argue that climate change, environmental pollution and energy shortages in the 20th century will extremely limit the possibilities for rebuilding future mobility and using the energy necessary to avoid the "societal collapse" of the kind that the Roman Empire or civilization Maya due to the development of their internal contradictions. Mobile life for millions can be a short-lived phenomenon. Over the course of a century, until the contradictions have fully manifested themselves, the rich world has gone wild, and as a result, in the 21st century, when societies will have a hard time, people and machines will have a much slower inheritance in their hands" (Urri, 2012, p. 62).

Due to the low density of settlements (increased average distances between settlements), the increased cost of transporting fuel in the Arctic today is such a model of the "society of expensive oil" (despite the fact that now the price of oil is low). In other words, the *Arctic today is already a laboratory of a possible future for all of humanity in the "after mobility" era*, and its "recipes" can be potentially recipes for the adaptation of mankind to low mobility conditions.

What are these recipes? There is, for example, the practice of complex, multifunctional trips, when a trip is used to the maximum to perform many tasks at the same time (treatment, rest, shopping, collecting information, etc.). Another example is the expanded role of stocks and the corresponding warehouse infrastructure, a strategy that is characteristic of the Arctic and completely opposite to the just-in-time strategy prevailing under milder conditions. It is possible that these exotic strategies will turn out to be the mainstream of the future development of mankind making it worth looking at the Arctic as an experimental training ground for survival strategies.

The Arctic strategy is a strategy of large stocks of equipment and spare parts. The unreliability of transport routes, the untimely delivery, the instability of aircraft schedules and blocking of the road due to weather conditions are absolute realities of modern, and not a hundred years ago, Arctic life. Today we can study this "reserve" not as the past but as, quite possibly, the future of mankind.

These examples disprove the traditional notion that innovations in the Arctic can be of interest to the rest of the world only in its basic extractive industries. The Arctic cities of Russia, (e.g. Norilsk) have been developing for many decades unique competencies for the collective survival of hundreds of thousands of people in conditions of extreme cold discomfort and extreme instability of permafrost. We are talking about new technologies for Arctic multi-story construction on permafrost, about geological research, and in general about the formation of a whole range of specific Arctic science-intensive business services that are demanded by resource corporations which in their essence are no longer industrial, but from a post-industrial era.

19.7 Conclusion

Our multi-faceted study of Arctic innovations is aimed at straightening out the bias that has developed in the social sciences in recent decades between numerous studies of innovations in large urban agglomerations of the world and an almost complete absence of such research in the Arctic territories. Since the 1970s, the Arctic was presented to researchers as a natural "research laboratory" for the analysis of socio-economic processes due to the sparseness of the infrastructure and the relative simplicity of the links within socio-economic systems.

Continuing this tradition, our work can be understood as a logical step for developing a methodological base for a better understanding of the nature of innovative development in the remote regions of Russia and the world. The scientific algorithms, methods, and research methodology worked out using the relatively simple economic structure of the Arctic zone as an example can subsequently be used constructively and make a contribution to the study of peripheral innovation systems in other parts of the world.

The emphasis on innovation provides a new interpretation of the usual phenomena of Arctic life (e.g. remoteness, cold discomfort, energy and food security) from the standpoint of the "dramaturgy" of the struggle between the new and the old in the Arctic.

The innovation process always transforms the status quo that existed before. The peculiarity of the Arctic is that here it is usually forced to have a more radical, more revolutionary character. In a poorly developed and settled social environment, any innovations cause a very noticeable and visible transformation.

The paradox is that the Arctic, which gave birth to the concept of sustainable development for the whole world due to the imperative of finding ways to balance conflicting environmental, social and economic goals, is itself often an example of non-equilibrium development. The regional innovation system and the innovation process further reinforce this disequilibrium, but give it a constructive and creative rather than a catastrophic character.

This exploratory behavior is aimed at overcoming the effects of exhaustion and stagnation, which in the Arctic are not only destructive, but threaten the very continued existence of man-made urban, economic and social systems. This resultant innovative search, which is a forced feature of both the natural and social systems of the Arctic, provides an opportunity for a new dynamic beginning of the process of economic development, which always at the first stage provides an attractive tone to development. The difference between the Arctic and other parts of the world is that it is constantly ready for such an innovative reformatting.

It is the Arctic innovation system that materializes the completely new role of science in the development of the Arctic. Science is a key factor in informed decisionmaking, as a guarantor of the formation of the common interests of influential actors and as an effective institution that ensures the resilience of the Arctic territories under conditions of rapid social and natural changes.

References

Asheim, B. T., & Gertler, M. S. (2005). The geography of innovation: Regional innovation systems. In J. Fagerberg, D. C. Mowery, & R. Nelson (Eds.), *The Oxford handbook of innovation* (pp. 291–317). Oxford University Press.

Asheim, B. T., Isaksen, A., & Trippl, M. (2019). *Advanced introduction to regional innovation systems* (146p). Edward Elgar.

Baburin, V. L., & Zemtsov, S. P. (2017). Innovatsionnyy potentsial regionov Rossii. KDU. In Russ [Baburin, V. L., & Zemtsov, S. P. *The innovative potential of the regions of Russia*. KDU].

Bathelt, H., Malmberg, A., & Maskell, P. (2004). Clusters and knowledge: Local buzz, global pipelines and the process of knowledge creation. *Progress in Human Geography, 28*, 31–56.

Boorstin, D. J. (1958). *The Americans: Colonial experience* (434 pp). Random House.

Boschma, R. (2005). Proximity and innovation: A critical assessment. *Regional Studies, 39*(1), 61–74.

Carrol, L. (1973). Alice through the looking glass.

Dawley, S. (2014). Creating new paths? Offshore wind, policy activism, and peripheral region development. *Economic Geography, 90*(1), 91–112.

Dosi, G. (1982). Technological paradigms and technological trajectories. *Research Policy, 11*, 147–162.

Feldman, M. (1994). *The geography of innovation*. Springer.

Feldman, M. P., & Audretsch, D. B. (1999). Innovation in cities: Science-based diversity, specialization and localized competition. *European Economic Review, 43*(2), 409–429.

Ferrucci, L., & Porcheddu, D. (2006). An emerging ICT cluster in a marginal region. The Sardinian experience. In *Regional development in knowledge economy* (pp. 203–226). Routledge.

Florida, R. (2008). *Who's your city?* (374p). Random House Canada.

Fujita, M., Krugman, P., & Venables, A. (1999). *The spatial economy: Cities, regions, and international trade*. MIT Press.

Healy, A. (2017). Innovation in circumpolar regions: New challenges for smart specialization. *Northern Review, 45*, 11–32.

Iammarino, S., & McCann, P. (2013). *Multinationals and economic geography: Location, technology and innovation*. Edward Elgar.

Isaksen, A., & Karlsson, J. (2016). Innovation in peripheral regions. In R. Shearmur, C. Carrincazeaux, & D. Doloreux (Eds.), *Handbook of the geographies of innovation* (482pp). Edward Elgar.

Karlsen, J., Isaksen, A., & Spilling, O. (2011). The challenge of constructing regional advantage in peripheral areas: The case of marine biotechnology in Thomson's, Norway. *Entrepreneurship and Regional Development, 23*(3), 235–257.

Krugman, P. (1991). *Geography and trade*. MIT.

Kryukov, V. A. (1998). Institutsional'naya struktura neftegazovogo sektora. Problemy i napravleniya transformatsii (280s). IEiOPP. In Russ [Kryukov, V. A. *Institutional structure of the oil and gas sector. Problems and directions of transformation* (280p.). IEiOPP].

Matveyev, O. A. (2011). Innovatsionnoye razvitiye regionov Severa: sostoyaniye i perspektivy (320s). Sovremennaya ekonomika i pravo . In Russ [Matveev, O. A. *Innovative development of the regions of the North: State and prospects* (320p). Modern Economics and Law].

Pelto, P. (1987). *The snowmobile revolution. Technology and social change in the Arctic*. Prospect Heights/Waveland Press.

Perroux, F. (1950). Economic space. Theory and applications. *The Quarterly Journal of Economics, 4*(1), 89–104.

Petrov, A. N. (2008, June). Talent in the cold? Creative capital and the economic future of the Canadian North. *Arctic, 61*(2), 162–176.

Petrov, A. N. (2011). Beyond spillovers. Interrogating innovation and creativity in the peripheries. In H. Bathelt, M. Feldman, & D. Kogler (Eds.), *Beyond territory. Dynamic geographies of knowledge creation, diffusion, and innovation* (pp. 168–190). Routledge.

Pilyasov, A. N. (1993). Trest Dal'stroy kak superorganizatsiya. Kolyma 8–11 (8, s. 34–37, 9–10, s. 37–41, 11, s. 28–33). In Russ [Pilyasov, A. N. *Dalstroy trust as a super-organization*. Kolyma, 8–11. (No. 8, p. 34–37, No. 9–10, pp. 37–41, No. 11, pp. 28–33)].

Pilyasov, A. N. (2018). Arkticheskaya diagnostika: plokh ne metr – yavleniye drugoye. Sever i rynok: formirovaniye ekonomicheskogo poryadka. Tom *61*(5), 35–56. In Russ [Pilyasov, A. N. Arctic diagnostics: Not a bad meter – Another phenomenon. *North and Market: The Formation of an Economic Order*. *61*(5), 35–56].

Pilyasov, A. N., & Yadryshnikov, G. N. (1997). Nauka kak faktor sotsial'no-ekonomicheskogo razvitiya rossiyskogo Severa (37s). SVKNII. In Russ [Pilyasov, A. N., & Yadryshnikov, G. N. *Science as a factor in the socio-economic development of the Russian North* (37p). SVKNII. 1997. 37p.].

Stammler, F. (2009). Mobile phone revolution in the tundra? Evolution in the tundra? Technological change among Russian reindeer nomads. *Folklore, 41*. http://www.folklore.ee/folklore/vol41/stammler.pdf

Torre, A. (2008). On the role played by temporary geographical proximity in knowledge transmission. *Regional Studies, 42*(6), 869–889.

Trippl, M. (2013). Islands of innovation as magnetic centres of star scientists? Empirical evidence on spatial concentration and mobility patterns. *Regional Studies, 47*(2), 229–244.

Urri, D. (2012). Mobil'nosti. Predisloviye k russkomu izdaniyu. Praktis (In Russ.). [Urry, J. *Mobility. Preface to the Russian edition*. Praktis].

Urry, J. (2007). *Mobilities* (336 p). Polity Press.

Virkkala, S. (2007). Innovation and networking in peripheral areas -a case study of emergence and change in rural manufacturing. *European Planning Studies, 15*(4), 511–529.

Zamyatina, N., & Goncharov, R. (2018). Arctic urbanization: Resilience in a condition of permanent instability. The case of Russian Arctic cities. In K. Borsekova & P. Nijkamp (Eds.), *Resilience and urban disasters surviving cities* (New horizons in regional science series) (pp. 136–154). Edward Elgar Publishing. https://doi.org/10.4337/9781788970105

Zbeed, S. O., & Petrov, A. N. (2017). Inventing the New North: Patents & knowledge economy in Alaska. In Arctic yearbook (pp. 1–18).

Zemtsov, S., Muradov, A., Wade, I., & Barinova, V. (2016). Determinants of regional innovation in Russia: Are people or capital more important? *Foresight-Russia, 2*, 29–42.

Chapter 20
(Action): Future Arctic Business

Annika Olsen

The Faroe Islands are 18 beautiful islands located centrally in the Arctic. We are a traditional society but still a modern society with a booming economy. There has never been more focus on our country than right now. From our neighbours and form the world's superpowers. What we emphasise more than anything is a good, quality infrastructure. Domestically and internationally. Therefore, we are investing heavily in infrastructure in order to prepare for new conditions in the region.

We have connected 88% of the people together with good roads, bridges, embankments, tunnels, and underwater tunnels. This is the very basis for a thriving industry and for our future competitiveness.

We are currently expanding our harbour in Tórshavn. The only airport in the Faroe Islands has been extended and we have the world's best internet. These are all key barometers for us when preparing for the future in Arctic.

Education and entrepreneurship is very important as well. Tórshavn has a newly established innovation house and in the same environment, we have gathered businesses and the University in the Faroe Islands.

The salmon company Bakkafrost is the largest enterprise in the Faroe Islands and between the five largest salmon businesses in the world. This is only possible because of a good infrastructure.

Tourism experiences a double-digit growth and the hotel capacity will double in summer 2020. This is key infrastructure. Beside this, we have focus on a sustainable development in tourism. The Closed for Maintenance campaign is a good example of how the Faroe Islands have managed to combine tourism and responsible growth. Furthermore, we are emphasising business tourism, particularly in the arrangement and hosting of meetings, conferences, and other business events.

A. Olsen (✉)
Tórshavn, The Faroe Islands
e-mail: hogni@torshavn.fo

P. A. Berkman et al. (eds.), *Building Common Interests in the Arctic Ocean with Global Inclusion, Volume 2*, Informed Decisionmaking for Sustainability,
https://doi.org/10.1007/978-3-030-89312-5_20

Our goal is to have a 100% green society in 2030 meaning that energy solutions are solely renewable. This will have significant impact on our brand as a green country in the Arctic.

The Northern Sea Route gives the Faroe Islands a unique opportunity to become a key maritime service and educational hub. The Faroe Islands are the natural port and hub into the Arctic and the Northern Sea Route. Shipping traffic has already increased and will increase further. This will open new opportunities to provide vessels with high quality and competitive maritime services.

Looking at new industries in the Arctic that will be more profitable in the future, tourism is going play a major role. It will affect societies and local communities. The Arctic is an untouched tourist destination and there is huge potential in enhancing the tourism sector. The tourism industry is on a rise and it can be as profitable as traditional industries. Tourism is now the single largest industry in Iceland and more profitable than the traditional fishing industry. It is important to listen to the local people and to secure them ownership and a part of the growing tourism. It needs to be growth that benefits local people.

The potential of Arctic resources is enormous. Nevertheless, it is important that all Arctic countries have a peaceful approach when we exploit these resources. For many small communities exploring for oil and gas can affect their societies to a large extant. We should invest potential oil money in green growth and renewable energy.

I often get the question if there are limitations for activity in the Arctic. Technological progress will widen the possibilities in the area. The weather conditions are extreme compared to what companies are used to so there are of course limitations. We also need to know why people move from peripheral areas and how new activities will affect the labour market. These are important questions that need research and that politicians need to address.

We also need a management regime for ensuring sustainable development for natural resources in the Arctic. We do not have the necessary rules. The economic viability of fisheries and aquaculture, which are the very basis of the economy in the Faroe Islands, depends on clean and productive seas. I firmly believe that the Arctic Council will set the norms necessary for a sustainable future. There are also areas in the Arctic that we cannot utilize until we have secured the environment. Other areas need protection and not to be exploited at all as we do not have sufficient management tools to develop natural resources in a sustainable way.

Like other nations and communities in the Arctic, the Faroe Islands society, economy and way of life is very close to, and dependent on, nature. Exploring for oil and gas can affect many small communities to a large extent. Potential money from the oil industry should be invested in green solutions and sustainable energy. There is no doubt that utilizing the resources in Arctic needs heavy funding from businesses. This calls for cooperation between research institutes, businesses, governments and the Arctic Council.

The future needs new solutions so we are also finding new solutions in the Faroe Islands. We are investing in windmills, we are researching in tidal energy and a biogas station will open in February. We will spend 1 billion EUR in green energy within the next 10 years. We will also invite international companies to test new technologies in the Faroe Islands.

Chapter 21
(Action): Future Arctic Business

Geir Seljeseth

It is good to be back in Tromsø. I have lived in Brussels for some months now. Last time I lived abroad I lived in Moscow and I had a short trip home in the spring of 2010 and then also a volcano erupted in Iceland, and I ended up staying four weeks in Tromsø. It was nice and we could stay here for some weeks and have fun without being disturbed. I was once upon a time an old history teacher so I will start with the Eufrat and the Tigris and the Mesopotamia. When we go back there, the only thing we have learned from then until now is that the only way to build societies is to work and to do business. All the way, we can do all kinds of decisions, but we end up with that resilient societies is made from work and from business. That's how it is in the Arctic today and that's how it will be in the Arctic in the future.

The Arctic is very many different things, so it ends up with, there is no one button you can push or one solution that will fit all. There are different solutions for different areas. And that depends on where you are - you have different Arctics, from the Russian Arctic, to the European Arctic to the American Arctic. And, if your main focus is the nature, the preservation of it, that means an Arctic with fewer people in the future. We have to have activity to make an Arctic active and livable.

It is going to be very interesting this spring or this year, because the EU is now building a new Arctic policy. And I sense already that there is a totally different discussion now in the capital of Europe than it was the last time when it was a lot about preservation. Now it is almost as much about security measure and how to make sure that EU has an influence in the Arctic area. That will be important to see whether were going to have an active Arctic with people living there, because there has been a lot of discussions about the climate challenges for the Arctic – how it is now. The other challenge that we face, just as big is the demographic challenge. If

G. Seljeseth (✉)
Head of Europe Office – Industri & energi, Brussels, Belgeium
e-mail: geir.seljeseth@industrienergi.no

P. A. Berkman et al. (eds.), *Building Common Interests in the Arctic Ocean with Global Inclusion, Volume 2*, Informed Decisionmaking for Sustainability,
https://doi.org/10.1007/978-3-030-89312-5_21

we should take the Faroe Islands into the Arctic, and we are an inclusive family, so let's take them in, they are the ones that has a rising population. All the other areas have a decline in population. And that is the huge challenge. I want an Arctic with societies that are active, with a lot of people. I want to have an Arctic growing. I was born in the Arctic. I was raised in the Arctic. I lived most of my life in the Arctic. And I hope that also my daughter and my grandchildren will see the same. With an active Arctic with big societies thriving.

Thank you all.

Chapter 22
(Action): Future Arctic Business

Anders Oskal

Arctic Opportunities as Opportunities for All?
Global Change, Business and the Realities of Nomadic Indigenous Peoples

The Arctic is changing in ways unprecedented in the long histories of the Indigenous peoples of the north. The Paris Agreement from 2015 seeks to limit global warming to below 2, preferably to 1.5 °C, compared to pre-industrial levels. Indigenous reindeer herders in the Arctic however face another reality: In the Sámi village of Guovdageaidnu in Arctic Norway, there has already been more than 3 °C warming of reindeer spring pastures since the 1930s. And in Chersky in northeastern Siberia reindeer herding communities have now experienced spring warming of more than 6 °C the 30 last years. This reality affects Indigenous peoples' economy, health and wellbeing.

Accumulated knowledge about climate variability is high among Arctic Indigenous peoples. Reindeer herding peoples have always known that humans cannot control nature, but that one needs to come to terms with it, and constantly cope and adapt to change. Land encroachment linked with an explosion of increased human activity in the Arctic, with irreversible fragmentation of reindeer pastures and migration routes, are negatively affecting reindeer herders' ability to adapt to climate change. Fennoscandia now faces the most serious situation in terms of cumulative loss of reindeer pastures. As also pointed out by IPCC, protection of grazing lands

This work was in part supported by RCN Rievdan and H2020 PolarNet II.

A. Oskal (✉)
Association of World Reindeer Herders (WRH), International Centre for Reindeer Husbandry (ICR), Guovdageaidnu, Norway
e-mail: oskal@reindeercentre.org

P. A. Berkman et al. (eds.), *Building Common Interests in the Arctic Ocean with Global Inclusion, Volume 2*, Informed Decisionmaking for Sustainability,
https://doi.org/10.1007/978-3-030-89312-5_22

represents the most important adaptive strategy for reindeer herders under climate change. As the Arctic is now quickly becoming an integrated part of the global economy, reindeer herders are additionally facing highly varying socio-economic conditions and effects of assimilation past and present.

Indigenous peoples often get entangled in the usual debates about the negative impacts of change, somewhere between being seen as victims of change, as the «canary in the coal mine», to being portrayed as greedy for compensations. The latter seems to be a preferred tactic from certain developers – but by no means all. Here one can certainly question how the small nomadic reindeer herding communities end up as the «greedy» part in such discourses, merely defending their ancestral homelands from often very resourceful industrial newcomers on their territories.

In principle however, change means both challenges and opportunities. But again, the realities of nomadic Indigenous peoples are often somewhat different: Most of the time, one has to spend so much resources, time and energy on the negatives that one is not really in position to effectively exploit the opportunities Arctic change brings.

Balance of opportunity is sometimes perceived as virtually impossible. As one young Sámi herder described their struggle against a multinational company: «*...It cannot be right that one side gets all the benefits and the other is struck with all the problems*». So things needs to be done differently for Indigenous reindeer herders to also actually benefit from Arctic change.

Here is imperative that reindeer herding peoples are able to *pursue their own opportunities* for their small nomadic communities. It is necessary to call for culturally-anchored development and entrepreneurship, building Indigenous economies and societies from within. Fair trade arrangements, friendly investments, joint ventures and assistance for entrepreneurship and innovation are all useful ways by which mainstream business could assist Indigenous youth and their societies develop their own economic base. Clearly, governmental and educational support is also needed, including development of Indigenous innovation systems.

Following are some observations on certain prejudices or myths that sometimes are observed concerning different parts of the Arctic. In the west it seems sometimes we have an understanding of being somehow «superior», where a reality assumption seems to be that the west is *per definition* better than the east on Indigenous issues. While good and bad examples can be found everywhere, however, this notion is in our experience not true. In Yamalo-Nenets AO industrial development represents a difficult prospect. These lands have been used by nomadic Indigenous Nenets, Khanty and Mansi reindeer herders from time immemorial, and challenges do remain. But there is a progressive legal framework in Yamalo-Nenets AO in the regional legislation, with a specific Act on reindeer herding aimed at securing the legal, economic, natural and social basis of reindeer herders. Also industrial companies attempt to have dialogue and adapt their operations to the needs of reindeer herders. We have experienced western companies not fulfilling the same standards, for instance a larger Norwegian company was unwilling to even have a meeting with Indigenous reindeer herders in Russia directly affected by their developments. Hence, we are not inclined to readily adopt any black-and-white understanding of reality, not in general nor under influence of any current east-west dichotomies and tensions.

The unique way of life of reindeer herders has formed the very foundations of their cultures, their traditional Indigenous knowledge, and their unique original Arctic civilization. To continue this particular human heritage and identity, one needs to identify development possibilities that are based on reindeer herders' *own premises*.

Arctic Indigenous reindeer herding peoples have ancient food systems and knowledge, that represents the least explored part of the culinary world today. This holds a great potential for local economic development; Our traditional Indigenous food book «EALLU» was presented as an Arctic Council deliverable to the Arctic Council Ministerial Meeting in 2017, subsequently awarded the *23rd Gourmand Awards Best Foodbook of the World*. It was followed up by the EALLU II report to the Arctic Council in 2019. Indeed, examples from the food sector shows that premium exclusive food products like Parma ham, Serrano ham and Parmesan cheese are not products invented by the food industry; They are hand-crafted products based on long traditions, adapted to modern contexts, and today collecting the highest price premiums in the most demanding food markets globally. Connections to Indigenous cultures is a strong motivation for consumers in choosing Indigenous food products and paying price premiums. Innovation is also about combining existing knowledge in new ways, and revitalizing traditional food products for modern markets is a forceful way of increasing local value added, innovation and entrepreneurship. This can be done within and on the terms of Indigenous peoples, their societies and communities.

While food security is a real and legitimate concern for parts of the Arctic, many reindeer herding regions have the capacity of producing surplus food resources. Under the conditions of encroachment of ancestral reindeer pasturelands, it is anyway necessary to get more from the living resources at reindeer herders' disposal to maintain their economies.

A central response strategy from reindeer herders is engagement of our new generations by education and exchange opportunities for Indigenous youth. We have developed courses for youth in food innovation and entrepreneurship together with UArctic, Nord University in Norway and universities in Russia and Canada. Together with Harvard University Kennedy School we have also developed a joint resilience leadership program for Arctic Indigenous youth in the name of late Harvard Professor James J McCarthy, the former co-chair of IPCC and Arctic Council ACIA. Our efforts in this regard will continue, including new initiatives underway.

Facing Arctic change, Indigenous peoples must be proactive and manage to create economic development driven from within their societies, reinforcing Indigenous cultures and ways of life. This can be achieved through partnerships, and business-sector support. It will contribute pathways to maintain reindeer herders' original nomadic civilization for the future, on reindeer herders' own terms, in face of unprecedented Arctic change.

Our Indigenous youth must be made capable of managing their own destinies. This way, an opportunity of a changing Arctic can truly be an opportunity for all.

Part IV
Informed Decisionmaking Tools and Approaches for the Arctic

Framing Questions

1. How can analysts and policymakers collaborate to achieve informed decisionmaking?
2. What is the role of observation systems in producing data and evidence for informed decisionmaking?
3. How do narratives affect the conduct of science and the contributions of science to informed decisionmaking?

Chapter 23
(Research): Sea Ice Hazard Data Needs for Search and Rescue in Utqiaġvik, Alaska

Dina Abdel-Fattah, Sarah Trainor, Nathan Kettle, and Andrew Mahoney

Abstract Sea ice dynamics, such as sea ice convergence and landfast break-out events, can cause individuals and vessels to be trapped or beset in the ice, posing major hazards for Arctic maritime operators and first responders. Search and Rescue (SAR) for these events can be challenging, depending on weather and maritime conditions such as low visibility and large wave action. This research investigated the use of radar, satellite, and other tracking data for sea ice and weather conditions in maritime-related SAR operations in Alaska. Specifically, we looked into how sea ice and weather data and models can help support emergency responders by analyzing a case study of a SAR event for a missing small vessel offshore from Utqiaġvik (formerly, Barrow) in July 2017. This research consisted of: (1) an archival analysis of the SAR communication email threads and official U.S. Coast Guard case file associated with the SAR event and (2) an analysis of interviews with individuals involved in the SAR event. We analyzed themes related to the timeline of the event, the use of scientific products in decisionmaking, challenges to data use, and lessons learned for future SAR events. Interviews were conducted over the course of fall 2017 and spring 2018 to explore how this SAR event unfolded and also SAR data needs more broadly. In this study, the data needs of information users were defined as those related to supporting an emergency response. This research holds implications for future use and uptake of modeling data in local SAR operations in Utqiaġvik specifically and potentially across the Arctic. For example, one of the main findings from this research is that while there exists a breadth of data sources that could potentially be applied in a SAR context, many of these resources are not known to SAR operators. Furthermore, many of these resources were created for a specific scientific purpose and are not readily available for a SAR situation. Given

D. Abdel-Fattah (✉)
UiT – The Arctic University of Norway, Harstad, Norway

University of Alaska Fairbanks, Fairbanks, AK, USA
e-mail: dina.abdel-fattah@uit.no

S. Trainor · N. Kettle · A. Mahoney
University of Alaska Fairbanks, Fairbanks, AK, USA

P. A. Berkman et al. (eds.), *Building Common Interests in the Arctic Ocean with Global Inclusion, Volume 2*, Informed Decisionmaking for Sustainability,
https://doi.org/10.1007/978-3-030-89312-5_23

that local SAR operators are predominantly the first line of response to maritime emergencies in Northern Alaska, the ability to share and provide a set of resources to support SAR operators is critical, particularly in a rapidly changing Arctic.

23.1 Introduction

Arctic sea ice cover and extent has seen dramatic decreases in recent years. Recent analysis has shown a 14% per decade decline in summer sea ice extent due to record-high decreases in sea ice extent since 2012 (Stroeve & Notz, 2018). It is predicted that there will be an ice-free Arctic Ocean during summer months before the year 2050 (Stroeve & Notz, 2018). Although sea ice retreat translates into more open water, sea ice still serves as a formidable hazard. As the seasonal Arctic sea ice extent has diminished and thinned over the past few decades, research has found that sea ice motion is increasingly more responsive to changes in geostrophic wind as well as the mean ocean current (Kimura & Wakatsuchi, 2000; Kwok et al., 2013). Furthermore, the stability of landfast ice is compromised due to increasingly warmer air temperature and varying snowfall rates, which lead to thinner landfast ice, increasing the risks for those who traverse sea ice for hunting, fishing, and exploration (Dumas et al., 2005; Eicken & Mahoney, 2015; Flato & Brown, 1996).

Understanding sea ice conditions and making decisions regarding the use of and travel across sea ice is an integral part of Indigenous communities in the Arctic as well as other vested sea ice users such as the maritime and shipping industry (Inuit Circumpolar Council, 2008; Durkalec et al., 2015; Pizzolato et al., 2014). Sea ice conditions are changing in a multitude of ways beyond just its diminishing extent. These changes have different implications for different animals and people that use Arctic sea ice. For example, the continued loss of multiyear ice constitutes a loss of habitat for marine mammals who rely on this specific type of sea ice, for example, ringed seals (Kovacs et al., 2011). Beyond ecosystem impacts, changes in Arctic marine mammal populations and distributions also affect subsistence hunting, which directly affect Indigenous communities (Lovvorn et al., 2018).[1]

The later onset of sea ice formation, particularly landfast ice, has contributed to less stable and less predictable sea ice, which impacts Indigenous communities across the Arctic who rely on sea ice for hunting and travel during the winter (Mahoney et al., 2014). Delays in and more variable sea ice formation also exposes coastal communities to high wave action, notably during wintertime, as well as later freezeup and earlier breakup, which poses hazards and threats to communities, such as increased coastal erosion and flooding and hazardous maritime conditions (Barber

[1] It is important to note regional differences exist regarding marine mammal populations. Changes in a marine mammal population in one area is not necessarily generalizable across the Arctic; for example, recorded declining polar bear population in the Hudson Bay area compared to increases seen in the polar bear population in the Alaskan Beaufort Sea area during similar periods of monitoring (Kovacs et al., 2011).

et al., 2018; Overeem et al., 2011; Petrich et al., 2012). These changes also impact industries – such as the offshore oil and gas industry – and therefore economies in the region, which rely on sea ice forecasts and measurements to assess access to and conduction of offshore drilling operations (Galley et al., 2013).

There is a clear need for data on changing sea ice conditions in the Arctic, for both the nearshore and maritime scale as well as daily to seasonal scale, to not only understand the sea ice changes taking place, but to also address the breadth of communities, ecosystems, and services these changes impact (Kettle et al., 2019). Traditional local knowledge (TLK) in conjunction with weather, water, ice, and climate data can provide a valuable resource for sea ice-related decisionmaking (Jeuring et al., 2019; Tremblay et al., 2006). There has been an increase in recent years in Arctic information providers that extend well beyond the national sea ice and meteorological services across the region (Knol et al., 2018). The development of various decision support tools to provide information on weather, water, ice, and climate data has spurred research and discussions on not only the operability and interoperability of these tools but more importantly the need for co-production with local and end users in the development of these tools (Jeuring et al., 2019; Knol et al., 2018). Co-production in the context of this chapter is defined as developing a service or product in an equal and reciprocal relationship between information providers and people using a service or product such that both sets of actors become effective agents of change (Fenwick, 2012). Co-production of decision support tools, particularly among local decisionmakers and scientists as well generally in an Arctic context, fosters tools that are better equipped to meet users' needs, bridging the science-to-policy interface and research-to-operations gap (Jeuring et al., 2019; Robards et al., 2018).

Beyond understanding different user needs, understanding different user contexts is important in the development of decision support tools. Decision support tools help complement and contribute to the broader concept of informed decisionmaking, which is defined to operate across a "continuum of urgencies," from short-term to long-term, independent of scale (Young et al., 2020). Decision support tools are a medium that can support decisionmakers and information users make informed decisions, by providing relevant information and knowledge in a synthesized or centralized manner. There is a diverse and complex constellation of stakeholder groups that rely on or are impacted by Arctic sea ice; their interpretations and responses to drivers of sea ice change and exposure to hazardous conditions are therefore based off their unique backgrounds. These different user groups thus experience variable vulnerability, exposure, and adaptive capacity to hazardous situations (Bennett et al., 2016). Decision support tools can therefore help increase a user's adaptive capacity, particularly if they are developed in a co-production manner and take into account various and different user needs.

In this chapter, we explore the use of TLK, various decision support tools, and weather, water, ice, and climate data by different user groups during a Search and Rescue (SAR) incident off the coast of Utqiaġvik, Alaska in July 2017. The six main user groups interviewed in this study were the U.S. Coast Guard (USCG), commercial operators, the North Slope Borough, research/data providers, local SAR, and subsistence hunters. This SAR event presented a novel opportunity to analyze the use of different information resources, particularly university-developed information products, by different user groups in an Arctic Alaska SAR context, which, to the best of our knowledge, has not previously been done. The different types of information resources analyzed in this study ranged from operational SAR products, to scientific information products, to TLK, to weather information products. We wanted to understand the role each information resource played in an Arctic Alaska SAR decision support case study, to better understand information user needs as well as how they are, or potentially are not, met. Therefore, we analyzed which information resource each user group used during the SAR event and why these resources were chosen. We compared which information sources were used by user groups, taking into account a user group's needs and contexts based on their decisionmaking related to temporal, spatial, and organizational cultural scales.

In this chapter, we discuss the reasons why various information resources are or are not used. In addition, we discuss the implications of our findings, potential limitations of this study as well as its potential applications. Given that this research was an analysis of a SAR event, it holds potential implications for the successful continued execution of the 2011 Arctic Council SAR Agreement, particularly Articles 7 and 9, which respectively refer to the conduct of SAR operations and SAR-related cooperation, including information sharing, among the Arctic Council parties (Arctic Council, 2011).

23.1.1 Case Study Background

There are three main types of sea ice hazard events in Utqiaġvik, Alaska: landfast breakout events, sea ice convergence events, and speed events (Eicken et al., 2018). Landfast breakout refers to when grounded pressure ridges become ungrounded such that the ice detaches from the shoreline and drifts off (Jones et al., 2016). Sea ice convergence occurs when ice is pushed together by winds or currents such that internal ice stress keeps the ice from moving (Dynamics | National Snow and Ice Data Center, n.d.). Speed events are when anomalously high rates of ice drift cause ice to collide with and damage vessels (Eicken et al., 2018). Between September 2006 and September 2017, findings from the Utqiaġvik Sea Ice Radar revealed that 245 sea ice convergence events occurred near Utqiaġvik (Kettle et al., 2018). These types of hazardous sea ice events have a consistent presence in the Utqiaġvik area and can pose significant hazards to maritime operators, particularly small vessels. Search, rescue, and response to these events can also be challenging, due to harsh and inclement weather and maritime conditions such as low visibility and fast currents. This specific case study investigates a SAR event in which a missing vessel was caught in a high-speed current event amongst drifting packed ice.

At approximately 2:30 am local time (GMT-8) on July 19, 2017, a fourteen-foot open top motor-powered boat went dead in the water approximately ten miles north of Point Barrow, Alaska among a drift ice pack. The boat had on board one adult and

Fig. 23.1 Summary of main events during July 19, 2017 SAR event.
The main events of the July 19, 2017 SAR event analyzed in this study. At ~2:30 am local time on July 19, 2017, a 14 foot open egg top shell boat, powered by a 90 horse power Yamaha motor, went dead in the water approximately 10 miles north of Point Barrow, Alaska among a drift ice pack. The boat had on board one adult and four children under the age of 13. Volunteers, the North Slope Borough Search and Rescue team, and the U.S. Coast Guard (USCG) conducted a search to find the missing vessel and persons. Their efforts were supported by several university scientists who helped determine the vessel's drift from its point of last known location. All five persons aboard were safely found at 12:00 am local time July 20, 2017

four children under the age of thirteen. Volunteers, the North Slope Borough SAR team, and the USCG conducted a search to find the missing vessel and persons. Several university scientists who helped determine the vessel's drift from its point of last known location also supported the SAR effort. All five people aboard were safely found at 12:00 am (GMT-8) July 20, 2017. Figure 23.1 outlines some of the main events that took place during the SAR event.

23.2 Methods

23.2.1 Materials and Methodology

The main objective of this research was to understand why seven different information products were or were not used by each user group during this specific SAR event as well as identify the differences, if any, between different user groups and the information products they used. These user groups represent some of the main user groups that may be involved in an Arctic maritime SAR context such that the findings could hold potential implications for broader understanding about Arctic SAR.

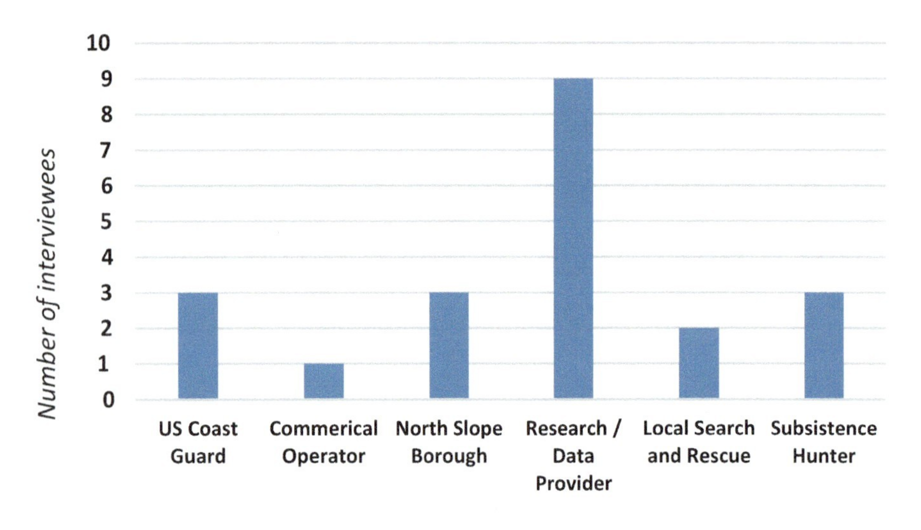

Fig. 23.2 Primary interviewee stakeholder group, self-identified (n = 17)
Different stakeholder groups identified via an archival analysis of the SAR event and interviewed (n = 17) for this study. Some interviewees spanned multiple stakeholder groups. For the purposes of this study, we used their perspectives regarding each group, hence the larger n (21) via the figure

This research investigated seven different information products used by six different user groups (Fig. 23.2) during the SAR event. We engaged a mixed methodology to achieve the research objectives which involved (1) archival analysis of the SAR communication email threads and the official USCG case file associated with the SAR event and (2) conducting semi-structured interviews (n = 17) with different individuals involved directly or indirectly with the SAR event. Both methods are described in further detail below. We identified seven different information resources and products that were provided and/or used during the SAR event. Table 23.1 lists and provides a detailed description of each identified information resource and product.

23.2.2 Method 1: Archival Analysis

We analyzed the e-mail communication (16 e-mails) among 16 individuals involved in the SAR event, which were obtained in August 2017. This SAR event holds potential implications for other Arctic marine coastal areas or communities given the offshore nature of the event. Offshore Arctic SAR events could potentially be on the rise in years to come, due to increased sea ice retreat. A university Arctic scientist who was in Utqiaġvik at the time of the event, initiated email communication with

Table 23.1 Weather-and sea ice-related information provided for the July 19, 2017 SAR effort. Seven different weather and sea ice-related information resources and products that were investigated as part of this case study research. Information on the data provider for each information resource or product as well as the information provided, how it can be accessed, its refresh rate, and general format are provided below

Information Product or Resource	Information Provider	Information Provided	Access	Refresh Rate	Format	Notes
International Arctic Buoy Programme (IABP)	University of Washington	Buoy (ID, skin temperature when available), ice concentration, sea level pressure, sea surface temperature	Online (requires internet)	Daily (every morning), monthly (once a month)	Map (dynamic, online) or table (.dat)	Buoys report on the hour but can be changed to every 15 min
High-Resolution Ice-Ocean Modeling and Assimilation System (HIOMAS)	University of Washington	Hindcast and forecast of Arctic Ocean currents, sea ice, and change	Online (requires Internet)	6 h	Via Arctic ERMA mapping tool and AOOS Ocean data explorer	Hindcasts and forecasts can be optimized and created in an hour
Chukchi Sea Surface Currents High Frequency (HF) Radar	University of Alaska Fairbanks	12-h animation as well as daily maps	Online (requires Internet)	Every hour (during open water season)	Map (dynamic, online) or raw data files (KML, PNGs, and KMZ)	Though the radar does not overwinter, ice drifting buoys have data during winter season (send GPS data every hour) Radar data goes to National HFR Data Center and is ingested into SAROPS
Windy.com	Windy	Wind, rain, snow, temperature, and cloud movement data	Online (requires internet)	Between 6 and 12 h depending on data source	Online webpage and phone application (Android and iPhone)	Ocean current information not available for Arctic Alaska
Traditional and local knowledge (TLK)	North Slope Borough local community	Information related to travel, ocean, and sea ice conditions	In-person, can be shared online or via distance communication			

(continued)

Table 23.1 (continued)

Information Product or Resource	Information Provider	Information Provided	Access	Refresh Rate	Format	Notes
Ice tracking drifters mapping tool	University of Alaska Fairbanks	Ice drifter location (from deployment, to current, and end destination)	Online (requires internet)	Consistently (precise interval not yet identified)	Online map	Drift rates can also be provided as a text file
Search and Rescue Optimal Planning System (SAROPS)	U.S. Coast Guard (USCG)	Drift probability, probability grid for search area, probability of success metric	Only accessible via USCG intranet	Consistently (precise interval not yet identified)	Charts, maps, text files, shapefiles, and KMZ files	SAROPS exports can be created by USCG Headquarters

other Arctic researchers regarding the SAR event, given the urgency of the event and their desire to help support the local SAR effort. Specifically, the researcher reached out to a number of different Arctic researchers for support in estimating the location of the missing vessel based off potential drift. The review of the email communication was used to identify the interviewees for the interview portion of this research, identify the data products that were shared regarding the SAR event, and create a preliminary timeline of events (Fig. 23.2). In addition to the email communication, the USCG case file[2] was analyzed to help reconstruct the timeline of events as well as provide information on the USCG's role in the SAR event and the information products used by the USCG during the event.

23.2.3 *Method 2: Stakeholder Interviews (N = 17, 100% Response Rate)*

Potential interviewees were selected based on a snowball sample, which began with the list of individuals identified in the archival analysis (Handcock & Gile, 2011). Prior to conducting interviews, we obtained Institutional Review Board approval for research with human subjects.[3] Interview participants spanned different stakeholder groups, specifically Arctic researchers, North Slope Borough employees, subsistence hunters, maritime commercial operators, North Slope Borough and volunteer local SAR responders, and the USCG (Fig. 23.2). It is important to note that many interviewees spanned multiple stakeholder groups; for example, some subsistence hunters, North Slope Borough employees, and North Slope Borough and volunteer local SAR responders spanned multiple stakeholder groups. For the purpose of this study, we used perspectives from each of their self-identified groups.

There was a larger number of interviewees from the researcher / data provider group, due to the number of different data products analyzed in this study. Given the small community in Utqiaġvik, having a comparable number of interviewees for the other stakeholder groups was not possible. A semi-structured interview protocol was used to ask the interviewees about: the SAR event, data provision – what type of data was provided and why, data use – what type of data was used in the SAR event and why, and the enabling and challenging factors in data provision and data use in a general SAR context.

Notes from the interviews were coded twice by one coder, approximately 1 year apart, to identify information on the aforementioned themes. Given the sensitive nature of the case study, detailed notes were taken during the interviews (n = 17) instead of being recorded and transcribed in order to increase comfort for interviewees. Grounded theory analysis was used to analyze the transcriptions and notes from the interviews and develop a coding structure (Charmaz & Belgrave, 2012).

[2]MISLE reference number: 1089697, provided by USCG District 17 SAR Coordinator.

[3]Institutional Review Board Project Number: 1123309-1.

Codes were developed to identify which information products were utilized by each user, any reasoning as to why information products were or were not used by a user, as well as challenges to utilizing information products, particularly in a SAR context. Interviews were analyzed using qualitative content analysis for themes related to the timeline of the event, the use of scientific products in decisionmaking, challenges to data use, and lessons learned for future SAR events (Mayring, 2004). Interviews were then coded to identify themes related to enabling and challenging factors in data sharing and use in an Arctic SAR context. Ten factors, five enabling factors and five challenging factors, were identified from the interviews. Each factor needed to have been mentioned and coded in at least two interviews in order to be included in the factors list, which is found in Table 23.6.

23.3 Findings

Based on the archival analysis and stakeholder interviews regarding the July 19, 2017 SAR event, the following three sections summarize the major findings from our analysis.

23.3.1 What Information Was Provided and/or Accessed and Why?

TLK of wind and ocean conditions was provided by the local SAR team to help pinpoint the initial search radius for the local SAR effort that began the morning of July 19. Information from windy.com (a weather forecast application, which provides visualizations of outputs from different weather models) was also accessed by the local SAR team to help determine wind direction. The USCG Search and Rescue Optimal Planning System (SAROPS), the system used across USGS nationwide, was accessed to plan the USCG SAR response. Lastly, information from the following four university-developed data and modeling tools were provided to the local SAR team over the course of the SAR event, once the group of university Arctic researchers was notified of the incident (Table 23.2).

23.3.2 What Information Was Used and Why

Of the six different stakeholder groups we interviewed, we identified two groups as information users for, and therefore decisionmakers in, this specific SAR event: local SAR (which includes members of the subsistence hunter group) and the USCG (Table 23.3). Since commercial operators were not involved in this specific SAR event, they are not included in the analysis of the SAR event. Two of the identified information user groups are also information providers, namely, subsistence hunters

Table 23.2 University-developed data and modeling tools provided in SAR event. Description of the four university-developed data and modeling tools that were provided in this SAR event, focusing on the type of information each product provides

University-developed data and modeling tools	Information provided
University of Washington International Arctic Buoy Programme	Provides real-time meteorological and oceanographic data from Arctic buoys
University of Washington High-Resolution Ice-Ocean Modeling and Assimilation System (HIOMAS)	Provides surface ocean velocity and ice drift prediction
University of Alaska Fairbanks (UAF) Chukchi Sea Surface Current High Frequency Radar	Measures surface currents
UAF ice-tracking drifters	Provides data on the speed and direction of ice

Table 23.3 Information users and information providers during SAR event by stakeholder group. Identification of which of the five main stakeholder groups involved in the analyzed SAR event were information users, information providers, or were both during the SAR event

Stakeholder Group	Information User	Information Provider
Local Search and Rescue	X	
North Slope Borough		X
Subsistence hunter	X	X
Research/data provider		X
U.S. Coast Guard	X	X

due to the wealth of TLK within this community as well as the USCG, since they are providers of the SAROPS platform. The local SAR team predominantly consists of individuals and community members that are extremely well versed in TLK. Many members of the local SAR team are subsistence hunters themselves, as well as well-established, respected members of the community, who therefore can, and do, serve as information providers.

Of the seven different information products and resources provided or accessed during the SAR event, only three were explicitly mentioned as being utilized during this SAR event (Table 23.4). TLK, windy.com, and SAROPS were primarily used due to the SAR responders' familiarity and positive previous experiences with these information resources and products, as well as due to the organizational culture and procedural mandates in utilizing these products in some cases. However, several interviewees mentioned that although the four university products that were shared with the SAR responders were not explicitly used by the SAR team (i.e. accessed directly by the SAR team), information was generated from these products and provided to the SAR team. This information helped to ascertain, inform, and confirm decisions that were being made by the SAR responders, specifically with the local SAR response team. This was done via communication between the university researchers (predominantly via email but also phone calls and text messages) and members of the local SAR response team. This indirect use of the four university products is accounted for in Table 23.4 below.

Table 23.4 Information resources and products used during SAR event by information users. Identification of which information resources and products were used by the information users (Local Search and Rescue and the U.S. Coast Guard) identified for this specific SAR event via our stakeholder interviews. Products that are mentioned below as being *indirectly* used refer to how, in several instances, we were told these products helped support the SAR event. However, since we did not have the ability to talk with the specific SAR responders who potentially utilized these products, we cannot determine if information derived from these products was simply passed along to the relevant SAR responders or if the SAR responders directly accessed the products themselves

	IABP	HIOMAS	HF Radar	Windy. com	TLK	Ice Drifters	SAROPS
Local Search and Rescue	*Indirect*	*Indirect*	*Indirect*	X	X	*Indirect*	
U.S. Coast Guard				X			X

As one interviewee mentioned, there were unusual wind patterns during this event, which made estimating the missing vessel's location particularly challenging. The offshore nature of this SAR event is what triggered the request, sent out by a university researcher who was in Utqiaġvik at the time, for modeling the missing vessel's location via university products. None of the interviewees "regretted" reaching out to the broader scientific community. In fact, one interviewee mentioned this SAR event showed the "untapped potential of the science community in Utqiaġvik." As one interviewee stated, "traditional knowledge works very well on land and in the near shore area but farther out to sea, a lot of the guys are still building their knowledge base, so when it comes to offshore SAR, we find that capacity in general hasn't expanded to the degree of sea ice retreat." One of the main reasons the university products were not explicitly used in this SAR event was the amount of time it took to get the information to and from the scientists. As one of the scientists involved in the SAR event mentioned, "if we had gotten the message out to the science community earlier in the day, we could have had a chance to better help."

We were not able to determine whether the USCG team used information from the four university-developed tools since we were not able to speak with the specific USCG team members involved in this SAR event, due to the inability to track down the specific USCG team that was involved.

23.3.3 Other Identified Information Resources and Products

Over the course of our interviews, several other information resources and products were identified. In addition to the seven products identified for this study, each information user group identified at least two additional information resources and products they utilize for sea ice hazard awareness and/or response. Three products in particular were used by more than one information user group, as seen in Table 23.5 below, which shows the additional information resources identified by interviewees.

Table 23.5 Other utilized information resources and products by stakeholder group, identified during interviews.
Other information resources and products used by each stakeholder group, that stakeholders indicated they use over the course of our interviews. The products outlined in bold below denote products that were identified by more than one stakeholder group. In total, four products were reported as being used by more than one stakeholder group. It is important to note stakeholders indicated they use these products in general and not specifically for the analyzed SAR event

Commercial operators	*Local Search and Rescue*
Canadian ice navigators (when travelling in Canadian waters) Communication with other vessels that travel the same routes **Satellite imagery from NASA MODIS website** **NOAA ASIP** Marine Exchange of Alaska Flying in personnel to physically assess landfast ice stability	Community members **University of Alaska Fairbanks Barrow Sea Ice Radar** Alaska Eskimo Whaling Commission **Satellite imagery from NASA MODIS website**
Subsistence hunters	*U.S. Coast Guard*
Facebook **University of Alaska Fairbanks Barrow Sea Ice Radar** Barrow Whaling Captain Commission	NOAA buoys **NOAA ASIP**

Regarding the additional information resources and products that were identified, commercial operators mentioned that the Marine Exchange of Alaska offered the ability to see information posted by other vessels, which is particularly useful when those vessels have recently been through ice. Commercial operators also noted that when Shell was operating in Northern Alaska, they would share a lot of information on ice conditions with the commercial maritime community, as well as advise the USCG on occasion of ice conditions. As the interviewee put it, "there was a lot of communication about sea ice when Shell was around."

Interviews with local SAR mentioned that the Barrow Sea Ice Radar was utilized quite frequently when it was operational. It has not been in operation since 2017 and therefore was not used in this SAR event. However, it was utilized in 2014 to help track the drift of missing persons on a landfast ice breakout floe, the first time in which the Sea Ice Radar was utilized for a SAR search. However, the limiting factor of the Sea Ice Radar is its range, up to 6 miles from the coast of Utqiaġvik, which as one interviewee stated, "is barely past what we can already see with our eyes." Another interviewee stated that "a lot of cruise ships are just out of the radar's range," which in the event of a cruise ship-related SAR event, an unlikely but potentially calamitous SAR scenario, would render it unhelpful. These findings were also affirmed in our interviews with subsistence hunters.

The importance of community and individual interactions was highly stressed in interviews with both local SAR and subsistence hunters. One interviewee mentioned that roughly 90% of information for SAR missions comes from knowing which people have knowledge of certain areas; "I already know who I want to talk to and what I want to know from them depending on where a search is taking place."

23.3.4 The Enabling and Challenging Factors in Data Sharing and Data Use

We identified five factors (E1–E5) that enabled information product use and five factors (C1–C5) that created challenges in information product use (Table 23.6). Familiarity with information or information provider (E1) and previous positive experiences with information and information provider (E2) were found to be enabling factors among all the interviewee groups. E2, E3, and E5 are linked to each other. Trust (E3) and previous positive experiences (E5) working with an information product or provider, such as windy.com and specific university researchers, increased an information user's perceived reliability of a specific information product or provider. In some cases, some information users have institutional mandates to utilize specific information products (E4), such as the USCG in terms of using SAROPS to plan a SAR mission, which automatically generates a search grid for the USCG responders. Local SAR also has long-standing experience relying on TLK (E4), among themselves but also among the community, when planning a SAR response.

Commercial operators and USCG were not readily aware of some the university-based information products (C1); local SAR and subsistence hunters were aware of some of the products, if not directly, then indirectly via contact with university researchers or the North Slope Borough. Satellite imagery data and even weather data is not provided in near-enough real-time where it can be used for real-time decisionmaking, especially in the Arctic, where weather can change very rapidly within a matter of minutes; in some cases, even hourly data is not good enough, from a reliability perspective (C2). One interviewee mentioned that though it is great to see new information always coming online, it is hard to truly rely on products (C3), especially when you have had experiences where a "website changes and where we once had reliable information to go to, the link is now broken and we cannot find where it went." Other accessibility challenges (C3) that were mentioned involved

Table 23.6 Enabling and challenging factors in data sharing and data use in Arctic SAR context. Identified enabling and challenging factors regarding data sharing and data use in a general Arctic SAR context. These factors were identified via the stakeholder interviews using grounded theory analysis, in which the stakeholder interviews were initially coded to identify broad themes and then subsequently coded (1 year later) to further refine and develop the ten factors, five enabling and five challenging, listed and discussed below

Enabling factors	Challenging factors
E1. Familiarity with information or information provider	C1. Lack of awareness of information resource or product's existence
E2. Previous positive experiences with information or information provider	C2. Infrequent information update or refresh
E3. Trust in information or information provider	C3. Inaccessibility of information resource or product
E4. Mandate to, or legacy of, using an information resource or product	C4. Procedural policies or organization culture barriers
E5. Reliability of information or information provider	C5. Information product not designed or readily applicable for a SAR context

information products that are Internet-based; Internet connectivity is increasingly limited the further you go out to sea. Any information product with imagery or consistent Internet connectivity is difficult, or impossible to use, due to bandwidth limits. Accessibility problems however are not just limited to the sea. As one interviewee said, "the Internet is really slow here in Utqiaġvik, especially for Facebook."

Procedural policies and organizational culture can also inhibit the use of information resources and products (C4), especially new ones. As one interviewee mentioned, "I have trust in the science and the broader system – I use it on a daily basis – but some of the others might be hesitant" while another interviewee stated that instruments used by "operational agencies often become stagnant due to the tendency to buy the same thing over and over because it 'works' but meanwhile, technology has evolved."

Lastly, regarding how an information product is not designed or readily applicable for a SAR context (C5), Fig. 23.3 shows the plotted location estimates provided by HIOMAS, the Chukchi Sea Surface Current HF Radar, and the ice-tracking drifters. The estimates from the Chukchi Sea Surface Current HF Radar group

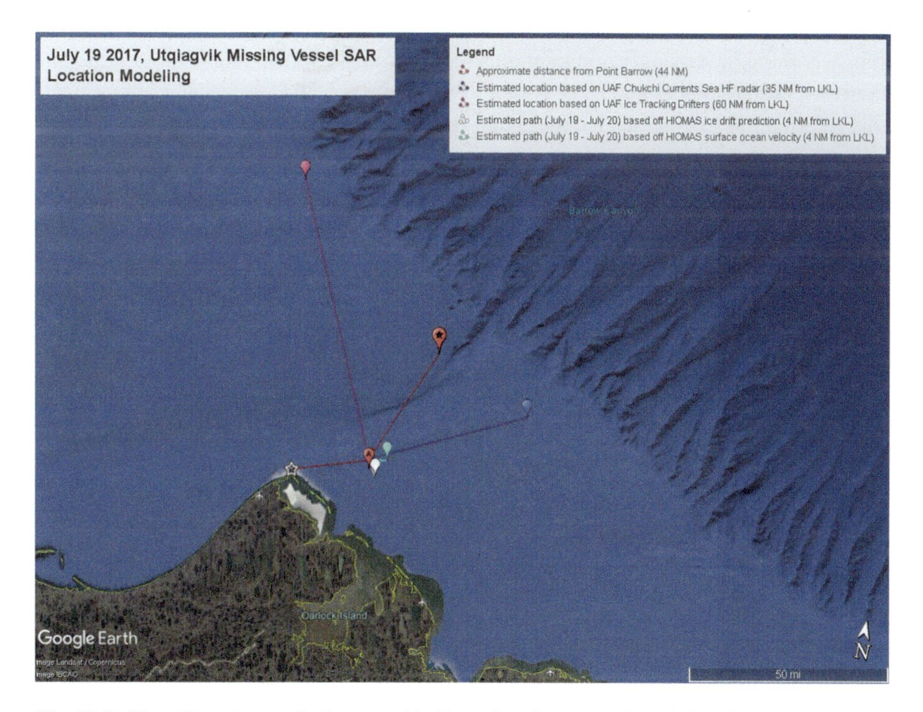

Fig. 23.3 Plot of location predictions provided by university researchers during SAR event, against point of last known location for the missing vessel as well as its retrieval location.
Analysis of the location predictions provided via the four university-developed products analyzed in this research, against the point of last known location (LKL) for the missing vessel (point A, red balloon) as well as its ultimate retrieval location (starred point, red balloon). Though none of the product estimated the retrieval point, they helped to indicate ocean current direction and a general search grid, despite having very little information to work off (e.g. only one point of LKL, almost 12 hours prior to when the models were run (see Fig. 23.2)

were in closest proximity (23 nautical miles) to where the missing vessel was retrieved. Though the surface velocity data from the HIOMAS model was only 24 nautical miles from where the missing vessel was retrieved, the estimate was more southward in comparison to the data from the Chukchi Sea Surface Current HF Radar group. It is important to note that the information from the ice-tracking drifter was provided less as a means to locate the missing vessel but rather to understand surface current speed and wind direction. Given that none of these information products were developed to provide estimated locations of a single object, based off past data, the fact that it was indeed possible to modify each product to generate this information is notable and commendable, in a short period of time (see SAR event timeline in Methods section). However, there is a need for optimizing and automating some of these processes, should they be used for other SAR events, where time and accuracy are of the essence.

23.4 Discussion

It is becoming increasingly apparent on a local, national, and an Arctic-wide scale that there is a need to model and forecast sea ice hazards, as well as a need to communicate this information to both sea ice users and rescue operators (Bridges, 2017). The SAR community in particular has recognized the necessity to actively pursue sea ice hazard research as pack ice and fast ice can limit the mobility of rescue vessels (Clark & Ford, 2017; Ford & Clark, 2019; Smith, 2017). Information on both current conditions (real-time to 2-day forecasts) and future conditions (1–2 month forecasts) in the Arctic Ocean is important to get a sense of both current and projected sea ice-related hazards. However, looking at sea ice melt and retreat is not enough to understand its implications on the Arctic maritime environment. Information on ice thickness, landfast ice stability, ice velocity (e.g. ice drift), as well as weather-related conditions such as atmospheric pressure, wind direction and speed, and ocean currents are important to get a holistic understanding of changes to sea ice and its potential impact on animal, human, and vessel movement in the Arctic (Eicken & Mahoney, 2015). Within this overarching context of information needs for sea ice-related SAR, we summarize the three main findings from our case study below.

23.4.1 Different Data Needs and Decision Contexts for Each Information User Group

The main information needs for a coastal sea ice decisionmaking context are identified in Fig. 23.4. These information needs spanned across the four identified information user groups in this study. However, depending on an information user's knowledge and familiarity of the area, familiarity with and previous use of an information product, as well as the situational context itself, some information

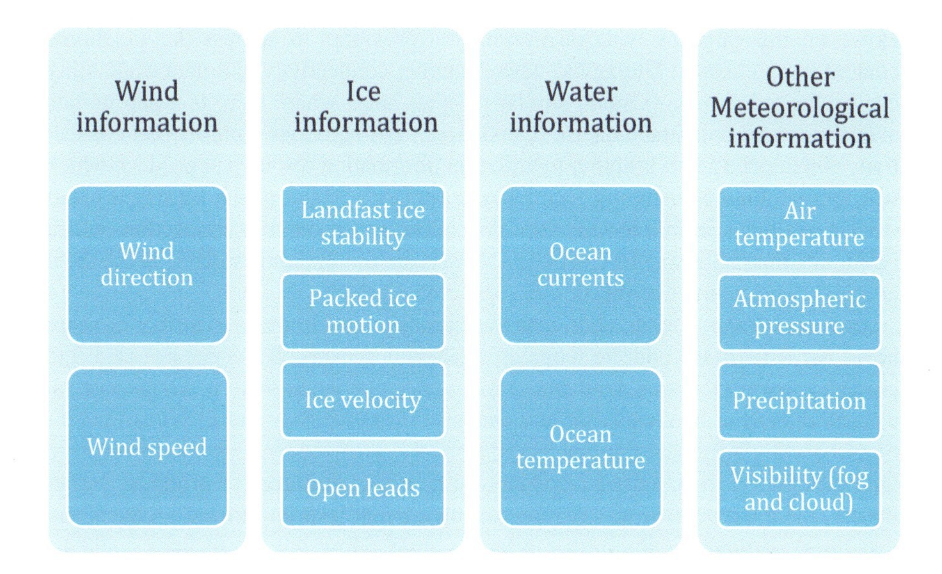

Fig. 23.4 Information needs regarding sea ice decisionmaking in a SAR context. Summary of the main information needs, identified via the stakeholder interviews for this case study research, that individuals need to assess for sea ice-related decisionmaking. It is important to note that all of these information needs may not always be relevant depending on the situation

needs can be inferred or are not relevant to a specific situation. For example, in the case of the SAR event analyzed in this study, information on landfast ice stability and open leads was irrelevant since not only was the missing vessel stuck in a drifting ice pack over 30 nautical miles from shore, but ice concentration in general was low. Thus, though Fig. 23.4 provides an overview of the main information needs for sea ice decisionmaking, actual information needs vary and depend on each SAR context.

23.4.2 Single Information Resource Cannot Meet All Informational Needs

We found in the context of this case study that a single resource cannot meet the needs of all sea ice information user groups. A multitude of different informational products and resources need to exist to meet each respective user needs, which include varying spatial and temporal information needs. This finding is in-line with existing literature on information resource use and development (Jeuring et al., 2019; Knol et al., 2018; Petrich & Eicken, 2009; Tremblay et al., 2006). Therefore, access to a suite of resources, including TLK, scientific information, as well as operational resources, can potentially better serve SAR needs. There is clear benefit though from generating information from a number of different resources (Knol et al., 2018).

However, this puts the onus on information providers to address the situational context in and around Utqiaġvik, where Internet connectivity and data availability can be limited or non-existent, similar to other areas where there is a gap in data availability and potential data use (Dinku et al., 2014). Issues with Internet connectivity can impede a user's ability to utilize an information resource or product, which was an identified challenging factor in our research. Many of the identified information resources in this research are dependent on an Internet connection, which constrains their ability to be used when vessels are out in Arctic waters, where there is limited / no Internet connection (Larsen et al., 2016).

Beyond Internet connectivity, data availability is also limited for further offshore areas, particularly beyond the scope of radars and trackers. Some ocean-based data products also may not be available in the wintertime when there is ice present; all data products can be subject to data outages due to weather and inclement conditions. Existing information products and resources are used by SAR operators due to their familiarity, trust, and/or mandate to use certain resources or products. Nonetheless, there are gaps in information for this current suite of utilized products and resources, particularly for further offshore areas. However, these gaps could be potentially addressed with the additional use of other information. There is merit therefore to leveraging different products to validate, interpolate, and extrapolate information against one another, in order to create a comprehensive suite of sea ice hazard decisionmaking data, especially for further offshore SAR events where there is limited data available.

23.4.3 Available Resources Not Used Due to Lack of Familiarity, Developed Trust, and Perceived Reliability

A number of different information resources and products are available to help meet information needs; our research identified and investigated the use of seven different information resources and products, as well as at least a dozen other information resources relevant for the community of Utqiaġvik. There is a breadth of potential data sources that could potentially be applied in a SAR context (Tables 23.1 and 23.5). However, based on our interview findings, many of these resources are not known to SAR operators, particularly those developed by university and other research institutions. This was one of the challenging factors we identified in our findings. Only three out of the seven identified information products or resources were used in this specific SAR event, which was primarily due to information user's lack of familiarity with these products. However, another challenging factor is that many of these resources were created for a specific purpose; they are not ready to be applied in a SAR situation as-is. All of the information derived from university-based data produced for this event required some level of pre- or post-processing, or separate individual analysis by university researchers, in order to be adapted for this specific SAR need; in this case, estimating the potential drift of a missing vessel.

23.4.4 Options

A targeted and systematic way for SAR operators to utilize and draw upon the various sea ice hazard information products and resources available can help support the local SAR community in future SAR events, especially when immediate information is necessary. Future SAR efforts can benefit from a more streamlined process of data sharing and engagement, in order to minimize response time. For example, an emergency email listserv and/or phone or text alert system that could be used to send requests for data support from data providers could be one way to help support SAR efforts as well as the preparation of a number of existing data resources to be "SAR ready," should they need to be utilized in a SAR effort. Data products, specifically those developed by universities, were used in ways for which they were not designed during this SAR effort. Yet, they were adapted to meet the demand. Optimizing some of these products for future use could be a highly applicable and useful contribution to further improving SAR responses moving forward. This can be done by incorporating different stakeholder decision contexts in current products and leveraging existing information networks to integrate additional information (Kettle et al., 2019).

Beyond creating SAR-relevant data, the following quotes (Fig. 23.5) were taken from a number of different interviews to show the varied need for different data and data formats for the variety of sea ice information users in Arctic Alaska. The need for continued and comprehensive sea ice research and sea ice data development is critical to furthering our understanding of the changing Arctic. However, there is also a clear benefit for the co-production of these resources with local communities. It is very important to note that distinct and sustained efforts need to be undertaken by all actors to integrate TLK with science products. Our findings have shown that TLK plays a critical role in understanding sea ice conditions as well as monitoring the changes happening to sea ice over time. Braiding TLK with science products will ensure not only their relevancy to local communities across the Arctic but also serve to honor the important knowledge and understanding to be gained from this very valuable and venerated information source found across the Arctic. Furthermore, research has indicated that products developed using co-production not only increases their relevancy but also their usability (Dilling & Lemos, 2011; Kettle et al., 2019).

23.4.5 Limitations and Potential Research Applications

This was a qualitative study due to our small sample size (n = 17) of our interviews; no statistical analyses were conducted. Furthermore, while we spoke with the general local SAR and USCG teams, we were not able to track down and therefore speak with the actual SAR responders for this event from both teams, which limited our ability to say which information was or was not used by the response teams.

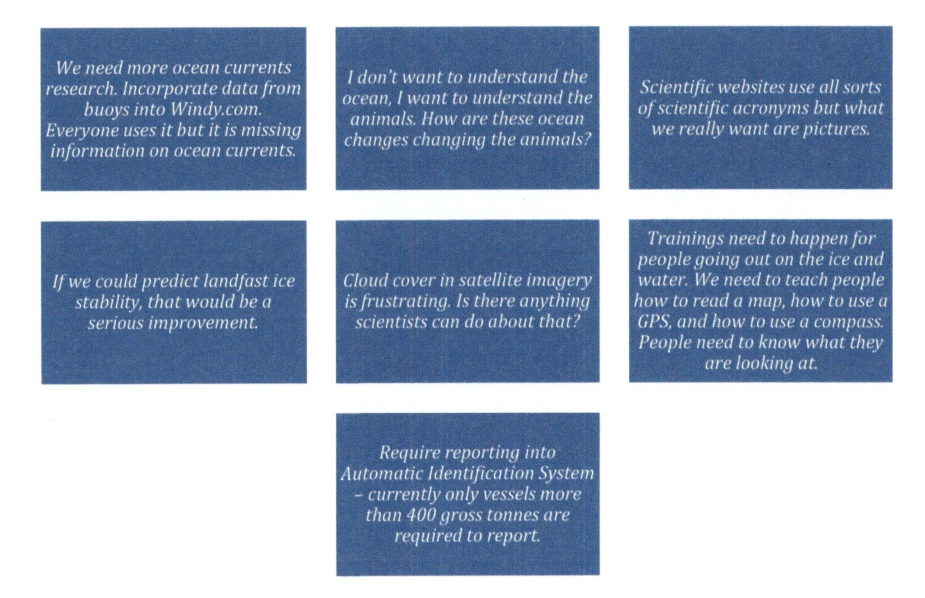

Fig. 23.5 Recommendations for future sea ice hazard research.
Several quotes that were garnered from the stakeholder interviews that were conducted for this case study research, about how to develop better sea ice hazard-related information products as well as bolster better sea ice hazard decisionmaking

To respect the privacy and anonymity of our respondents, given the small size of the community in Utqiaġvik, we refrained from providing any detailed information on the interviewees. Lastly, this research was a pilot study to investigate the use of the seven identified information products in a specific SAR event. Over the course of the interviews, we identified additional information products that are used by information users. Future research could conduct additional interviews to analyze systematically why and how these additional information products are used.

This study was an analysis of a single case study (Fig. 23.1). The analysis of a historical SAR event provided insight and helped to improve our understanding of a SAR response. In particular, the seven different information resources analyzed in this research could potentially be of use and applied in other Arctic SAR incidents, to help common-interest building regarding Arctic SAR. In particular, the use of multiple information sources in this case study highlights a need for continued information sharing and development, which can be a point for collaboration and cooperation between the Arctic Council parties, local responders, local data providers, and operational programs, as part of the 2011 Arctic Council SAR Agreement. Although our findings were illuminative regarding this case study, our findings cannot be generalized without further research on other SAR events related to sea ice. Further analysis of other SAR events, in Utqiaġvik, in Arctic Alaska, as well as the broader Arctic, is important to see if these findings can extend to other contexts. Expanding this research to a broader set of events can be especially helpful in investigating whether there have been any changes to SAR data needs over time. Specific events should be identified in conjunction with local communities as well as emergency responders and operators in order to ensure the findings are pertinent to those the research affects the most.

In addition, further research on the relationship between information providers and information users in an Arctic SAR context can help to expand upon some of the findings from this work. Particularly given that TLK can both be provided and used in a SAR context, this opens up the opportunity to explore the ensuing cyclical relationship between information providers and information users but also between information users. In general, there is a move away from the "traditional" linear relationship between information providers and information users, particularly in the weather, water, ice, and climate information realm, which calls for further research into how information is both shared and used in this sector (Beck, 2011; Haavisto et al., 2020; Jeuring et al., 2019; Knol et al., 2018).

Lastly, this research looked at the role of scientific information and TLK in SAR decisionmaking and response. However, other parameters can factor into SAR decisionmaking response, such as missing vessel characteristics, communication systems (both for SAR responders and the missing vessel), and SAR agreements and protocols. Further research on the role of these different parameters in SAR decisionmaking is imperative, in order to ensure a holistic understanding of the factors that contribute to a successful SAR response.

23.5 Conclusions

As more sea ice-related risks arise for individuals and vessels, this has direct implications for maritime operators, including subsistence hunters, and rescue operators, such as national Coast Guards and local SAR (U.S. Coast Guard, 2015). Research is currently underway to understand these risks better and their effects on the maritime environment. For example, the broader project this research is a part of, the U.S. Arctic Domain Awareness Center (ADAC)-funded project, *Developing Sea Ice and Weather Forecasting Tools to Improve Situational Awareness and Crisis Response in the Arctic,* seeks to create an early notification system for sea ice-related hazards (Kettle et al., 2019). The early warning system is designed to support USCG and local SAR operators in the North Slope of Alaska.

However, there is both need and merit to work with other Arctic communities and contexts on such a topic, as there are lessons learned, similarities, and differences across the Arctic that when looked at together, can only help to increase our understanding of the changing sea ice landscape (Hovelsrud et al., 2011). This is particularly important from both a pan-Arctic geopolitical and local operational perspective. A better understanding of the changing sea ice landscape holds implication for the successful fulfillment of political agreements such as the 2011 Arctic Council SAR Agreement as well as addressing local operational and SAR needs in Arctic communities. From a local perspective, co-production research of Arctic weather and sea ice services is well underway in Canada and Scandinavia (Armitage et al., 2011; Dale & Armitage, 2011; Jeuring et al., 2019). There is opportunity therefore to increase collaboration with other researchers and research institutes on a pan-Arctic level, such as the conduction of an Arctic-wide sea ice hazards assessment, as well as a study of the various weather and sea ice services that are the most beneficial to information users. This study helped to demonstrate the potential for common interest building by synthesizing SAR information needs for a specific case study in Arctic Alaska. Furthering and expanding upon this research in other Arctic contexts would not only be a form of collaboration but also a strong demonstration of science diplomacy and common interest building within the changing Arctic (Berkman et al., 2017; Bertelsen, 2020).

References

Arctic Council. (2011). *Agreement on cooperation on aeronautical and maritime search and rescue in the Arctic*. Arctic Council.

Armitage, D., Berkes, F., Dale, A., Kocho-Schellenberg, E., & Patton, E. (2011). Co-management and the co-production of knowledge: learning to adapt in Canada's Arctic. *Global Environmental Change, 21*(3), 995–1004. https://doi.org/10.1016/j.gloenvcha.2011.04.006

Barber, D. G., Babb, D. G., Ehn, J. K., Chan, W., Matthes, L., Dalman, L. A., Campbell, Y., Harasyn, M. L., Firoozy, N., Theriault, N., Lukovich, J. V., Zagon, T., Papakyriakou, T., Capelle, D. W., Forest, A., & Gariepy, A. (2018). Increasing mobility of high Arctic Sea Ice increases marine hazards off the East Coast of Newfoundland. *Geophysical Research Letters, 45*(5), 2370–2379. https://doi.org/10.1002/2017GL076587

Beck, S. (2011). Moving beyond the linear model of expertise? IPCC and the test of adaptation. *Regional Environmental Change, 11*(2), 297–306. https://doi.org/10.1007/s10113-010-0136-2

Bennett, N. J., Blythe, J., Tyler, S., & Ban, N. C. (2016). Communities and change in the anthropocene: understanding social-ecological vulnerability and planning adaptations to multiple interacting exposures. *Regional Environmental Change, 16*(4), 907–926.

Berkman, P. A., Kullerud, L., Pope, A., Vylegzhanin, A. N., & Young, O. R. (2017). The Arctic science agreement propels science diplomacy. *Science, 358*, 6363.

Bertelsen, R. G. (2020). *Science diplomacy and the arctic*. Routledge Handbook of Arctic Security.

Bridges, R. (2017). Risks and damages caused in ice navigation. In *Encyclopedia of maritime and offshore engineering* (pp. 1–12).

Charmaz, K., & Belgrave, L. (2012). Qualitative interviewing and grounded theory analysis. *The SAGE Handbook of Interview Research: The Complexity of the Craft, 2*, 347–365.

Clark, D. G., & Ford, J. D. (2017). Emergency response in a rapidly changing arctic. *Canadian Medical Association Journal, 189*(4), E135–E136. Canadian Medical Association. https://doi.org/10.1503/cmaj.161085

Council, I. C. (2008). *The sea ice is our highway: an Inuit perspective on transportation in the artic*. Inuit Circumpolar Council (Canada).

Dale, A., & Armitage, D. (2011). Marine mammal co-management in Canada's Arctic: Knowledge co-production for learning and adaptive capacity. *Marine Policy, 35*(4), 440–449. https://doi.org/10.1016/j.marpol.2010.10.019

Dilling, L., & Lemos, M. C. (2011). Creating usable science: Opportunities and constraints for climate knowledge use and their implications for science policy. *Global Environmental Change, 21*(2), 680–689.

Dinku, T., Block, P., Sharoff, J., Hailemariam, K., Osgood, D., del Corral, J., Cousin, R., & Thomson, M. C. (2014). Bridging critical gaps in climate services and applications in africa. *Earth Perspectives, 1*(1), 15. https://doi.org/10.1186/2194-6434-1-15

Dumas, J., Carmack, E., & Melling, H. (2005). Climate change impacts on the Beaufort shelf landfast ice. *Cold Regions Science and Technology, 42*(1), 41–51. https://doi.org/10.1016/j.coldregions.2004.12.001

Durkalec, A., Furgal, C., Skinner, M. W., & Sheldon, T. (2015). Climate change influences on environment as a determinant of Indigenous health: Relationships to place, sea ice, and health in anInuit community. *Social Science & Medicine, 136–137*, 17–26. https://doi.org/10.1016/j.socscimed.2015.04.026

Dynamics | National Snow and Ice Data Center. (n.d.). Retrieved November 28, 2019, from https://nsidc.org/cryosphere/seaice/processes/dynamics.html

Eicken, H., & Mahoney, A. R. (2015). Sea ice: hazards, risks, and implications for disasters. In *Coastal and marine hazards, risks, and disasters* (pp. 381–401). Elsevier.

Eicken, H., Mahoney, A., Jones, J., Heinrichs, T., Broderson, D., Statscewich, H., Weingartner, T., Stuefer, M., Ravens, T., & Ivey, M. (2018). Sustained observations of changing Arctic coastal and marine environments and their potential contribution to Arctic maritime domain awareness. *Arctic, 71*, 1–15.

Fenwick, T. (2012). Co-production in professional practice: A sociomaterial analysis. *Professions and Professionalism, 2*(2), 1–16.

Flato, G. M., & Brown, R. D. (1996). Variability and climate sensitivity of landfast Arctic Sea ice. *Journal of Geophysical Research, Oceans, 101*(C11), 25767–25777. https://doi.org/10.1029/96JC02431

Ford, J., & Clark, D. (2019). Preparing for the impacts of climate change along Canada's Arctic coast: the importance of search and rescue. *Marine Policy, 108*. https://doi.org/10.1016/j.marpol.2019.103662

Galley, R. J., Else, B. G. T., Prinsenberg, S. J., Babb, D., & Barber, D. G. (2013). Summer sea ice concentration, motion, and thickness near areas of proposed offshore oil and gas development in the Canadian Beaufort Sea—2009. *Arctic*, 105–116.

Haavisto, R., Lamers, M., Thoman, R., Liggett, D., Carrasco, J., Dawson, J., Ljubicic, G., & Stewart, E. (2020). Mapping weather, water, ice and climate (WWIC) information providers in polar regions: who are they and who do they serve? *Polar Geography*, 1–19. https://doi.org/10.1080/1088937X.2019.1707320

Handcock, M. S., & Gile, K. J. (2011). Comment: On the concept of snowball sampling. *Sociological Methodology, 41*(1), 367–371.

Hovelsrud, G. K., Poppel, B., Van Oort, B., & Reist, J. D. (2011). Arctic societies, cultures, and peoples in a changing cryosphere. *Ambio, 40*(SUPPL. 1), 100–110. https://doi.org/10.1007/s13280-011-0219-4

Jeuring, J., Knol-Kauffman, M., & Sivle, A. (2019). Toward valuable weather and sea-ice services for the marine Arctic: Exploring user–producer interfaces of the Norwegian Meteorological Institute. *Polar Geography*, 1–21. https://doi.org/10.1080/1088937X.2019.1679270

Jones, J., Eicken, H., Mahoney, A., Rohith, M. V., Kambhamettu, C., Fukamachi, Y., Ohshima, K. I., & George, J. C. (2016). Landfast Sea ice breakouts: Stabilizing ice features, oceanic and atmospheric forcing at Barrow, Alaska. *Continental Shelf Research, 126*, 50–63.

Kettle, N. P., Abdel-Fattah, D., Mahoney, A. R., Eicken, H., Brigham, L. W., & Jones, J. (2019). Linking Arctic system science research to decision maker needs: Co-producing sea ice decision support tools in Utqiaġvik, Alaska. *Polar Geography*, 1–17. https://doi.org/10.1080/1088937X.2019.1707318

Kimura, N., & Wakatsuchi, M. (2000). Relationship between sea-ice motion and geostrophic wind in the northern hemisphere. *Geophysical Research Letters, 27*(22), 3735–3738. https://doi.org/10.1029/2000GL011495

Knol, M., Arbo, P., Duske, P., Gerland, S., Lamers, M., Pavlova, O., Sivle, A. D., & Tronstad, S. (2018). Making the Arctic predictable: the changing information infrastructure of Arctic weather and sea ice services. *Polar Geography, 41*(4), 279–293. https://doi.org/10.1080/1088937X.2018.1522382

Kovacs, K. M., Lydersen, C., Overland, J. E., & Moore, S. E. (2011). Impacts of changing sea-ice conditions on Arctic marine mammals. *Marine Biodiversity, 41*(1), 181–194. https://doi.org/10.1007/s12526-010-0061-0

Kwok, R., Spreen, G., & Pang, S. (2013). Arctic Sea ice circulation and drift speed: Decadal trends and ocean currents. *Journal of Geophysical Research, Oceans, 118*(5), 2408–2425. https://doi.org/10.1002/jgrc.20191

Larsen, L.-H., Kvamstad-Lervold, B., Sagerup, K., Gribkovskaia, V., Bambulyak, A., Rautio, R., & Berg, T. E. (2016). Technological and environmental challenges of Arctic shipping—a case study of a fictional voyage in the Arctic. *Polar Research, 35*(1), 27977. https://doi.org/10.3402/polar.v35.27977

Lovvorn, J. R., Rocha, A. R., Mahoney, A. H., & Jewett, S. C. (2018). Sustaining ecological and subsistence functions in conservation areas: eider habitat and access by native hunters along landfast ice. *Environmental Conservation, 45*(4), 361–369.

Mahoney, A. R., Eicken, H., Gaylord, A. G., & Gens, R. (2014). Landfast Sea ice extent in the Chukchi and Beaufort Seas: The annual cycle and decadal variability. *Cold Regions Science and Technology, 103*, 41–56. https://doi.org/10.1016/J.COLDREGIONS.2014.03.003

Mayring, P. (2004). Qualitative content analysis. *A Companion to Qualitative Research, 1*, 159–176.

Overeem, I., Anderson, R. S., Wobus, C. W., Clow, G. D., Urban, F. E., & Matell, N. (2011). Sea ice loss enhances wave action at the Arctic coast. *Geophysical Research Letters, 38*(17), L17503. https://doi.org/10.1029/2011GL048681

Petrich, C., & Eicken, H. (2009). Growth, structure and properties of Sea Ice. In *Sea Ice* (pp. 23–77). Wiley-Blackwell. https://doi.org/10.1002/9781444317145.ch2

Petrich, C., Eicken, H., Zhang, J., Krieger, J., Fukamachi, Y., & Ohshima, K. I. (2012). Coastal landfast sea ice decay and breakup in northern Alaska: Key processes and seasonal prediction. *Journal of Geophysical Research, Oceans, 117*(C2), n/a-n/a. https://doi.org/10.1029/2011JC007339

Pizzolato, L., Howell, S. E. L., Derksen, C., Dawson, J., & Copland, L. (2014). Changing sea ice conditions and marine transportation activity in Canadian Arctic waters between 1990 and 2012. *Climatic Change, 123*(2), 161–173. https://doi.org/10.1007/s10584-013-1038-3

Robards, M. D., Huntington, H. P., Druckenmiller, M., Lefevre, J., Moses, S. K., Stevenson, Z., Watson, A., & Williams, M. (2018). Understanding and adapting to observed changes in the Alaskan Arctic: Actionable knowledge co-production with Alaska native communities. *Deep-Sea Research Part II: Topical Studies in Oceanography, 152*, 203–213. https://doi.org/10.1016/j.dsr2.2018.02.008

Smith, T. (2017). *Search and Rescue in the Arctic: is the US prepared?* RAND.

Stroeve, J., & Notz, D. (2018). Changing state of Arctic Sea ice across all seasons changing state of Arctic Sea ice across all seasons. *Environmental Research Letters, 13*, 103001. https://doi.org/10.1088/1748-9326/aade56

Tremblay, M., Furgal, C., Lafortune, V., Larrivée, C., Savard, J.-P., Barrett, M., Annanack, T., Enish, N., Tookalook, P., & Etidloie, B. (2006). Communities and ice: Bringing together traditional and scientific knowledge. In *Climate change: Linking traditional and scientific knowledge* (p. 289). Aboriginal Issues Press.

U.S. Coast Guard. (2015). *Arctic strategy implementation plan*. U.S. Coast Guard.

Chapter 24
(Research): Maritime Ship Traffic in the Central Arctic Ocean High Seas as a Case Study with Informed Decisionmaking

Paul Arthur Berkman, Greg Fiske, Jacqueline M. Grebmeier, and Alexander N. Vylegzhanin

Abstract This chapter applies the baseline satellite record of maritime ship traffic in the Central Arctic Ocean (CAO) High Seas from 1 September 2009 through 31 December 2018 as a case study with informed decisionmaking to operate across a 'continuum of urgencies'. Starting with questions to generate data as stages of research, the geospatial analyses herein involve cloud-based innovations with the space-time cube and binned queries to interpret the dynamics of maritime ship traffic based on the vessel flag states, types and sizes within the CAO High Seas and surrounding Exclusive Economic Zones (EEZ). These 'big data' are being transformed into evidence for decisions in view of the institutions that produce governance mechanisms and built infrastructure. With science diplomacy, the next level of action is to introduce options (without advocacy), which can be used or ignored explicitly, contributing to informed decisionmaking by the institutions short-to-long term. Objective integration with satellite sea-ice records further reveals ship-ice dynamics in the CAO High Seas – where the highest number and diversity

P. A. Berkman (✉)
Science Diplomacy Center, EvREsearch LTD, Falmouth, MA, USA

Science Diplomacy Center, MGIMO University, Moscow, Russian Federation

United Nations Institute for Training and Research (UNITAR), Geneva, Switzerland

Program on Negotiation, Harvard Law School, Harvard University, Cambridge, MA, USA
e-mail: paul.berkman@unitar.org; pberkman@law.harvard.edu

G. Fiske
Woodwell Climate Research Center, Falmouth, MA, USA

J. M. Grebmeier
Chesapeake Biological Laboratory, University of Maryland Center for Environmental Science, Solomons, MD, USA

A. N. Vylegzhanin
International Law School, MGIMO University, Moscow, Russian Federation

P. A. Berkman et al. (eds.), *Building Common Interests in the Arctic Ocean with Global Inclusion, Volume 2*, Informed Decisionmaking for Sustainability,
https://doi.org/10.1007/978-3-030-89312-5_24

of ships are entering from the Pacific Ocean side – introducing urgent questions to generate informed decisions across the Bering Strait Region south to the Aleutian Islands and northward. The holistic (international, interdisciplinary and inclusive) analyses herein of Arctic Ocean satellite records complement the intent of the *"precautionary approach"* embodied in international law, as provided by the 2018 *Agreement to Prevent Unregulated High Seas Fisheries in the Central Arctic Ocean* that entered into force on 25 June 2021 with ten Arctic and non-Arctic States. In the CAO High Seas, as an area beyond national jurisdictions, maritime ship traffic is highlighted with global inclusion under the Law of the Sea, where all Arctic states and Indigenous peoples *"remain committed"* as they shared in their 2013 *Vision of the Arctic*. These next-generation Arctic marine shipping assessments reflect socioeconomic drivers of change in the Arctic Ocean, as revealed by the ecology of maritime ship traffic in all EEZ and High Seas areas north of the Arctic Circle, with global lessons from the CAO High Seas about balancing national interests and common interests.

24.1 Introduction

24.1.1 Observing Pan-Arctic Maritime Ship Traffic with Satellites

Maritime ship traffic underscores the socioeconomic dynamics of commercial, scientific and other forms of human presence in the Arctic Ocean, which was the overarching rationale to design *Next-Generation Arctic Marine Shipping Assessments* with satellite Automatic Identification System (AIS) records (Table 24.1). AIS signal transmission (NAVCEN, 2016) is mandated by the International Maritime Organization (IMO, 2020a) for ships larger than 300 gross tonnes engaged on international voyages, as implemented globally through the *International Convention for the Safety of Life at Sea,* 1974, as amended (SOLAS, 2020). Accelerating from the Arctic Marine Shipping Assessment report approved by the Arctic Council (AMSA, 2009), with the satellite record of ship movements north of the Arctic Circle from 2009 forward (Berkman et al., 2020a) – this chapter builds on the baseline satellite record of Pan-Arctic maritime ship traffic from 1 September 2009 to 31 December 2016, which can be accessed through the Arctic Data Center supported by the US National Science Foundation (Berkman et al., 2020b). In this chapter, additional satellite AIS data are included through 31 December 2018 for the region north of the Arctic Circle. These maritime ship traffic data provide the framework to generate an objective assessment of the socioeconomic system (associated with science, technology and innovation) coupled with the biophysical system (associated with environmental factors and biological productivity) in the Central Arctic Ocean (CAO) High Seas beyond national jurisdictions under Law of the Sea (see Chap. 1 in this book).

Satellite AIS records enable synoptic patterns, trends and processes with maritime ship traffic to be interpreted on a Pan-Arctic scale objectively in relation to complementary satellite records with the biophysical system, including with sea-ice

Table 24.1 Next-Generation Arctic Marine Shipping Assessments (AMSA)[a]

	Arctic Marine Shipping Assessments (AMSA)	
Attribute	AMSA (2009)	Next-Generation AMSA
Sampling period	2004	2009-present
Data sources	Arctic states individually and with the Arctic Council	Diverse government and commercial Automatic Identification System (AIS) sources
Observation coverage	Point, Regional	Point, Regional and Pan-Arctic
Observation scope	Ground-based	Ground-based and Satellite
Observation frequency	Inconsistent over space and time	Synoptic and continuous (from minutes to decades)
Ship-type designations	Variable national designations	Standardized international designations
Individual ship attributes	Inconsistent and incomplete	Consistent and comprehensive
Analytical capacity	Limited granularity and questions	Open-ended granularity and questions
Science-diplomacy contributions	Scenarios and negotiated recommendations	Holistic evidence and options (without advocacy)
Informed decisionmaking[b]	Governance mechanisms	Operations, Built Infrastructure and Governance Mechanisms

[a]Updated from Berkman et al. (2020a), involving Automatic Identification System (AIS) data collected by polar-orbiting satellites
[b]Informed decisions operate across a 'continuum of urgencies' short-to-long term (Berkman et al., 2020c), as elaborated subsequently (Berkman, 2020a, b)

coverage (NSIDC, 2020) and ocean color patterns as a representation of primary production (Comiso et al., 2020). These maritime ship traffic analyses complement the Synoptic Arctic Survey (Anderson et al., 2018; Ashjian et al., 2019. Paasche et al. (2019) that is underway with international and interdisciplinary inclusion to *"generate a comprehensive dataset that allow for a complete characterisation of Arctic hydrography and circulation, carbon uptake and ocean acidification, tracer distribution and pollution, and organismal and ecosystem functioning and productivity."*

For example, as the sea-ice has been diminishing, the centroid of Arctic maritime ship traffic has shifted 300 kilometers north-eastward based on the continuous satellite AIS record from 2009 to 2016 (NASA Earth Observatory, 2018). This observation enhances the monthly interpretation of satellite AIS data from 2010 to 2014 north of the Arctic Circle (Eguíluz et al., 2016), with Pan-Arctic ship traffic predominating in the Norwegian and Barents Seas. Relationship between sea-ice and ship traffic similarly has been interpreted with the Arctic Ship Traffic Database (ASTD) through the Protection of the Arctic Marine Environment (PAME) working group of the Arctic Council (PAME, 2020a), revealing a 75% increase in the distance sailed by all ships from 2013 to 2019 in the area of the Polar Code (IMO,

2017a), which largely excludes areas in the Norwegian and Barents Seas because they are perennial open water areas. The observed maritime traffic increase appears to be related to destinational shipping, for example, associated with Liquid Natural Gas (LNG) in the Yamal Peninsula and associated logistic chains prior to the COVID-19 pandemic.

Additional assessments with satellite AIS records of Arctic maritime ship traffic also are emerging, as with models of ship emission inventories (Winther et al., 2014) and intercalibration with land-based AIS records (Wright et al., 2019). The goal of this chapter is to demonstrate the fundamental necessity of next-generation Arctic marine shipping assessments (Table 24.1) to implement "precautionary" approaches with decisionmaking for Arctic Ocean management (see Chap. 1 in this book), as established with entry into force of the *Agreement to Prevent Unregulated High Seas Fisheries in the Central Arctic Ocean* on 25 June 2021 (CAO High Seas Fisheries Agreement, 2018).

24.1.2 Methodology of Informed Decisionmaking

Assessments of maritime ship traffic as well as any other system parameters in the Arctic Ocean – or elsewhere at local-global scales – involves data to answer questions. Diverse methods may be applied to generate the data, including from the natural sciences and social sciences as well as Indigenous knowledge, considering science in an holistic (international, interdisciplinary and inclusive) manner as the 'study of change' (Berkman et al., 2020c). Questions create capacities to consider change short-term to long-term – to make "informed decisions" that operate across a 'continuum of urgencies' (Berkman et al., 2016; Berkman, 2020a, b). For example, the underlying questions with Arctic maritime ship traffic in this chapter relate to patterns of diminishing sea ice in the Arctic Ocean (Thoman et al., 2020), which may be non-linear (Eisenmann & Wettaufer, 2009).

Progressing from questions to data represents stages of research in the *Pyramid of Informed Decisionmaking*, where the apex goal is an informed decision (see Chap. 1 in this book). However, to produce an informed decision requires evidence, which are distinct from data because decisionmaking institutions are involved (Donnelly et al., 2018). The distinction is that data are generated with diverse methods to answer questions with research whereas evidence is for decisions with action, integrating the data in the context of the decisionmaking institutions in a purposeful manner (Berkman et al., 2020a):

$$\text{Data} + \text{Institution} = \text{Evidence} \tag{24.1}$$

Importantly, evidence is insufficient for decisions, only compelling decisionmaking institutions to act, if they so choose. Beyond evidence – with science diplomacy – options (without advocacy), which can be used or ignored explicitly, are required for informed decisionmaking (Berkman et al., 2016, 2020c;

Berkman, 2020a, b). In this sense, evidence and options represent stages of action, informing decisions about governance mechanisms and built infrastructure as well their coupling to achieve progress with *"sustainable development and environmental protection,"* which are the *"common Arctic issues"* established by the eight Arctic states and six Indigenous Peoples Organizations with the high-level forum of the Arctic Council (Ottawa Declaration, 1996).

In the Arctic Ocean as elsewhere, the challenge is to operate with research and action, building common interests across the data-evidence interface to produce informed decisions. The basic objective of this chapter is to illustrate how satellite AIS data can be integrated into evidence for informed decisionmaking (see Chap. 1 in this book), applying the CAO High Seas Fisheries Agreement as an institutional case study (Vylegzhanin et al., 2020).

24.2 Arctic Ocean Ship Traffic within Law of the Sea Zones

24.2.1 Synoptic Geospatial Analyses with Satellite Big-Data

This chapter continues to elaborate as well as utilize geospatial methodologies with the baseline of satellite AIS data from the Arctic Ocean, involving cloud computing and binned solutions with the space-time cube – based on user-defined polygons – as described with regional lessons from the Bering Strait and Barents Sea in Volume 1 of the Informed Decisionmaking for Sustainability book series (Berkman et al., 2020a). Briefly, the same standardized methods and satellite AIS data are applied herein from 1 September 2009 through 31 December 2018 north of the Arctic Circle with 21,005 ships in total, as interpreted from the Maritime Mobile Service Identity (MMSI) of each unique vessel across more than 173,000,000 AIS records. The cloud-based methods with Google Big Query enable queries to be run across the entire dataset within seconds at $5 USD per terabyte processing costs and $0.02 USD per gigabyte storage costs (Google, 2020).

These cloud-based methodologies accentuate the geospatial questions that can be addressed with user-defined scalability about maritime ship traffic changes over time and space in the Arctic Ocean, applying satellite AIS records north of the Arctic Circle. The framework question to illustrate in this chapter involves the Law of the Sea zones (see Chap. 1 in this book) across the entire Arctic Ocean with its centrality at 90° North latitude, considering the North Pole as a "Pole of Peace" (Gorbachev, 1987):

What is the distribution of maritime ship traffic in the Exclusive Economic Zones (EEZ) of the Arctic states and the High Seas that exist beyond national jurisdictions in the Arctic Ocean (i.e., north of the Arctic Circle)?

Answering this framework question provides the first rendering of maritime ship traffic within, between and beyond national jurisdictions north of the Arctic Circle comprehensively (Fig. 24.1). In addition, this synoptic profile of maritime ship traffic within jurisdictional zones highlights regional granularity in a Pan-Arctic

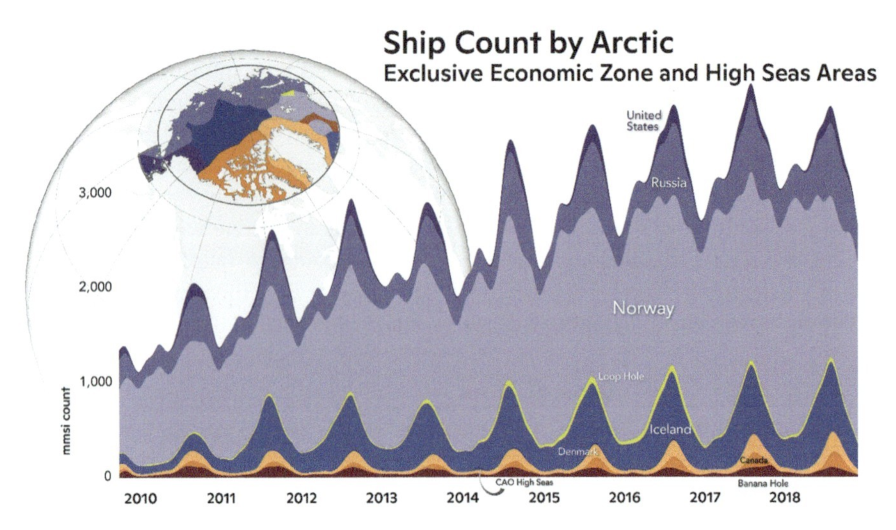

Fig. 24.1 Pan-Arctic Ecosystem of Maritime Ship Traffic among Law of the Sea zones in the Arctic Ocean derived from satellite Automatic Identification System (AIS) big-data with synoptic circumpolar coverage within the Exclusive Economic Zones (EEZ) of Arctic coastal states as well as High Seas areas beyond national jurisdictions from 1 September 2009 through 31 December 2018 north of the Arctic Circle, including the Bering Strait Region as analyzed previously (Berkman et al., 2020a). These data represent more than 173,000,000 AIS records with 21,005 unique ships during the 2009–2018 observation period. Longitudes range from 0°East-West in the Barents Sea with surrounding Norwegian and Russian EEZ to 180°East-West through the Bering Strait with surrounding United States and Russian EEZ. Additional mapping of High Seas areas north of the Arctic Circle is shown in Harrison et al. (2020) for the Banana Hole in the Norwegian Sea and Loop Hole in the Barents Sea as well as the Central Arctic Ocean

context that can be interpreted with new satellite AIS observations, providing an indicator of socioeconomic change that can be integrated objectively with biogeophysical changes continuously across seasons and years in the Arctic Ocean.

The satellite AIS data from 2009 to 2018 reveal seasonality of maritime ship traffic within the EEZ of Arctic coastal states (Canada, Denmark, Iceland, Norway, Russian Federation, United States) as well as within the three High Seas areas north of the Arctic Circle ("Banana Hole" in the Norwegian Sea, Central Arctic Ocean and the "Loop Hole" in the Barents Sea). The high number of ships within the Barents Sea is well known as this region is largely open water throughout the year, further explaining the relatively low-amplitude seasonal variation in maritime ship traffic within the Norwegian EEZ.

As a socioeconomic indicator within Law of the Sea zones, maritime ship traffic reflects the relative change of human presence and interests across the Arctic Ocean. Increasing trends of ship traffic are suggested in all jurisdictional regions, but more clearly in those jurisdictions where there are larger numbers of unique ships (Fig. 24.1). These analyses further reveal the relative dimensions and rates of change with ship traffic across these Arctic maritime jurisdictions from 2009 to 2018 (Table 24.2), noting all regions have increasing maritime ship traffic with the highest increases in the Norwegian EEZ.

Table 24.2 Regional trends of maritime ship traffic within jurisdictions defined by the international framework of the law of the sea, derived monthly from Fig. 24.1

Law of the Sea zone	Arctic ocean area	Monthly number of unique ship days[a]	Regression line[b] [y = rate of change (year) ± constant]
Exclusive Economic Zone (EEZ)	**(areas within national jurisdictions)**		
	Canada	2541	y = 0.012x−474.48 (r^2 = 0.086)
	Denmark	7563	y = 0.169x−638.22 (r^2 = 0.129)
	Iceland	40,644	y = 0.109x−4181.8 (r^2 = 0.322)
	Norway	176,048	y = 0.420x−15928.0 (r^2 = 0.813)
	Russian Federation[c]	43,950	y = 0.088x−3279.1 (r^2 = 0.246)
	United States[c]	6836	y = 0.010x-333.4 (r^2 = 0.023)
High Seas	**(areas beyond national jurisdictions)**		
	Banana Hole	6426	y = 0.012x + 7.7 (r^2 = 0.001)
	Central Arctic Ocean	494	y = 0.002x−89.0 (r^2 = 0.076)
	Loop Hole	3275	y = 0.011x−447.6 (r^2 = 0.306)

[a]Derived with satellite Automatic Identification System (AIS) data north of the Arctic Circle from the monthly totals of unique ships in each area during a daily observation period from 1 September 2009 through 31 December 2018 (i.e., combination of 111 monthly totals), as a measure of relative maritime ship traffic across jurisdictional zones in the Arctic Ocean
[b]Derived from the monthly number of unique ships for 111 months, as shown daily during the observation period (Fig. 24.1), noting the same unique ships may appear in multiple months
[c]The Bering Strait Region with the Russian Federation and United States includes area south of the Arctic Circle (Fig. 24.1), as analyzed and defined previously (Berkman et al., 2020a)

The focus herein with the CAO High Seas involves the jurisdictional zone where there is the slowest increase in maritime ship traffic to date (Table 24.2). Nonetheless, with precaution, the CAO High Seas underlies the potential for a trans-Arctic shipping route when there is open water across the North Pole (Smith and Stephenson, 2013; Stevenson et al., 2019), introducing all manner of questions about *"logistical, geopolitical, environmental, and socioeconomic impacts"* (Bennett et al., 2020).

24.2.2 International Maritime Ship Traffic Patterns and Trends in the CAO High Seas

With maritime ship traffic in an ecological context – studying the home ('eco') – individual ships can be considered as representatives of 'ship species' with known attributes (e.g., flag state, type and size). Similarly, aggregations within a ship species underscore the dynamics of 'maritime ship traffic populations,' which are interacting among 'maritime ship traffic communities' characterized by their diversities within bounded habitats. These habitats are illustrated regionally by Law of the Sea zones that can be interpreted objectively from satellites over time within the 'Pan-Arctic ecosystem of maritime ship traffic' (Fig. 24.1). In an economic context – managing the home – the patterns, trends and processes associated with the Pan-Arctic ecosystem of maritime ship traffic become fundamental to informed decisionmaking about operations, governance mechanisms and built infrastructure in the Arctic Ocean (Table 24.1).

While there are relatively few ships in the CAO High Seas (Fig. 24.1), this jurisdictional region is globally important because it illustrates balancing between national interests and common interests (Berkman & Young, 2009; Berkman & Vylegzhanin, 2013; Berkman et al., 2020c; Berkman, 2010, 2014). This jurisdictional balancing is highlighted by the 2018 CAO High Seas Fisheries Agreement, which is the first North Polar agreement with Arctic and non-Arctic states involving an official translation in an Asian language.

From more than 173,000,000 AIS records with 21,005 unique ships across the Arctic Ocean, in the CAO High Seas there were 185 vessels during the 2009–2018 observation period (Fig. 24.1). As the corpus for the subsequent analyses of this chapter, these vessels were cross-validated in view of their identities and operational characteristics (IMO, 2020b) as well as further confirmed in relation to their transit histories (MyShipTracking, 2020). This dataset of IMO-registered vessels with Class-A transponders (NAVCEN, 2019) is interpreted herein with vessel locations and metadata from 2009 to 2018 (Table 24.3) to generate the first comprehensive assessment about the socioeconomic dynamics of the CAO High Seas, where maritime ship traffic represents human activities, impacts and interests.

The composite maritime traffic pattern in the CAO High Seas from 2009–2018 is shown in relation to vessel flag states (Fig. 24.2), as one of several attributes to quantify ship species' diversity, providing the granularity to assess the dynamics of the Pan-Arctic ecosystem of maritime ship traffic (Fig. 24.1). Other attributes that are considered herein include ship types (e.g., research, cargo, fishery and enforcement vessels) and their sizes (e.g., tonnage classes). These ship attributes are analyzed individually, but can be combined to address user-defined questions with international and interdisciplinary inclusion. The spatial distribution of ships from all nations is circumpolar, but national activities of Arctic coastal states do seem to predominate adjacent to their respective jurisdictions, notably in parallel with Canada and Russia. Higher diversity of flag states is shown in the CAO High Seas with vessels in the vicinity of the Beaufort and Chukchi Seas.

Different types of ship movements are indicated in Fig. 24.2, as with direct transit lines to the North Pole, where the 'Barneo Ice Camp' operated seasonally from 2002 to 2018 (Barneo, 2020). Various shipping patterns (e.g., rectangular zig-zag across extended region, tight zigzag in confined region or two-ship parallel transits) also are revealed, relating to types of maritime activities, as with research or fishing that could be further quantified (Visalli et al., 2020). Moreover, transits of individual ships can be investigated over time as with the 2009–2016 voyages of the German

Table 24.3 Maritime ship traffic attributes to interpret socioeconomic dynamics in the Central Arctic Ocean (CAO) High Seas[a] with surrounding Exclusive Economic Zones (EEZ) shown in Fig. 24.1

Unique ship designation[b]			Ship metadata attribute[c]			CAO High Seas regional visit	
MMSI[d]	Ship name[e]	IMO[f]	Flag[g]	Type[h]	Size[i]	Dates in CAO[j]	Longitudinal Positions[k]

[a]Summary of the satellite Automatic Identification System (AIS) data for the CAO High Seas is available through the Arctic Data Center (https://arcticdata.io/) in conjunction with baseline dataset from September 1, 2009 through December 31, 2016 north of the Arctic Circle (Berkman et al., 2020a), derived from the Aprize satellite constellation launched by SpaceQuest Ltd. (Berkman et al., 2020b); [b]From AIS data file; [c]Selected AIS metadata attributes from among those available (NAVCEN, 2019); [d]Mobile Maritime Service Identity (MMSI) as the unique ship identifier, which is redacted with the Arctic Ship Traffic Database (ASTD) that anonymizes records with access Levels 2 and 3 (PAME, 2020b); [e]Ship names (which may change) were noted, but MMSI (which remains with each ship) was used to identify unique ships; [f]International Maritime Organization (IMO) registered ships with Class-A transponders were used to validate the AIS record; [g]Nation (which may change) at time of each CAO visit; [h]Designation of ship type directly from the AIS data file (Marine Traffic, 2018), recognizing there is a different IMO schema of ship types (IHS Markit, 2017); [i]tonnage size-classes; [j]During period; [k]Longitudinal positions in the CAO High Seas

Polarstern (Berkman et al., 2020a), with its epic MOSAiC (*Multidisciplinary drifting Observatory for the Study of Arctic Climate*) expedition in the CAO High Seas during September 2019 to September 2020 (MOSAiC, 2020).

In addition to patterns of vessel flag states over the CAO High Seas (Fig. 24.2) – across ice-covered and open-water areas with different extents annually (NSIDC, 2020) – the number of nations operating in this international space has been trending upward (Fig. 24.3). Further elaboration of the 30 flag states among the 185 vessels in the CAO High Seas from 2009–2018 are shown in Fig. 24.4, raising questions about the relative number of ships from Arctic and non-Arctic States.

24.2.3 *Socioeconomic Trends and Characteristics in the CAO High Seas*

The diverse international presence of ships (Figs. 24.3 and 24.4) underlies investments with institutions that enabled their operation in the CAO High Seas. There also are associated questions about risk-management that accompany the decisionmaking. With additional granularity for decisionmaking about built infrastructure (Berkman et al., 2020c; Berkman, 2020a, b), it is clear the number of ship types (Table 24.3) also has been increasing annually in the CAO High Seas (Fig. 24.5), noting a jump in 2014 among the two dozen vessel types recorded from 2009 to 2018 (Fig. 24.6). Independent ASTD analyses (Jon Arve Røyset personal communication October 2020) indicate that many of the unspecified ships are research vessels of different types. The importance of consistent

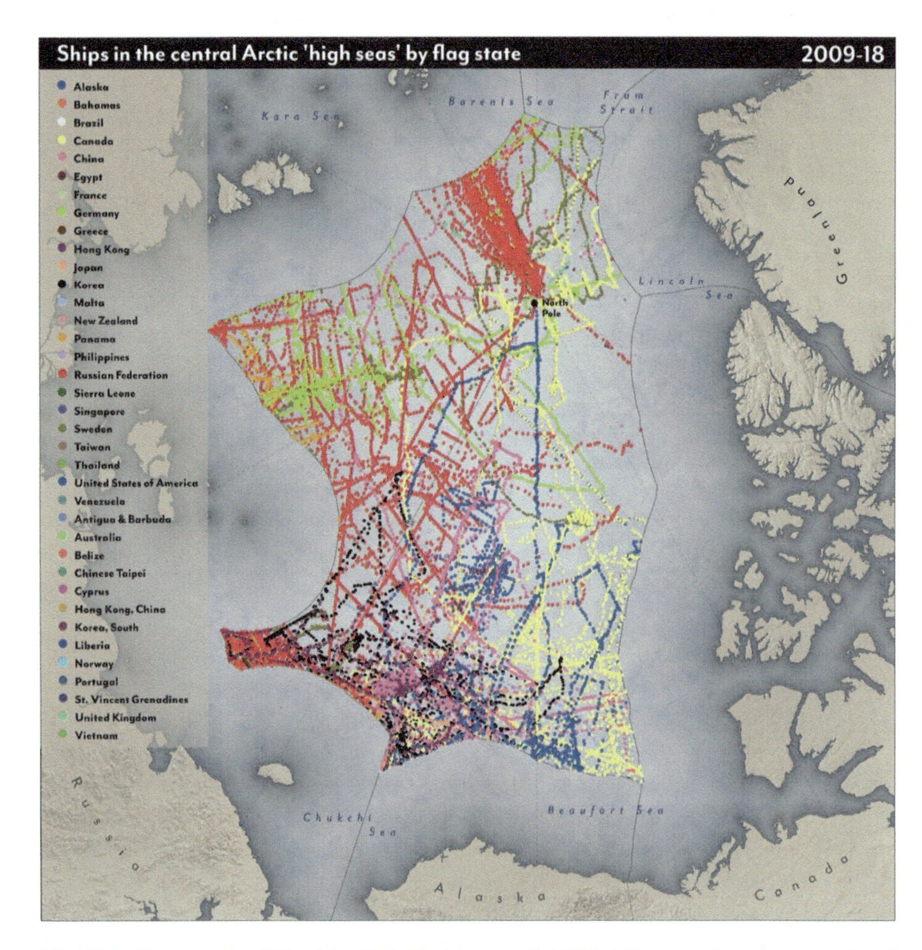

Fig. 24.2 Community of Maritime Ship Traffic in the CAO High Seas based on the composite of vessel flag states (Table 24.3) with distinct ship tracks from 1 September 2009 to 31 December 2018 (see legend). These data have been cross-checked with the Arctic Ship Traffic Database (ASTD) to confirm, for example, that Norwegian flagged vessels were absent in the CAO High Seas until 2019. See Fig. 24.1 for additional East-West orientation around Arctic Ocean longitudes

international strategies with ship-type designations, which is recognized to be a complex challenge (IHS Markit, 2017), are herein highlighted for regional and inter-annual comparisons that contribute to informed decisionmaking.

The socioeconomic dimensions, capacities and dynamics in the CAO High Seas (as elsewhere across the Arctic Ocean) are reflected by ship characteristics (Figs. 24.5 and 24.6) and their national relationships (Figs. 24.3 and 24.4), noting there are "flags of convenience" that complicate any assessments attributed to national activities. It is further noted that additional financial, geopolitical and logistic analyses will be required to produce rigorous socioeconomic interpretations with next-generation Arctic marine shipping assessments (Table 24.1), as interpreted in view of opening of the Transpolar Sea Route (Bennett et al., 2020).

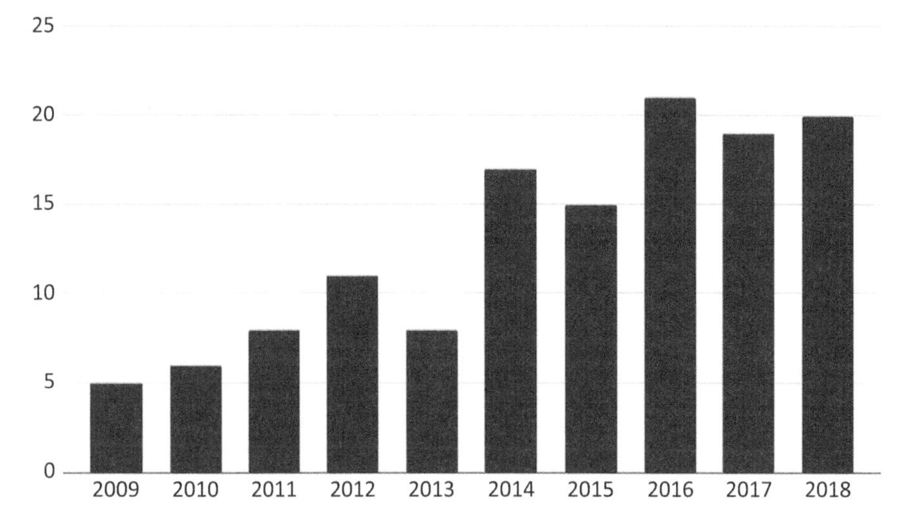

Fig. 24.3 International Presence and Dynamics of the maritime ship traffic community in the CAO High Seas (Fig. 24.2) based on the number of flag states among the 185 vessels (Table 24.2) annually from 1 September 2009 to 31 December 2018. These data have been cross-checked and are in close agreement with independent data collected for the Arctic Ship Traffic Database (ASTD). The Y-axis is the number of ships

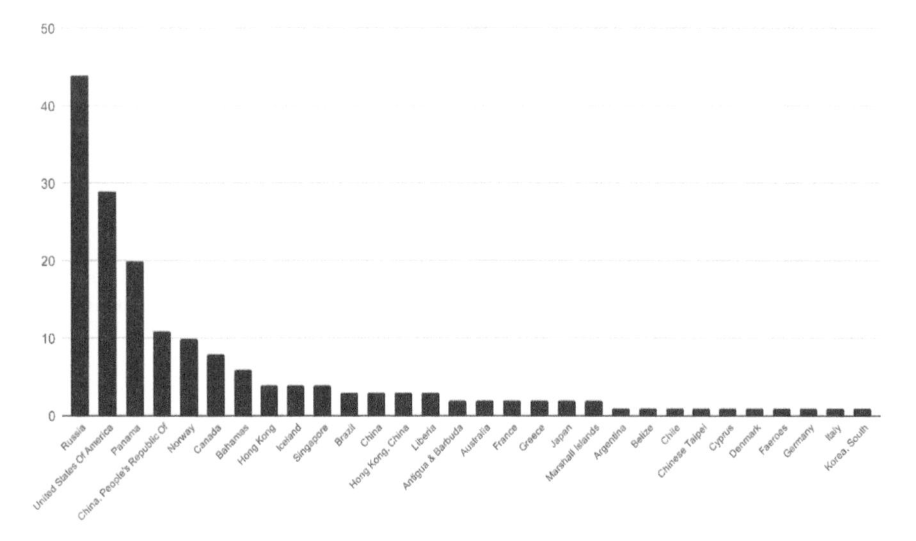

Fig. 24.4 International Characteristics of the maritime ship traffic community in the CAO High Seas (Fig. 24.2) based on the diversity of flag states (Table 24.2) among the 185 vessels across the period from 1 September 2009 to 31 December 2018. The Y-axis is the number of ships

A fundamental ship type for the Arctic Ocean is the icebreaker with its various classes, involving an international fleet size of 94 vessels in 2017 (USCG, 2017), indicating about a third of the world icebreaker fleet was operating in the CAO High

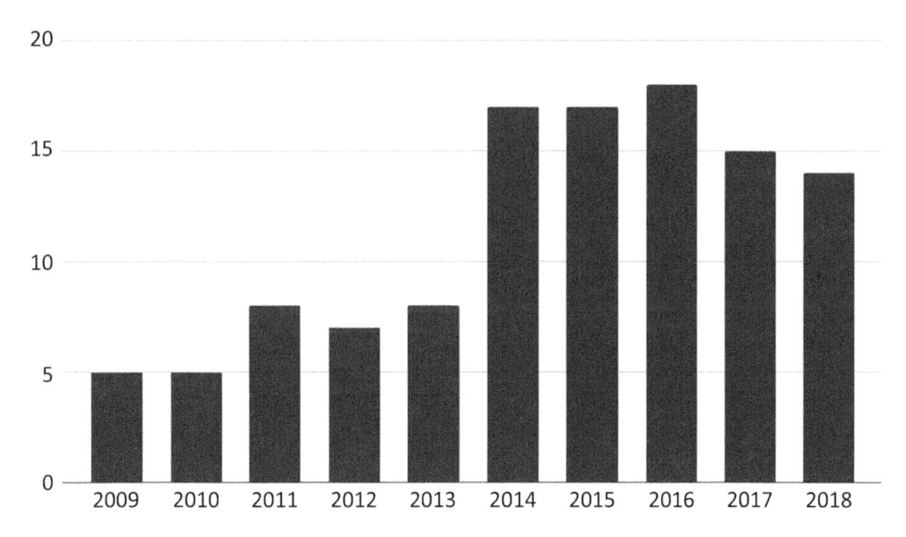

Fig. 24.5 Socioeconomic Trends of the maritime ship traffic community in the CAO High Seas (Fig. 24.2) based on the number of ship types (Table 24.3) annually from 1 September 2009 to 31 December 2018. The Y-axis is the number of ships

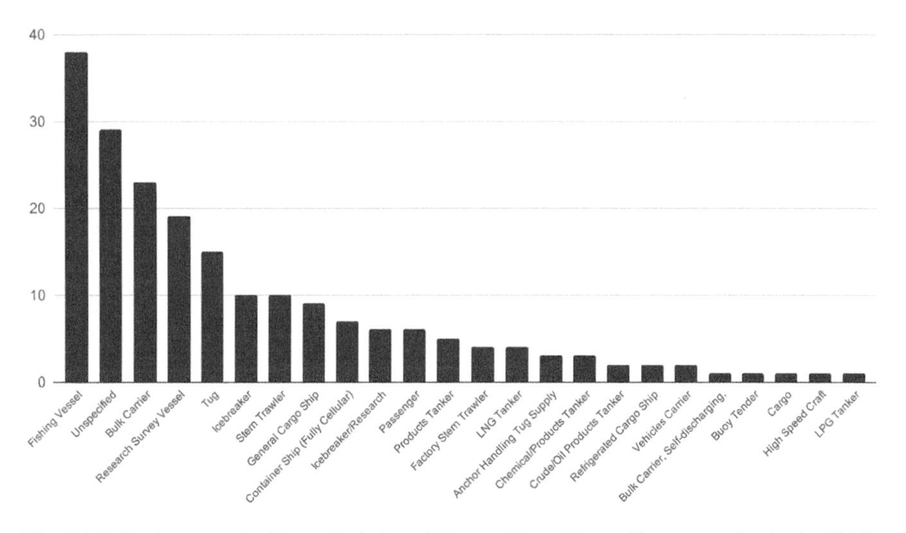

Fig. 24.6 Socioeconomic Characteristics of the maritime ship traffic community in the CAO High Seas (Fig. 24.2) based on the diversity of ship types (Table 24.3) among the 185 vessels across the period from 1 September 2009 to 31 December 2018. The Y-axis is the number of ships

Seas during the observation period (Fig. 24.6). Distinct from ice-strengthened vessels, icebreakers are designed for operations that include escorts, search-and-rescue and other emergency responses as well as maritime domain awareness. As the most seaworthy vessels for the Arctic Ocean, can this international icebreaker fleet be better coordinated to implement the emergency-response agreements in force with all of the Arctic states in the Arctic Ocean? Specifically, this question applies to the 2011 *Agreement on Cooperation on Aeronautical and Maritime Search and Rescue in the Arctic* (Arctic SAR Agreement, 2011) and 2013 *Agreement on Cooperation on Marine Oil Pollution Preparedness and Response in the Arctic* (Arctic MOPP Agreement, 2013)? Addressing this question also is an example of where data can be integrated into evidence (Eq. 24.1) for decisions in view of relevant institutions in the CAO High Seas as well as elsewhere in the Arctic Ocean.

Satellite AIS data facilitate holistic integration with diverse user-defined questions to transform data into evidence, stimulating research and action that contribute to informed decisionmaking. While it is beyond the scope of this chapter, synoptic analyses based on the characteristics of the vessels and their movements could contribute to informed decisions with next-generation Arctic marine shipping assessment (Table 24.1), identifying questions of common concern to address: black-carbon production; ship strikes on marine mammals; noise pollution; introduction of invasive species; or the effectiveness of existing international agreements generally. Importantly, framing such questions with holistic integration would contribute to common-interest building in the Arctic Ocean, moving beyond self-interests that commonly limit progress with decisionmaking.

In this regard, the CAO High Seas offers a potent case study, as reflected by ambassadorial dialogues on *Building Common Interests in the Arctic Ocean* with the ambassadors of six then twelve nations in 2015 and 2016, respectively (Ambassadorial Panel, 2015, 2016; Pan-Arctic Options Project, 2016). These inclusive dialogues serve as stimulus for this second volume in the Informed Decisionmaking for Sustainability book series, enabled by questions to build common interests beyond the *"concern about a unilateral declaration of five states regarding prevention of unregulated commercial fishing in the Central Arctic Ocean"* (Alfreðsdóttir, 2016). The lesson is that questions of common concern build common interests among allies and adversaries without being prescriptive to enable progress with sustainable development (United Nations, 1987, 2015), which is a "common" Arctic issue (Ottawa Declaration, 1996).

24.3 CAO Ship Traffic Coupling with Sea Ice

24.3.1 Ship-Ice Patterns and Trends in the CAO High Seas

Satellite sea-ice data from the National Snow and Ice Data Center (NSIDC, 2020), covering the same region and period as the satellite AIS data in the CAO High Seas (Fig. 24.2), were integrated into the space-time cube (see above) to analyze ship-ice interactions (Berkman et al., 2020a). These ship-ice interactions represent ship occurrences within 4 km^2 bins that contain ice, quantified on a daily basis. Complementing overall trends with maritime ship traffic north of the Arctic Circle from 2009 to 2016 (Berkman et al., 2020a), ship-ice interactions during this same period increased toward higher latitudes just in the CAO High Seas (Fig. 24.7).

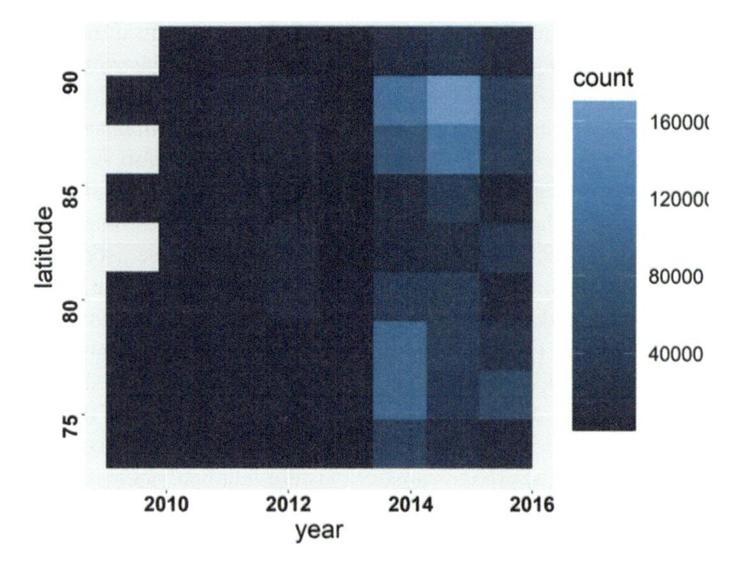

Fig. 24.7 Ship-ice interactions associated with the dynamics of the maritime ship traffic community in the CAO High Seas (Fig. 24.2) assessed within 4 km^2 grids daily from 1 September 2009 to 31 December 2016 based on satellite sea-ice and ship-traffic data, as analyzed previously for the entire maritime region north of the Arctic Circle (Berkman et al., 2020a)

These satellite sea-ice and ship-traffic data further reflect a jump in 2014 (Fig. 24.5) with the coupled biophysical and socioeconomic dynamics of the CAO High Seas system. The three-dimensional pattern of ship-ice interactions in the CAO High Seas (Fig. 24.7) is a space-time representation of the flag-state track lines shown above (Fig. 24.2).

24.3.2 Testing the 'Ship-Ice Hypothesis' in the CAO High Seas

A central contribution of this chapter is applying the CAO High Seas as a regional test of the 'ship-ice hypothesis' that Arctic ship traffic is increasing as sea-ice is diminishing (Berkman et al., 2020a). Without falsifying the hypothesis, assessment in the CAO High Seas (Fig. 24.7) suggests a trend of increasing ship traffic toward higher latitudes in the Arctic Ocean over time, as has been predicted (Smith & Stephenson, 2013; Norwegian Environment, Agency, 2014; Stephenson & Smith, 2015; Stephenson et al., 2018).

However, with the CAO High Seas, the East-West directionality of maritime ship traffic also can be assessed within longitudinal sectors in a circumpolar context surrounding the North Pole. More specifically, the CAO High Seas offers a unique regional test of the ship-ice hypothesis because diminished sea-ice and open-water predominate only in the Beaufort Sea and Chukchi Sea sectors (Thompson et al., 2016; Armitage et al., 2020), adjacent to the 180°East-West meridian. Consequently, a corollary of the 'ship-ice hypothesis' is that maritime ship traffic (i.e., socioeconomic activity) in the CAO High Seas will predominate from the Pacific Ocean rather than from the Atlantic Ocean sectors, even though vessels north of the Arctic Circle predominate in the EEZ connected to the North Atlantic (Fig. 24.1, Table 24.2).

Test of the 'ship-ice hypothesis' is characterized by vessel numbers and diversities within adjacent polygons to reveal 30° sectoral trends during the 2009–2016 period. Within the area of the CAO High Seas, international presence predominates in the Pacific Arctic sectors (Fig. 24.8), centering along the 180° East-West meridian, adjacent to the Bering Strait. This maritime-traffic directionality literally is 180° offset from the majority of shipping north of the Arctic Circle, which is in the Barents Sea (Fig. 24.1), where there is open water, as noted above in view of the *Polar Code* implementation. Concentrated international maritime ship traffic in the Pacific Arctic sectors of the CAO High Seas also is independent of national origin.

The Bering Strait is particularly important as the choke point of maritime ship traffic into and out of the Arctic Ocean (Rothwell, 2017), where the north-south transit gap is only 47 kilometers wide at its narrowest point in the Pacific Arctic

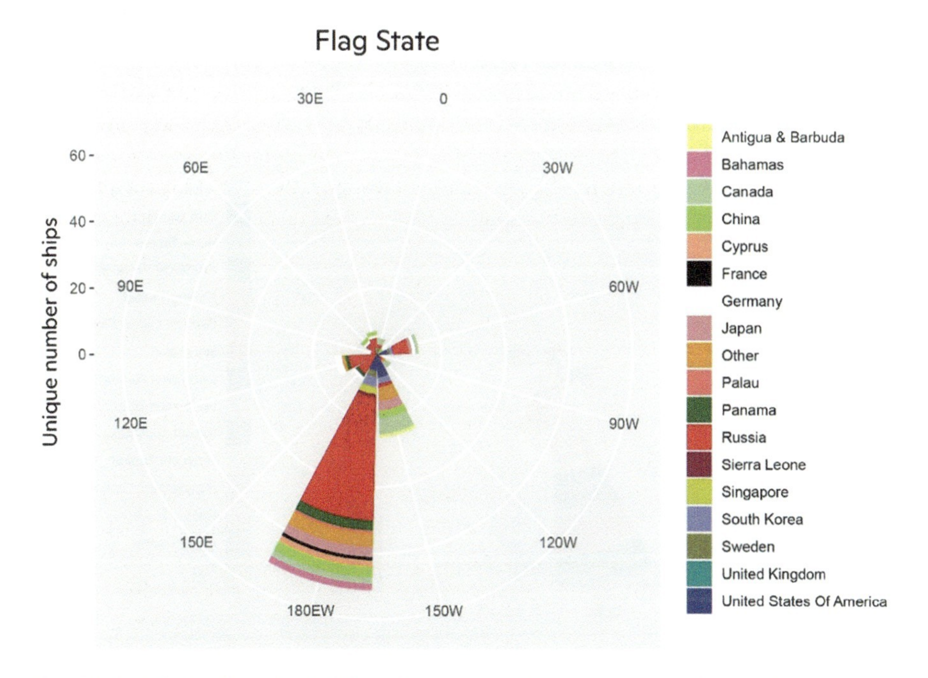

Fig. 24.8 'Ship-Ice Hypothesis' Test (Flag States) with maritime ship traffic populations in the CAO High Seas (Fig. 24.2) based on the distribution of ship flag states (Figs. 24.3 and 24.4) from MMSI records (Table 24.3) across 30° meridional sectors surrounding the North Pole. See Fig. 24.1 for East-West orientation around Arctic Ocean longitudes with 0°East-West in the Barents Sea to 180°East-West through the Bering Strait

sectors along the 180° East-West meridian (WWF, 2020). Along this maritime boundary region with the Russian Federation and United States (Berkman et al., 2016; Young et al., 2020), "two-way routes" and "precautionary areas" have been established for ship traffic (IMO, 2017b). Implications of maritime ship traffic dominating in the Pacific Arctic sectors of the CAO High Seas (Fig. 24.8) also relates to implementation of the "precautionary approach" (Pan & Huntington, 2016; Harrison et al., 2020) intended with the CAO High Seas Fisheries Agreement:

> *"precautionary conservation and management measures as part of a long-term strategy to safeguard healthy marine ecosystems and to ensure the conservation and sustainable use of fish stocks."*

Transforming these data into evidence (Eq. 24.1) relates to the CAO High Seas Fisheries Agreement (2018) as well as ship-traffic governance mechanisms and built infrastructure that are being considered specifically for the Bering Strait Region (CMTS, 2019).

As shown in a circumpolar context (Fig. 24.9), icebreaker movements exist across all sectors of the CAO High Seas, as would be expected because they are designed to move in ice-covered areas. Conversely, less ice-worthy vessels would be expected to be more restricted in their movements, where sea ice is diminished, which is the case in the CAO High Seas sectors in the vicinity of the Beaufort and

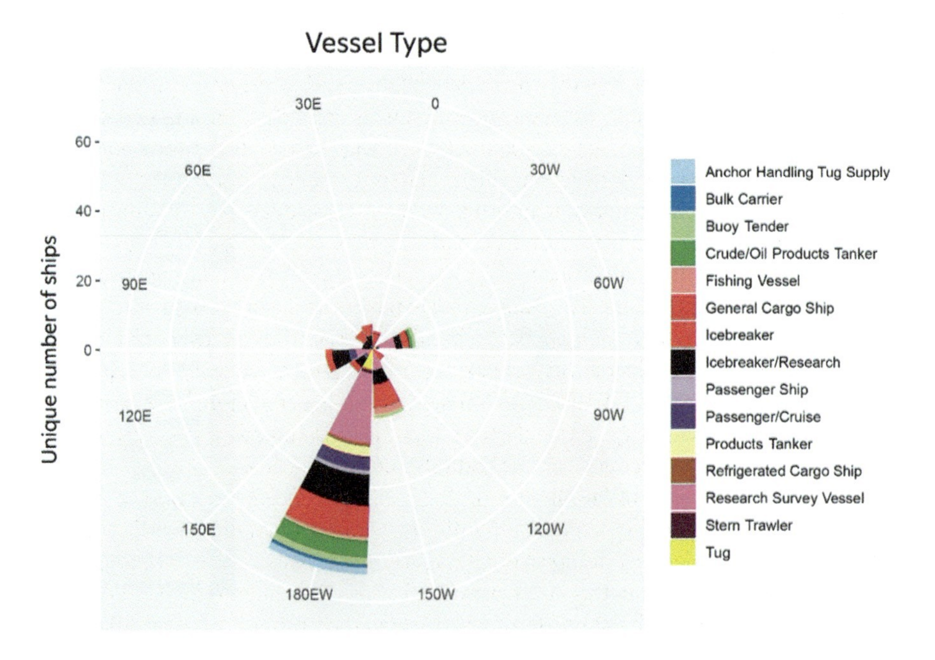

Fig. 24.9 'Ship-ice hypothesis' Test (Ship Types) with maritime ship traffic populations in the CAO High Seas (Fig. 24.2) based on the distribution of ship types (Figs. 24.5 and 24.6) from MMSI records (Table 24.3) across 30° meridional sectors surrounding the North Pole. See Fig. 24.1 for East-West orientation around Arctic Ocean longitudes with 0°East-West in the Barents Sea to 180°East-West through the Bering Strait

Chukchi Seas (Fig. 24.9), supporting the 'ship-ice hypothesis' and its corollary above. Moreover, with commercial considerations of harvesting living resources in the CAO High Seas, it also would be expected that fishing vessels may be present in the open water areas, even for exploratory purposes as shown. Ship sizes additionally reveal directionality with small tonnage ships only appearing in the Beaufort Sea region of the CAO High Seas (Fig. 24.10).

Together, ship densities and diversities among meridional sectors (based on the characteristics of the maritime ship traffic) increase with diminishing sea ice in the CAO High Seas surrounding the North Pole (Figs. 24.8, 24.9 and 24.10). As a practical outcome, testing the 'ship-ice hypothesis' connects the socioeconomic and biophysical systems of the Arctic Ocean. With such integration, next-generation Arctic marine shipping assessments (Table 24.1) will continue to reinforce the application of a *"precautionary approach"* to produce informed decisions across a

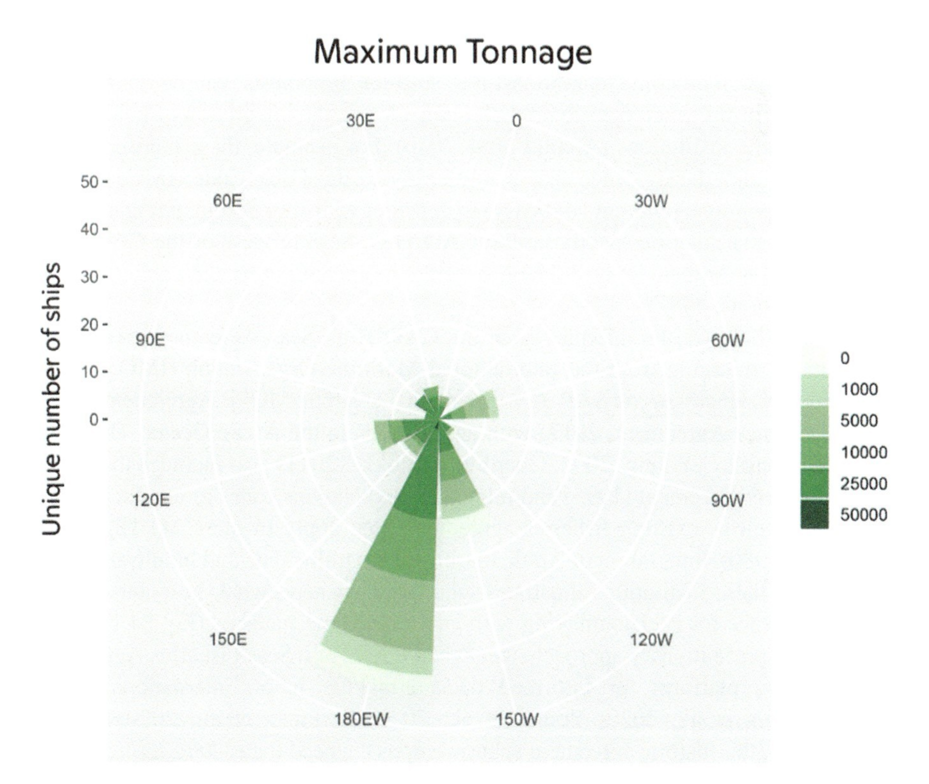

Fig. 24.10 'Ship-Ice Hypothesis' Test (Size-Classes) with maritime ship traffic populations in the CAO High Seas (Fig. 24.2) based on the distribution of ship size-classes from MMSI records (Table 24.3) across 30° meridional sectors surrounding the North Pole. See Fig. 24.1 for East-West orientation around Arctic Ocean longitudes with 0°East-West in the Barents Sea to 180°East-West through the Bering Strait

'continuum of urgencies' with common-interest building (see Chap. 1 in this book) in the CAO High Seas surrounding the North Pole as a "Pole of Peace" (Gorbachev, 1987; Berkman, 2009b, 2012).

24.4 Global Inclusion in the CAO High Seas

24.4.1 Informed Decisionmaking in the CAO High Seas

Understanding the system dynamics of species applies to marine living resources as well as ships. In this ecological context (Crowder & Norse, 2008), ships are analogous to individual fish, which have populations of the same species, involving diverse interactions within communities and ecosystems. Such ship species' interactions are represented, in part, by their feedback and intended interplay with governance mechanisms.

As a research outcome, data to test the 'ship-ice hypothesis' can be transformed into action for informed decisionmaking, considering the integration of evidence in view of Arctic institutions (Arctic Portal, 2020). For example, these maritime ship-traffic data underlie evidence that would apply to the Polar Code (IMO, 2017a), introducing options (without advocacy) to consider with ship design, navigation and monitoring that may be specific to the CAO High Seas in view of the CAO High Seas Fisheries Agreement or the *United Nations Convention on the Law of the Sea* (United Nations, 1982).

As noted above, the interplay with the CAO High Seas Agreement extends to institutions emerging from the International Maritime Organization (IMO, 2017a) and Arctic Council (Arctic SAR Agreement, 2011; Arctic MOPP Agreement, 2013; Arctic Science Agreement, 2017) with applications to the Arctic Ocean. The institutional interplay (Young, 2002, Oberthür & Stokke, 2011) also includes the Straddling Stocks Agreement (1995) and related United Nations codes of conduct (FAO, 1995) as well as existing fisheries agreements that apply to the CAO High Seas (NEAFC, 1980). Integration of Arctic maritime ship traffic data and biophysical data in view of these institutions illustrates who, when, where, what, how and why to create evidence for decisionmaking with governance mechanisms (Eq. 24.1).

With its precautionary approach, the signed CAO High Seas Fisheries Agreement represents a platform for informed decisionmaking in an international space (Vylegzhanin et al., 2020; Young et al., 2020; Berkman et al., 2020a). More specifically, this historic agreement acknowledges the need for a *"long-term strategy to safeguard healthy marine ecosystems,"* addressing *"long-term conservation and sustainable use of living marine resources and in healthy marine ecosystems in the Arctic Ocean."*

Informed decisionmaking in the CAO High Seas involves science broadly as the 'study of change' with biophysical and socioeconomic dynamics interpreted with natural and social sciences as well as Indigenous knowledge, as stated in the CAO High Seas Fisheries Agreement, desiring *"to promote the use of both scientific knowledge and indigenous and local knowledge."* Key natural and social science organizations are involved in the CAO High Seas Fisheries Agreement, as had been suggested (Van Pelt et al., 2017), appreciating Indigenous knowledge is being included (Schatz, 2019). Importantly, since 2016, the International Council for the Exploration of the Sea (ICES) and North Pacific Marine Science Organization (PICES) along with PAME have been coordinating the Working Group on Integrated Ecosystem Assessment for the Central Arctic Ocean (WGICA). The ICES/PICES/PAME efforts have been generating continuous progress to interpret the rapidly changing biophysical dynamics of the CAO system (WGICA, 2016, 2017, 2018, 2019, 2020). Implications of the 'precautionary approach' with the CAO High Seas are global, especially with precedents that will contribute to sustainable management of biodiversity beyond national jurisdictions (BBNJ, 2019; De Santo et al., 2019). With the CAO High Seas Fisheries Agreement and related institutions, the *"precautionary"* approach or principle (see Chap. 1 in this book) with short-to-long term consideration exemplifies informed decisionmaking under international law.

24.4.2 Common-Interest Building in the CAO High Seas

The *Convention on the High Seas* (1958) established the first international space ever on a planetary scale, promoting peace after the second world war (Berkman, 2009a). Emerging from cooperation among allies and adversaries alike at the height of the cold war – the *Convention on the High Seas* now is awakening lessons from the CAO High Seas that have relevance for humanity, which still is in its infancy as a globally-interconnected civilization (Berkman, 2020a,b), learning to balance national interests and common interests at local-global levels across the spectrum of subnational-national-international jurisdictions (Berkman 2019).

Lessons include socioeconomic dynamics, which can be revealed across the entire Arctic Ocean in relation to maritime ship traffic with objectivity and synoptic scope (Tables 24.1, 24.2, and 24.3; Fig. 24.1), enabling cooperation, coordination and consistency. As an option (without advocacy), next-generation Arctic marine shipping assessments (Table 24.1) can be treated as a fundamental indicator of socioeconomic dynamics in the Pan-Arctic maritime ecosystem, as illustrated with CAO High Seas (Figs. 24.2, 24.3, 24.4, 24.5, 24.6, 24.7, 24.8, 24.9, and 24.10). With the CAO High Seas Fisheries Agreement, these socioeconomic data will help to implement a *Joint Program of Scientific Research and Monitoring* to address questions short-to-long term (Balton & Zagorski, 2020), complementing Pan-Arctic research that is underway with the Synoptic Arctic Survey to understand the biophysical system *"beyond the scope of any single nation"* (Anderson et al., 2018).

At the top of the Earth, the CAO High Seas is unambiguously an area beyond national jurisdictions under the international framework of the Law of the Sea. Building on the initiative of the five surrounding Arctic coastal states (Ilulissat Declaration, 2008), the eight Arctic states and six Indigenous peoples organizations

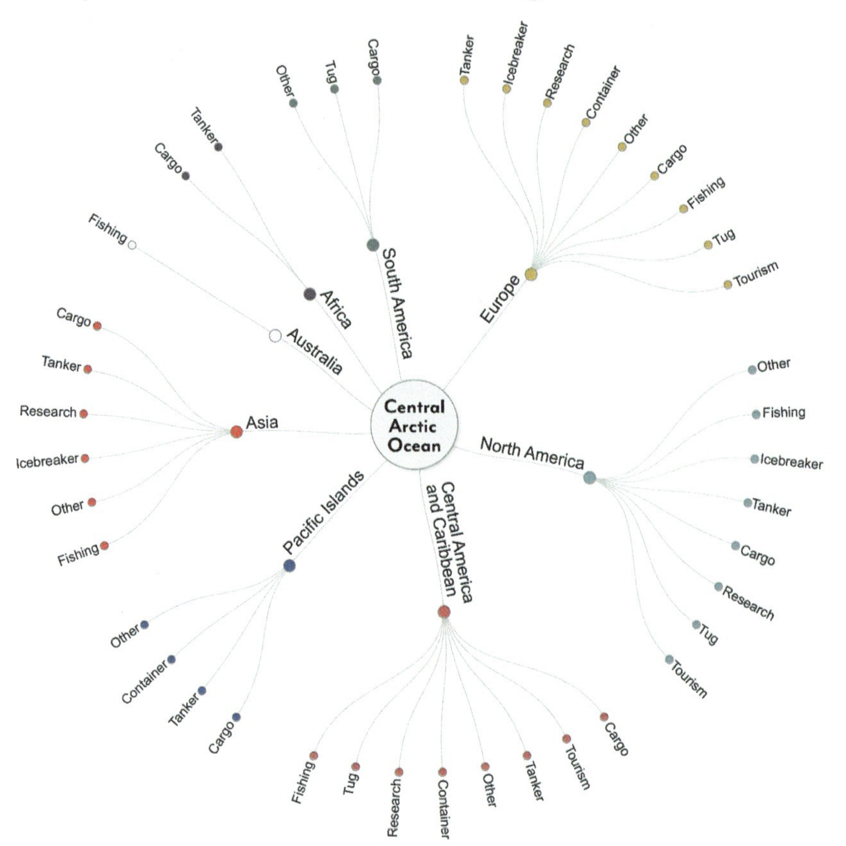

Fig. 24.11 **Complex of Attributes** (Table 24.3) with the maritime ship traffic community in the CAO High Seas (Fig. 24.2), as an area beyond national jurisdictions (ABNJ), reflecting global inclusion based on ship types flagged from all continental regions on Earth from 2009 through 2018 (Figs. 24.2, 24.3, 24.4, 24.5, 24.6, 24.7, 24.8, 24.9, and 24.10)

together *"remain committed"* to this international legal framework (Arctic Council, 2013). The product of their leadership is global inclusion in the CAO High Seas (Fig. 24.11), where the world has shared rights and responsibilities.

With science diplomacy as an holistic process involving the skills, methods and theory of informed decisionmaking (see Chap. 1 in this book), there is a local-global opportunity to frame questions that build common interests in the CAO High Seas, recognizing the starting point determines the journey of cooperation or conflict. As an option (without advocacy), the journey of humanity in the CAO High Seas can be characterized as an 'Index of Global Inclusion' with hope and inspiration for the benefit of all on Earth across generations.

Acknowledgements This article is a product of the Science Diplomacy Center, coordinated through EvREsearch LTD (previously through The Fletcher School of Law and Diplomacy at Tufts University), with support from the United States National Science Foundation (Award Nos. NSF-OPP 1263819, NSF-ICER 1660449, NSF-OPP 1917434 and NSF-ICER 2103490). We gratefully acknowledge the thoughtful reviews by Lawson W. Brigham and by Jon-Arve Røyset, who provided independent data cross-checking with the Arctic Ship Traffic Database (ASTD) coordinated by the Protection of the Arctic Marine Environment (PAME) Working Group of the Arctic Council.

References

Alfreðsdóttir, L. (2016). *Second Annual Ambassadorial Panel on the Arctic High Seas, Pan Arctic Options and the Reykjavik University, Reykjavik 6 October 2016.* Opening Address. Minister for Foreign Affairs of Iceland. https://www.stjornarradid.is/library/04-Raduneytin/Utanrikisraduneytid/PDF-skjol/Arctic-pre-event%2D%2D-raeda-ra%CC%81dherra.pdf.

Ambassadorial Panel. (2015). *Building common interests in the Arctic Ocean.* University of Reykjavik, Reykjavik. https://en.ru.is/news/arctic-high-seas-oct15.

Ambassadorial Panel. (2016). *Building common interests in the Arctic Ocean.* University of Reykjavik, Reykjavik. https://en.ru.is/news/building-common-interests-in-the-arctic-ocean-1.

AMSA. (2009). *Arctic Marine Shipping Assessment report.* Protection of the Arctic Marine Environment (PAME) Working Group of the Arctic Council, Akureyri. https://pame.is/index.php/projects/arctic-marine-shipping/amsa.

Anderson, L.G., Ashjian, C., Azetsu-Scott, K., Bates, N.R., Carmack, E., Chierici, M., Cho, K.H., Deming, J., Edelvang, K., Gerland, S., Grebmeier, J., Hölemann, J., Itoh, M., Ivanov, V., Kang, S.H., Kassens, H., Kikuchi, T., Lien, V., Mathis, J., Novikhin, A., Olsen, A., Paasche, Ø., Schlosser, P., Swift, J., Stedmon, C., Sørensen, L.L., Titov, O., Tyrrell, T., Wilkinson, J., & Willams, B. (2018). *Synoptic Arctic Survey – A Pan-Arctic research program.* Science and Implementation Plan. 29 June 2018. https://synopticarcticsurvey.w.uib.no/science-plan/.

Arctic Council. (2013). *Vision for the Arctic.* Arctic Council Secretariat, Kiruna, Sweden. 15 May 2013.

Arctic Council. (2020. Exploring the Arctic Ocean: The Agreement that Protects an Unknown Ecosystem. *Arctic Council News.* https://arctic-council.org/en/news/exploring-the-arctic-ocean-the-agreement-that-protects-an-unknown-ecosystem/.

Arctic MOPP Agreement. (2013). *Agreement on Cooperation on Marine Oil Pollution Preparedness and Response in the Arctic.* Signed: Kiruna, 15 May 2013 by the 8 Arctic States; Entry into Force: 25 March 2016. https://oaarchive.arctic-council.org/handle/11374/529.

Arctic Portal. (2020). *Arctic Policies Database.* https://arcticportal.org/arctic-governance/arctic-policies-database.

Arctic SAR Agreement. (2011). *Agreement on Cooperation on Aeronautical and Maritime Search and Rescue in the Arctic* Signed: Nuuk, 12 May 2011 by the 8 Arctic States; Entry into Force: 19 January 2013. https://oaarchive.arctic-council.org/handle/11374/531.

Arctic Science Agreement. (2017). *Agreement on Enhancing International Arctic Scientific Cooperation.* Signed: Fairbanks, 11 May 2017 by the 8 Arctic States; Entry into Force: 23 May 2018. https://oaarchive.arctic-council.org/handle/11374/1916.

Armitage, T. W. K., Manucharyan, G. E., Petty, A. A., Kwok, R., & Thompson, A. F. (2020). Enhanced Eddy activity in the Beaufort Gyre in response to sea ice loss. *Nature Communications, 11,* 761. https://www.nature.com/articles/s41467-020-14449-z

Ashjian, C. et al. (2019). *Synoptic Arctic Survey.* Report of the Open Planning Workshop, May 15–16, 2019. Woods Hole Oceanographic Institution, Woods Hole, Massachusetts, USA, 37 pp. https://web.whoi.edu/sas2019/wp-content/uploads/sites/130/2019/09/2019_SAS_Workshop_WoodsHole.pdf.

Balton, D. A., & Zagorski, A. (2020). *Implementing marine management in the Arctic Ocean.* Russian International Affairs Council, Moscow and Wilson Center, Washington, DC. https://www.wilsoncenter.org/publication/implementing-marine-management-arctic-ocean.

Barneo. (2020). *Barneo Ice Camp.* http://campbarneo.com/.

BBNJ. (2019). *Revised draft text of an agreement under the United Nations convention on the law of the sea on the conservation and sustainable use of marine biological diversity of areas beyond national jurisdiction.* United Nations General Assembly, New York. https://undocs.org/en/a/conf.232/2020/3.

Bennett, M. M., Stephenson, S. R., Yang, K., Bravo, M. T., & De Jonghe, B. (2020). The opening of the Transpolar sea route: Logistical, geopolitical, environmental, and socioeconomic impacts. *Marine Policy.* In press. https://doi.org/10.1016/j.marpol.2020.104178.

Berkman, P. A. (2009a). International spaces promote peace. *Nature, 462*, 412–413. https://www.nature.com/articles/462412a

Berkman, P. A. (2009b). North Pole as a pole of peace. *The Circle, 1*, 14–17. https://wwfeu.awsassets.panda.org/downloads/thecircle0109.pdf

Berkman, P. A. (2010). Integrated Arctic Ocean governance for the lasting benefit of all humanity. In R. Wolfrum (Ed.), *New challenges and new responsibilities in the Arctic Region* (pp. 187–194). Max Planck Institute.

Berkman, P. A. (2012). Our common future in the Arctic Ocean. *The Round Table, 101*(02), 123–135. https://www.tandfonline.com/doi/abs/10.1080/00358533.2012.661527

Berkman, P. A. (2014, June). Stability and peace in the Arctic Ocean through science diplomacy. *Science & Diplomacy, 2014*, 26–35. https://www.sciencediplomacy.org/perspective/2014/stability-and-peace-in-arctic-ocean-through-science-diplomacy.

Berkman, P.A. (2019). Evolution of Science Diplomacy and Its Local-Global Applications. Special issue, '*Broadening Soft Power in EU-US Relations*', European Foreign Affairs Review 24:63–79. (https://www.ingsa.org/wp-content/uploads/2019/09/Evolution-of-Science-Diplomacy-andits-Local-Global-Applications_23JUL19.pdf).

Berkman, P. A. (2020a). Science diplomacy and it engine of informed decisionmaking: Operating through our global pandemic with humanity. *The Hague Journal of Diplomacy, 15*, 435–450. https://brill.com/view/journals/hjd/15/3/article-p435_13.xml

Berkman, P. A. (2020b, November 13). *The pandemic lens*: Focusing across time scales for local-global sustainability. *Patterns, 1*(8), 4p. https://www.cell.com/action/showPdf?pii=S2666-3899%2820%2930195-1

Berkman, P. A., & Vylegzhanin, A. N. (2013). Conclusions: Building common interests in the Arctic Ocean. In P. A. Berkman & A. N. Vylegzhanin (Eds.), *Environmental security in the Arctic Ocean* (pp. 371–404). Springer.

Berkman, P. A., & Young, O. R. (2009). Governance and environmental change in the Arctic Ocean. *Science, 324*, 339–340. https://science.sciencemag.org/content/324/5925/339

Berkman, P. A., Vylegzhanin, A. N., & Young, O. R. (2016). Governing the Bering Strait Region: Current status, emerging issues and future options. *Ocean Development and International Law, 47*(2), 186–217. https://www.tandfonline.com/doi/full/10.1080/00908320.2016.1159091

Berkman, P. A., Fiske, G., Røyset, J. A., Brigham, L., & Lorenzini, D. (2020a). Chapter 11: Next-generation Arctic marine shipping assessments. In O. R. Young, P. A. Berkman, & A. N. Vylegzhanin (Eds.), *Informed decisionmaking for sustainability* (Volume 1. Governing Arctic seas: Regional lessons from the Bering Strait and Barents Sea) (pp. 241–268). Springer. https://link.springer.com/content/pdf/10.1007%2F978-3-030-25674-6_11.pdf

Berkman, P. A., Fiske, G., & Lorenzini, D. (2020b). *Baseline of next-generation Arctic Marine Shipping Assessments – Oldest continuous Pan-Arctic Satellite Automatic Identification System (AIS) data record of maritime ship traffic, 2009–2016.* Arctic Data Center. https://arcticdata.io/catalog/view/doi%3A10.18739%2FA2TD9N89Z.

Berkman, P. A., Young, O. R., & Vylegzhanin, A. N. (2020c). Book series preface: Informed decisionmaking for sustainability. In O. R. Young, P. A. Berkman, & A. N. Vylegzhanin (Eds.), *Informed decisionmaking for sustainability* (Volume 1. Governing Arctic seas: Regional lessons from the Bering Strait and Barents Sea) (pp. v–xxv). Springer. https://link.springer.com/content/pdf/bfm%3A978-3-030-25674-6%2F1.pdf

CAO High Seas Fisheries Agreement. (2018). *Agreement to prevent unregulated high seas fisheries in the Central Arctic Ocean*. Signed: Ilulissat, 3 October 2018. Entry into force: pending. https://www.mofa.go.jp/mofaj/files/000449233.pdf.

CMTS. (2019). *A ten-year projection of maritime activity in the U.S. Arctic Region, 2020–2030*. U.S. Committee on the Marine Transportation System. September 2019. https://www.cmts.gov/downloads/CMTS_2019_Arctic_Vessel_Projection_Report.pdf

Comiso, J. C., Frey, K., Stock, L. V., Gersten, R. A., & Mitchell, H. (2020). *Satellite visualization data for the Distributed Biological Observatory (DBO)*. https://earth.gsfc.nasa.gov/cryo/data/distributed-biological-observatory.

Convention on the High Seas. (1958). Signed: Geneva, 29 April 1958. Entry into Force: 30 September 1962. https://www.gc.noaa.gov/documents/8_1_1958_high_seas.pdf.

Crowder, L., & Norse, E. (2008). Essential ecological insights for marine ecosystem-based management and marine spatial planning. *Marine Policy, 32*, 772–778. https://mcbi.marine-conservation.org/publications/pub_pdfs/Norse_MarinePolicy_2008.pdf

De Santo, E. M., Áeirsdóttir, Á., Barros-Platiau, A., Biermann, F., Dryzek, J., Gonçalves, L. R., Kim, R. E., Mendenhall, E., Mitchell, R., Nyman, E., Scobie, M., Sun, K., Tiller, R., Webster, D. G., & Young, O. R. (2019). Protecting biodiversity in areas beyond national jurisdiction: An earth system governance perspective. *Earth System Governance, 2*, 1–7. https://www.sciencedirect.com/science/article/pii/S258981161930028X

Donnelly, C. A., Boyd, I., Campbell, P., Craig, C., Vallance, P., Walport, M., Whitty, C. J. M., Woods, E., & Wormald, C. (2018). Four principles for synthesizing evidence. *Nature, 558*, 361–364. https://www.ingsa.org/wp-content/uploads/2018/07/4-principles-Nature-June-2018.pdf

Eguíluz, V. M., Fernández-Gracia, J., Irigoien, X., & Duarte, C. M. (2016). A quantitative assessment of Arctic shipping in 2010–2014. *Nature, 6*(30682), 1–6. https://www.nature.com/articles/srep30682

Eisenmann, I., & Wettlaufer, J. S. (2009). Nonlinear threshold behavior during the loss of Arctic Sea ice. *Proceedings of the National Academy of Sciences of the United States of America, 106*, 28–32. https://www.pnas.org/content/pnas/106/1/28.full.pdf

European Commission. (2018). *EU and Arctic partners enter historic agreement to prevent unregulated fishing in high seas*. European Commission, Brussels. 3 October 2018. https://ec.europa.eu/fisheries/eu-and-arctic-partners-enter-historic-agreement-prevent-unregulated-fishing-high-seas_en.

FAO. (1995). *Code of conduct for responsible fisheries*. Food and Agriculture Organization of the United Nations, Rome. http://www.fao.org/3/v9878e/V9878E.pdf.

Google. (2020). *BigQuery costs*. https://cloud.google.com/bigquery/pricing.

Gorbachev, M. (1987). *Speech at the ceremonial meeting on the occasion of the presentation of the order of Lenin and the Gold Star to the City of Murmansk*. 1 October 1987 (English translation prepared by the Press Office of the Embassy of the Soviet Union, Ottawa 1988).

Harrison, P., Shin, H. C., Huntington, H. P., Balton, D., Benton, D., Min, P., Fujio, O., Peiqing, G., Grebmeier, J. M., Highleyman, S., Jakobsen, A., Meloche, M., Romanenko, O., & Zilanov, V. K. (2020). How non-government actors helped the Arctic fisheries agreement. *Polar Perspectives, 2*(October 2020) https://www.wilsoncenter.org/publication/polar-perspectives-no-2-how-non-government-actors-helped-arctic-fisheries-agreement

IHS Markit. 2017. *StatCode 5 Shiptype coding system. A categorisation of ships by type-cargo carrying ships*. https://cdn.ihs.com/www/pdf/Statcode-Shiptype-Coding-System.pdf.

Ilulissat Declaration. (2008). *Declaration from the Arctic Ocean conference*, 28 May 2008. Ilulissat. https://arcticportal.org/images/stories/pdf/Ilulissat-declaration.pdf.

IMO. (2017a). *International code for ships operating in polar waters (Polar Code).* Marine Environmental Protection Committee, MEPC 68/21/Add. 1, Annex 10. International Maritime Organization. Entry into Force 1 January 2017. http://www.imo.org/en/MediaCentre/HotTopics/polar/Pages/default.aspx.

IMO. (2017b). *Routeing measures and mandatory ship reporting systems. Establishment of two-way routes and precautionary areas in the Bering Sea and Bering Strait.* Submitted by the Russian Federation and the United States. Sub-Committee on Navigation, Communications and Search and Rescue. International Maritime Organization (NCSR 5/3/7, 17 November 2017). https://www.navcen.uscg.gov/pdf/imo/ncsr_5_3_7.pdf.

IMO. (2020a). *AIS Transponders.* International Maritime Organization. http://www.imo.org/en/OurWork/Safety/Navigation/Pages/AIS.aspx.

IMO. (2020b). *Global integrated shipping information system.* International Maritime Organization. https://gisis.imo.org/Public/Default.aspx.

Marine Traffic. (2018). *What is the significance of the AIS Shiptype number?* https://help.marinetraffic.com/hc/en-us/articles/205579997-What-is-the-significance-of-the-AIS-Shiptype-number-.

MOSAiC. (2020). *Multidisciplinary drifting observatory for the study of Arctic climate.* https://mosaic-expedition.org/.

MyShipTracking. (2020). https://www.myshiptracking.com/.

NASA Earth Observatory. (2018). *Shipping responds to Arctic Ice Decline.* National Aeronautics and Space Administration. https://earthobservatory.nasa.gov/images/91981/shipping-responds-to-arctic-ice-decline.

NAVCEN. (2016). Navigation Center: Automatic identification system. *How it Works.* United States Coast Guard. https://www.navcen.uscg.gov/?pageName=AISworks.

NAVCEN. (2019). Navigation Center: Automatic identification system. *AIS Messages.* United States Coast Guard. https://www.navcen.uscg.gov/?pageName=AISMessages.

NEAFC. (1980). *Convention on future multilateral cooperation in North-East Atlantic Fisheries.* Signed: London, 18 November 1980; Entry into Force. 17 March 1982. https://www.neafc.org/system/files/Text-of-NEAFC-Convention-04.pdf.

Norwegian Environment Agency. (2014). *Specially designated marine areas in the Arctic High Seas.* Report No./DNV Reg No.: 2013–1442 / 17JTM1D-26 Rev 2. https://www.pame.is/index.php/document-library/shipping-documents/arctic-marine-shipping-assessment-documents/recommendation-iid/343-specially-designated-marine-areas-in-the-arctic-part-ii/file.

NSIDC. (2020). *Arctic Sea ice news and analysis.* National Snow and Ice Data Center. http://nsidc.org/arcticseaicenews/.

Oberthür, S., & Stokke, O. S. (Eds.). (2011). *Managing regime complexity: Regime interplay and global environmental change.* MIT Press.

Ottawa Declaration. (1996). *Declaration on the establishment of the Arctic Council.* Signed 19 September 1996, Ottawa. https://oaarchive.arctic-council.org/handle/11374/85.

Paasche, Ø., Olsen, A., Årthun, M., Anderson, L.G., Wängberg, S. A., Ashjian, C. J., Grebmeier, J. M., Kikuchi, T., Nishino, S., Yasunaka, S., Kang, S. H., Cho, K. H., Azetsu-Scott, K., Williams, W. J., Carmack, E., Torres-Valdés, S., Tyrrell, T., Edelvang, K., He, J., & Kassens, H. M. (2019). Addressing Arctic challenges requires a synoptic ocean survey. *Transactions of the American Geophysical Union (EOS),* 19 November 2019. https://eos.org/opinions/addressing-arctic-challenges-requires-a-synoptic-ocean-survey.

PAME. (2020a). *Arctic Ships Status Report (ASSR) #1. The increase in Arctic shipping 2013–2019.* Protection of the Arctic Marine Environment Working Group, Akureyri. https://www.pame.is/projects/arctic-marine-shipping/arctic-shipping-status-reports/723-arctic-shipping-report-1-the-increase-in-arctic-shipping-2013-2019-pdf-version/file.

PAME. (2020b). *ASTD Data.* The Cooperative Agreement among the Arctic States Regarding Arctic Ship Traffic Data (ASTD) sharing. Protection of the Arctic Marine Environment Working Group, Akureyri. https://pame.is/images/03_Projects/ASTD/ASTD_Data_Document.pdf.

Pan, M., & Huntington, H. (2016). A precautionary approach to fisheries in the Central Arctic Ocean: Policy, science, and China. *Marine Policy, 63*, 153–157. https://doi.org/10.1016/j.marpol.2015.10.015

Pan-Arctic Options Project. (2016). *Arctic High Seas*. https://www.panarcticoptions.org/category/arctic-high-seas/.

Rothwell, D. R. (2017). *Arctic Ocean shipping: Navigation, security and sovereignty in the north*. Brill.

Schatz, V. (2019). The incorporation of indigenous and local knowledge into Central Arctic Ocean fisheries management. *Arctic Review Law and Politics, 10*, 130–134. https://arcticreview.no/index.php/arctic/article/view/1630/3173

Smith, L. C., & Stephenson, S. R. (2013). New trans-Arctic shipping routes navigable by midcentury. *Proceedings of the National Academy of Sciences, 110*(13), E1191–E1195. https://www.pnas.org/content/110/13/E1191

SOLAS. (2020). *Annex 17*. Automatic Identification System. Convention on the Safety of Life at Sea. http://solasv.mcga.gov.uk/Annexes/Annex17.htm.

Stephenson, S. R., & Smith, L. C. (2015). Influence of climate model variability on projected Arctic shipping futures. *Earth's Future, 3*, 331–343. https://doi.org/10.1002/2015EF000317

Stephenson, S. R., Wang, W., Zender, C. S., Wang, H., Davis, S. J., & Rasch, P. J. (2018). Climatic responses to future trans-Arctic shipping. *Geophysical Research Letters, 45*, 9898–9908. https://doi.org/10.1029/2018GL078969

Stevenson, T. C., Davies, J., Huntington, H. P., & Sheard, W. (2019). An examination of trans-Arctic vessel routing in the Central Arctic Ocean. *Marine Policy, 100*, 83–89. https://www.sciencedirect.com/science/article/pii/S0308597X18307334?via%3Dihub

Straddling Stocks Agreement. (1995). *Agreement for the implementation of the provisions of the United Nations convention on the law of the sea of 10 December 1982 relating to the conservation and management of straddling fish stocks and highly migratory fish stocks of 4 August 1995.*

Thoman, R. L., Richter-Menge, J., & Druckenmiller, M. L. (Eds.). (2020). *Arctic Report Card 2020*. United States National Oceanic and Atmospheric Administration, Rockville. https://arctic.noaa.gov/Portals/7/ArcticReportCard/Documents/ArcticReportCard_full_report2020.pdf.

Thompson, J., Fan, Y., Stammerjohn, S., Stopa, J., Rogers, W. E., Girard-Ardhuin, F., Ardhuin, F., Shen, H., Perrie, W., Shen, H., Ackley, S., Babanin, A., Liu, Q., Guest, P., Maksym, T., Wadhams, P., Fairall, C., Persson, O., Doble, M., ... Bidlot, J. (2016). Emerging trends in the Sea State of the Beaufort and Chukchi Seas. *Ocean Modelling, 105*, 1–12. https://doi.org/10.1016/j.ocemod.2016.02.009

United Nations. (1982). *United Nations convention on the law of the Sea*. (Signed: Montego Bay, 10 December 1982; Entry into Force: 16 November 1994). https://www.un.org/depts/los/convention_agreements/texts/unclos/unclos_e.pdf.

United Nations. (1987). *Our common future: From ONE EARTH TO ONE World*. Report Transmitted to the General Assembly as an Annex to Resolution A/RES/42/187. United Nations, World Commission on Environment and Development, New York. https://sustainabledevelopment.un.org/content/documents/5987our-common-future.pdf.

United Nations. (2015). *Transforming our world: The 2030 Agenda for sustainable development*. United Nations General Assembly, New York. https://sustainabledevelopment.un.org/post2015/transformingourworld/publication.

USCG. (2017). *Major icebreakers of the world*. Office of Waterways and Ocean Policy (CG-WWM). United States Coast Guard. https://www.dco.uscg.mil/Portals/9/DCO%20Documents/Office%20of%20Waterways%20and%20Ocean%20Policy/20170501%20major%20icebreaker%20chart.pdf?ver=2017-06-08-091723-907.

Van Pelt, T. I., Huntington, H. P., Romanenko, O. V., & Mueter, F. J. (2017). The missing middle: Central Arctic Ocean gaps in fishery research and science coordination. *Marine Policy, 85*, 79–86. https://doi.org/10.1016/j.marpol.2017.08.008

Visalli, M. E., Best, B. D., Cabral, R. B., Cheung, W. W. L., Clark, N. A., Garilao, C., Kaschner, K., Kesner-Reyes, K., Lam, V. K. Y., Maxwell, S. M., Mayorga, J., Moeller, H. V., Morgan, L., Crespo, G. O., Pinsky, M. L., White, T. D., & McCauley, D. L. (2020). Data-driven approach for highlighting priority areas for protection in marine areas beyond national jurisdiction. *Marine Policy*. in press. https://www.sciencedirect.com/science/article/pii/S0308597X19309194?via%3Dihub.

Vylegzhanin, A. N., Young, O. R., & Berkman, P. A. (2020). The Central Arctic Ocean fisheries agreement as an element in the evolving Arctic Ocean governance complex. *Marine Policy, 118*, 1–10. https://doi.org/10.1016/j.marpol.2020.104001

WGICA. (2016). *First interim report of the ICES/PAME Working Group on Integrated Ecosystem Assessment for the Central Arctic Ocean (WGICA)*. International Council for the Exploration of the Sea, Copenhagen. http://www.ices.dk/sites/pub/Publication%20Reports/Expert%20Group%20Report/SSGIEA/2016/WGICA/WGICA%202016.pdf.

WGICA. (2017). *Interim report of the ICES/PICES/PAME Working Group on Integrated Ecosystem Assessment (IEA) for the Central Arctic Ocean (WGICA)*. International Council for the Exploration of the Sea, Seattle. http://ices.dk/sites/pub/Publication%20Reports/Expert%20Group%20Report/SSGIEA/2017/WGICA/WGICA%202017.pdf.

WGICA. (2018). *Interim report of the ICES/PICES/PAME Working Group for Integrated Ecosystem Assessment of the Central Arctic Ocean (WGICA)*. International Council for the Exploration of the Sea, Newfoundland. https://www.ices.dk/sites/pub/Publication%20Reports/Expert%20Group%20Report/IEASG/2018/WGICA/WGICA%202018.pdf.

WGICA. (2019). *ICES/PICES/PAME working group on Integrated Ecosystem Assessment (IEA) for the Central Arctic Ocean* (Vol. 2, issue 33). http://ices.dk/sites/pub/Publication%20Reports/Expert%20Group%20Report/IEASG/2019/WGICA%20report%202019.pdf.

WGICA. (2020). *ICES/PICES/PAME Working Group on Integrated Ecosystem Assessment (IEA) for the Central Arctic Ocean (WGICA)* (Vol. 2, issue 79). http://www.ices.dk/sites/pub/Publication%20Reports/Forms/DispForm.aspx?ID=36908.

Winther, M., Christensen, J. H., Plejdrup, M. S., Ravn, E. S., Eriksson, O. F., & Kristensen, H. O. (2014). Emission inventories for ships in the arctic based on satellite sampled AIS data. *Atmospheric Environment, 91*, 1–14. https://doi.org/10.1016/j.atmosenv.2014.03.006

Wright, D., Janzen, C., Bochenek, R., Austin, J., & Page, J. (2019). Marine observing applications using AIS: Automatic Identification System. *Frontiers in Marine Science, 6*, 1–7. https://doi.org/10.3389/fmars.2019.00537

WWF. (2020). *Safety at the Helm: A plan for smart shipping through the Bering Strait*. WWF, Anchorage. https://c402277.ssl.cf1.rackcdn.com/publications/1314/files/original/WWF_Bering_Straits_Shipping_Report_UPDATE.pdf?1588034355.

Young, O. R. (2002). Institutional interplay: The environmental consequences of cross-scale interactions. In E. Ostrom (Ed.), *The drama of the commons*. National Academy Press.

Young, O. R., Berkman, P. A., & Vylegzhanin (Eds.). (2020). *Informed decisionmaking for sustainability* (Volume 1. Governing Arctic Seas: Regional lessons from the Bering Strait and Barents Sea). Springer. 358p. https://www.springer.com/gp/book/9783030256739

Chapter 25
(Research): Science for Management Advice in the Arctic Ocean: The International Council for the Exploration of the Sea (ICES)

Alf Håkon Hoel

Abstract The International Council for the Exploration of the Sea was established in 1902 and is one of the oldest marine science institutions in the world. It has aged well – today it provides scientific advice for the management of the marine environment and the natural resources there to governments and regional commissions for fisheries and environment in the Northeast Atlantic. It has 20 member nations and a network of 6000 scientists and 700 institutes as the foundation of its activities, spanning from basic marine science via data management to the provision of scientific advice on marine management. The purpose of this chapter is to provide an overview of the ICES organization and its functions, discuss its provision of scientific advice and thereby its role at the science-policy interface in the North Atlantic and the Arctic, including how this role is changing with the development of integrated, ecosystem based management of the oceans. The final part of the chapter addresses the current governance of Arctic marine science and its science – policy interfaces.

25.1 Introduction

The fate of the oceans and their governance is a major issue of our times. A number of international commissions, task forces, and expert groups have assessed the state of marine environments, identified problems, and proposed solutions (Independent Commission on the Oceans, 1998; Global Ocean Commission, 2016; High Level Panel, 2020). A critical issue running through these initiatives is how scientific knowledge can be effectively communicated to policymakers and put to use in marine management. Also, the UN General Assembly has proclaimed a Decade of

A. H. Hoel (✉)
UiT – The Arctic University of Norway, Tromsø, Norway
e-mail: alf.hakon.hoel@uit.no

P. A. Berkman et al. (eds.), *Building Common Interests in the Arctic Ocean with Global Inclusion, Volume 2*, Informed Decisionmaking for Sustainability,
https://doi.org/10.1007/978-3-030-89312-5_25

Ocean Science[1] during 2021–2030 to address the UN 2030 agenda and support the Sustainable Development Goals.

The Arctic Ocean can be loosely defined as the central Arctic ocean (CAO) above the continents and the marginal seas surrounding it (see map). This is a huge area – the Arctic above the Arctic Circle is some 20 million km^2. The CAO alone is more than seven million km^2 – an area almost three times the size of the Mediterranean - consisting of the waters of the five coastal states (USA, Russia, Norway, Denmark/Greenland, and Canada) as well as an area beyond national jurisdiction. The area covered by sea ice in the wider Arctic Ocean is about 15 million km^2 at its maximum in early spring, and less than five million km^2 in early fall.[2]

While there is little human activity in the CAO, the surrounding marginal seas (the Bering Sea, the Barents Sea, waters around Iceland and Greenland, Russia's and Canada's northern waters),[3] have substantial economic activities in fisheries (Hoel, 2018), shipping (Hildebrand et al., 2018), petroleum development (Baker, 2020), aquaculture, and others. Over time, the amount and diversity of human activity is also increasing (ACIA, 2005), a trend that is expected to continue with declining sea ice in the CAO and warming waters. All of these issues, including climate change, and their combined effects call for massive investments in science for societies to understand and adapt to these on-going changes. The science-policy interface is therefore of particular interest when discussing the future of Arctic marine stewardship.

The science-policy interface is an important aspect of environmental politics (Andresen et al., 2000; SAPEA, 2019), not least in the marine realm. In the case of ICES, the science-policy relationship has evolved over more than a 100 years (Holland & Pugh, 2010), and science is a critical factor in decisionmaking in international fora dealing with marine issues (Miles, 1987) as well as at the domestic level of governance (Sakshaug et al., 2009).

The International Council for the Exploration of the Sea – ICES – is one of the oldest and most important marine science organizations in the world, with 20 member countries and a network of some 6000 scientists and 700 marine science institutes.[4] It has a special role in the Arctic, as all Arctic coastal states are members, a large part of its work is related to Arctic and sub-arctic marine ecosystems, and its scientific advisory function is critical to marine governance in the Northeast Atlantic part of the Arctic in particular and increasingly also Arctic-wide.

The purpose of this chapter is first to describe the role and functions of ICES in marine science and its advisory functions, including an account of how the organization has evolved to address ecosystem-based and integrated oceans management.

[1] https://oceandecade.org/assets/The_Science_We_Need_For_The_Ocean_We_Want.pdf

[2] See the Sea Ice Index of the National Snow and Ice Data Center: https://nsidc.org/data/seaice_index/

[3] See the Arctic Ocean Review phase I report for a discussion of which ocean areas that constitute the Arctic Ocean.

[4] http://www.ices.dk/about-ICES/who-we-are/Pages/Who-we-are.aspx

We then proceed to discuss its role in Arctic marine management, drawing on the author's experience, conversations with colleagues and practitioners, and academic publications as well as grey literature.

25.2 ICES History and Organization

ICES was established in 1902, following international conferences in 1899 and 1901 on promotion of international cooperation in marine science (Nature, 1902). It has evolved considerably over the years (Rozwadowski, 2002),[5] becoming a formal international intergovernmental organization (IGO) with the adoption of the ICES convention in 1964.

The ICES convention sets out the purposes of ICES to promote and encourage research for the study of the marine systems, "particularly those relating to the living resources", to draw up programs for this purpose, and to disseminate the results of research.[6] It also defines its geographical scope to encompass the Atlantic Ocean and its adjacent seas, provides for relations with other organizations, and commits member countries to supply ICES with the information needed to fulfill its mission. The convention furthermore sets out organizational arrangements.

Following the increasing uses and pressures on the oceans, developments in international ocean law and other initiatives as well as the increasing use of ICES advice in fisheries management, the Copenhagen Declaration on Future ICES strategy was adopted on its 100 year anniversary in 2002.[7] The declaration reaffirms a commitment to maintain ICES as an independent science organization and stresses the need for ICES to strengthen relations with the users of marine science.[8] Since then, ICES has developed its strategy through several cycles, the latest being adopted in 2019 (ICES, 2019a). Over time the ICES strategy has placed increasing emphasis on an ecosystem approach to the study and management of the oceans. The 2019 strategic plan represents a further step in this direction, reflecting also an increasing concern with the human dimension. Another long-term development in the evolution of the organization is increased attention to the needs and wishes of the users of ICES advisory products.

The mission of the current ICES organization is to "advance and share scientific understanding of marine ecosystems and the services they provide and to use this knowledge to generate state-of-the-art advice for meeting conservation, management, and sustainability goals" (ICES, 2019a). The 2019–2024 ICES Strategic Plan sets out science priorities (see below) and outlines steps to address them.

[5] See also the ICES website at http://www.ices.dk/about-ICES/who-we-are/Pages/Our-history.aspx

[6] The convention text can be found here: http://www.ices.dk/about-ICES/who-we-are/Documents/ICES_Convention_1964.pdf

[7] http://www.ices.dk/about-ICES/who-we-are/Documents/CPH_declaration_2002.pdf

[8] http://www.ices.dk/about-ICES/who-we-are/Documents/CPH_declaration_2002.pdf

The ICES Council is the organization's key decisionmaking body, led by an elected president and consisting of two representatives from each member country. A Bureau acts as the executive committee of the Council and a Finance Committee oversees the organization's budget and finances. The 60+ person secretariat, led by a General Secretary, is located in Copenhagen, Denmark.[9]

The scientific work of ICES is governed through its Science Committee (SCICOM) and Advisory Committee (ACOM). The SCICOM drives the ICES science program, links science, data and advice, and organizes the annual science conference as well as meetings and workshops.[10] A number of steering groups under SCICOM address "broad and enduring areas of science and advice" such as aquaculture, fisheries resources, and integrated ecosystem assessments, drawing on the work of around 200 expert groups. The expert groups address a wide range of issues, including a strategic initiative on the human dimension.[11]

The ICES annual science conferences are major events in the world of marine science, gathering participants from all over the world. Also, in cooperation with other marine science organizations, such as the Pacific Marine Science Organization (PICES), ICES organizes global conferences on topical issues such as climate change.[12] In keeping with its Convention, ICES is also engaged in the dissemination of marine research and hosts the *ICES Journal of Marine Science*, a prominent marine science journal published by Oxford University Press.[13]

Another important part of ICES is its role in the management of marine data in the North Atlantic and the Arctic. The ICES Data Centre[14] is directed by a Data and Information Group which works to ensure the alignment of data policies and processes. Data are collected by its members, and work on methods, quality checks, and submission of data to ICES also relies on inputs from members. ICES has established a data pipeline from collection of data to advice products, supported by a set of best practices intended to ensure quality and consistency as well as transparency. Data services are delivered via various web services; AI, cloud services, and machine learning are increasingly important in this respect (ICES, 2019b).

While ICES has been in existence for more than 100 years, a number of other regional organizations concerned with the marine environment and associated natural resources have emerged over the last decades. ICES has established working relationships through MoUs or similar documents with, among others, the Northeast Atlantic Fisheries Commission (NEAFC), the North Atlantic Marine Environment Organization (OSPAR), the HELCOM which addresses the marine environment in the Baltic, the North Atlantic Marine Mammals Commission (NAMMCO) as well as

---

[9] The ICES convention specifies its location to be in Copenhagen.

[10] http://www.ices.dk/community/groups/Pages/SCICOM.aspx

[11] http://www.ices.dk/community/groups/Pages/SIHD.aspx

[12] Cfr the Effects of Climate Change on the World's Oceans quadrennial conference series. https://meetings.pices.int/meetings/international/2018/climate-change/Background

[13] https://academic.oup.com/icesjms

[14] http://admin.ices.dk/Submissions/index.aspx?t=1

the European Union. The relationship with NEAFC is particularly close, as the NEAFC convention requires NEAFC to seek scientific advice from ICES.[15]

Also, relationships are developing beyond the North Atlantic. ICES has recently been granted observer status in the UN General Assembly, providing it with the opportunity to participate in oceans- and science-related meetings there. It also participates in other UN bodies such as the Intergovernmental Oceanographic Commission of UNESCO. And it has a working relationship with Arctic Council working groups PAME and AMAP as well as its sister organization in the North Pacific, the North Pacific Marine Science Organization PICES.

25.3 ICES Science

Use of ocean space and its natural resources requires an understanding of the nature and dynamics of marine ecosystems. Nowhere is this more evident than in fisheries management, where assessments of abundance, distribution, and other characteristics of fish stocks subject to harvest are critical for decisionmakers to be able to establish regulatory measures to ensure a sustainable harvest (Pitcher & Hart, 1982). This was recognized as a key function of ICES already at the outset,[16] and remains a central area of work.

The science agenda of ICES has evolved considerably since then, in response to scientific developments as well as to increasing uses of and pressures on the oceans. The current ICES Strategic Plan outlines seven interrelated science priorities (ICES, 2019a):

Ecosystem science
Impacts of human activities
Observation and exploration
Emerging techniques and technologies
Seafood production
Conservation and management science
Sea and society

Each of the science priorities is accompanied by a set of tasks designed to address it. The ecosystem science priority is foundational in the sense that it addresses the need to understand the dynamics, structure, and functions of marine ecosystems as a basis for scientific advice as well as for marine management (Wilson, 2009; Walther & Møllmann, 2013). It also reflects a long-term evolution in ICES's organizational

[15] Article 14 of the NEAFC Convention. https://www.neafc.org/system/files/Text-of-NEAFC-Convention-04.pdf

[16] In a report of the first meeting of ICES in 1902, it was noted that the funding from governments was conditioned by the need for knowledge for fisheries management. "Practical results of direct value to the fisheries are sought for" (Nature, 1902).

focus (Ballesteros et al., 2018). Similarly, with increasing impacts if climate change, pollution and human uses of the oceans, understanding the impacts of human activities, including cumulative effects, becomes critical.

The next two priorities relate to collection of data and technologies for monitoring and analysis of data. This is a rapidly expanding field where our capacity to collect and assimilate data is increasing exponentially (European Marine Board, 2020). The seafood production priority is a traditional core area for ICES, providing scientific advice for marine capture fisheries as well as aquaculture. A more recent priority is conservation and management science, which is concerned with providing options for managers to set and meet objectives for management. The final priority on sea and society reflects ICES's intent to address issues relating to culture, recreation and human livelihoods.

All these endeavors rely on the science institutions in member states and beyond to provide the data and the resources needed to address the priorities.

25.4 ICES Advice

Providing scientific advice for fisheries has been a raison d'être for ICES since its inception. Its work in this respect has contributed significantly to the evolution of fisheries management, a development that was reinforced with the law of the sea negotiations (UNCLOS III) during the 1970s. UNCLOS III resulted in the 1982 UN Convention on the Law of the Sea,[17] conferring sovereign rights over natural resources in 200 nautical mile Exclusive Economic Zones on coastal states (Churchill & Lowe, 1989). This provided coastal states with a strong incentive to invest in marine science and to manage fisheries through regulating access to and utilization of resources (Juda, 1996).

The convention proved deficient when it came to regulating fisheries beyond national jurisdiction (Burke, 1994), and another UN conference produced the 1995 UN fish stocks agreement (Balton, 1996).[18] Both treaties have a number of provisions regarding marine science. The 1982 convention defines maximum sustainable Yield (MSY) as an objective of fisheries management, which has proved important to subsequent developments in marine science (Hoel, 2017). By making the application of a precautionary approach mandatory under international law, the 1995 agreement spurred a significant change in how scientific advice was to be developed and communicated to policymakers (Kvamsdal et al., 2016). The 1995 agreement also requires states to address ecosystems and biodiversity in their management of fisheries, which requires additional scientific inputs. The 1995 agreement has wide ranging requirements regarding data collection and

[17] https://www.un.org/Depts/los/convention_agreements/texts/unclos/closindx.htm

[18] https://www.un.org/Depts/los/convention_agreements/texts/fish_stocks_agreement/CONF164_37.htm

transparency. These provisions have had a significant impact on the work and practices of ICES (Lassen et al., 2013).

While the initial mission as well a large part of ICES history has been weighted towards fisheries science and advice, this has changed in recent years, bringing changes to how its advisory functions are organized and work. Three advisory bodies for fisheries (Advisory Committee on Fisheries Management), environment (Advisory Committee for the Marine Environment), and ecosystems (Advisory Committee on Ecosystems) were collapsed into the Advisory Committee for Ocean Management (ACOM) in 2008, reflecting the increasing emphasis on understanding marine ecosystems and addressing them holistically (Strange et al., 2012).

The current mission of ACOM, therefore, is to translate "...ICES science into advice on the sustainable use and protection of marine ecosystems".[19] ACOM has a representative from each of the member countries. It provides scientific advice to its clients who are the member countries, the European Union, and regional commissions such as OSPAR and NEAFC.

An advisory plan (ICES, 2019c) sets forth the framework for advice, where various types of requests for advice from clients are addressed through a process starting with the formulation of a request. Requests in many cases will be recurrent, as in the scientific advice provided on the status of fish stocks and options for management including total quotas. A second step in the advisory framework is the role of expert groups synthesizing knowledge syntheses using data that conforms to ICES standards.[20] The products of expert groups are subject to independent peer review (the third step), before ACOM formulates the actual advice as the fourth and final step. The advice is published on the ICES Website.[21]

Fisheries-specific advice is supported by other advisory products such as Ecosystem Overviews and Fisheries Overviews, intended to complement and provide context for fisheries-specific advice. Such overviews are based on the ICES ecoregions.

A new framework for ICES advice – a "more appropriate framework that incorporates the ecosystem approach in all sectors" – specifically addresses EBM. The framework was adopted in 2020, reflecting a further evolution in the organization's thinking about the science-policy interface and its emphasis on an ecosystem approach. The Guide explains how ICES provides advice based on ten principles:

1. Document openly
2. Formulate request iteratively
3. Clarify objectives & risks
4. Deliver knowledge timely
5. Use best available science
6. Apply data FAIR principles

[19] http://www.ices.dk/community/groups/Pages/ACOM.aspx

[20] https://www.ices.dk/sites/pub/Publication%20Reports/Advice/2020/2020/Guide_to_ICES_Advice.pdf

[21] https://www.ices.dk/advice/Pages/default.aspx

7. Undergo peer review
8. Develop clear & consistent advice
9. Agree by consensus
10. Explain without advocacy

Principles 1–3 are guidelines for advice and refer to the first step in the framework for advice. Principles 4–6 deal with the second step of the framework (the knowledge syntheses), and 7 refers to peer review. The fourth step is addressed by principles 8–10 and focuses on the formulation of advice.

A pertinent question is what happens after ICES advice is provided. It is widely recognized that this advice is not always acted upon and that disentangling the causal path from scientific advice to policy outcomes is complex (Stokke, 2012). A case in point is the situation with regard to pelagic species in the Norwegian Sea, where controversies related to allocation of fish quotas have prevented lasting agreement on management (Bjørndal & Ekerhovd, 2014). It is beyond the scope of this chapter to address this issue in depth.

25.5 ICES and the Arctic

ICES has a long standing engagement with the Arctic. Its Arctic Fisheries Working Group (its oldest expert group) has existed for more than 50 years. This group plays a critical role in developing the scientific basis for management advice for the fisheries of the Barents Sea (Kovalev & Bogstad, 2011), a globally significant fishing ground with the world's largest cod fisheries. The recipient of advice in this case is the Norway-Russia Joint Fisheries Commission, which manages five shared fish stocks in the Barents Sea.[22]

The Northeast Atlantic has a large number of fish stocks that are shared between two or more countries and/or extend into waters beyond national jurisdiction. ICES therefore also provides advice to a number of other sub-Arctic cooperative arrangements, including those between Norway and Iceland, Norway and Greenland, Norway and the Faroes, and Norway and the EU.[23] It also coordinates scientific cooperation on Norwegian Sea surveys of pelagic fish stocks. ICES provides advice directly to coastal states and the EU, and this is the basis for management of the fish stocks in the waters of Greenland, Iceland, Norway and Russia.

The Northeast Atlantic has three areas of waters extending beyond EEZs: in the Norwegian Sea and the Barents Sea, in the sub-Arctic, and in the central Arctic Ocean. The sub-Arctic waters are home to significant fisheries, while the European wedge of the high seas portion of the CAO is ice-covered and does not have any

[22] Cod, haddock, capelin, Greenland halibut, and redfish.

[23] These agreements are subject to annual review in reports to Parliament in Norway.

fisheries.[24] These areas are Regulatory Areas of the Northeast Atlantic Fisheries Commission.[25] NEAFC regulations apply to all regulatory areas, including the European wedge of the high seas in the CAO, specifically the scheme on control and enforcement, protection of vulnerable marine ecosystems, deep sea fisheries, and annual regulations on a series of fish stocks.[26]

While most other regional fisheries management organizations (RFMOs) have an in-house mechanism to provide for scientific advice (FAO, 2020), NEAFC relies on ICES for this purpose. The 1980 NEAFC convention explicitly requires that it "... shall seek information and advice from the International Council for the Exploration of the Sea."[27] NEAFC and ICES have established an MoU for this arrangement.[28] Thus, NEAFC gets scientific advice that is independent of NEAFC and its members.

ICES involvement in the Arctic is also significant in the context of the 2018 agreement to prevent unregulated fishing in the high seas portion of the Central Arctic Ocean. This ten-party[29] agreement, which establishes a 16-year moratorium on fishing (Balton, 2019), has been more than a decade in the making. A number of scientific meetings since 2011 have been important to its development and conclusion. ICES contributed substantively to these meetings by providing advice on how to organize the functions of a science mechanism to be established when the agreement enters into force (Hoel, 2020)[30]. When such a mechanism eventually is set up,[31] ICES is likely to be important by virtue of its central role in scientific cooperation and provision of scientific advice in the Northeast Atlantic, the fact that all coastal states are ICES members, and its special relationship with NEAFC, which has a Regulatory Area in the European wedge of the high seas portion of the Central Arctic Ocean.

ICES also has working relationships with the Arctic Council[32] and with ICES's sister organization PICES in the North Pacific. This is the basis for a 3-way cooperation on developing an integrated ecosystem assessment of the central Arctic Ocean. Integrated ecosystem assessments (IEAs) are critical elements in the

[24] The world's northernmost fisheries are on the Northern flank of Norway's Svalbard archipelago, well inside its 200-mile zone.

[25] NEAFC has seven contracting parties: The European Union, the Faroe Islands, Iceland, Greenland, Norway, the Russian Federation, and the United Kingdom. The UK became a contracting party in 2020 following Brexit.

[26] Statement by NEAFC regarding the conclusion of the negotiations on the Agreement to Prevent Unregulated High Seas Fisheries in the Central Arctic Ocean. https://www.neafc.org/system/files/NEAFC-statement_Central-Arctic-Ocean-Agreement.pdf

[27] NEAFC Convention article 14.

[28] https://www.neafc.org/system/files/ices_mou-2019.pdf

[29] Canada, Denmark/Greenland, Norway, the Russian Federation, USA, China, the Republic of Korea, Japan, Iceland and the EU.

[30] The agreement requires that all 10 parties have ratified for it to enter into force. By the end of 2020 nine out of 10 signatories have ratified.

[31] Currently, a Provisional Scientific Coordination Group has held one meeting.

[32] Observer status, as well as functions in relation to working groups.

development of ecosystem-based management (Levin et al., 2009), and ICES is currently engaged in producing several such assessments in the Arctic and sub-Arctic, including for the Central Arctic Ocean, the Barents Sea, and the Norwegian Sea. The working group established for the conduct of an integrated ecosystem assessment of the Central Arctic Ocean (WGICA) has met since 2016, and has recently embarked on its second three-year mandate period.[33] The first WGICA IEA report and the first Ecosystem Overview of the CAO will be published by ICES in 2021.

Ecosystem Overviews are priority action areas for ICES have become advisory products along with Fisheries Overviews,[34] complementing the regular scientific advice for fisheries management. Ecosystem Overviews follow a human activity – pressures – states conceptual scheme,[35] and are already published for the subarctic ecoregions[36] in the Barents Sea, the Norwegian Sea, the Greenland Sea, and Icelandic waters. ICES ecoregions are the spatial units for ecosystem-based scientific advice.[37]

With the increasing impacts of climate change in the Arctic and its ramifications affecting Arctic marine ecosystems (Haug et al., 2017), along with increasing human activity, the role of science and scientific advice for management will become ever more important. ICES is not the only game in town. An assessment of its future role in the Arctic needs to factor in other organizations and initiatives and how they relate to each other.

25.6 The Wider Context of Science and the Arctic Ocean

While commercial activities are the dominant human presence in the sub-Arctic, marine scientific research is probably the most significant human activity in the Central Arctic Ocean today. The conduct of marine scientific research in the Arctic Ocean is governed by global norms as well as regional and domestic institutions.

As for the global norms, the 1982 UN Convention on the Law of the Sea provides the legal framework for all activities in the oceans globally, including science (Churchill & Lowe, 1989). Within national jurisdictions, marine scientific research by entities from other nations requires the consent of the coastal state. In areas beyond national jurisdiction, marine scientific research is one of the freedoms of the

[33] http://www.ices.dk/community/groups/Pages/WGICA.aspx . The mandate of WGICA is here: http://www.ices.dk/community/Documents/Science%20EG%20ToRs/IEASG/2019%20-%202020/WGICA%20resolution%202019-2021.pdf

[34] http://www.ices.dk/advice/Fisheries-overviews/Pages/fisheries-overviews.aspx

[35] https://www.ices.dk/advice/ESD/pages/preview.aspx?diagramid=52

[36] http://www.ices.dk/advice/ESD/Pages/Ecosystem-overviews.aspx

[37] http://www.ices.dk/advice/ICES%20ecoregions%20and%20advisory%20areas/Pages/ICES-ecosystems-and-advisory-areas.aspx

high seas. The Intergovernmental Oceanographic Commission (IOC) of UNESCO is the global marine science body tasked with promoting marine science and implementing global marine science programs[38]; it is the coordinator and driver of the 2021–2030 UN Decade of Ocean Science for Sustainable Development.[39]

A number of regional organizations and arrangements are engaged in Arctic marine science. The International Arctic Science Committee (IASC) was established in 1990 to encourage and facilitate cooperation in all aspects of Arctic research (Rogne et al., 2015).[40] IASC, an NGO, has members from 23 countries and can be viewed as the science community's own organization, relying on bottom-up processes to identify cutting-edge research topics (Smieszek, 2015). It has a marine working group dealing with basic science.

Another, more recent, regional initiative features the Arctic Science Ministerial Meetings, which have been held in 2016, 2018 and 2021. The 2018 meeting was attended by 26 countries. The main goal of the ministerial meetings is to shape the course of future Arctic research. The outcomes of the meetings is a set of conclusions setting out priorities for research, such as increased international cooperation.[41] The main themes for cooperation are observations and data, regional and global dynamics, and vulnerability and resilience.

Still another regional arrangement is the 2018 Agreement on Preventing Unregulated Fishing in the High Seas Portion of the Central Arctic Ocean. It is the outcome of negotiations, first among the five coastal states, subsequently expanded to include potential distant water fishing nations (Japan, China, Republic of Korea, the EU, and Iceland). Interactions over several years between science and policy actors was critically important for the conclusion of the agreement, which contains provisions for the establishment of a Joint Program of Scientific Research and Monitoring. Given that a 16-year moratorium will commence when the agreement enters into force, to be continued beyond the initial 16 years in five-year increments as long as no party objects, scientific research is likely to constitute a large part of this body's agenda in the coming years (Hoel, 2020).

The Arctic Council was established in 1996 as a high-level intergovernmental forum to provide a means for promoting cooperation, coordination and interaction among the Arctic States (Young, 2010). While not a scientific body, its various working groups (AMAP,[42] CAFF,[43] PAME,[44] EPPR,[45] ACAP,[46] and SDWG[47]) are

---

[38] https://ioc.unesco.org

[39] https://oceandecade.org

[40] https://iasc.info

[41] https://www.arcticscienceministerial.org/files/ASM2_Joint_Statement.pdf

[42] Arctic Monitoring and Assessment Programme.

[43] Conservation of Arctic Flora and Fauna.

[44] Protection of the Arctic Marine Environment.

[45] Emergency Prevention and Preparedness and Response.

[46] Arctic Contaminants Action Programme.

[47] Sustainable Development Working Group.

users of scientific research, focusing on monitoring and assessments. A significant legacy of the Arctic Council is therefore that our understanding of the Arctic is greater than ever before, resulting inter alia from the Arctic Climate Impact Assessment (ACIA) (ACIA, 2005), the Snow, Water, Ice and Permafrost in the Arctic (SWIPA) (AMAP, 2017), the State of the Arctic Marine Biodiversity Report 2017 (CAFF, 2017), the Arctic Marine Shipping Assessment (PAME, 2009), and the Arctic Human Development Report II (SDWG, 2015). ICES acquired observer status in the Arctic Council in 2017, a relationship that encourages enhanced collaboration.

Under the auspices of the Arctic Council, an agreement on international scientific cooperation in the Arctic was signed in 2017 and entered into force in 2018. The purpose of this agreement is to enhance cooperation in scientific activities in order to improve scientific knowledge about the Arctic by providing access to areas, data, and infrastructure (Smieszek, 2017). While it has the potential to boost cooperation (Berkman et al., 2017), so far there appears to have been little activity under this agreement.

With respect to Arctic-wide coordination of observations and data, the Arctic Council and the International Arctic Science Committee established the Sustaining Arctic Observations Network (SAON) in 2011 in the wake of the International Polar Year (2007–2009).[48] The mission of SAON is to strengthen pan-Arctic observing, and its 2018–2028 strategy sets out the principles for this.[49] ICES is a SAON partner organization.

In a larger perspective on Arctic marine stewardship, a significant recent development is the establishment of the SAO-based Marine Mechanism (SMM) in the Arctic Council, aiming to provide a high-level coordination and steering function for the marine activities of the Arctic Council. The outcome of a 2015–2018 Task Force on Arctic Marine Cooperation, the SMM will likely be an important arena for discussion of marine scientific research. Its first meeting took place in October 2020, and ICES contributed with an introduction to ecosystem-based management. With the establishment of the SMM, the Arctic Council has created a focal point for Arctic marine issues at a strategic level, an important development when viewed in a wider, global perspective and in relation to the on-going negotiations of an international legally binding instrument for the conservation and use of biodiversity in the areas beyond national jurisdiction (Balton, 2019).

[48] https://www.arcticobserving.org/images/pdf/Board_meetings/5th_tromso/nuuk_declaration_final.pdf

[49] https://www.arcticobserving.org/images/pdf/Strategy_and_Implementation/SAON_Strategy_2018-2028_version_16MAY2018.pdf

25.7 Discussion

The International Council for the Exploration of the Sea (ICES) has a long history; it was established more than a century ago, at the beginning of the twentieth century. With a network of 6000 scientists and 700 institutes, it has a large pool of intellectual capital for marine science. It is an Intergovernmental Organization (IGO) based on its 1964 convention and has evolved from a body mainly concerned with fisheries science and advice to a broad-based marine science organization now having marine ecosystems as its organizational focus. In this respect it represents a broad, international development over recent decades where ecosystem science and the need for integrated ocean management has become broadly accepted (Winther et al., 2020), if not yet widely implemented.

The advisory function of ICES is unique. It is the only international marine scientific organization with such a strong mandate for provision of scientific advice to its clients – coastal states and regional marine management organizations in the North Atlantic. Its sister organization PICES in the North Pacific has a similar mandate but does not perform the same advisory functions.

ICES is also an independent scientific organization, and an IGO in its own right where scientific integrity is valued highly and where a number of safeguards and procedures are in place to protect the scientific work from undue political influence. With the growing threats to the oceans and their resources, the growth in interest in marine issues, and the proliferation of private initiatives to influence marine governance, the need for impartial scientific advice for marine management is more important than ever.

With the onset of the International Decade on Ocean Science[50] in 2021, a pertinent question is whether and how the ICES can represent a model for the organization of marine science and provision of scientific advice to other regions in the world. A number of features of ICES are of interest in this respect, such as the organization of its work, the data pipelines and their governance, and the protection of scientific integrity.

While ICES as an organization represents a cutting-edge approach to the provision of scientific advice to management authorities, an important question is "what happens next"? Is the advice listened to and followed by its clients? A full answer to that question is beyond the scope of this chapter and would require a major effort to address fully. Also, the quality and role of scientific advice in marine management is but one of several factors explaining the status of marine ecosystems and the natural resources they encompass. The overall development in the status of fish stocks in the Northeast Atlantic is however generally improving, as can be seen for example in the Barents Sea or the North Sea (Hilborn et al., 2020) and indeed in regions where modern fisheries management plans are implemented (Melnychuk et al., 2021). Increases in the quality and influence of scientific advice obviously play some role in this development. Regarding the current preoccupation with ecosystem-based

[50] https://www.oceandecade.org

management and advice, it could be asked how ICES can advance the implementation of the ecosystem approach to fisheries management, so long as clients primarily ask for advice on single species management (Ramirez-Monsalve et al., 2021).

As regards the Arctic specifically, ICES has a mandate for the Northeast Atlantic up to the North Pole. Its historical as well as current engagement in the Arctic includes both its traditional preoccupation with fisheries and its more recent emphasis on ecosystem science. The first is amply illustrated by its critical role in providing scientific advice for fisheries management in the Northeast Atlantic, a role that is set to become more important as climate change drives fish stocks north (Hastings et al., 2020, Fossheim et al., 2015). As regards ecosystem science and management, the cooperation with PICES and the Arctic Council's working group PAME on an integrated ecosystem assessment for the Central Arctic Ocean (WGICA)[51] is a significant initiative in several ways. It represents a new and important foundation for subsequent development of ecosystem advice; it is also a novel model of cooperation among key scientific bodies in that region (ICES, 2020). ICES performs such integrated ecosystem assessments also in the seas surrounding the Central Arctic Ocean, such as the Barents Sea and the Norwegian Sea.[52] In addition, ICES is connected to other recent initiatives of importance to marine science in the Arctic, such as the Sustaining Arctic Observation Networks (SAON) and the scientific work under the 2018 agreement to prevent unregulated fishing.

The Arctic marine science landscape is a complex work in progress. Also, it evolves in the context of increasing geopolitical tensions (Stavridis, 2017), a development that could have major repercussions for science (Nature, 2020). Still, it seems safe to conclude that ICES will remain a core part of the fabric of Arctic marine science in the future.

Acknowledgement I am grateful to Kåre Nolde Nielsen and two anynomous reviewers for comments to an earlier version of this chapter.

References

ACIA. (2005). *The Arctic climate impact assessment*. Cambridge University Press.

AMAP. (2017). *Snow, water, ice, and permafrost*. SWIPA 2017 report. https://swipa.amap.no

Andresen, S., et al. (2000). *Science and politics in international environmental regimes*. Manchester University Press.

Arctic Ocean Review phase I report.: https://www.researchgate.net/publication/308519139_The_Arctic_Ocean_Review_-_phase_1_report_

Baker, B. (2020, November). Arctic overlaps: The surprising story of continental shelf diplomacy. *Polar Perspectives, 3*.

[51] https://www.ices.dk/community/groups/Pages/WGICA.aspx

[52] https://www.ices.dk/news-and-events/news-archive/news/Pages/A-healthy-ocean.aspx

Ballesteros, M., et al. (2018). Do not shoot the messenger: ICES advice for an ecosystem approach to fisheries management in the European Union. *ICES Journal of Marine Science, 75*(2), 519–530. https://doi.org/10.1093/icesjms/fsx181

Balton, D. (1996). Strengthening the law of the sea: The new agreement on straddling and highly migratory fish stocks. *Ocean Development and International Law, 26*(1–2), 125–151.

Balton, D. (2019). What will the BBNJ agreement mean for the Arctic fisheries agreement. *Marine Policy*.

Berkman, P., et al. (2017). The Arctic science agreement propels science diplomacy. *Science, 358*(6363), 596–598.

Bjørndal, T., & Ekerhovd, N. A. (2014). Management of Pelagic fisheries in the Northeast Atlantic. *Marine Resource Economics, 29*(1), 69–83.

Burke, W. (1994). *The new international law of fisheries*. Oxford University Press.

CAFF. (2017). *State of the Arctic Marine biodiversity report*. https://oaarchive.arctic-council.org/bitstream/handle/11374/1945/SAMBR_Scientific_report_2017_FINAL_LR.pdf

Churchill, R., & Lowe, A. (1989). *The law of the sea*. Manchester University Press.

European Marine Board. (2020). *Big data in marine science*. Future Science Brief No 6 April 2020. https://marineboard.eu/sites/marineboard.eu/files/public/publication/EMB_FSB6_BigData_Web_v4_0.pdf

FAO. (2020). *Regional fisheries management organizations and advisory bodies*. FAO fisheries and aquaculture technical paper 651, Rome.

Fossheim, M., Primicerio, R., & Climate, E. J. N. (2015). Recent warming leads to a rapid borealization of fish communities in the Arctic. *Nature Climate Change*. https://doi.org/10.1038/nclimate264

Global Ocean Commission. (2016). *The future of our ocean*. https://www.some.ox.ac.uk/wp-content/uploads/2016/03/GOC_2016_Report_FINAL_7_3.low_1.pdf

Hastings, R. A., et al. (2020). Climate change drives poleward increases and equatorward declines in marine species. *Current Biology, 30*, 1–6.

Haug, T., Bogstad, B., Chierici, M., Gjøseter, H., Hallfredsson, E., Høynes, E., Hoel, A. H., Ingvaldsen, R., Jørgensen, L. L., Knutsen, T., Loeng, H., Naustvoll, L. J., Røttingen, I., & Sunnanå, K. (2017). Future harvest in the Arctic Ocean north of the Nordic and Barents Seas: A review of possibilities and constraints. *Fisheries Research, 188*, 38–57.

High Level Panel for a Sustainable Ocean Economy. (2020). *Transformations for a sustainable ocean*. https://www.oceanpanel.org/ocean-action/files/transformations-sustainable-ocean-economy-eng.pdf

Hilborn, R., et al. (2020). Effective fisheries management instrumental in improving fish stocks status. *PNAS, 117*(4), 2218–2224.

Hildebrand, L. P., Brigham, L. W., & Johansson, T. M. (eds.). (2018). *Sustainable shipping in a changing Arctic* (World Maritime University studies in maritime affairs 7). Springer.

Hoel, A. H. (2017). The importance of marine science in sustainable fisheries: The role of the 1995 UN fish stocks agreement. In M. H. Nordquist, J. N. Moore, & R. Long (Eds.), *Legal order in the world's oceans: UN convention on the law of the sea*. Brill/Nijhoff.

Hoel, A. H. (2018). Northern fisheries. In M. Nuttall (Ed.), *The Routledge handbook of the polar regions*. Routledge.

Hoel, A. H. (2020). Ch. 11: The evolving management of fisheries in the Arctic. In: K. Scott & D. VanderZwaag (Eds.), *Research handbook on Polar Law* (pp. 199–216). https://www.e-elgar.com/shop/gbp/research-handbook-on-polar-law-9781788119580.html

Holland, G., & Pugh, D. (Eds.). (2010). *Troubled waters: Ocean science and governance*. Cambridge University Press.

ICES. (2019a). *Strategic plan*. https://doi.org/10.17895/ices.pub.5470.

ICES. (2019b). *ICES user handbook – Best practice for data management*. https://www.ices.dk/sites/pub/Publication%20Reports/User%20Handbooks/uh-best-practice-data-management.pdf

ICES. (2019c). *ICES advisory plan*. https://issuu.com/icesdk/docs/ices_advisory_plan

ICES. (2020). ICES/PICES/PAME working group on integrated ecosystem assessment for the Central Arctic Ocean (WGICA). ICES Scientific Reports, 2, 79, 144 pp. https://doi.org/10.17895/ices.pub.7454.

Independent World Commission on the Oceans. (1998). *The ocean – Our future*. Cambridge University Press.

Juda, L. (1996). *International law and ocean use management*. Routledge.

Kovalev, Y. A., & Bogstad, B. (2011). The scientific basis for management. In T. Jakobsen & V. K. Ozhigin (Eds.), *The Barents Sea – Ecosystem, resources, management* (pp. 621–646). Tapir.

Kvamsdal, S. F., Eide, A., Ekerhovd, N.-A., Enberg, K., Gudmundsdottir, A., Hoel, A. H., Mills, K. E., Mueter, F. J., Ravn-Jonsen, L., Sandal, L. K., Stiansen, J. E., & Vestergaard, N. (2016). Harvest control rules in modern fisheries management. *Elementa: Science of the Anthropocene, 4*, 000114. https://doi.org/10.12952/journal.elementa.000114

Lassen, H., et al. (2013). *ICES Advisory framework 1977–2012. From Fmax to precautionary approach and beyond.*

Levin, P., et al. (2009). Integrated ecosystem assessments: Developing the scientific basis for ecosystem-based management of the ocean. *PLoS Biology, 7*(1), 2009.

Melnychuk, M., et al. (2021). Identifying management actions that promote sustainable fisheries. *Nature Sustainability*. https://doi.org/10.1038/s41893-020-00668-1

Miles, E.L. (1987). *Science, politics, and international ocean management*. Policy papers in International Affairs. Institute of International Studies/University of California.

Nature. (1902, August 7). The first meeting of the International Council for the Exploration of the Sea. *Nature*.

Nature (editorial). (2020, October 1). Arctic science cannot afford a new cold war. *Nature, 586*.

PAME. (2009). https://www.pame.is/index.php/projects/arctic-marine-shipping/amsa

Pitcher, T. J., & Hart, P. J. (1982). *Fisheries ecology*. Croom Helm.

Ramirez-Monsalve, et al. (2021). Pulling mechanisms and pushing strategies: How to improve ecosystem advice for fisheries management advice within the European Union's common fisheries policy. *Fisheries Research, 233*. https://doi.org/10.1016/j.fishres.2020.105751

Rogne, O., Rachold, V., Hacquebord, L., & Corell, R. (2015). *IASC after 25 Years*. http://iasc25.iasc.info

Rozwadowski, H. (2002). *The sea knows no boundaries: A century of marine science under ICES*. University of Washington Press.

Sakshaug, E., et al. (Eds.). (2009). *Ecosystem Barents Sea*. Tapir Academic.

SAPEA. (2019). *Making sense of science for policy*. https://www.sapea.info/topics/making-sense-of-science/

SDWG. (2015). *Arctic human development report*. http://norden.diva-portal.org/smash/record.jsf?pid=diva2%3A788965&dswid=-4440

Smieszek, M. (2015). 25 Years of the International Arctic Science Committee (IASC). *Arctic Yearbook 2015*. https://www.arcticyearbook.com/briefing-notes2015/173-25-years-of-the-international-arctic-science-committee-iasc

Smieszek, M. (2017). The agreement on enhancing international Arctic Scientific Cooperation: From paper to practice. In L. Heininen, H. Exner-Pirot, & J. Plouffe (Eds.), *Arctic yearbook 2017*. . Northern Research Forum. Retrieved from https://arcticyearbook.com/images/yearbook/2017/Briefing_Notes/6_The_Agreement_on_Enhancing_International_Arctic_Scientific_Cooperation.pdf

Stavridis, J. (2017). *Sea power – The history and geopolitics of the world's oceans*. Penguin Press.

Stokke, O. S. (2012). *Disaggregating international regimes*. MIT Press.

Strange, K., et al. (2012). *Managing organizational change in an international scientific network: A study of ICES reform processes*.

Walther, Y., & Møllmann, C. (2013). Bringing ecosystem assessments to real life: A scientific framework for ICES. *ICES Journal of Marine Science, 71*(5), 1183–1186. https://doi.org/10.1093/icesjms/fst161

Wilson, D. C. (2009). *The paradoxes of transparency – Science and the ecosystem approach to fisheries management in Europe* (304pp). MARE/Amsterdam University Press. https://www.jstor.org/stable/j.ctt46mxkb

Winther, J.-G., Dai, M., Rist, T., Hoel, A. H., Li, Y., Trice, A., Morrissey, K., Juinio-Meñez, M. A., Fernandes, L., Unger, S., Scarano, F. R., Halpin, P., & Sandra Whitehouse. (2020). Integrated oceans management for a sustainable ocean economy. *Nature Ecology and Evolution*. https://www.nature.com/articles/s41559-020-1259-6

Young, O. R. (2010). Arctic governance – Pathways to the future. *Arctic Review of Law and Politics, 1*(2), 164–185.

Chapter 26
(Research): The Arctic Is What Scientists Make of It: Integrating Geopolitics into Informed Decisionmaking

Sebastian Knecht and Mathias Albert

Abstract The present chapter seeks to bridge the two worlds of geopolitics and science. It starts from the perspective of science in order to show how the geopolitical narratives in which the work, methods and approaches of scientists and experts are embedded can shape the form and function of science diplomacy. The chapter assumes a perspective in the tradition of 'critical geopolitics' in order to show how the seemingly simple assumption and equations of geopolitics in fact are based on a wealth of complex narratives and practices certainly not limited to, but definitely also including science. It then continues to establish science geopolitics as a new heuristic for the study of geopolitical imaginations and spatial constructions in world politics. Making both a conceptual argument and using the scientific construction of Arctic large marine ecosystems (LMEs) under the Arctic Council's (AC) Protection of the Arctic Marine Environment Working Group (PAME WG) as an empirical illustration, we demonstrate how science engages in the making of spatial realities against the backdrop of an imaginary Arctic space that provide the scientific community with formative power in the pyramid of informed decisionmaking.

26.1 Introduction

In the present chapter, we engage in an exercise to bridge the two worlds of geopolitics and science. The history of the Arctic is paved with examples of how science has been used and abused to serve geopolitical ideas, interests and performances, from developing and underscoring national stakes in the early days of exploration of the 'Far North' (Levere, 1993), over helping to preserve economic, security and strategic motives of Arctic states during the Cold War period (Doel et al., 2014), to serving as a tool for trust-building by non-Arctic stakeholders seeking involvement in Arctic governance (Su & Mayer, 2018) or legitimating

S. Knecht · M. Albert (✉)
Faculty of Sociology, Bielefeld University, Bielefeld, Germany
e-mail: sebastian.knecht@uni-bielefeld.de; mathias.albert@uni-bielefeld.de

P. A. Berkman et al. (eds.), *Building Common Interests in the Arctic Ocean with Global Inclusion, Volume 2*, Informed Decisionmaking for Sustainability,
https://doi.org/10.1007/978-3-030-89312-5_26

extended continental shelf claims under the United Nations Convention on the Law of the Sea (Riddell-Dixon, 2017). We start from the other side of this allegedly strained relationship, the perspective of science, to show how the geopolitical narratives in which the work, methods and approaches of scientists and experts are embedded can shape the form and function of science diplomacy. This is particularly important in a burgeoning field such as Arctic science and research that is supposed to feed data, evidence and options into processes of informed decisionmaking for sustainability under conditions of scarce financial and material resources, high specialization and immense time pressure. For consistency's sake, we apply the editors' understanding of science diplomacy 'as an international, interdisciplinary, and inclusive (holistic) process, facilitating informed decisionmaking to balance national interests and common interests for the benefit of all on Earth across generations' (Berkman et al., 2020, p. v). In what the editors of this volume further propose to constitute a pyramid of informed decisionmaking starting from asking questions of common concern to accumulating data and generating evidence and in the final step formulating options to policy-makers, science diplomacy can lead to common-interest building and collective action that might otherwise not be possible (ibid., pp. xiii–xv). In our view, much then depends on finding common ground and broader agreement between various geopolitical worldviews held in scientific communities as a key factor for facilitating both process and outcome of science diplomacy.

We start by giving a brief review of geopolitical thought and place the role of science in it. In doing this, we assume a perspective in the tradition of 'critical geopolitics' in political geography that shows how the seemingly simple assumption and equations of geopolitics in fact are based on a wealth of complex narratives and practices certainly not limited to, but definitely also including science. We then continue to establish science geopolitics as a new heuristic for the study of geopolitical imaginations and spatial constructions in world politics. Making both a conceptual argument and using the scientific construction of Arctic large marine ecosystems (LMEs) under the Arctic Council's (AC) *Protection of the Arctic Marine Environment* Working Group (PAME WG) as an empirical illustration, we demonstrate how science engages in the making of spatial realities against the backdrop of an imaginary Arctic space that provide the scientific community with formative power in the pyramid of informed decisionmaking. We conclude with a number of research vistas that emerge from our case study, particularly highlighting that we need to take a closer look into how geopolitics is infused in science itself.

26.2 The Complexities of Geopolitics

Much of what goes on in the Arctic is about 'geopolitics', that is generally the construction and articulation of space-related political interests, strategies, and worldviews. Frequently, references are made to geopolitical interests that are at play in defining the relevant actors' interests in and on the Arctic, and the Arctic is depicted as some kind of playground in a geopolitical game.[1] In order to get a grasp on the importance of geopolitics regarding the Arctic, and in order to establish the link to the process of informed decisionmaking later, it is useful to differentiate between 'classical' and 'critical' geopolitics. Both originally refer to different modes and traditions of thought regarding the importance of geography for international politics. 'Classical' geopolitics sees political interests and the ensuing power struggles between states to be at least partially determined by geographical location. International politics then to a large degree appears as the politics of controlling space. 'Critical geopolitics' emerged as a mode of thought in political geography and, as the name implies, as criticism of the tradition of 'classical' geopolitical thought (cf., for example, Agnew, 1998; Ó Tuathail, 1996). The main rationale here is the idea that there are no 'natural' geographical features that would somehow pre-determine state interests, but that spaces and their relevance are always the result of social constructions, and strongly linked to the formation of identities, the contingent drawing of borders, and the on-going reproduction of political orders (cf. in overview: Albert et al., 2014, also Albert et al., 2001).

Although also referring to the respective academic discussions, in the present context we take classical and critical geopolitics to generally mean modes of thought that structure and influence how actors think about, frame and spatialize, and how they consequentially act in and on the Arctic (Knecht & Keil, 2013; Wegge & Keil, 2018). The main difference between the two modes of thought then would be that in classical geopolitics many of the main features that define the Arctic in political terms are simply 'there', they exist as basic facts that cannot be changed, while in critical geopolitics these features are socially constructed and they can change (although that in itself says nothing about how stable these features are and about what it takes to change them). In classical geopolitics, international politics is basically seen in terms of political realism, with states pursuing their interests defined solely in terms of power. Geographical space in that context is both an object of political interests (either regarding the direct control of territory, or of land-, sea-, or air-spaces in terms of control and influence), as well as a source of power. Ultimately, 'power' here boils down to 'hard power', with military capabilities ranking first, although in direct relation to it most notably economic and financial capabilities play an important role. In the world of classical geopolitics, science would feature only indirectly in the sense of supporting these capabilities. In classical geopolitical thought, the Arctic is not something that is 'made' or 'imagined', but rather something that is 'there'. It is an area in which powers (i.e. states) compete for control and influence, be it in terms of strategic positions or control over natural resources.

[1] To cite but a random recent example: 'How the global battle for the Arctic became the new Cold War' (*New Statesman*, 28 August 2019). For a fuller statement, see for example Sale and Potapov (2010), and Huebert (2019) for a succinct statement of the thesis that there can be no 'new' Cold War in the Arctic because 'the old one never ended'. For a highly nuanced reading of the 'geopolitics' discourses regarding both the Arctic and the Antarctic, see Dodds and Nuttall (2016).

Put simply, approaches from 'critical geopolitics' do not question that interest, power, and territory/space play important roles in the system of world politics. However, they point out that interests result from social practices and conventions, that power is not objectively given, but a relational social category, and territory and spaces in their importance in and for the social world are not unchanging expressions of geomorphological characteristics. In short, understanding the geopolitics of something is not simply an issue of finding out what this geopolitics *is*, but how it is *made* in the context of social reality. Identifying, delineating, and naming a specific space is probably the most basic operation of 'making' geopolitics. This operation in itself is a highly complex process, composed of many different actors, strategies, images, narratives, and practices; it is not a process that in its outcomes can easily be ascribed to specific original intentions; and it often takes a long time. What is important, though, is that it is a process that basically works on different scales which themselves are discursively constructed, and that thus works from the meta-geographies (cf. Lewis & Wigen, 1997) of continents to all types of constructing regions (cf. Albert, 2021). The interesting question then is which of these constructions stabilize over time and with some endurance, and the degree to which these constructions are contested. It is in that sense that answering the questions of what and where the Arctic is and where its boundaries are, as well as struggling over and establishing specific answers as reference points, is a highly political exercise. It is not a purely 'academic' exercise whether the Arctic is defined as the area of the Earth's surface north of the Arctic Circle, as the area north of the 10 °C July isotherm, or as the Arctic Ocean and its littoral areas. Rather, such definitions have direct consequences for legitimizing political and legal claims. Ironically, this is probably best demonstrated by China's self-identification as a 'near-Arctic state' in its published Arctic Policy (The State Council, 2018) exactly because it cannot claim to be included in any conceivable acceptable definition of the Arctic (see also Dodds & Nuttall, 2016 on the 'polar orientalism' involved in claiming stakeholdership in the Arctic).

Obviously, constructing a region is not only a matter of its designation and delimitation, but also a matter of its *inclusion* particularly in narratives of national identity (see, for example, Wehrmann, 2018; Medby, 2018). The Arctic is a particularly interesting case in this respect, woven not only into the identity constructions for specific national identities, but also, for example, into the collective identity constructions among Nordic countries in the form of an 'Arctic Norden'. It is here that through a dense interplay of scientific exploration and exhibition, political claims and projects, as well as cultural practices, constructions of 'Norden' draw from references to the Arctic, while vice versa the Arctic is designated as a space lived by Norden (and Nordeners) (see Sörlin, 2013). Of course, such constructions of the Arctic through narratives not only pertain to contemporary practices of identity-construction related to the nation-state, but build on long traditions of 'Narrating the Arctic' particularly also based on scientific practices of both colonial powers and of nation-states, with often quite significantly 'different versions of Arctic narration, often involving complex relations between them: scientific, nationalist, imperial, evolutionary, religious' (Bravo & Sörlin, 2002, p. 20).

It is in this sense that defining the Arctic is equally a political exercise and an exercise in establishing imaginaries (cf. Steinberg et al., 2015). Those same definitions of the Arctic that serve as the geographical basis for collective action in international governance are deeply rooted in scientific understandings of what the Arctic is as developed in different scientific disciplines. The ongoing scientific disagreements regarding the boundary of the Arctic demonstrates that the Arctic as a geographical region and governance object eludes precise measurement and a fixed definition, and that it is thus socially constructed. To paraphrase Alexander Wendt's (1992) classic constructivist argument that the seemingly natural anarchical condition in world politics is the result of interaction processes between actors, not system structure, one could say that the Arctic is 'what scientists make of it'.[2] As Ó Tuathail (1996, p. 1) remarks, 'the geography of the world is not a product of nature but a product of histories of struggle between competing authorities over the power to organize, occupy, and administer space'. Given their central and often authoritative role in processes and institutions of Arctic governance, particularly in the Arctic Council, science communities have joined this 'struggle' over spatial constructions of what constitutes the Arctic, and hence how it is to be governed.

26.3 Science Geopolitics as a Fourth Geopolitical Heuristic

It is beyond the scope of this chapter to give a full and systematic account of the different linkages through which science is interconnected with geopolitics as an analytical device. The ways in which specific perceptions of place, space and political orders are constructed through politicians and government representatives (*practical geopolitics*), institutions like think-tanks and academia (*formal geopolitics*), and mass media as well as popular culture and everyday practices (*popular geopolitics*), has long been at the heart of critical geopolitical research (Kuus, 2010). However, while there is little in science as an institution and a method of generating knowledge and interpretation that would lead us to expect geopolitical imaginaries to be entirely absent in its preconceptions, practices, routines or outcomes, we argue that neither practical, formal nor popular geopolitics can fully capture the relevant dynamics at play in the case of science. Rather, we argue that these three heuristics of geopolitical reasoning can and need to be supplemented by a fourth which we describe as *science geopolitics*.

Where the four heuristics differ is not only in the agency of who constructs geopolitical narratives and images, but also how, in what social context, and by what means they are (re)produced. All four of these types draw on specific systems of knowledge that feed the geopolitical discourse. The practical geopolitics of 'practitioners of statecraft' (Ó Tuathail & Agnew, 1992, p. 194) is concerned with the study

[2] For a somewhat similar argument to Wendt's but based on Wittgenstein's philosophy of language, see Medby, 2019.

of 'how foreign policy decisionmakers make sense of international crises, how they construct stories to explain these crises, how they develop strategies for handling these crises as political challenges, and how they conceptualize "solutions" to these crises' (Ó Tuathail, 2002, p. 603). This can be contrasted with the 'citizen statecraft' (Pinkerton & Benwell, 2014) of popular geopolitics often rooted in shared understandings of cultural imaginings, symbolisms and conventional wisdoms, but also prejudices and stereotypes. The realm of science geopolitics certainly has the most overlap with formal geopolitics, though in our understanding the two rest on different knowledge categories that allow for their separation as analytical heuristics. For the sake of a widest possible, albeit simplified and ideal-type juxtaposition, formal geopolitics can be said to lean towards 'orientation knowledge' – 'belief systems, values, cultural traditions, worldviews, ideologies, religions, moral positions, mindsets, action-guiding norms [. . .], and reflection about the ethical conduct of one's life [. . .] providing actors and social systems with a moral compass, ideologies, goals, values, a cultural memory, and a collective identity' (Meusburger, 2015, p. 27), whereas science geopolitics has its basis more in 'factual knowledge', i.e. knowledge 'perceived, discovered, or learned by means of a methodically well-regulated procedure bound to justification, truth, and verification' (ibid., p. 24).

From this epistemological point of view, the terminology of science geopolitics appears to be an oxymoron; it is hard to think of any other way of knowledge-generation that is less geopolitical than the scientific method of producing reliable, valid, rationalized knowledge based on hard data, empiricist methodologies and conclusive evidence. Where the production of geopolitical narratives and images is most pronounced is in the realm of practical geopolitics in constant need of making sense of the world to shape, legitimize or redirect political decisions (Ó Tuathail & Agnew, 1992, p. 194). In contrast, scientists who map, measure and survey the world neither require, nor intentionally promote, geopolitical imaginings. *Informed* decisionmaking essentially describes both a process as well as an outcome in which science can serve as a major substitute for narrative authority as a source of legitimate decisionmaking. In doing so, science can shift the policy focus from national interests and underlying geopolitical conceptions towards questions of common concern and governance (Berkman et al., 2020, p. xiii).

The high expectation of informed decisionmaking as a holistic approach to balance national and common interests for sustainable development rests on the assumption of a pure, objective and unerring science 'speaking truth to power'. Indeed, it is the central role of science as the fundament of the pyramid of informed decisionmaking that puts it in both a crucial and vulnerable position in the holistic process. Particularly in governance settings marked by high complexity, uncertainty and urgency and crossing local-global scales, for which the Arctic is a prime example, science is indispensable as a source of good governance and as a consequence of which it has not only an informative, but very much a formative function. Much of what is taken for granted or seems commonsensical today is a result of the social construction not of politics, but of science. The history of the discursive and political construction of the Arctic is one that is well-recorded (e.g. Keskitalo, 2004; English, 2013), while its scientific construction as a spatial entity and research object often predates and then shapes political initiatives aimed at region-building (Knecht, 2013).

Under conditions of complexity, uncertainty and a 'continuum of urgencies' (Berkman et al., 2020, p. xi), the epistemic authority of science communities can be both anchor and engine of informed decisionmaking (Haas, 1992; Mitchell et al., 2006). But complexity, uncertainty and urgency are challenging contextual conditions for science as a knowledge-system as well. The scientific process is by far not a rigid system of neutral and objective knowledge-accumulation, but at best a close approximation that is nevertheless influenced by prior knowledge, unknowns, assumptions, research conventions, judgements and other factors to be found in the individual researcher, a scientific community or an entire discipline and which may themselves be of non-scientific origin. This is not to say that science is per se political – its methods shall guarantee it is based on rationality and reason – but that it would be a misconception to think of science to work like clockwork under a dome and in isolation from outside events and particularly its own research objects.

26.4 PAME and the Arctic LME Delineation Process

In order to provide a brief illustration of our argument thus far, we trace the evolution of the scientific delimitation of the network of Arctic large marine ecosystems (LMEs) in the Arctic Council's *Protection of the Arctic Marine Environment* Working Group and related Expert Groups. The Council had adopted ecosystem-based management (EBM) in the 2004 *Arctic Marine Strategic Plan* (AMSP) as a key instrument to 'achieve the sustainable development of the Arctic marine environment' (PAME, 2004, p. 1), one of the two pillars upon which the Arctic Council was founded in 1996, the other being environmental protection. We take the delineation process of the Arctic LME network as a succinct example of how science un/makes geographical boundaries against the backdrop of an imaginary Arctic space and with concrete implications for informed decisionmaking in circumpolar governance.

Saying that science is based on rationality and reason does not mean it will automatically lead to consensus between scientists or entire disciplines about concepts, methods and research results. Variable measurement is as inherent to science as it is to other systems that are based on processing knowledge, and it is the search for consensus and widely accepted definitions and findings that makes scientific progress possible in the first place. Different disciplines have defined the Arctic differently, and scientists have struggled to make sense of the polytheism of what the Arctic 'is'. The 1998 report on *Arctic Pollution Issues* by the Arctic Council's Arctic Monitoring and Assessment Programme (AMAP) WG mentions no less than five different conceptualizations of the Arctic derived from different scientific disciplines referring to a geographical boundary as developed in physical geography (the 'Arctic Circle' definition at 66°32′N), a temperature boundary from meteorology (the 'isotherm' definition of a mean July temperature of 10 °C), a permafrost boundary from climatology, a vegetation boundary from ecology (the 'treeline' definition north of which trees do not grow), and a marine boundary from oceanography 'along the convergence of cool, less saline surface waters from the Arctic Ocean and warmer, saltier waters from oceans to the south' (AMAP, 1998, p. 10). In addition to the cacophony of Arctic definitions based on physical conditions come those set by state authorities following legal, political and administrative considerations under national and international law as well as those 'invented' by epistemic communities usually along functional lines to address specific governance sectors. The International Maritime Organization (IMO) in its *Guidelines for Ships Operating in Polar Waters* (IMO, 2010, p. 9), the Arctic Human Development Report (AHDR), and, from the Arctic Council, AMAP and the Emergency Prevention, Preparedness and Response (EPPR) and Conservation of Arctic Flora and Fauna (CAFF) WGs have all adopted varying and only partly overlapping definitions of what they consider to be the 'Arctic' for their respective work (see Fig. 26.1).

The variable geometry of boundary-making and their reproduction in maps has two direct implications for processes of spatialization in Arctic governance. First, boundaries require broad consensus to be both legitimate and effective tools for policy planning and governance, but often their shape and/or the process and methods that led to their definition are contested. For instance, early on in the negotiations on the IMO's *International Code for Ships Operating in Polar Waters* (Polar Code) the definition of Arctic waters for the Code, the same as in the previous IMO Guidelines, was criticized in strong words by an NGO alliance as 'arbitrary and

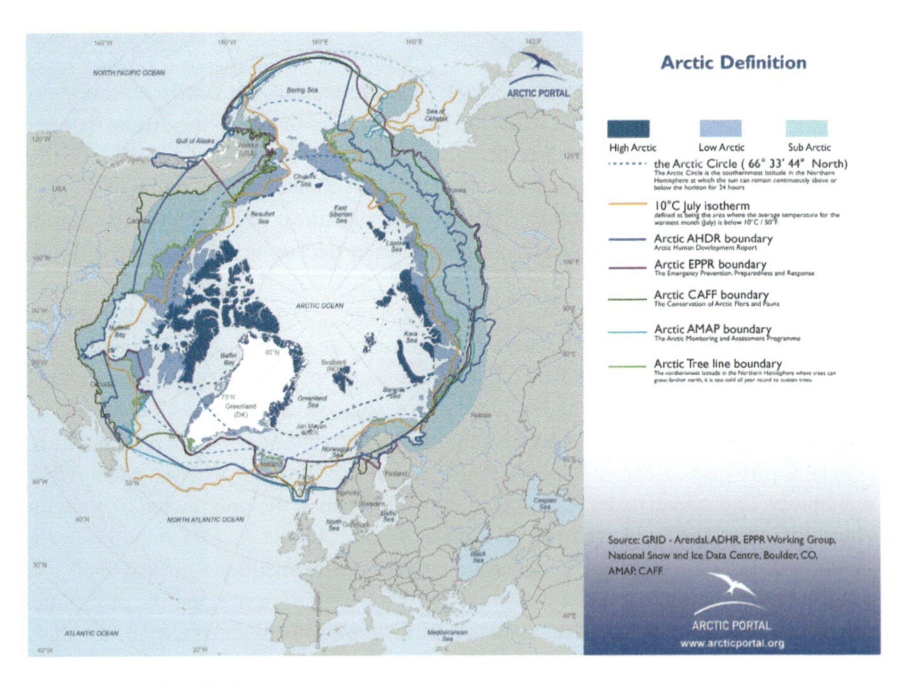

Fig. 26.1 Arctic definitions map

inadequate' (IMO, 2011a, p. 5). The alliance of Friends of the Earth International (FOEI), the International Fund for Animal Welfare (IFAW), the World Wide Fund for Nature (WWF) and Pacific Environment demanded – unsuccessfully – that the application area be extended into the North Pacific (ibid.) as well as further into the North Atlantic and Barents Sea (IMO, 2011b). In their view, sea ice extent should have been the most important parameter for delineating the Polar Code boundary. Further citing a 'strong scientific foundation', the group called for the LME network developed under PAME to be taken into consideration, warning that '[f]oregoing LMEs is inconsistent with modern scientific environmental management, impairs regional efforts to meet marine resource goals (e.g., the Arctic Council's Arctic Marine Strategic Plan), and diverges from the practices of other U.N. bodies, intergovernmental organizations, and national governments'(IMO, 2011a, p. 4).

Second, designating and delineating geographical areas of operation is often instrumental but not necessarily prescriptive for these institutions and communities, or for the governance and decisionmaking that they wish to inform. Contested social constructs that they are, designated areas of operation are subject to configuration, change, and continuous reinterpretation by both scientific and policy communities. Yet, how- and wherever set, a boundary does not only necessarily establish a more or less clear-cut geographical distinction between an inside Arctic and an outside non-Arctic, but also inevitably cuts through the issues and transboundary problems these same boundaries intend to demarcate for the Arctic governance complex. As CAFF, addressing issues of biodiversity and species and habitat management in the Arctic, warns with regard to its defined Arctic boundary, '[t]here is virtually no place on the earth that is not connected to the Arctic by migratory species. The Arctic is a distinctive region of the planet, and as such is worthy of a dedicated review of its environment, but its conservation and its future are tied to what happens throughout the world' (CAFF, 2001, p. 14). The main purpose of geographical boundary-drawing in the Arctic then is less to raise artificial distinctions between otherwise interconnected spaces, but to make a certain area identifiable and actionable for scientific research, policy planning and governance. As Dodds argues with regard to the 2018 *Agreement to Prevent Unregulated High Seas Fisheries in the Central Arctic Ocean* (CAOF agreement) that applies the precautionary principle to fisheries in a designated area of the 'Central Arctic Ocean', the agreement works towards 'the mobilization of geographical imaginaries and territorial practices being brought to bear on the Arctic Ocean. As a referent object, the CAO becomes a space of and for geopolitical interest, legal intervention and resource management' (Dodds, 2019, p. 549).

The same can be said of large marine ecosystems – 'areas of coherent ecological and geophysical processes [that] provide an appropriate scale for assessing the structural and functional integrity of ecosystems, including the separate and cumulative impacts of human activities' (PAME, 2019, p. 6) – that the PAME WG is working on to support the Arctic Council in its efforts to implement ecosystem-based management. As PAME notes, 'LMEs represent the appropriate and primary units for applying the ecosystem approach to management of the marine environment' (PAME, 2014, p. 3). EBM, in turn, was defined by the Council's *Expert*

Group on Ecosystem-Based Management (EBM EG) as 'the comprehensive, integrated management of human activities based on best available scientific and traditional knowledge about the ecosystem and its dynamics, in order to identify and take action on influences that are critical to the health of ecosystems, thereby achieving sustainable use of ecosystem goods and services and maintenance of ecosystem integrity' (Arctic Council, 2013, p. 12). The EBM EG sees the concept's main benefits in introducing 'flexible and adaptive management approaches in the Arctic that recognize cultural, governmental/legal and sub-regional differences, apply an integrated and interdisciplinary approach to understanding and managing these ecosystems, and ultimately maintain the resilience of Arctic ecosystems and communities' (Arctic Council, 2013, p. 10).

The case of the Arctic LME network constructed through the work of PAME is a striking example of how science geopolitics can result in variable geometries of the Arctic for four reasons. First, in contrast to the respective Arctic boundaries agreed by AMAP, CAFF and EPPR, PAME has refrained from defining a delineated 'referent object' of what it conceives to be 'Arctic' for its purposes. In PAME's *Arctic Offshore Oil and Gas Guidelines*, first published in 1997 and updated in 2002 and 2009, the WG makes clear that the definition of the geographical Arctic for the WG 'is left for Arctic states to determine' (see PAME, 2009a, p. 1 in the latest version). With each state deciding for itself how far south to stretch the boundary, the Arctic states indeed produced what can be labelled a distinct 'PAME area' (PAME, 1997, p. 8, 2002, p. 9, 2009a, p. 5), although this area only applies to the *Guidelines*, not the work of PAME as a whole. With regard to LMEs, no equivalent boundary has been constructed by PAME or the Arctic states. Second, the reluctance to boundary-making by PAME can be seen to run counter to initiating and implementing EBM, which is a deeply geographical/geopolitical concept. The EBM EG outlined altogether nine core principles to guide a common understanding of EBM across the Arctic Council, which later were approved by Arctic ministers at the Ministerial Meeting in Kiruna in 2013, and among which is the recommendation to recognize that 'EBM is place-based, with geographical areas defined by ecological criteria, and may require efforts at a range of spatial and temporal scales (short-, medium- and long-term)' (Arctic Council, 2013, p. 13). Third, the holistic approach to place-based EBM built on LMEs defined by four ecological criteria – bathymetry, hydrography, productivity and trophic relations – cuts through the sectoral approach found in other Arctic Council WGs and makes the determination of a designated area of operation for the separation of Arctic LMEs from non-Arctic LMEs in an interlinked global system of LMEs (see Sherman & Hempel, 2009) more difficult and necessarily arbitrary.

Fourth and finally, science is at the heart of EBM and the determination of LMEs. LME delineation is supposed to follow from the scientific assessment of the four ecological criteria to ensure the 'LME approach is applied within geographical management areas which are based on distinctive ecosystems rather than political boundaries' (PAME, 2009b, p. 1). The *Ecosystem Approach Expert Group* (EA EG) established under PAME in 2007 has developed a six-step implementation framework for EBM that sets identification of the geographical extent of the ecosystem as the very basis, followed by five more steps: description of the ecosystem, definition of ecological objectives, assessment of the ecosystem, valuation of the ecosystem, and management of human activities in the ecosystem (PAME, 2017, pp. 6–8). LME determination is a crucial step towards informed decisionmaking in Arctic EBM because '[t]he geographic extent of the LME, its coastal zone and contributing basins constitute the place-based area for assisting countries to focus on ecosystem-based strategies to recover completed fisheries, reduce coastal pollution, and restore damaged habitats' (PAME, 2009b, p. 4).

Taking those four aspects together, the basic formula that EBM 'is a science-based, place-based and adaptive approach to management of ecosystems' (PAME, 2016, p. 2) is not without inner tensions and only holds true if one accepts a considerable degree of variety in underlying basic, 'factual' understandings. Thus, for example, when PAME first addressed the issue of Arctic LME delineation in 2005, the initial list contained 17 potential LME candidates producing a spatially extended network of Arctic LMEs whose outer boundary reached into the North Pacific as well as North Atlantic Ocean areas.[3] Existing regional definitions used in the Arctic Council or other WGs did not serve as a delimiting tool for spatial orientation in that regard. The initial list included three LMEs located south of any of the physical Arctic definitions or those applied in other Arctic Council WGs: the Sea of Okhotsk LME in the Northwest Pacific between the Russian Federation and the Japanese island of Hokkaido, the Gulf of Alaska LME in the Northeast Pacific, and the North Sea LME bordered by the United Kingdom, France, Belgium, the Netherlands and Germany, in addition to Norway, Sweden and the Kingdom of Denmark (cf. Sherman & Hempel, 2009, pp. 573–580). Of those three, only the Sea of Okhotsk LME was ultimately kept, while the other two were removed, resulting in a tentative list of 15 Arctic LMEs (PAME, 2005, p. 15). Then again, the Faroe Plateau LME in the Northeast Atlantic, covered in whole or in part by the Arctic definitions of the AHDR and AMAP and CAFF WGs (cf. Figure 26.1), was not yet included. These decisions have little to do with questions of the methodology or accuracy of ecological assessments to determine the boundary of individual LMEs. They are deliberate choices about the spatial extent of what in the view of scientists the 'Arctic' is or should be. The discussion within scientific communities about the inclusion/exclusion of LMEs inside/outside of established Arctic boundaries indicates that the geographical 'Arctic' is a negotiable concept.

A year later, the total number of Arctic LMEs was brought back to 17 when two new LMEs, the Baffin Bay/Davis Strait LME and the Arctic Archipelago LME, were

[3]We restrict our analysis to the outer boundary of the Arctic LME network and do not consider the many and often substantial boundary adjustments made between adjacent LMEs inside that network over time. Though based on regular reassessment of the ecological criteria, these too are deeply geopolitical processes that involve constant deliberation and negotiation between, on the one hand, scientific communities in the Arctic Council and on the national level, and, on the other, administrations in neighboring Arctic states.

designated off the coast of Canada (PAME, 2009b, p. 7). As a PAME report notes, the resulting compilation of Arctic LMEs still 'includes the Sea of Okhotsk LME, which lies outside the Arctic area as used by the Arctic Council, but it does not include the Faroe Plateau LME which is included in the Arctic Council work' (PAME, 2013, p. 1). The map, adopted at the PAME WG meeting in Oslo in March 2006 and endorsed at the Arctic ministerial meeting in Salekhard later that year, was treated explicitly as a 'working map' in due consideration of the need for constant reassessment of the ecological state of the ecosystems in the future.

The working map did not stay within the confines of the PAME WG, however, but spread to inform other Arctic Council WGs as well, not least as a result of emerging cooperation with AMAP, CAFF and SDWG under the EA EG since 2007. For instance, a substantially adjusted version of PAME's LME map was used by AMAP in the 2007 *Assessment of Oil and Gas Activities in the Arctic* (OGA) (AMAP, 2010). The two maps are available for visual comparison in PAME's report *Large Marine Ecosystems (LMEs) of the Arctic area* (PAME, 2013, pp. 2–3). The OGA map, in turn, was used in 2013 by AMAP, CAFF and SDWG for the identification of Arctic marine areas of heightened ecological and cultural significance in 'AMSA IIc' (AMAP/CAFF/SDWG, 2013), a follow-up report to the 2009 *Arctic Marine Shipping Assessment* (AMSA) (PAME, 2009c). This process of incremental 'map diffusion' between scientific communities of PAME and other Arctic Council WGs caused some debate inside the Council and had larger consequences for the further delineation process of Arctic LMEs. AMAP had modified the 'working map' for the purposes of its assessment by removing the Sea of Okhotsk LME and adding the Faroe Plateau LME that PAME had originally excluded, as well as a large area off the Canadian province of Newfoundland and Labrador, designated as the Southern Labrador Shelf LME (PAME, 2013, p. 1). At a WG meeting in September 2007, those modifications raised objections by members of the PAME WG who pointed out that the altered map used by AMAP, in contrast to theirs, had not been endorsed by Arctic ministers (PAME, 2007, p. 11). AMAP's response came in the form of an outright rejection of both the competence and responsibility of PAME to even conduct assessment work with regard to LMEs, claiming lack of experience on the side of the WG (AMAP, 2008, pp. 23–24). Instead, the view was held that 'AMAP is the AC WG responsible for assessments and it is important that scientific experts are responsible for conducting such assessments' (ibid., p. 11).

This short, rare and rather unusual episode of a turf war between Arctic Council WGs was resolved in 2011 with an agreement to intensify cooperation under the PAME-led EA EG, which was broadened to a Joint Expert Group with participation from AMAP, CAFF and SDWG. The expansion of the EG coincided with a planned revision of the Arctic LME map which was the central objective in the first of a series of EA workshops held since 2011. The outcome of this revision process was a list of 18 LMEs resulting from the addition of the Faroe Plateau and Labrador-Newfoundland LMEs, on the one hand, and the segmentation of formerly two LMEs in the Bering Sea region (East Bering Sea and West Bering Sea LMEs) into three by designating the Aleutian Archipelago LME, on the other (PAME, 2013, p. 5) (see Fig. 26.2). In turn, PAME maintained the removal of the Sea of Okhotsk

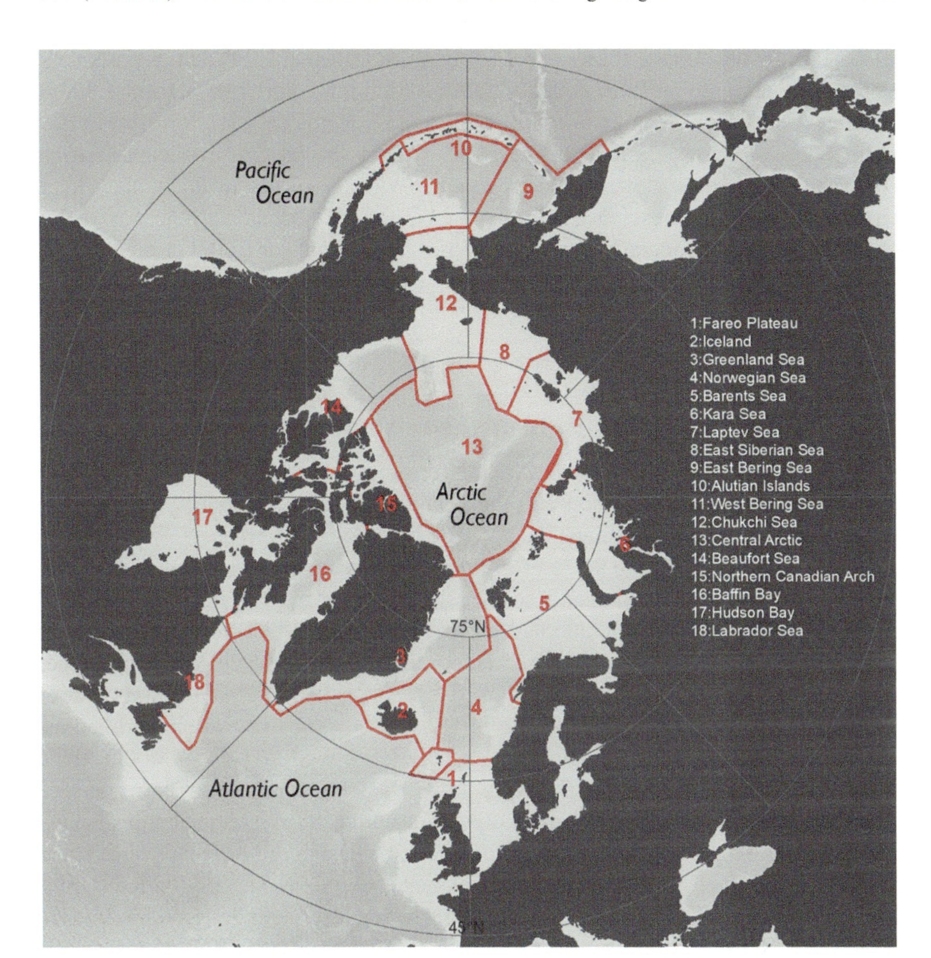

Fig. 26.2 PAME LME map 2013

LME, and restructured formerly four LMEs between Canada and Greenland into three. As one of the reasons for the adjustments, PAME cited inaccuracy of the ecosystem boundaries of the 2006 Arctic LMEs after being used in OGA and AMSA IIc, 'when it was observed that boundaries were cutting through important ecological features to split them somewhat arbitrarily in two parts, in the absence of an actual natural ecological or geophysical discontinuity' (ibid., p. 3).

Constant reassessments of the ecological state of LMEs, and the map modifications that result from it, constitute a core challenge to proper and timely implementation of EBM measures in the Arctic (cf. PAME, 2017). As the first of six steps in the implementation process developed by the Arctic Council, the spatial process of identifying, delineating and possibly reconstructing LMEs is deeply intertwined with institutional processes of Arctic Ocean governance. When new scientific data and information becomes available or ecological changes in the LME occur, reassessment becomes a necessary component for effective EBM across spatial and temporal scales to ensure policy affects the 'right' unit. But those scientific reassessments also steadily put pressure on the governance system to reevaluate and adapt implementation objectives and methods in order to avoid mismatches between ecosystems and their management. What scientists make of the Arctic influences processes of science coordination between WGs inside the Arctic Council and international cooperation between Arctic states whenever LMEs cross maritime boundaries of adjacent states, and even more so where they reach into areas beyond national jurisdiction.

26.5 Implications for Informed Decisionmaking

What does this short excursion into the field of geopolitical thought imply for the 'international, interdisciplinary, and inclusive (holistic) process' of informed decisionmaking that forms the overarching theme of the present volume? We have argued that the formative power of science geopolitics as a fourth heuristic should be given more careful consideration in the process of informed decisionmaking. In complex and challenging areas such as Arctic governance, science communities enjoy a high degree of epistemic authority that makes them a fundamental pillar of international cooperation.

We have shown for the case of the Arctic LME network developed in the Arctic Council that factual knowledge is not 'objective' knowledge to inform science diplomacy, but can also be, and quite often is, a source of disagreement and dispute. What counts as 'factual' knowledge about the same designated reference object, and the criteria derived from it for decisionmaking, obviously not only varies significantly between institutions and fora wide apart, but also potentially within science communities, as for example in our case between different AC WGs. The main point here, however, is that all the observed practices of generating and processing knowledge are implicated in what we have termed 'science geopolitics', that is the construction and 'making' of spatial images ('regions'). It is futile to argue that the geographical, temperature, permafrost, vegetation and marine boundaries of the Arctic, as determined in different scientific disciplines, are 'factually' wrong, or that one of them is more accurate than the other. Each of them individually constitutes a valid and justified construction of the Arctic, while together they form a set of competing claims to a spatial truth.

However, the possibility remains that disagreements about research objects, scientific methods, and findings remain unresolved and become part of the process of informed decisionmaking, where they complicate the provision of conclusive evidence for the formulation of viable policy options, or are used by political interests to undermine the validity and legitimacy of the scientific method. Needless to say, of course, this is anything but a one-way street. An essential part of informed decisionmaking lies in the political system (and the relevant actors) being able to exercise sound judgement on the possibilities and limits of scientific expertise.

Informed decisionmaking, in other words, depends on being able to at least partially reflect on the methodological disagreements mentioned, so as not to slip into the extremes of dismissing scientific expertise because of its inbuilt uncertainties, on the one hand, or unfiltered belief in science, on the other hand. This is not an easy task. In fact, it is extremely difficult. If anything, the reactions to the Covid-19 pandemic have highlighted on a grand scale the extreme difficulties that are always associated with the translation process between two social systems, i.e. science and politics, that necessarily always primarily operate according to their own logics. While designating and delineating the Arctic as a region in many ways obviously differs from reactions to a global pandemic in scale, it does not do so in kind when it comes to the difficulties involved with the translation process mentioned. Nonetheless, it would seem that in some sense the (relatively) small is beautiful here. As our examples have demonstrated, we are dealing here with contexts and settings that with some effort can be reconstructed, and the processes of 'construction' therein can be reflected upon. It is such a reflection on a systematic and continuous level that we deem to be a necessary ingredient to turn any kind of 'informed' decisionmaking into '*well*-informed decisionmaking'.

Science exists in, and plays an important role for, geopolitics. This is to say that quite obviously it never has been, and never can be, the pure quest for knowledge on which cooperative structures can be built to counter the 'power politics' of geopolitics. All this very often, and probably mostly, has nothing to do with the intentions of scientists. It is mostly about uses of science. At this point it should but serve as a reminder that what is certainly not implied is that the science diplomacy side is about the uses and involvement of science, while the geopolitics side is about the absence of science. What makes the relation between science and politics so difficult is not only that in many respects both exist in worlds of their own. What makes it so difficult is that despite this separation they remain deeply linked to, and dependent on each other.

References

Agnew, J. A. (1998). *Geopolitics: Re-visioning world politics*. Routledge.

Albert, M. (2021). Regions in the system of world politics. In J. Paul & N. Godehardt (Eds.), *The multidimensionality of regions in world politics* (pp. 59–74). Routledge.

Albert, M., Jacobson, D., & Lapid, Y. (Eds.). (2001). *Identities, Borders, orders. Rethinking international relations theory*. University of Minnesota Press.

Albert, M., Reuber, P., & Wolkersdorfer, G. (2014). Critical geopolitics. In S. Schieder & M. Spindler (Eds.), *Theories of international relations* (pp. 321–335). Routledge.

AMAP. (1998). *AMAP assessment report: Arctic Pollution issues*. Arctic Monitoring and Assessment Programme (AMAP). Available at https://oaarchive.arctic-council.org/handle/11374/924. Last accessed 16 Dec 2020.

AMAP. (2008). *Minutes of the 22nd meeting of the Arctic Monitoring and Assessment Programme (AMAP) Working Group, Quebec, Canada, 7–10 December 2008*. Arctic Monitoring and Assessment Programme (AMAP). Available at https://www.amap.no/work-area/document/61. Last accessed 16 Dec 2020.

AMAP. (2010). *Assessment 2007: Oil and gas activities in the Arctic – Effects and potential effects. Volume One*. Arctic Monitoring and Assessment Programme (AMAP). Available at https://www.amap.no/documents/doc/assessment-2007-oil-and-gas-activities-in-the-arctic-effects-and-potential-effects.-volume-1/776. Last accessed 16 Dec 2020.

AMAP/CAFF/SDWG. (2013) *Identification of Arctic Marine areas of heightened ecological and cultural significance: Arctic Marine Shipping Assessment (AMSA) IIc*. Arctic Monitoring and Assessment Programme (AMAP). Available at https://www.amap.no/documents/doc/identification-of-arctic-marine-areas-of-heightened-ecological-and-cultural-significance-arctic-marine-shipping-assessment-amsa-iic/869. Last accessed 16 Dec 2020.

Arctic Council. (2013). *Ecosystem-based management in the Arctic*. Report submitted to Senior Arctic Officials by the Expert Group on Ecosystem-Based Management (Arctic Council). Available at https://oaarchive.arctic-council.org/handle/11374/122. Last accessed 16 Dec 2020.

Berkman, P. A., Young, O. R., & Vylegzhanin, A. N. (2020). Book series preface: Informed decisionmaking for sustainability. In O. R. Young, P. A. Berkman, & A. N. Vylegzhanin (Eds.), *Governing Arctic seas: Regional lessons from the Bering Strait and Barents Sea, volume 1* (pp. v–xxv). Springer.

Bravo, M., & Sörlin, S. (2002). Narrative and practice – An introduction. In M. Bravo & S. Sörlin (Eds.), *Narrating the Arctic. A cultural theory of Nordic scientific practices* (pp. 3–32). Science History Publications/USA.

CAFF. (2001). *Arctic flora and fauna: Status and conservation*. Conservation of Arctic Flora and Fauna (CAFF). Available at https://www.caff.is/assessment-series/167-arctic-flora-and-fauna-status-and-conservation. Last accessed 16 Dec 2020.

Dodds, K. (2019). "Real interest"? Understanding the 2018 agreement to prevent unregulated high seas fisheries in the Central Arctic Ocean. *Global Policy, 10*(4), 542–553.

Dodds, K., & Nuttall, M. (2016). *The scramble for the poles: The geopolitics of the Arctic and Antarctic*. Polity Press.

Doel, R. E., Friedman, R. M., Lajus, J., Sörlin, S., & Wråkberg, U. (2014). Strategic Arctic science: National Interests in building natural knowledge – Interwar era through the cold war. *Journal of Historical Geography, 44*, 60–80.

English, J. (2013). *Ice and water: Politics, peoples, and the Arctic council*. Penguins.

Haas, P. M. (1992). Introduction: Epistemic communities and international policy coordination. *International Organization, 46*(1), 1–35.

Huebert, R. (2019). A new cold war in the Arctic?! The old one never ended! In L. Heininnen, H. Exner-Pirot, & J. Barnes (Eds.), *The Arctic yearbook 2019* (pp. 75–78). Akureyri.

IMO. (2010). *Guidelines for Ships operating in Polar waters*. International Maritime Organization (IMO). Resolution A.1024(26). Available at https://www.imo.org/en/OurWork/Safety/Pages/polar-code.aspx. Last accessed 16 Dec 2020.

IMO. (2011a). *Development of a mandatory code for ships operating in Polar waters. Polar Code Boundaries for the Arctic and Antarctic*. Submitted by FOEI, IFAW, WWF and Pacific Environment to the Sub-Committee on Ship Design and Equipment, 14 January 2011. DE 55/12/8. Available at https://webaccounts.imo.org. Last accessed 16 Dec 2020.

IMO. (2011b). *Development of a mandatory code for ships operating in Polar waters. Polar Code Boundaries for the Atlantic side of the Arctic*. Submitted by FOEI, IFAW, WWF and Pacific Environment to the Sub-Committee on Ship Design and Equipment, 28 January 2011. DE 55/12/17. Available at https://webaccounts.imo.org. Last accessed 16 Dec 2020.

Keskitalo, E. C. H. (2004). *Negotiating the Arctic: The construction of an international region*. Routledge.

Knecht, S. (2013). Arctic regionalism in theory and practice: From cooperation to integration? In L. Heininen (Ed.), *Arctic yearbook 2013* (pp. 164–183). Northern Research Forum.

Knecht, S., & Keil, K. (2013). Arctic geopolitics revisited: Spatialising governance in the circumpolar north. *The Polar Journal, 3*(1), 178–203.

Kuus, M. (2010). Critical geopolitics. In R. A. Denemark (Ed.), *Compendium of international studies* (pp. 683–701). Blackwell.

Levere, T. J. (1993). *Science and the Canadian Arctic. A century of exploration 1818–1918*. Cambridge University Press.

Lewis, M. W., & Wigen, K. E. (1997). *The myth of continents: A critique of metageography*. University of California Press.

Medby, I. A. (2018). Articulating state identity: "Peopling" the Arctic state. *Political Geography, 62*, 116–125.

Medby, I. A. (2019). Language-games, geography, and making sense of the Arctic. *Geoforum, 107*, 124–133.

Meusburger, P. (2015). Relations between knowledge and power: An overview of research questions and concepts. In P. Meusberger, D. Gregory, & L. Suarsana (Eds.), *Geographies of knowledge and power* (pp. 19–74). Springer.

Mitchell, R. B., Clark, W. C., Cash, D. W., & Dickson, N. M. (Eds.). (2006). *Global environmental assessments: Information and influence*. MIT Press.

Ó Tuathail, G. (1996). *Critical geopolitics: The politics of writing global space*. Routledge.

Ó Tuathail, G. (2002). Theorizing practical geopolitical reasoning: The case of the United States' response to the war in Bosnia. *Political Geography, 21*(5), 601–628.

Ó Tuathail, G., & Agnew, J. (1992). Geopolitics and discourse: Practical geopolitical reasoning in American foreign policy. *Political Geography, 11*(2), 190–204.

PAME. (1997). *Arctic offshore oil and gas guidelines*. Protection of the Arctic Marine Environment (PAME). Available at https://www.pame.is/index.php/document-library/resource-exploration-and-development/arctic-offshore-oil-and-gas-guidelines. Last accessed 29 Aug 2020.

PAME. (2002). *Arctic offshore oil and gas guidelines*. Protection of the Arctic Marine Environment (PAME). Available at https://www.pame.is/index.php/document-library/resource-exploration-and-development/arctic-offshore-oil-and-gas-guidelines. Last accessed 16 Dec 2020.

PAME. (2004). *Arctic Marine Strategic Plan*. Protection of the Arctic Marine Environment (PAME). Available at https://oaarchive.arctic-council.org/handle/11374/72. Last accessed 16 Dec 2020.

PAME. (2005). *Working group meeting report no. I-2005, Copenhagen, Denmark, 22–23 February 2005*. Protection of the Arctic Marine Environment (PAME). (available at https://www.pame.is/document-library/pame-reports-new/pame-working-group-meeting-reports. Last accessed 16 Dec 2020.

PAME. (2007). *Working group meeting report no. II-2007, Reykjavik, Iceland, 26–27 September 2007*. Protection of the Arctic Marine Environment (PAME). Available at https://www.pame.is/document-library/pame-reports-new/pame-working-group-meeting-reports. Last accessed 16 Dec 2020.

PAME. (2009a). *Arctic offshore oil and gas guidelines*. Protection of the Arctic Marine Environment (PAME). Available at https://www.pame.is/index.php/document-library/resource-exploration-and-development/arctic-offshore-oil-and-gas-guidelines. Last accessed 16 Dec 2020.

PAME. (2009b). *PAME progress report on the ecosystem approach to Arctic Marine Assessment and Management 2006–2008*. Protection of the Arctic Marine Environment (PAME). Available at https://www.pame.is/projects/ecosystem-approach/ea-documents-and-workshop-reports. Last accessed 16 Dec 2020.

PAME. (2009c). *Arctic Marine Shipping Assessment 2009 Report*. Protection of the Arctic Marine Environment (PAME). Available at https://pame.is/index.php/projects/arctic-marine-shipping/amsa. Last accessed 16 Dec 2020.

PAME. (2013). *Large Marine Ecosystems (LMEs) of the Arctic area. Revision of the Arctic LME Map* (2nd ed.). Protection of the Arctic Marine Environment (PAME). Available at https://www.pame.is/projects/ecosystem-approach/arctic-large-marine-ecosystems-lme-s. Last accessed 16 Dec 2020.

PAME. (2014). *The ecosystem approach to management of Arctic marine ecosystems: Concept paper*. Protection of the Arctic Marine Environment (PAME). Available at https://www.pame.is/projects/ecosystem-approach/ea-documents-and-workshop-reports. Last accessed 16 Dec 2020.

PAME. (2016). *The ecosystem approach to management of Arctic marine ecosystems: Brochure*. Protection of the Arctic Marine Environment (PAME). Available at https://www.pame.is/projects/ecosystem-approach/ea-documents-and-workshop-reports. Last accessed 16 Dec 2020.

PAME. (2017). *Status of implementation of the ecosystem approach to management in the Arctic. Report from the Joint Ecosystem Approach Expert Group as delivered through PAME to Senior Arctic Officials*. Protection of the Arctic Marine Environment (PAME). Available at https://www.pame.is/document-library/pame-reports-new/pame-ministerial-deliverables/2017-10th-arctic-council-ministerial-meeting-fairbanks-alaska-usa. Last accessed 16 Dec 2020.

PAME. (2019). *EA Guidelines: Implementing an ecosystem approach to management of Arctic marine ecosystems*. Protection of the Arctic Marine Environment (PAME). Available at https://www.pame.is/document-library/ecosystem-approach-to-management-documents/ea-guidelines. Last accessed 16 Dec 2020.

Pinkerton, A., & Benwell, M. (2014). Rethinking popular geopolitics in the Falklands/Malvinas sovereignty dispute: Creative diplomacy and citizen statecraft. *Political Geography, 38*, 12–22.

Riddell-Dixon, E. (2017). *Breaking the ice: Canada, Sovereignty, and the Arctic extended continental shelf*. Dundurn.

Sale, R., & Potapov, E. (2010). *The scramble for the Arctic: Ownership, exploitation and conflict in the far north*. Frances Lincoln Publishers.

Sherman, K., & Hempel, G. (2009). *The UNEP large marine ecosystem report: A perspective on the changing conditions of the LMEs of the world's regional seas*. United Nations Environment Programme.

Sörlin, S. (Ed.). (2013). *Science, geopolitics and culture in the polar region*. Ashgate.

Steinberg, P. E., Tasch, J., Gerhardt, H., Keul, A., & Nyman, E. A. (2015). *Contesting the Arctic: Politics and imaginaries of the circumpolar north*. I.B. Tauris.

Su, P., & Mayer, M. (2018). Science diplomacy and trust building: "Science China" in the Arctic. *Global Policy, 9*(S3), 23–28.

The State Council, People's Republic of China. (2018). *China's Arctic Policy*. Available at http://english.www.gov.cn/archive/white_paper/2018/01/26/content_281476026660336.htm. Last accessed 16 Dec 2020.

Wegge, N., & Keil, K. (2018). Between classical and critical geopolitics in a changing Arctic. *Polar Geography, 41*(2), 87–106.

Wehrmann, D. (2018). *Critical geopolitics of the polar regions. An inter-American perspective*. Routledge.

Wendt, A. (1992). Anarchy is what states make of it: The social construction of power politics. *International Organization, 46*(2), 391–425.

Chapter 27
(Action): Powered by Knowledge

Anita L. Parlow

Thank you to Øle, Lina and great Arctic Frontiers Team.

Whether the *Power of Knowledge* – the subject of our remarkable – and remarkably important – week in Trømso – or *Powered by Knowledge*, the subject of this panel, the central issue is the question of who unearths, shapes and controls the knowledge that gives definition to the Arctic region.

Who establishes the rules of the road, sets the priorities and implements them is vital to the outcome. For whom does science serve? Almost prescient for our current conversation was Michel Foucault', the twentieth century French philosopher, who grappled with this question in a far less complex, or globalized, world. Foucault warned that power, knowledge and the science that enables it, was too frequently used for social control.

It might not be a step too far, today, to note that capitalism and its knowledge systems, itself, is in crisis. National populist movements – be they in places like Argentina, the Middle East, or direct democracies or republics such as Italy, Britain or the US – or, directed capital nations such as China – the world's-people are informing their leaders that that continued marginalization from a globalization that does not serve their individual or community interests – will no longer be tolerated.

In what some have called a *Faustian Bargain* – the triumph of market-promised riches including its broader freedoms cost a price that has, far too often, been the social, economic and cultural marginalization of individuals, communities, or a stranding of entire nations. The combination of inequality, and, the capacity of the Earth to sustain its oceans, plant and animal life – as well as human beings, are in play. What questions cry out for scientific inquiry and knowledge to either reverse current practices or build greater resilience? What knowledge encourages greater equity into new paradigms?

A. L. Parlow (✉)
Fulbright Scholar, Iceland
e-mail: anita.parlow@kellogg.oxford.ac.uk; http://www.sustain-the-globe.com

P. A. Berkman et al. (eds.), *Building Common Interests in the Arctic Ocean with Global Inclusion, Volume 2*, Informed Decisionmaking for Sustainability,
https://doi.org/10.1007/978-3-030-89312-5_27

The engines of commerce, the velocity of money and questions of sustainability are increasingly focusing on the Arctic region – both from within and without. It is vital to prevent the recreation of globalization's past negative behaviors.

How science is deployed – how marginalized stakeholders are meaningfully engaged in shaping the questions of scientific research, policy-making and practices is a paramount question for global civilization, and the Arctic region specifically.

My Fulbright Scholar research on Iceland and Norway's scrapped offshore petroleum development in the *Jan Mayen* – taught a best-practices approach to include and balance competing commerce, development and environment interests. The inclusive approach, "the Norwegian model," – labor, – NGOs, environment and community stakeholders – offers an effective approach for a more inclusive government-corporate-citizen-environment dynamic.

Including, the right to say no.

As Øle Øvretveit said, the process must be "co-produced" – for both questions and implementation – thus, less likely to perpetuate the power imbalances that have marginalized so many and, continues to destroy the planet earth.

The various western liberal models and China's state-ownership approach to capital will advance, clash or find accommodation in the twenty-first century. It's basically up to us – rich or poor Native or non-Native, powerful or disempowered – to ask questions, along with the power to ensure they count.

Advocacy, public demonstrations worldwide and voluntary standards designed to equalize power is shaping the global debate. But, as a series of lawsuits move forward, an interesting view of plaintiffs is that even unsuccessful suits are effective in encouraging government and business leaders to improve practices particularly, regarding energy production.

According to the London School of economics, over 1,500 climate-related lawsuits have been brought in 28 countries between 2007 and 2020. Norway's Court of Appeals in Oslo rendered a judgment in the 2020 Greenpeace lawsuit against the Norwegian government on the grounds that the Constitution upholds the right to a healthy environment under the Constitution that guarantees a right to an environment that is conducive to health.

While the Norwegian Court did not invalidate the drilling licenses, it did acknowledge Constitutional limits to drilling for oil. At this writing, an appeal is likely – and, the process sets the stage for what can be achieved when new questions are raised.

Perhaps a bit of "greenwash" is visible in the Arctic Corporate Shipping Pledge recently supported by companies like Nike, shipping companies like Evergreen and others who took the pledge, encouraging others to not ship in the Arctic region – despite that, in the future, it might cost less CO2 emissions crossing the Central Arctic Ocean

Decisions by powerful interests, with its broad implications, highlight the need for sufficient knowledge of the truths and equitability of the questions asked. And the knowledge gained. The Central Arctic Ocean, likely to becomes as large as the Mediterranean by mid-century, might cost less in CO2 emissions than the longer existing routes for which the shippers are better equipped to operate.

The range of Arctic voices and salient questions that originate in the Arctic must be more amplified to the world at large. In that respect, the launching of the *Arctic Mayors Forum* – a new Arctic-wide forum devoted to municipal issues – is vital for a voice from the front lines of the climate initiatives that will define our planet into the distant future. It is the Mayors who are on the front lines, can balance the competing interests and, in that respect, give definition to geopolitics, commerce, the next generation and, indeed, a direction to scientific inquiry.

The Arctic must define its own future – no matter how chaotic, varying and, contentious it might be. The region's people, institutions, experience and long histories are inextricably connected to the region, and its residents know their own priorities. Their collective breadth of information, knowledge, and wisdom, must be incorporated into scientific inquiry for the knowledge that emerges to have both meaning – and, a fundamentally non-hierarchical respect– for the Arctic's people and the environment.

Chapter 28
(Action): Powered by Knowledge

Annika E. Nilsson

To address the cause of climate change, the Arctic, like the rest of the world, must move away from fossil fuels, not only their *use* but also the *production* of gas and oil. This will not be easy, given the national interests at stake. But I will not talk about geopolitics and national interests in fossil fuel today and instead focus on knowledge and decisionmaking related to non-fossil energy system.

Non-fossil energy also requires resources: The expansion of wind energy in the Arctic creates demand for access to land, sometimes in ways that can preclude other uses of the land. Many so-called green energy technologies rely on minerals – including rare earth minerals – that require mining. This has increased the interest in opening new mines and created a push for speeding up decisionmaking processes related to mining. At the same time, protests are mounting against new mining activities. The reason is that mines, similar to the expansion of wind power, create conflicts over land use. The Arctic is not an empty space and land already has value for other activities, such as reindeer herding and tourism, and for safeguarding essential ecosystem functions.

Knowledge will be essential for making sure that the necessary push away from fossil fuels does not lead to energy production systems with negative side effects on the sustainability of development. But whose knowledge should count when assessing the consequences? Who should decide what is important when the decisions are made? If we want to smooth decisionmaking, the knowledge production and decisionmaking processes must be seen as legitimate also by those who will be directly affected by new industrial activities.

A. E. Nilsson (✉)
KTH Royal Institute of Technology, Stockholm, Sweden

Nordland Research Institute, Bodø, Norway
e-mail: annika.nilsson@vetani.se

P. A. Berkman et al. (eds.), *Building Common Interests in the Arctic Ocean with Global Inclusion, Volume 2*, Informed Decisionmaking for Sustainability,
https://doi.org/10.1007/978-3-030-89312-5_28

Today, the processes used for assessing the impacts of new industrial ventures are often criticized. One reason is that not all knowledge is included. Another is that assessments are made piecemeal project by project, without attention to cumulative impacts. While the knowledge input in the details might be very good, systems perspectives of multiple pressures and erosion of resilience are missing.

We know from the past that national and global interests in Arctic resources have played a fundamental role in shaping many northern societies. Investments have created path dependencies with long-term consequences related to economic, social and technical structures. We know that climate change will have profound impacts on Arctic environments and societies. The implications of a new energy infrastructures could be just as profound.

We have been encouraged to end with a question for the panel discussion and my question is:

How do we ensure that knowledge processes and impact analyses are considered legitimate by all relevant parties and include attention to potential long-term and large-scale systems changes that could profoundly affect the sustainability of the transition to non-fossil energy systems?

Chapter 29
(Action): Powered by Knowledge

Paul Arthur Berkman

We are entering a world with 8 billion people this decade, inhabiting a globally-interconnected civilization aligned with changes on a planetary scale, recognizing that human-population size has skyrocketed over 400% just since World War I – one century ago, in the lifetimes of our oldest living relatives. Crossing thresholds unlike any in human history – considering *Our Common Future* – there is great responsibility for decisions that operate in the face of change.

With such responsibility that belongs to all of us – **science diplomacy** has been accelerating as an **international, interdisciplinary and inclusive (holistic) process**, involving **informed decisionmaking** to balance national interests and common interests for the benefit of all on Earth across generations.

To restate the obvious – at the levels of peoples, nations and our world there is a **'continuum of urgencies'** to address from **security time scales** (mitigating the risks of political, economic and cultural instabilities that are immediate) to **sustainability time scales** (balancing environmental protection, economic prosperity and societal well-being across generations). With this observation – introducing definition to avoid jargon – *informed decisions operate across a 'continuum of urgencies.'*

Children and even young adults living today – our children and grandchildren – will be alive in the 22nd century. As a scalable proposition – for each of us as individuals, the 'continuum of urgencies' is like driving on any road, constantly adjusting to the surrounding vehicles and circumstances while being alert to the red lights ahead and traffic behind.

P. A. Berkman (✉)
Science Diplomacy Center, EvREsearch LTD, Falmouth, MA, USA

Science Diplomacy Center, MGIMO University, Moscow, Russian Federation

United Nations Institute for Training and Research (UNITAR), Geneva, Switzerland

Program on Negotiation, Harvard Law School, Harvard University, Cambridge, MA, USA
e-mail: paul.berkman@unitar.org; pberkman@law.harvard.edu

P. A. Berkman et al. (eds.), *Building Common Interests in the Arctic Ocean with Global Inclusion, Volume 2*, Informed Decisionmaking for Sustainability,
https://doi.org/10.1007/978-3-030-89312-5_29

The Arctic that we can observe and measure is crossing environmental thresholds where the boundaries of systems are changing. In the ocean and on land – with the sea ice and permafrost, respectively – these environmental state-changes are unambiguous. In the Arctic today, now, without projecting into the future – the risks of instabilities are inherent with these marine and terrestrial ecosystems.

Fortunately, we have science, which can be characterized as the 'study of change,' with international and interdisciplinary inclusion involving the natural sciences and social sciences as well as Indigenous knowledge – all of which characterize patterns, trends and processes (albeit with different methodologies) that become the bases for decisions.

Informed decisionmaking starts with questions, where each of us has the capacities to contribute as both observers and participants, convening dialogues among allies and adversaries alike to build common interests. There also is opportunity to champion the momentum of the **"common Arctic issues"** of "sustainable development and environmental protection" established by the eight Arctic states and six Indigenous Peoples Organizations with the Arctic Council from their 1996 *Ottawa Declaration*.

Arctic sustainability involves governance mechanisms, highlighted by the three binding agreements that have emerged from Arctic Council task forces since 2009. With the Arctic Economic Council, there also is recognition about the fundamental importance of built infrastructure, which is characterized by technology and capitalization with goods, services and markets. Together, the sustainability of the Arctic as elsewhere on Earth involves effectively coupling between the governance mechanisms and built infrastructure.

The Arctic is special for humanity, introducing hope and inspiration with the North Pole as a *"pole of peace"*, as envisioned by Mikhail Gorbachev is his famous 1987 Murmansk speech. Our common journey – to operate across a 'continuum of urgencies' – fundamentally involves next-generations on a global scale, stimulating holistic educational initiative like the University of the Arctic; evolving as a globally-interconnected civilization with knowledge and lessons of the 20th century that nationalism breeds global conflict. Together, we can contribute to informed decisionmaking, building a sustainable Arctic for the benefit of all on Earth across generations.

Chapter 30
(Action): Resilient Arctic Communities

Aileen Campbell

Ladies and gentlemen,

God dag.

It is a pleasure to be with you at Arctic Frontiers today and it is timely – Tomorrow, Shetland, our archipelago that sits as far north as Cape Farewell in Greenland, celebrates its annual Up Helly Aa Viking fire festival – a very explicit and lively celebration of Shetland's and Scotland's long and intertwined historic links with Norway.

And this gathering also comes just a couple of days after we celebrate the poetry and song of our national Bard Robert Burns. The poetry, and songs that he penned spoke of love, fairness and egalitarianism. But he also espoused an internationalist outlook that remains as crucial as ever for Scotland as we seek to continue to build new relationships with countries around the world and cement and celebrate old partnerships that have been integral to our story.

I mention Burns because those themes of fairness, kindness, solidarity speak to the values that underpin our National Performance Framework. Our Framework has been developed jointly by ourselves in national government along with our partners in local government and embraced by much of public life. It articulates the type of country that we want Scotland to be – a fairer country with opportunity for all. And it is important in helping us shift from judging the success of Scotland by GDP alone to instead also focusing on the wellbeing of our people.

We join many of the countries represented here today in that wellbeing journey as we attempt to rebalance our economies for the betterment of the people we all care so much about whether we are politicians, policy makers or professionals.

And that is why this gathering is so important. It provides a valuable opportunity to share ideas, test new approaches, and learn from one another. To work in

A. Campbell (✉)
Scottish Government, Edinburgh, UK
e-mail: info@parliament.scot

P. A. Berkman et al. (eds.), *Building Common Interests in the Arctic Ocean with Global Inclusion, Volume 2*, Informed Decisionmaking for Sustainability,
https://doi.org/10.1007/978-3-030-89312-5_30

partnership where our pooled endeavours can maximise our potential and help us unlock barriers and tackle common challenges.

30.1 Shared Challenges

Scotland's population is just over 5.4 million, similar to Norway, Finland and Denmark. Like Arctic nations, most of our landmass is classified as rural – as much as 98%. We have 96 inhabited islands, with local populations often in single figures.

Indeed, with only 9 people per square kilometre, our Highlands and Islands region is one the most sparsely populated areas of Europe.

We have an ageing population and an economy that – while performing strongly – still sees too many young people struggle to thrive where they were born and raised.

Scotland and the Arctic therefore have much in common when it comes to ensuring that – no matter how rural – all of our communities are resilient, successful and have access to high-quality services.

By working together and pooling our expertise, we can make sure that *remote* doesn't become a byword for *marginal*.

That is why last September we published *Arctic Connections*, Scotland's first Arctic Policy Framework.

Our document puts communities at the heart of a renewed Scottish-Arctic dialogue, celebrating the many existing links while highlighting opportunities for even greater cooperation.

Scotland is eager to learn from our Arctic neighbours and we also want to be an active contributor into developing approaches that help improve the resilience of communities across the region.

30.2 Case Studies

The Scottish Government's approach is to empower communities to take the lead in delivering local projects. We do that by providing funding and support, and removing barriers, sometimes ourselves, which stand in the way of progress. Our Place principle, which means being guided by the needs of the community and the place asks that local partners – both private and the public sector – cooperate to develop local solutions to local issues. Not doing things to folk, not assuming we know best, but instead co-producing sustainable and more transformative positive outcomes.

One example of successful community-led regeneration is the Isle of Eigg on the Inner Hebrides. Before the community took ownership, Eigg faced problems like high unemployment and poor access to services, which led to population decline. Eigg had no mains electricity, poor housing and no proper pier for ferries.

The Heritage Trust – a group of local residents – became the Island's owners 22 years ago. Since then:

- the Island's population has increased by 60%;
- Eigg runs its own housing association; and
- between 85 and 95 per cent of its energy comes from renewable resources.

But it's not only our rural and island communities that are leading change. In one of our most deprived areas of Glasgow, the community and public agencies have a created a partnership – Clyde Gateway – to deliver improvements to housing, environment and employment opportunities. Clyde Gateway's approach to regeneration is successful because it acknowledges that the experts are the people who live there.

So we want to see more of this, and it is clear that when we listen, work with, and trust our communities, that incredible things can happen. But it also points to a need to shift the power balance of Scotland. And 20 years on from the establishment of that parliament, it is right that we think about what is next for governance in Scotland, and further strengthen our approaches to community empowerment and right to buy schemes.

30.3 Community Climate Action

Building resilience in our local communities, especially on islands and in coastal areas, will also be critical in terms of preparing for the global threat of climate change.

We want communities to drive solutions, so we created the Climate Challenge Fund in 2008. The Fund has supported community-led organisations to run projects that reduce local carbon emissions, cut waste, grow local food and lower energy use in homes and community buildings.

30.4 Arctic Frontiers in Scotland

Many of the shared challenges we face, environmental, economic, and demographic require partnership working and cooperation. And that is why we in Scotland value this opportunity to continue that dialogue, discourse and discussion with our Arctic partners in order that we can emerge through those challenges in stronger positions and with successful communities.

And that is why I am also delighted to announce that later this year Scotland will host Arctic Frontiers Abroad, which will help us further reflect on how we can increase our cooperation further.

So Scotland looks forward to welcoming you and I look forward to continued working with you to make our shared aspiration of successful and resilient places becomes the reality we all seek.

So thank you for letting us be part of this conference, and I look forward to the panel discussion and your questions. Thank you very much.

Chapter 31
(Action): Resilient Arctic Communities

Joel Clement

The top down or bottom-up question is riveting. One of the primary findings from the Arctic Resilience Report is that ideas and innovation and leadership can, should, and often do come from local communities, particularly from Indigenous leaders who have a relationship with place and time and change that surpasses that of any other people on earth. The problem is that the change is happening too fast. At this point no one is off the hook in terms of making resilience work. And the number one finding of the report was that local empowerment is the most important factor for building resilience. That means that governments at all scales need to take off the handcuffs. There are laws and planning processes and consultation rules that are in place that don't work well and that often constrain resilience. So, on one hand, the governance that comes from outside needs to be fixed to support resilience, and those handcuffs need to come off.

But the big constraint we always run into when we talk about resilience is that we can have all these great ideas, they can come from the bottom or the top, but no one is putting money into them. So it comes down to the investment question. That's where the rest of civil society, I think, can engage more fully. We need to stop wondering and vexing about where the big money is going to come from and start experimenting and moving forward on resilience ideas, even if it is on a shoestring. We are having conversations here with a lot of people on how they do that. We can talk about that more throughout this conversation, but I do think that's where the dam breaks. That's when you start to see the proliferation of projects attract notice and, ideally, investment.

As I mentioned, often the number one indicator of resilience is fate control or empowerment and the degree of empowerment. But also important is the diversity of voices within the community, and that is why I think that youth and gender issues are

J. Clement (✉)
Arctic Initiative, Belfer Center, Harvard Kennedy School, Cambridge, MA, USA
e-mail: joel_clement@hks.harvard.edu

P. A. Berkman et al. (eds.), *Building Common Interests in the Arctic Ocean with Global Inclusion, Volume 2*, Informed Decisionmaking for Sustainability,
https://doi.org/10.1007/978-3-030-89312-5_31

very important. There is more we can all do to enable women and youth to have strong, impactful voices. It's a form of empowerment that must go beyond merely giving youth and women the room to operate, and within communities, this can really be a great indicator of success.

If your goal is long term resilience you are not welcoming extractive boom and bust industries such as oil and gas. Despite the short-term economic gains, it's not worth the long-term stress on communities. Historically we have seen that such extractive economies are not good for either the communities or the ecosystems they depend upon, so it creates an automatic tension between people that are coming in to work in those industries and those who are of that place. It has a negative effect on resilience unless there is an immense amount of respect and investment in shared long-term goals.

Chapter 32
(Action): Resilient Arctic Communities

Mikhail Pogodaev

Dear Arctic friends,

First of all let me thank the organizers of this conference for inviting me! It is the honor for me to be at this stage and speak about perspectives of our region! Thanks to Norway and Tromsø for Arctic cooperation!

I also would like to thank Sami people for hospitality and leadership in the Arctic cooperation for all Indigenous Peoples of the world!

I am originally Even reindeer herder from Sakha Republic Yakutia which is a biggest region of the Russian Arctic with a territory of more than 3 million sq. Km. It is 8 times of Norway. Inhabited by almost 1 million people including Indigenous Peoples. From time immemorial Indigenous Peoples like Even, Evenki, Yukagirs, Chukchi and Dolgan inhabited this land doing reindeer herding, fishing and hunting. I want you to understand that it is a challenge to manage territory of this size. Our challenges are huge distances, sparsely populated areas, lack of infrastructure, limited access to markets, extreme climate conditions.

As we say Yakutia is a fourth Pole on our planet. You know we have North Pole, we have South Pole, we have Himalaya as a third Pole and we have Yakutia – the fourth Pole of the Cold. Climate is extreme, in the Pole of the Cold Oimyakon was registered lowest temperature −72 °C and the difference between winter and summer temperatures is more than 100°

Almost all the territory of Sakha Republic is covered by Permafrost in some areas it is 1500 m deep.

Today we talk about the Power of Knowledge. How people could survive and thrive in this harsh climate conditions? The answer to this question is Knowledge accumulated and transferred for thousands of years from one generation to another.

M. Pogodaev (✉)
Ministry for Arctic Development and Indigenous Peoples Affairs, The Government of Sakha Republic (Yakutia), Yakutsk, Russian Federation
e-mail: pogodaev.ma@sakha.gov.ru

P. A. Berkman et al. (eds.), *Building Common Interests in the Arctic Ocean with Global Inclusion, Volume 2*, Informed Decisionmaking for Sustainability,
https://doi.org/10.1007/978-3-030-89312-5_32

Indigenous knowledge let our people to utilize natural resources in a sustainable way for millennia, they created their unique circumpolar civilization, which was unique way of adaptation to harsh climate conditions and they live in a harmony with Nature.

Modern Yakutia is developing. And we would like to develop our region, improve the quality of life, develop education, social services, living conditions, preserve and sustain traditional culture and economic activities. Economy of Yakutia is based on natural resources, first of all – diamonds, gold, silver, coal, oil and gas etc. So the biggest challenge for our region is to find an answer to the question: how to keep the balance between economic development and preservation of fragile nature and Indigenous Peoples livelihoods since they keep very close connection to the Nature.

In our region we make it in our way, we have a special law on Ethnological Expert Review. And Yakutia is only a region in Russia where we have this kind of law. It means that any development project to be realized on Indigenous lands, before it starts, in addition to Environmental and Social Impact Assessment, must be investigated by an Ethnological Expert Assessment. In other words they have to find out how this particular project will impact Indigenous Peoples and their livelihoods. This is the way to protect Indigenous people's rights. Companies must compensate all negative impacts to Indigenous peoples. Another tool to protect Indigenous lands is the law on territories of traditional nature use (TTNU). It is special protected areas reserved for preservation of traditional livelihoods of Indigenous peoples. In Yakutia we have established 62 TTNU and they cover more than 60% of the territory of Sakha Republic. I also must admit that in Yakutia 37% of its territory is protected areas. 37%! You remember I told you more than 3 million sq.km. Think about it.

But I will get back to the theme of this conference – the Power of Knowledge.

As I said knowledge let our people to thrive on our land with extreme conditions for millennials. But today we face major challenges with climate change and globalization. Thawing permafrost put on a risk all infrastructure, houses, roads, releases methane etc. More flooding affect many remote settlements, we already have climate refugees. Winter roads, food security, energy infrastructure, transport, reindeer herding. Taiga and tundra wild fires represent huge risk for security of people and loss of pastures. Every year we observe increase of number of wild fires in our region and this challenge will be accelerated in the future.

Because of globalization Indigenous knowledge is disappearing. And therefore we need to take actions to preserve this knowledge, transfer it to young generations and let people use their own knowledge. Science is important, but in combination with Indigenous knowledge it can give more comprehensive answers to the society and to be better prepared to meet global challenges. In this regard we are very happy to say that cooperation between researchers from Norway and Yakutia is developing. I would like to thank Nordforsk for support of the project led by UArctic EALAT Institute at the International Centre for Reindeer Husbandry and focused on Co-production of knowledge between scientists and Indigenous peoples.

We also have another project which aims to preserve Indigenous knowledge. It is a joint initiative of the Russian SAO Nikolai Korchunov and the President of Sakha

Republic Aisen Nikolaev and they proposed this project to the Arctic Council among other project proposals from Russia. The title of the project is Digitalization of Linguistic and Cultural Heritage of Indigenous Peoples of the Arctic. This project is led by UNESCO Department of North-Eastern Federal University in Yakutsk. Project aims to engage youth to document, digitalize and create a portal to store linguistic and cultural heritage of Indigenous peoples of the Arctic. We appeal to Permanent Participants (PP's) and Arctic states to join and support this initiative.

Education is important, sometimes it can be used as a tool to assimilate Indigenous people, but it can be also a key to maintain Indigenous people's livelihoods if it will be also based on Indigenous languages and knowledge. But how to achieve this? Russia proposed a project in the Arctic Council called Children of the Arctic. It aims to provide a platform for exchange of experiences, identify best practices and let educators work together across all Arctic nations to develop new models of Arctic education. This is the Russian initiative proposed and supported by RAIPON – organization of Indigenous peoples of the Russian Federation. Now this project focuses on Preschool education, but we propose to extend this project also to school education. We also hope that this project will develop the concept of Nomadic schools, which was developed in Yakutia by our Even people researcher academician Vasily Robbek. We also propose our International Arctic School in Yakutsk as a platform for international cooperation in the Arctic on education.

It is not enough time to tell you about all our perspectives, but at least I told you some of them. We are interested to cooperate with Norway and all other Arctic nations. We need even more cooperation in the Arctic, we need to share our experiences, use the best available knowledge, new and traditional technologies, best practices. We need to keep Arctic Spirit of cooperation, to keep Peace and Stability in our Arctic home!

When I was leaving my home town Yakutsk it was −51 °C. It is warmer than it used to be when I was a child. But I hope that Arctic will stay cool!

Thank you!

Mikhail Pogodaev

Part V
Conclusion

Chapter 33
(Action): The State of the Arctic

Ine Eriksen Søreide

Speech/statement | Date: 27/01/2020.

By Minister of Foreign Affairs Ine Eriksen Søreide (Tromsø, 27 January).

Good morning,

It is a pleasure to attend the Arctic Frontiers conference again, and to be back here in Tromsø.

Ladies and gentlemen,

The Arctic today is a region characterized by peace, stability and international cooperation. This is no coincidence. It is a result of political choices. It is something we have worked hard to achieve. And it is something we will work hard to maintain.

33.1 Arctic Governance

I think it is fair to say that during the last years we have seen an increase in the attention to the Arctic region.

We have heard views that there is a legal vacuum in the Arctic – a kind of a "no man's land". Some have argued that a new "Great Game" is taking place in the Arctic, and that we need new structures of Arctic governance, including a new security policy forum for the Arctic.

First and foremost it is extremely important to underline that there is no legal vacuum in the Arctic. On the contrary: An extensive national and international legal framework already applies to the Arctic.

The Law of the Sea provides the basic architecture underpinning all ocean governance in the Arctic. It distributes jurisdiction and establishes clarity and

I. E. Søreide (✉)
Minister of Foreign Affairs, Oslo, Norway
e-mail: utenriksminister@mfa.no

P. A. Berkman et al. (eds.), *Building Common Interests in the Arctic Ocean with Global Inclusion, Volume 2*, Informed Decisionmaking for Sustainability,
https://doi.org/10.1007/978-3-030-89312-5_33

predictability as regards rights and obligations for all States in all ocean and sea areas.

The Law of the Sea also provides important rights and obligations regarding freedom of navigation, marine scientific research, protection of the marine environment, delineation of the outer limits of the continental shelf and other uses of the Arctic seas.

Furthermore, a series of treaties and legal instruments apply to the Arctic, as they do to other parts of the world. A concrete example is the new Agreement against unregulated fishing in the Central Arctic Ocean.

So to sum up, national and international mechanisms are already in place to clearly clarify both ownership and sustainable management for utilizing resources in the Arctic.

33.2 Arctic Council

Ladies and gentlemen,

Regarding governnance structures, the Arctic Council has been the primary arena for addressing international issues in the region for more than 20 years. This was reconfirmed by the Foreign Ministers in Rovaniemi, Finland last year. The Council – firmly supported by its member states, Indigenous peoples and observer countries – has been instrumental in setting the agenda, developing new knowledge and building trust across borders. I would argue that the Council has served all its members well in the current format discussing the current topics.

The Council provides a platform for addressing cross-border issues in the region. It has been instrumental in developing new knowledge on climate change in the Arctic. And it has been setting the agenda when it comes to discussing the opportunities and challenges in the Arctic.

The Arctic Council has also facilitated the negotiation of three important, legally binding agreements between the eight Arctic states. These agreements – on search and rescue, marine oil pollution preparedness and response, and scientific cooperation – highlight areas where cross-border cooperation is the way forward.

Under the steady guidance of Iceland's chairmanship, the Arctic Council will help to ensure that the working program of the Council can deal effectively with future demands and challenges.

33.3 Barents Euro-Arctic Council

In October last year Norway assumed the chairmanship of the Barents Euro-Arctic Council. This forum has been a cornerstone of regional cooperation in the Arctic since 1993. We will use our chairmanship to develope our relationship with our

Nordic neighbors and Russia. The close involvement of local and regional actors in the Barents region will be crucial to achieving these goals.

Together with Governor Magdalena Anderson of Västerbotten, we will later today discuss how the national and regional level can work better in the Barents region, and in particular new policies for the youth. I hope you will join us at that side event.

To deal with the challenges and to take advantage of the opportunities in the Arctic, Norway continues to believe in effective multilateral cooperation within the framework of international and national law and well-established governance structures.

33.4 Climate Change

Ladies and gentlemen,

It is here in the Arctic where we see the effects of climate change most dramatically.

The temperature rises more the further north you come. Last year, 3.3 degrees over average temperature at Svalbard, Norway's northernmost archipelago. That means 31 consecutive years with temperatures above the average at Svalbard.

Climate changes has implications for ice conditions, sea level and air temperature. However, it is important to underline that this is not directly linked to the activities in the Arctic, but predominantly due to activities and emissions outside the Arctic.

This is why the Paris Agreement is so important. It is the main legal vehicle for cooperation on reducing greenhouse-gas-emissions. Norway intends to do her part. Our current target under the Paris Agreement is to reduce emissions by at least 40% by 2030 compared with the 1990 level.

33.5 White Paper

Ladies and gentlemen.

The government will present a new white paper on the Arctic to our parliament this fall, which will outline our ambitions for a strong and innovative Arctic region. It will confirm our commitment to international cooperation as foreign and domestic policy are very much interconnected in the North. We need to follow closely global and local developments affecting the Arctic region, and we need to be well placed to meet them.

Ladies and gentlemen – in closing,

For Norway, the Arctic is not a remote place. Ten percent of our population live here north of the Arctic Circle. Some of our most innovative industrial areas are located here. Norwegians have lived by and off the sea for centuries. We will continue to utilize ocean resources and to strike the balance between sustainable use and protection. To ensure sustainable economic development in the Arctic, based on the best available knowledge and science. And based on the highest environmental standards.

I wish you all a pleasant and fruitful stay at here at Arctic Frontiers!

Thank you.

https://www.regjeringen.no/no/aktuelt/arctic_frontiers/id2688675/

Chapter 34
(Action): The State of the Arctic

Mike Sfraga

I do believe that the Arctic now is a global Arctic. It is no longer "emerging"—the Arctic has emerged. There certainly are new tensions among the Great Powers, as Secretary Pompeo's speech in Finland made clear. To understand how the Arctic nations approach the current circumstances of today's Arctic, it is important to understand their different perspectives. To me, it is as if they are playing different games. China plays the game *Go*. They take a long-term approach, and consider many parts of the world, Africa to the Arctic. Russia plays the game *Survivor*. They rely heavily on oil and gas development, have a declining population and an oligarchical structure. And the United States plays the game *Twister*. We have got one arm in the South China Sea, one in the Indian Ocean, one somewhere else. And now the United States is engaging in the new Arctic. That's a stretch of energy, time, money, navy. These different approaches are what is pressuring Arctic exceptionalism, the collaborative structures we have created for the Arctic. I'm not sure we need a new order as Bobo Lo suggests. What I would say is that we need a supplement and an innovative next step for the Arctic Council, for the IMO, for UNCLOS. We should strengthen these existing international orders.

If you want to see the internationalization of the Arctic, look no further than the Northern Sea Route. There is direct investment from China, and a massive build-up in oil and gas productivity along that Route destined for Chinese and other Asian markets. There are ice-reinforced oil tankers built in the shipyards of South Korea, financed by other nations. This is the new Arctic. The Arctic was an energy-rich region in the past, but now because of the melting and retreating sea ice, the whole area is more accessible. These dynamics are redefining the energy equation, supplementing President Putin's economy significantly, and reshaping geopolitics. I still believe that the current Arctic institutions and their traditions of cooperation

M. Sfraga (✉)
Wilson Center, Washington, DC, USA
e-mail: Michaela.Stith@wilsoncenter.org; mike.sfraga@wilsoncenter.org

P. A. Berkman et al. (eds.), *Building Common Interests in the Arctic Ocean with Global Inclusion, Volume 2*, Informed Decisionmaking for Sustainability,
https://doi.org/10.1007/978-3-030-89312-5_34

and collaboration, will ensure proper management of the Northern Sea Route and that the international regimes dictating shipping rules and regulations will hold. However, it is going to be up to the other nations to make sure it holds.

As far as the future of the Arctic goes, two things give me hope. First, today's young leaders—they get it. They are *iphoning* away and they know the future that they want. If they don't like the future that now seems to be coming, they will make it the one they want. I think it is our collective job to help them along. The second is industry, which is not the villain here. I believe that industry can be innovative and think outside the box, not just for the Arctic, but for the world. There is a way to provide innovative educational technology and opportunities, innovation to help our youth become leaders. There are also ways to let industry lead. Right now, I think there is a lot of money to be made in the Arctic. But the North could be a model for how to do it right. So, it seems like we have two really good things here, the youth equation and the innovation equation.

Chapter 35
(Research): Conclusions: Building Global Inclusion with Common Interests

Paul Arthur Berkman, Oran R. Young, Alexander N. Vylegzhanin, David A. Balton, and Ole Rasmus Øvretveit

Abstract The premise for Building Common Interests in the Arctic Ocean with Global Inclusion recognizes the Arctic is being transformed profoundly with immediate implications for the residents and our world. The Arctic Ocean is at the center of the Arctic region, which is home to Indigenous peoples for millennia as well as more recent arrivals. The Arctic Ocean also is a bellwether, reflecting the urgent need to produce informed decisions that operate short-to-long term. In the Arctic, the maturing focus on climate – as a *"common concern of humankind"* since the 1992 *United Nations Framework Convention on Climate Change* – exemplifies our quest for coordination and cooperation, locally, regionally and more broadly across our world, identifying essentials with the United Nations Sustainable Development Goals *"for the benefit of all on Earth across generations."*

P. A. Berkman (✉)
Science Diplomacy Center, EvREsearch LTD, Falmouth, MA, USA

Science Diplomacy Center, MGIMO University, Moscow, Russian Federation

United Nations Institute for Training and Research (UNITAR), Geneva, Switzerland

Program on Negotiation, Harvard Law School, Harvard University, Cambridge, MA, USA
e-mail: paul.berkman@unitar.org; pberkman@law.harvard.edu

O. R. Young
Bren School of Environmental Science & Management, University of California Santa Barbara, Santa Barbara, CA, USA
e-mail: oran.young@gmail.com

A. N. Vylegzhanin
International Law School, MGIMO University, Moscow, Russian Federation
e-mail: danilalvy@mail.ru

D. A. Balton
Polar Institute, Wilson Center and U.S. Arctic Steering Committee, Washington, DC, USA
e-mail: davebalton@comcast.net

O. R. Øvretveit
Initiative West Bergen, Bergen, Norway
e-mail: ole@initiativvest.no

P. A. Berkman et al. (eds.), *Building Common Interests in the Arctic Ocean with Global Inclusion, Volume 2*, Informed Decisionmaking for Sustainability,
https://doi.org/10.1007/978-3-030-89312-5_35

35.1 Building the Future

The premise for Building **Common**[1] Interests in the Arctic Ocean with Global Inclusion recognizes the Arctic is being transformed profoundly with immediate implications for the residents and our world. The Arctic Ocean is at the center of the Arctic region, which is home to **Indigenous peoples** for millennia as well as more recent arrivals. The Arctic Ocean also is a bellwether, reflecting the **urgent** need to produce informed decisions that operate short-to-long term. In the Arctic, the maturing focus on **climate** – as a *"common concern of humankind"* since the 1992 *United Nations Framework Convention on Climate Change* – exemplifies our quest for coordination and cooperation, locally, regionally and more broadly across the **Earth**, identifying essentials with the 17 United Nations **Sustainable Development Goals** (SDG) from 2015.

The primary conclusion of this book is that informed decisionmaking requires science as well as diplomacy with international, interdisciplinary and **inclusive** integration, noting inclusion is the biggest challenge. In the absence of inclusive considerations, informed decisionmaking is incomplete and sub-optimal in the complex global **system** that we now inhabit.

The inescapable truth is we now live in an interconnected world, but plagued with nationalism and the perpetual problems of systemic exclusion. We also are in the midst of a global pandemic, when lives and livelihoods of people are compromised everywhere, revealing once again that **survival** is a common interest among all of us. How can we build the future to address challenges and opportunities inclusively? Addressing this question is the outcome of this second volume in the initial trilogy of the book series on Informed Decisionmaking for **Sustainability**.

Anywhere can hold lessons for humankind, contributing insights for our world with nearly eight billion people today. Every moment also can hold lessons, especially since human populations began accelerating across billions of people, starting around 1800 at dawn of the Industrial Revolution. Inclusively in view of time and space – the Arctic Ocean presents a case study for humankind because it illustrates diverse perspectives with science[2] in a scalable manner, addressing **change** with **research** and **action** to produce informed decisions.

The **holistic** (international, interdisciplinary and inclusive) process with informed decisionmaking in the Arctic Ocean and elsewhere on Earth starts with **questions**.

---

[1] Search terms (bolded) were discovered comprehensively to reveal relevant chapters (see Table of Contents) with the KnoHow™ knowledge bank (https://knohow.co) for Volume 2. Building Common Interests in the Arctic Ocean with Global Inclusion, using the final PDF files for the initial book compilation to weave these conclusions with all chapters in many contexts, inclusively. The relevant chapters for each search term are indexed below in alphabetical order in the Chapter References (By Search Term).

[2] Science as the 'study of change' includes natural sciences, social sciences and Indigenous knowledge as complementary research systems that reveal patterns, trends and processes (albeit with different methodologies) that serve as the bases for decisions.

Questions also are core to any negotiation and all transdisciplinary dialogues, fueling the engine of **science diplomacy**[3] to address immediacies as well as eventualities with knowledge co-creation and co-production. Questions ultimately can give rise to **governance** mechanisms, and built **infrastructure** as well as their coupling with **economic**, societal and environmental considerations for sustainable development. In a world where everyone is looking for answers, questions are the differentiator to facilitate dialogues that build common interests, which are herein recognized as the key to being inclusive.

The goal of this concluding chapter is across the gamut of questions (as **fil rouge**) to reveal scalable elements of inclusion that can be illustrated with this book about building common interests in view of the Arctic Ocean as a test case for informed decisionmaking. The illustrations about the six elements of inclusion (i.e., fundamental questions) emerge from individual chapters and their juxtaposition, converging with content (who, what, when and where) and process (how and why) questions inclusively. Because they are interlinked, the proposition is that all of these questions are required to be inclusive, as a necessary step to both promote cooperation and prevent conflict. Testing this proposition broadly is among the **options** (without advocacy), which can be used or ignored explicitly, to facilitate inclusion with **respect** for the institutions as well as the decisionmakers and those reading forward.

Options (without advocacy) guide diplomacy, helping governments and others to navigate the winds of the present into the future with informed decisions, addressing urgencies that extend from **security** to sustainability time scales. The options transform **evidence** into decisions, transforming the **data** that arise to answer questions with research into actions by institutions. As a region raising local-global questions about inclusion, the Arctic Ocean is ripe for consideration to awaken informed decisionmaking for the benefit of all on Earth across **generations**.

35.2 Inclusion Element 1: Who? (Crossing Boundaries)

Inclusion extends across boundaries temporally, spatially, culturally, chemically, physically, economically, socially and naturally with any other ally that can be imagined. The boundaries define systems, such as our Solar system with the Earth system and the Arctic Ocean system embedded at different scales through to communities and people as well as the species and habitats on which they depend. Each system has its own internal dynamics, which in turn are influenced by external forces with varying intersections. The reference points to interpret change within and across these systems involves each of us as observers, asking and answering questions with diverse methodologies for research.

[3] See Chap. 1 about the theory, methods and skills of informed decisionmaking as the engine of science diplomacy to build common interests and enhance research capacities, transforming research into action with the apex goal of informed decisions that operate across a 'continuum of urgencies'.

The opportunity to be an observer is self-selected, but limited by boundaries across systems, where transparency and access are the challenges. Solutions for transparency certainly involve the development and application of **observing** systems and networks. However, the challenges to open doors of transparency and access are far more basic with systemic racism, exclusion, prejudice and injustice that continue to infect our world at all levels. With respect for the self-selected **identity** of individuals and institutions, the trick is to facilitate dialogues that build common interests, empowering observers to contribute as participants with capacities that create inclusion.[4]

Inclusion in this book is built with common interests, expressed with the insights of **youth** alongside others across society at local, national and international levels. As an example of institutional inclusion, the **Arctic Council** that was established in 1996 is a system in which the eight **Arctic States** and the six Indigenous **Permanent Participants** grant access to **Observers**. With the Arctic Council, informed decisionmaking is stimulated by *"sustainable development and environmental protection"* as *"common Arctic issues"* framed by the 1996 ***Ottawa Declaration***.

In the Arctic, Indigenous peoples arose from the first humans in the region with communities connected to the Arctic Ocean. Other Arctic residents are distributed within the boundaries of the eight States who have territories north of the Arctic Circle. In view of the Arctic Ocean, the **challenges** involve diverse **stakeholders**, **rights**holders and other actors.

Despite the flaws of humankind and our history, the richness of our world is its diversity. Looking across time, we are awakening to the necessity to act as stewards, with compassion for each other and all that surrounds us, short-to-long term. Seeking to be inclusive, any observer can raise questions. The trick is to facilitate dialogues that build common interests, moving observers into the realms of participants, transforming research into action with science diplomacy in an holistic process to deliver informed decisions at local-global levels.

35.3 Inclusion Element 2: When? (Past, Present and Future)

Time is the essence of change and also the biggest challenge to address at all levels. The meaning of short-to-long term depends on the questions, complementing the dimensions of informed decisions at all scales. With inclusion, weighing the past and future in view of the present underlies a fundamental source of inquiry as an egalitarian framework for lifelong learning.

In the Arctic, research stimulated by questions reveals insights about the drivers and impacts of change to address with **innovation** over security to sustainability time scales: **days**; **weeks**; **months**; **years**; **decades**; **centuries** and even **millennia**.

[4]See the *Informed Decisionmaking Pyramid* (Fig. 1.6) in Chap. 1.

Others are listening to the world's varying paces, trying to make sense of the rhythms of discourse and events now conveyed at the electronic speeds of **social media** with constantly flashing reactions.

Treating the chapters in this book as data, the most frequently cited unit of time is years, with more references to decades than months. These data suggest there is a tendency to address issues with longer time horizons than years, as with the United Nations Millennium Development Goals (2000–2015) and SDG (2015–2030) or the United Nations Decade on Ocean Science for Sustainable Development (UNDOS – see Appendix). This hypothesis is strengthened by the time series of International Decades since 1960 (Fig. 35.1), just after the International Geophysical Year of 1957–1958 and its preceding International **Polar Years** (IPY), revealing a step-change with common-interest building since the end of the Cold War, as a signal with informed decisionmaking into the future across the Earth.

The Arctic Council reinforced this observation about operating over longer periods in adopting its first long-term strategic plan at the biennial Arctic Council Ministerial Meeting in May 2021, a plan that will cover the ensuing decade.[5] Inclusion involves the sort of continuity that can only be achieved over significant periods of time. It takes time to generate informed decisions, recognizing that decisions are uniformed if they only operate at a particular moment, excluding considerations of either the present or the future. Lengthening the timeframes of initiatives (Fig. 35.1) is a key metric in assessing informed decisionmaking.

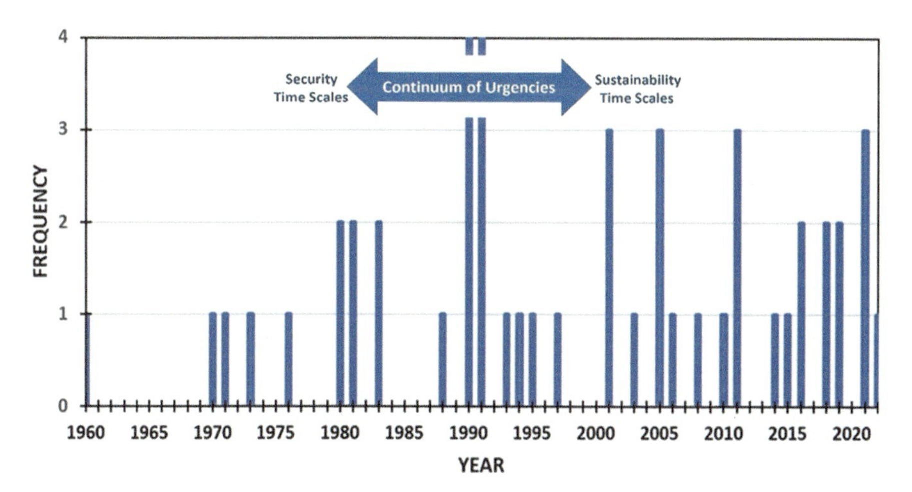

Fig. 35.1 Frequency of International Decades, based on their year of origin as compiled by the United Nations (https://www.un.org/en/observances/international-decades), with end of the Cold War in 1991, enhancing capacities of humanity to operate across a 'continuum of urgencies' with sustainability since the Second World War (Chap. 1)

[5]2021 Arctic Council Ministerial Meeting documents can be found at: https://oaarchive.arctic-council.org/handle/11374/2586

The challenge with time is to operate short-to-long term, whatever that means to you or anyone else, correctly setting expectations that progress takes time to mature, often more slowly than desired. Consequently, inclusion is the responsibility of all, highlighting the vital importance to enhance the common-interest building capacities of next-generation leaders, recognizing young adults today will be living into the twenty-second century. Whoever is involved, operating across time enhances the opportunity to transcend business as usual.

35.4 Inclusion Element 3: Where? (Marine and Terrestrial Ecosystems into Outer Space)

On Earth, space is generally easier to comprehend than time, largely because we actively can visualize the surface of our planet as well as peer from outer space with sub-meter **satellite** resolution. Across our home ('eco') with its diverse **ecosystems**, **water** is the fundamental driver of life in oceanic, continental and atmospheric areas of our planet. Humankind created **ecology** and economics for the study and **management** of our home systems, respectively. Understanding our home becomes increasingly vital as we venture across the curvature of spacetime into the universe, which is where humankind is headed, turning science fiction into science reality.

Like time, space is embedded: **centimeters**, **meters** and **kilometers** with bigger and smaller to explore. Across the physical dimensions of our globally-interconnected civilization are artificial boundaries imposed by humankind to protect and exclude interests, resulting in the eco**political** dynamics that we see at all scales. These ecopolitical scales are paralleled by **nations** as the basic jurisdictional unit since the 1648 *Treaty of Westphalia.* Subnational levels of governance range from **families** to **cities** and other governments across larger **regions**. **International** levels of governance include transboundary as well as global institutions affiliated with the **United Nations**.

The Arctic Ocean is a case-study with diverse spatial boundaries, both natural and anthropogenic, involving systems that are interconnected across the Earth. Progressing from the **North Pole** as a geographic point, there is the surrounding **sea ice** that is diminishing and beyond there is **open water** in the **Central Arctic Ocean** (CAO). Superimposed, there is the CAO **high seas** and surrounding Exclusive Economic Zones as well as other international maritime zones north of the Arctic Circle under **law of the sea**.

Surrounding the Arctic Ocean are land areas with **glaciers** and **permafrost** that also are diminishing, just as ice in the sea. Compared to the lower latitudes, Arctic marine and terrestrial areas are responding to climate warming with amplification, reflecting connections between these biogeophysical systems. Superimposed on the terrestrial areas are the jurisdictional boundaries of the eight Arctic States as well as the areas of the Arctic represented by the six Indigenous Peoples' Organizations.[6]

[6]See Fig. 1.1 in Chap. 1 as well as the book cover.

Beyond the Arctic are **non-Arctic** areas that also are included to interpret biogeophysical and socio-economic changes in the high north. The context of the Arctic and non-Arctic areas, inclusively, is the Earth. Beyond Earth is outer space, noting the Earth-Sun connections that have been explicitly researched on a global scale since the first IPY in 1882–1883, following the Little Ice Age that ended in the nineteenth century with humanity warming to the fact that global climate and local **weather** are connected.

35.5 Inclusion Element 4: What? (Natural Sciences, Social Sciences and Indigenous Knowledge)

In view of who, when and where to include – what are the issues, impacts and resources to consider and how do we measure them? Responses to this open-ended question can be considered inclusively, involving research to reveal patterns, trends and processes that ultimately become the bases for informed decisionmaking at any scale.

Building on the concept of systems that are bounded across space and time, internally there are the components that reflect their dynamics. Some of the components that move through systems occur naturally, such as **species** and water with its liquid, solid and vapor phases. Some components are introduced internally and externally from anthropogenic sources, as in the Arctic Ocean, where there are **chemicals**, **plastics** and other **pollutants** as well as **ships**. **Human** presence in the Arctic also involves exploitation of living and non-living **resources**.

Each of these Arctic system components can be measured to create data, which can be used to address questions with diverse methodologies that include hypotheses, values, ethics and cultural wisdom. The data come from the natural and social **sciences** along with **Indigenous knowledge**. Moreover, the data range with granularity from the metric system to the different Inuit words for snow, revealing system dynamics that underlie evidence for decisions, as actions by individuals and institutions. Further illustrating inclusion, seeking an umbrella framework, science broadly is the 'study of change' (symbolized by the Greek letter delta Δ) with basic and applied research that together contribute to informed decisions, especially in preparing next-generation decisionmakers.

Importantly, Arctic systems represent a special class of change, which happens when boundaries are altered, as with sea ice in the ocean and permafrost on land. Such environmental state-changes create new systems (e.g., there is a new Arctic Ocean without multiyear sea-ice predominating), representing inherent **risks** of instabilities with immediacies, which define security time scales (Fig. 35.1) that connect to consequent urgencies short-to-long term in view of sustainability time scales across generations.

35.6 Inclusion Element 5: How? (Governance, Infrastructure and Sustainability)

The proposition is sustainability operates across generations. If we think it, we can build it!

But, how do we operate at the time scale of generations, noting there are somewhere between five and six 20-year generations of people alive at any time? How do we convey urgencies across decades to centuries while extinguishing brushfires of the moment?

A key is to learn from the cultural wisdom of Indigenous peoples, revealing resilience across generations with grandparents, parents, yourself, children, grandchildren and great-grandchildren included. The Arctic offers a special example for our world, as the six Indigenous Peoples' Organizations share decisionmaking responsibilities about the destiny of the region with the eight Arctic States through the Arctic Council, addressing *"common Arctic issues"* as well as *"issues of common interest"* and *"common concern"* expressed in the 2021 *Reykjavik Declaration* along with *"common priorities"* through the *Arctic Council Strategic Plan 2021 to 2030*. In view of the Arctic Ocean as a case study, together these signatories of the 1996 *Ottawa Declaration* also *"remain committed to the framework of the Law of the Sea"*, as shared in their 2013 ***Vision for the Arctic*** with the *"Arctic region as a zone of peace and stability… at the heart of our efforts."*

These efforts progressed significantly at the 2009 Arctic Council Ministerial Meeting, when **peace** first was introduced into a declaration from the eight Arctic States. The 2009 *Tromsø Declaration* changed the dynamics of the Arctic Council, opening the door to task forces that would produce three binding agreements with all of the Arctic States across the following decade. Such convergence reflects common-interest building with knowledge co-production of governance mechanisms as an arena of informed decisionmaking.

Thinking short-to-long term, the 2011 and 2013 emergency response agreements anticipate issues and impacts with the changing Arctic. The 2011 *Agreement on Cooperation on Aeronautical and Maritime* ***Search and Rescue*** *in the Arctic* addresses questions of **safety** of life at sea. The 2013 *Agreement on Cooperation on Marine* ***Oil*** *Pollution Preparedness and Response in the Arctic* addresses sources and threats of **pollution** with impacts that are both acute and chronic. These agreements are complemented by the *International Code for Ships Operating in Polar Waters (**Polar Code**)* that entered into force in 2017 with the Arctic States and broader international community through the International Maritime Organization.

Effectiveness of these governance mechanisms depends on the **platforms** that exist for their implementation from paper to practice. In this direction, the Arctic States produced the 2017 *Agreement on Enhancing International Arctic* ***Scientific Cooperation***, supporting access with research, observing, communication and other information systems as elements of built infrastructure, which require **technology** plus **investment**.

In the Arctic Ocean, as elsewhere, there are questions about sustainable yields of **fish**. Fish are symbolic of all living resources harvested by humans, including considerations about species' recruitment and production across generations in view of their ecosystems. Most importantly, to create sustainable fisheries requires restraint to operate short-to-long term, being both tactical and **strategic**.

With CAO high seas fisheries, there is opportunity to learn about the species' dynamics in their changing ecosystems before any exploitation, remembering 1970s lessons of El Niño and the Peruvian anchovy with periodicities as well as impacts in the absence of informed decisions. To avoid another "Donut Hole" catastrophe, which happened with the pollock fishery in the high seas of the Bering Sea in the early 1990s, Arctic and Non-Arctic States signed the 2018 *Agreement to Prevent Unregulated High Seas Arctic Fisheries in the Central Arctic Ocean* that entered into force on 25 June 2021. The 16-year moratorium that is mandated with this agreement is an essential step to prevent unregulated commercial fishing activities in this Area Beyond National Jurisdiction (ABNJ), evolving a **precautionary approach** with worldwide precedent (See also the Appendix of Chap. 1 that elaborates international legal institutions with the precautionary approach and principle). Precaution provides time to:

- build common interests and raise the questions of common concern;
- generate the necessary data with appropriate methods to answer the questions of common concern;
- transform the data into evidence in view of the institutions that will make decisions about governance mechanisms and built infrastructure;
- couple decisions about governance mechanisms and built infrastructure to achieve progress with sustainable development; and
- ultimately reveal options (without advocacy) for humans to operate short-to-long term with informed decisionmaking.

The precautionary approach with research and action inclusively is an example of informed decisionmaking under international law – as illustrated by the CAO High Seas fisheries agreement as well as ongoing negotiations toward a global agreement on Biodiversity Beyond National Jurisdiction (BBNJ).

35.7 Inclusion Element 6: Why? (Balancing National Interests and Common Interests)

Inclusion involves **balance**. Over time, balance and resilience in the face of change produces sustainable development across generations. Systems that are out of balance are unstable, requiring processes to address diverse and often unknown urgencies over time (Fig. 35.1), which is why institutions and governments emerge with legacy responsibilities. A key feature of such processes is their scalability in an holistic manner.

With international, interdisciplinary and inclusive considerations – as illustrated with sustainable development in homes and villages to nations and the world – at all levels, there is urgency to balance economic, societal and environmental considerations. This is the gift of the SDGs, building common interests from the United Nations across nations into communities, necessitating progress upwards and downwards in both directions. Moreover, urgencies with sustainability are continuous short-to-long term, characterizing the ubiquitous need for informed decisionmaking.

The challenge remains to balance national interests with common interests that include each of us across the spectrum of subnational-national-international jurisdictions. Such balance is the objective of science diplomacy as a means of enhancing informed decisionmaking to promote cooperation and prevent conflict, recognizing that nations always will look after their national interests first and foremost.

Into this history of humanity, inclusion and balance are illustrated in the Arctic with science to build common interests as a necessary step, before it becomes possible to balance national interests. A high-level example is with the Arctic Council and its six scientific working groups, progressing with biennial declarations from the foreign ministers of all eight Arctic states, who declared again in 2021 their *"commitment to maintain peace, stability and constructive cooperation in the Arctic."* The 2017 Arctic Science Agreement uses the same language of peace, further emphasizing the common-interest building contributions of science among all Arctic States and Indigenous Peoples' Organizations, *"using the best available knowledge for decision-making."*

Global relevance of Arctic science is maturing with the Arctic Science Ministerial process that welcomes contributions from the six Indigenous Peoples' Organizations and non-Arctic States, with vision of human capacities to address climate change and other challenges. In particular, with global application, common-interest building is highlighted under law of the sea, surrounded by **Superpowers**, accentuating the North Pole as a **"pole of peace"**.

35.8 Lifelong Learning with Global Inclusion

This book series seeks to produce insights about Informed Decisionmaking for Sustainability that can be developed, applied, trained and refined, inclusively. We create the 'rules of the road' to steer a safe course into the future, maneuvering in view of the red lights ahead. Where the rules are exclusive, systems transform, like nations producing different constitutions. The underlying process has always worked, at least since origin of the Socratic method, starting with questions and research to inform decisions.

Once in a hundred generations – from stone to clay to papyrus to paper to digital – humankind invents a new medium to create and communicate knowledge. Today, with digital technologies, we can communicate across the world with information access that is effectively instantaneous and infinite, looking backward and forward across time inside and out of phenomena at all scales with unprecedented clarity. The consequences of our digital era are open-ended, involving artificial intelligence, cryptocurrencies, nanomedicine, renewable energy, robotics, social media, 3-D printing and all other manner of built infrastructure. With exponential growth of computing capacities across years to decades, science is expanding as a public good, opening the door for everyone to contribute as both an observer and participant with data, primed with transformational capacities at local-global levels.

However, the reality is humankind still is in its infancy to operate on a planetary scale. For example, there is still debate about our interconnections across the Earth as revealed by human population and atmospheric carbon dioxide increasing exponentially in parallel over decades to centuries since beginning of the Industrial Revolution. Our interconnections are even more evident over shorter periods, as harshly introduced by the COVID-19 pandemic with human infections and deaths increasing exponentially over months-years across the Earth.

With **hope** as much as certainty, COVID-19 impacts will decelerate, just as with the Spanish Flu a century ago and other plagues in human history. The advancement now is we have vaccines to hasten arrival of the global inflection point,[7] awakening the challenge and opportunity for great nations, especially the three Superpowers, to end the COVID-19 pandemic together. Such global inclusion is illustrated in the Arctic, where nations are balancing national interests and common interests, operating short-to-long term with informed decisionmaking.

Responsibilities to produce informed decisions extend especially to the people living today who will be alive in the twenty-second century. Such longevity includes month-years, years-decades and decades-centuries: across time scales with global impacts from humankind during the **Anthropocene**, raising questions across lifetimes about effective coupling between governance and infrastructure to achieve progress with sustainable development. Elaborating lessons about inclusion and common-interest building, the third volume in this trilogy will focus on Pan-Arctic Implementation of Coupled Governance and Infrastructure.

Inclusion is a matter of lifelong learning (Fig. 35.2), stimulated by curiosity, questioning who, when, where, what, how and why. The journey starts with **education** to introduce theory, methods and skills, revealing options that are available to each of us, like choosing whether the glass is half-full or half-empty. The options are further informed with research and leadership, generating synergies with knowledge co-production. Such convergence is facilitated by science diplomats who can operate inclusively, as brokers of dialogues across the data-evidence interface, transforming research into action to inform decisions.

With informed decisionmaking about governance, infrastructure and sustainability – there also is a basic choice to start from a position of conflict or common interests. This choice exists even among Superpowers, as illustrated in the **Antarctic** and Outer Space with continuous cooperation throughout the Cold War. This book and these conclusions highlight global inclusion as an outcome of common-interest building, with the Arctic as a case-study, revealing the scalable implications of informed decisionmaking *"for the benefit of all on Earth across generations."*

[7] For context, in May 2021, 170 million reported COVID-19 cases represent slightly more than 2% of the 7.9 billion people on Earth.

Fig. 35.2 Lifelong Learning with Global Inclusion, triangulating education, research and leadership to co-produce knowledge that empowers informed decisionmaking and inclusion short-to-long term with common interests

Acknowledgements This concluding chapter and those that preceded emerged with *Arctic Frontiers* 2020, building on the 2018 *Memorandum of Understanding* with the Science Diplomacy Center on behalf of the editors for the book series on Informed Decisionmaking for Sustainability. This chapter is a product of the Science Diplomacy Center through EvREsearch LTD, coordinating the *Arctic Options* and *Pan-Arctic Options* projects with support from the United States National Science Foundation (Award Nos. NSF-OPP 1263819, NSF-ICER 1660449 and NSF-ICER 2103490) along with the Fulbright Arctic Chair 2021–2022 awarded to P.A. Berkman by the United States Department of State and Norwegian Ministry of Foreign Affairs with funding from the United States Congress. These international projects include support also from national science agencies in Canada, China, France, Norway, Russia from 2013 to 2022 in coordination with the Belmont Forum, gratefully acknowledging the collaboration with the University of California Santa Barbara, MGIMO University, Université Pierre et Marie Curie, Norwegian Polar Institute, University of Alaska Fairbanks, University of Colorado Boulder, Carleton University and the Ocean University of China among other institutions throughout this period. We also thank the Polar Institute with the Wilson Center for their leadership and support of this contribution. Comprehensive integration of chapters in this book includes knowledge-discovery application of KnoHow™ (http://knohow.co), acknowledging support to EvREsearch LTD as a subawardee on the National Science Foundation project through the University of Colorado regarding *"Automated Discovery of Content-in-Context Relationships from a Large Corpus of Arctic Social Science Data"* (Award No. NSF-OPP 1719540).

Chapter References (By Search Term)

Arctic Council: Chapters 1, 5, 6, 7, 8, 9, 15, 18, 20, 22, 23, 24, 25, 26, 29, 33, 34 and Appendix
Arctic States: Chapters 1, 5, 7, 8, 9, 15, 18, 24, 25, 26, 29, 32, and 33
Balance: Chapters 1, 3, 4, 5, 7, 8, 10, 16, 18, 19, 22, 24, 26, 27, 29, 30, 32, and 33
Centimeters: Chapter 1
Central Arctic Ocean: Chapters 1, 5, 8, 15, 16, 18, 24, 25,26, 27, 33 and Appendix
Centuries: Chapters 1, 2, 15 and 33
Challenges: Chapters 3, 5, 6, 7, 8, 9, 10, 13, 15, 16, 17, 18, 19, 21, 22, 23, 24, 26, 30, 32, 33 and Appendix
Change: Chapters 1, 5, 6, 7, 8, 9, 10, 11, 12, 13, 14, 15, 16, 17, 18, 19, 22, 23, 24, 25, 26, 27, 28, 29, 30, 31, 32, 33 and Appendix
Chemicals: Chapters 1, 6, 7, 9 and Appendix
Cities: Chapters 1, 17, and 19
Climate: Chapters 1, 2, 5, 6, 7, 8, 9, 10, 11, 12, 13, 14, 15, 16, 17, 18, 19, 20, 21, 22, 23, 24, 25, 27, 28, 30, 32, and 33
Common: Chapters 1, 3, 5, 6, 7, 8, 9, 15, 16, 18, 19, 23, 24, 25, 26, 29, 30 and Appendix
Data: Chapters 1, 6, 7, 8, 9, 10, 16, 17, 18, 23, 24, 25, 26 and Appendix
Days: Chapters 3, 5, 8, 15, 26, and 30
Decades: Chapters 1, 6, 10, 19, 23, 24, 25 and Appendix
Earth: Chapters 1, 5, 6, 7, 8, 9, 23, 24, 26, 27, 28, 29, 31 and Appendix
Ecology: Chapters 1, 6, 9, 17, 24, 25, 26 and Appendix
Economic: Chapters 1, 5, 6, 7, 8, 10, 15, 17, 18, 19, 20, 22, 25, 26, 27, 28, 29, 30, 31, 32, 33 and Appendix
Ecosystems: Chapters 1, 6, 8, 9, 13, 23, 24, 25, 26, 29, 31 and Appendix
Education: Chapters 1, 4, 9, 16, 17, 18, 19, 20, 22, 32 and Appendix
Evidence: Chapters 1, 6, 8, 9, 18, 19, 24, 26 and Appendix
Families: Chapter 17
Fish: Chapters 1, 5, 6, 8, 9, 11, 12, 16, 18, 24, 25 and Appendix
Generations: Chapters 1, 16, 17, 22, 24, 26, 29, 32 and Appendix
Glaciers: Chapters 3, 8, 9 and Appendix
Governance: Chapters 1, 5, 7, 8, 15, 16, 17, 18, 24, 25, 26, 29, 30, 31, 33 and Appendix
High seas: Chapters 1, 5, 8, 15, 16, 18, 24, 25, and 26
Holistic: Chapters 1, 7, 16, 19, 23, 24, 26, and 29
Hope: Chapters 1, 3, 11, 13, 21, 24, 29, 32, 33, and 34
Human: Chapters 1, 5, 6, 7, 8, 9, 11, 12, 16, 18, 19, 22, 23, 24, 25, 26, 27, 29 and Appendix
Identity: Chapters 1, 2, 7, 16, 17, 22, 24, and 26
Inclusive: Chapters 1, 5, 7, 13, 16, 18, 21, 24, 26, 27, and 29
Indigenous knowledge: Chapters 1, 2, 9, 16, 22, 24, 29, 32 and Appendix
Indigenous peoples: Chapters 1, 4, 7, 8, 9, 15, 16, 18, 19, 22, 24, 29, 32 and 33
Infrastructure: Chapters 1, 5, 7, 8, 9, 14, 16, 17, 18, 19, 20, 23, 24, 25, 29, 32 and Appendix
Innovation: Chapters 1, 10, 17, 18, 19, 20, 22, 24, 31, and 34

Appendix

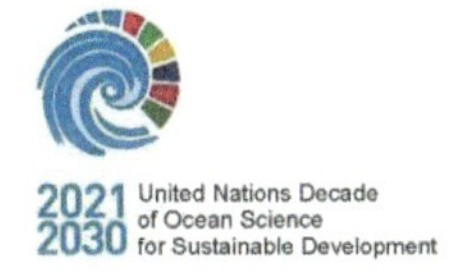

Arctic Ocean Decade Workshop: Policy-Business-Science-Dialogue

Full report from Tromsø, Norway, 29 January 2020

Inputs to the planning of UN Decade of Ocean Science for Sustainable Development (2021–2030)

P. A. Berkman et al. (eds.), *Building Common Interests in the Arctic Ocean with Global Inclusion, Volume 2*, Informed Decisionmaking for Sustainability,
https://doi.org/10.1007/978-3-030-89312-5

Contents

Introduction

This report summarises inputs given by participants at the Arctic Ocean Decade Workshop. The inputs are meant to feed into the planning of the UN Ocean Decade for Sustainable Development.

This was a one-day workshop covering four of the Decades six societal outcomes: a clean ocean, a healthy and resilient ocean, a predicted ocean and a sustainably harvested and productive ocean. For each topic a keynote introduced the topic, followed by break out group discussions giving input on the topic. All together there was five break out groups, all discussing the same topics. In this report, inputs from the groups are summarised "together".

100 participants from 15 countries attended. Approximately 60% Norwegians. About 70% from the research community and 30 % from business, policy and non-governmental organisations.

The report first sums up the opening remarks. Then the inputs given by the workshop groups are listed under each topic together with a short summary of the keynote talks introducing each topic. Many inputs are relevant to more than one topic and were discussed at various times and in various contexts during the workshop. At the end inputs concerning other topics than the four societal outcomes are listed.

The workshop provided a dialogue and discussions between representatives from policy, business and science. This report reflects the viewpoints of the participants, sometimes without concluding remarks.

Opening and Introduction Remarks

The Executive Director of the Division of Oceans, Energy and Sustainability, The Research Council of Norway, Fridtjof Unander, underlined the importance of ocean science and the special features of the Arctic Ocean in these times of rapid change. He noted that the huge interest to participate in the workshop with more than 100 participants from 15 countries showed that the Arctic Ocean issues engage many more than the Arctic states.

Ole Øvretveit, director of Arctic Frontiers underlined the importance of Arctic Ocean issues and international collaboration.

Ivet Petkova, representing Arctic Frontiers Young, told the participants how important the Decade of Ocean Science is to the young generation. The Arctic Frontiers Young program gathers and connects the next generation of upcoming scientists, to create opportunities for young people, to develop their inner curiosity, knowledge and skills. She reminded the audience that ocean science needs to be interdisciplinary and that a stable Arctic is essential for a stable world.

Peter Thomson, UN Secretary-General's special Envoy for the Ocean, greeted participants via video. The Decade is an opportunity to raise awareness of the ocean health and how important it is for food security and human wellbeing.

Vladimir Ryabinin, Executive Secretary Intergovernmental Oceanographic Commission of UNESCO, gave a presentation of the Decade of Ocean Science. He reminded the participants about the six societal outcomes of the Decade and that the Decade is not simply about ocean science, but ocean science for sustainable development. He noted also that predictions say ocean economy will develop faster than economy on land. The ongoing preparatory phase is to engage stakeholders to give IOC input. So far there are no detailed plans about allocation of resources for the Decade, but good plans open the door for resources.

Topic 1 – A Clean, Healthy and Resilient Arctic Ocean

A clean Ocean: whereby sources of pollution are identified, quantified and reduced and pollutants removed from the ocean

A healthy and resilient Arctic Ocean: whereby marine ecosystems are mapped and protected, multiple impacts, including climate change, are measured and reduced, and provision of ocean ecosystem services is maintained

Keynote speaker Anne-Christine Brusendorff, International Council for the Exploration of the Sea (ICES), introduced the topic.

Brusendorff informed that ICES is an international governmental organisation giving scientific advice on marine issues. One of their science priorities is advice on fish stocks in the Arctic ocean and adjacent seas where they give advice about many of the commercial important fish stocks.

The changing environment of the northern seas has a huge influence on species and habitats. She said it's a need to quantify the impacts on human activities in the Arctic Ocean. This includes also the impacts of pollution like harmful substances, nutrients run off and ocean acidification.

Her advice to the workshop was that for a clean ocean the discussion should focus on key pollution issues, which sectors are involved and if there are increasing or decreasing trends. For a healthy and resilient ocean, she said key issues are ecosystem science and Arctic Ocean resilience. Different kinds of knowledge should be considered and how western science and Indigenous knowledge can complement each other.

Knowledge Gaps and Input

We do know quite much about the Arctic Ocean and we often know enough to act. Often, we need to change the mind set of humans and especially the decisionmakers to act based on existing knowledge. Some of the ongoing changes in the Arctic can be limited or reversed (e.g. mitigation of emission of pollutants and greenhouse gases, regulation of the pace of industrial exploration). Other changes are inevitable and require plans for adaptation. But there is still much we do not know about the Arctic Ocean, the ongoing rapid changes and predictions for the future. Arctic is the geographical area on earth with highest temperature increase. The most profound is the loss of sea-ice.

Ecosystems

Vital for the Arctic Ocean is to understand ongoing changes and predicted future changes to the ecosystems. To know what has happened in the past is vital to understand the ongoing changes. There is a need to close knowledge gaps and understand the differences in the Arctic Ocean, because the Arctic Ocean is not uniform.

There is a need to know how commercial exploration of the Arctic Ocean could impact specific parts of ecosystems.

Research and monitoring of the seabed and subsurface seabed life should be prioritised and can tell what has happened in the past and thereby we can we learn from the past. Sediment core samples with analyses of a variety of substances, including contaminants, is important.

We need to improve our knowledge about the ecosystems of the Arctic Ocean. We have many different ecosystems and we need to establish a baseline for the different ecosystems. This will help in understanding the changes and the resilience of the ecosystems. Very little is known about the ecosystems in the Central Arctic Ocean basin, along the slopes and the effect of changes in sea ice. But changes are more intense, and biological productivity higher, along the margins of the Arctic ocean than in the Central Arctic Ocean and therefore there should be focus on both areas. While there is a range of ecosystem types in the Arctic Ocean, these systems are monitored in different degrees. We have good knowledge about some Arctic Ocean ecosystems and very little knowledge about others. If projecting knowledge from well-studied regions to poorly studied ones, we can learn what can be transferred from region to region and what knowledge is really missing. If we establish in depth knowledge about the Arctic Ocean ecosystems, we can predict much better how human activities in the Arctic (ship traffic, sea-bed activity, fisheries, etc) will affect the ecosystems. Also, environmental stressors may have different effects on different sub-populations (including of commercially harvested species); there is therefore a need for detailed studies. While modelling is extremely useful, it also has limitations. We need to be aware that while ongoing trends are likely to continue, there will also be surprises. Therefore, prediction is not extrapolation. This principle also applies more broadly for Arctic Ocean research, extrapolation from research from lower latitudes, or from local Arctic studies, cannot give the correct answer for other parts of the Arctic Ocean. The concept of connecting scales is important; for example, regional studies can often be combined with larger-scale modelling in order to consider the Arctic as a whole.

Focus on Arctic Ocean ecosystem understanding should be on the 18 Large Marine Ecosystems (LME) of the Arctic. Cooperation between ICES, PICES and the Arctic Council working groups on Integrated Ecosystem Assessment of the LMEs will identify knowledge gaps of each LME and be a basis for ecosystem-based management (EBM) of the Arctic Ocean. Ecology as a science is still a bit fragmented, and understanding of ecosystem functioning is still limited, e.g. the mechanism behind decline of sea bird species is not well understood. It is important during the Decade to improve interdisciplinary ecosystem approach which can help us understand the ongoing changes.

Pollution

A specific feature with the Arctic Ocean is that there are few local pollution sources, but global wind and ocean currents, and animal migration, transport pollutants from lower latitudes to the Arctic.

There is a knowledge gap about impacts of possible increase in local pollution sources as anthropogenic activity in the north increases.

The understanding of the combined effects of climate change and other stressors caused by human activities is the key research priority for the Arctic Ocean, along with the spatial and temporal distribution of pollutants.

Distribution, accumulation and possible adverse effects of micro- and macroplastics in the Arctic needs to be further investigated.

There are high concentrations of some contaminants, e.g. mercury and PCBs, in higher trophic levels with consequences for wildlife, a food safety aspect and for those Indigenous people who depend on hunting and fishing for subsistence. We need more knowledge about how contaminants, and especially a cocktail of contaminants, impact Arctic organisms which are often rich in fat. Further how contaminants are affected by climate change, e.g. remobilisation from melting sea ice and thawing of permafrost. Seasonal variations in fat content of the animals and susceptibility of contaminants are not understood. Time trends of contaminants are very important for science and management. Some legacy persistent organic pollutants show decreasing trends, but some emerging contaminants show increasing trends. The network of Arctic contaminant monitoring needs to be improved. There is often lack of data from Russia. Further development and systematisation of the Arctic Environmental Specimen Banks (ESBs) can, in the future, help detect time series of emerging contaminants. Boreal species migrate northwards, but we do not know trophic interaction, how contaminants affect fish, shrimp, and how trophic pathways change within ecosystem structure. Chemicals used in the industry are often tested, but not in terms of their influence on cold and dark environments. There is an urge to improve how to regulate waste management chemicals, which go into the Arctic Ocean, regulate new products on the market and test them before use.

Climate and Physical/Chemical Issues

Small organisms are very strongly controlled by their physical environment (advection etc.) and physical-biological coupling should be a focus area.

There are large gaps in our knowledge of the circulation of the Arctic Ocean, including in the inflow of Atlantic Water. 3D-modelling of the interior Arctic Ocean is crucial.

The impacts of increased anthropogenic noise are a knowledge gap.

Adaptations vary greatly between species. There should be more research on specific Arctic species in order to understand their resilience and adaptation. E.g., what the effects are of ocean acidification on lower trophic levels as well as commercially harvested species. Some studies indicate that Arctic ocean acidification can have a huge financial impact on e.g. commercial fisheries.

Geophysical understanding in an earth system view is also important to understanding the Arctic Ocean.

Reducing uncertainties in regional climate models should be prioritised as well as putting emphasis on communication of model uncertainties.

Land-Ocean Interactions

The Arctic Ocean is surrounded by land masses, as rapid changes are happening both in the ocean and on land. We need to increase the knowledge about land-ocean interactions. Freshwater fluxes have large impacts on many aspects relating to the Arctic Ocean, including carbon fluxes, coastal erosion, primary production, and ocean circulation. Riverine inputs and land runoff to the Arctic Ocean have increased, but there are major knowledge gaps about impacts and consequences of freshwater input to the Arctic Ocean. We still know too little about the cryosphere in general. How are the melting ice sheets (glaciers) affecting physical and chemical properties of the Arctic Ocean?

Coastal erosion is a major issue along parts of the Arctic coastline. There are many engineering challenges related to this and industrial research could help us find solutions. Scientific research also needs to focus on coastal erosion and help determine what areas will be affected. Scientific challenges relating to coastal erosion are multidisciplinary and include land-ocean interaction, freshwater, and sea ice decline.

Topic 2 – A Predicted Arctic Ocean

A predicted ocean: whereby society has the capacity to understand current and future ocean conditions, forecast their change and impact on human wellbeing and livelihoods.

Keynote speaker Heidar Gudjonsson, Arctic Economic Council, introduced the topic.

Arctic Economic Council is an independent organisation that facilitates Arctic business-to-business activities and responsible economic development. Three issues are as of today the main business activities in the Arctic Ocean; transport, energy, and fisheries.

In his talk, Gudjonsson put emphasise on doing more with less.

Transport is the foundation of business. The container evolution made transport costs drop 90 %. New icebreakers can break ice sideways which means more ice is broken with smaller icebreakers.

Fuel is changing from heavy fuel oil to lighter petroleum components and liquid natural gas. Arctic is estimated to hold ca. 30 % of world's undiscovered gas and

13 % of undiscovered oil. Ice class commercial ships can break ice themselves, reducing the need for icebreakers and thereby reduce fuel consumption.

Nowadays 97 % of the fish is utilised, compared to 40 % some decades ago which means many more products are made from the fish now than before. It is important for businesses to utilise the existing resources as much as possible.

Knowledge Gaps and Input

There are huge geographical differences related to knowledge of the Arctic Ocean. We know a lot about the Barents Sea and the Fram Strait, but there is much less knowledge about the Siberian seas and the Central Arctic Ocean.

There should be developed a management control system for ships (like air control), e.g. warning signs when ships cross into Marine Protected Areas (MPA) or areas of ecological heightened concern and which features the ship crew especially needs to be aware of in these areas. Similarly, it should be developed a detection system for ships to avoid whales. Improved understanding of ice dynamics is important to predict the movement of ice in shipping routes.

Further develop models for predictions of impacts after major emergencies due to human activity in the Arctic Ocean. Better predictions will help authorities to deal with oil spill effects.

Data Collection, Storage and Sharing

A key point to increase the understanding of the Arctic Ocean is to establish a good, international system for data collection, storage and sharing.

During the Decade it should be established programs and combined networks to collect data in a systematic way and prioritising the Central Arctic Ocean.

The industry can play a major part in developing a system for data collection, storage and sharing from the Arctic Ocean. Such a system should be one of the top priorities of the Decade in the Arctic Ocean. It is vital that by the end of the Decade there is an operating system for data handling and sharing. Data is collected both from research projects and trend monitoring programmes. Data from spatial and trend monitoring programmes are important information for research projects and is also crucial to understand Arctic Ocean changes.

Currently, there is a huge lack of in-situ observations from the Arctic Ocean, particularly long-term ocean observations. This is a major obstacle to the understanding of the Arctic Ocean. Observations are currently funded through (time-limited) research projects and some trend monitoring programmes – there is a need for a publicly financed, secure source of research funds and monitoring programmes.

Data collection could be through a pan-Arctic network of autonomous samplers/ instruments for data collection. In general, there is a need for better network to collect data across the Arctic Ocean because environmental status and changes are not uniform circum-Arctic.

Underwater robotics can be further developed with commercial actors. More cruise ships and other commercial vessels (ships of opportunity) can install monitoring devices.

As part of the systems with sharing data an Arctic regionally coordinated platform to discuss and share research plans, infrastructure and data should be established.

Monitoring for some parameters are good in some parts of the Arctic while there are some geographical areas without monitoring data. This makes it difficult to do e.g. pan-Arctic assessments. Monitoring surveys should integrate data on human activities and their effects. Monitoring should include not only instrumental observations, but also different sorts of other knowledge, including Indigenous knowledge in places where it is relevant. Industrial activities are an important data source in addition to research, and should be available to all, as well as citizen science data (community monitoring). There exists much data, but those are so far mainly from accessible/ice-free areas and there is lack of data from the ice-covered Central Arctic Ocean. It is not possible to do assessments based on data from ice-free areas projecting it on the entire Arctic Ocean.

One of the challenges with Arctic Ocean data collection is that funding is national and therefore the monitoring is national. Monitoring of the Arctic Ocean should be harmonised between the nations because what happens in the Arctic Ocean affects all Arctic nations. The Decade could be the platform for the Arctic nations, and others who perform monitoring in the Arctic, to join forces and find a common approach to monitoring surveys of the Arctic Ocean.

There is need for an infrastructure to streamline the data sharing process and develop an Arctic data portal. Data and observations from ocean, atmosphere, sea ice, extreme events, interaction between terrestrial and ocean environments (nutrients, sediments etc.) should be included. Major obstacles to effective data sharing include methodological differences between researchers, and varying degrees of willingness to share. It is important to identify successful examples of data management/sharing from elsewhere, and to learn from their experience.

FAIR data (Findability, Accessibility, Interoperability, Reusability) is also about best available informed decisionmaking. There is a huge need for access to data to be used in models, assessments, time trends, financing and many other issues. Someone needs to take the initiative to establish and maintain a data portal and add resources. Then it is easier for others to follow and add human and financial resources. One suggestion is that SAON (Sustaining Arctic Observing Network) should be the top outcome of the Decade. While there are problems with centralised archiving of actual data, one could instead aim for a catalogue or metadata archive which would make it easier to find out what data exists and where is can be obtained. Commercial actors could pay a tax and get access to all data.

Data Use and End Products

There should be improved integration between researchers, monitoring observations and various stakeholders and knowledge holders both in the planning process and in the use of results. Also, knowledge built by Arctic people, Indigenous communities and other local communities, should be integrated.

Stakeholders should be included in order to produce useful end products to predict the future Arctic Ocean. Their input is crucial in order to ensure that the products from e.g. modelling are useful.

Successful examples from meteorological forecasting are often a result of a clear picture of what needs to be delivered, and to whom. The outcome of such a process will then be more useful and valuable.

Topic 3 – A Sustainable Harvested and Productive Arctic Ocean

A sustainably harvested and productive ocean: ensuring the provision of food supply and alternative livelihoods.

Keynote speaker David Balton, Wilson Center, introduced the topic.

Balton talked about existing and possibly new management regimes for the Arctic Ocean.

Nine nations and EU have signed the arctic fishery agreement. The agreement covers an area the size of the Mediterranean Sea. The agreement bans commercial fishing in the Central Arctic Ocean until we have the appropriate science and management in place, not likely in the next 16 years. Seven countries have ratified the agreement.

He challenged the workshop participants that the existing agreements are not enough, and governance framework should be strengthened. These unmet needs could be met if Arctic Council changed their way of working. At the latest Arctic Council Ministerial there was no signed Ministerial Declaration for the first time. Arctic Ocean governance needs improvement. Effective Arctic governance is a combination of science function and management function.

Balton suggested that the ICES/PICES/PAME (Protection of the Arctic Marine Environment) Working Group on Integrated Ecosystem Assessment for the Central Arctic Ocean should finalise their work. He suggested to create a standalone science body and a management body for the Central Arctic Ocean. He also asked the workshop what more to do in the shipping space in addition to the polar code.

Knowledge Gaps and Input

One obvious challenge in the Arctic is the warming of the ocean, ocean acidification and migration of commercial fish stocks. A higher expected increase in food production in the Arctic Ocean as compared to other oceans, is expected, both in fisheries and possibly also aquaculture. Patterns of productivity and species distributions are changing. FAO estimates that high latitude areas have increased potential for food production. We need an efficient use of the resources harvested or produced, in the sea. One key word is circular economy, no waste, should be the goal.

Food supply is not only fish stocks but also harvesting of other species at lower trophic levels has been discussed, and marine bioprospecting for many purposes is increasing.

We need knowledge about how harvesting and production might impact the ecosystems. Food supply also includes marine mammals and seabirds, which are critical for Indigenous communities. Technology development for energy effectiveness of harvesting is needed and will play an important role in the future.

Human activity in the Arctic is increasing (ship traffic, seabed activity, fisheries, etc). Key questions are: What are the impacts from these activities, and how will they affect the ecosystems. This requires a deep and thorough understanding of how the ecosystems function. Arctic food webs are short, and different links have great impact on each other. We therefore need to understand the entire systems not just e.g. single fish stocks. If human activity also includes oil, gas and mineral exploration, science must evaluate the consequences and ask if food production would be damaged and such industries are wanted.

The key issue to deliver a sustainably harvested and productive ocean will be to make all citizens take ownership of/and responsibility for the importance of science in a changing world. In the context of a changing Arctic we need more knowledge about human behaviour and reaction to these changes like for example reaction to fish stocks migrating out of an area. We want discussions about the trade- offs of the emerging industrial activities (e.g. shipping and fishing) and the interests of local populations whose livelihood depends on local resources. A possibility is to develop scenarios for the Arctic Ocean that include not only economical values, but also cultural, historical, values of different generations, even family values. We need to develop methods to measure these "other" values. Is it right to maximise or optimise all activities, do we need to go into all areas? Do we need alternative models/theories compared to the usual economic growth models with less use of natural resources?

Management Tools

The key to delivering a sustainably harvested ocean is the creation of mutually supportive mechanisms by and amongst science, industry, and governance. Ensure the expansion of existing regional observation and management initiatives to cover Arctic region, coupled to a widespread ocean literacy activity to link science, policy and general public awareness. We need more knowledge on how the rapid changes will impact the function of the management mechanisms and agreements and to support appropriate area-based management tools (e.g. static or dynamic Marine Protected Areas). To inform management we need models to include climate science in stock assessment

We need research on how to manage trade-offs (and conflicting benefits) between the pillars of sustainable development for the Arctic Ocean. With boreal species moving north there is a need to coordinate science and evidence for decisionmakers to avoid conflict. A possibility is to build on existing national regulations, identify best practice (integrated management plans) and work towards international agreement.

Management Mechanisms and Bodies

The global legal framework applies to the Arctic Ocean as it does for other oceans. There are 70-80 international treaties that are applicable to the Arctic. There are several management bodies for the Arctic Ocean and different agreements include different countries. There are eight Arctic Council member states with 13 non-Arctic states as observers, there are five Arctic Ocean states and nine states and EU agreed on the fishery agreement.

In general there was agreement among the participants that management of the Arctic Ocean should build on established organisations (and possibly put more efforts into the existing ones) to coordinate integrated ecosystem based management of activities, communicate the importance of the ocean, and act now: show the power of knowledge. Although, it was also discussed the possibility to establish a new convention type organisation management body for the Central Arctic Ocean.

Even if the Arctic Council is not legally binding, it has initiated a number of Arctic specific agreements and has the potential role as the Central Arctic Ocean management body or a body to initiate Central Arctic Ocean issues agreed under other management bodies.

Knowledge about how the rapid Arctic changes will impact the function of the management mechanisms and agreements is important. There are concerns that the reaction time of existing mechanisms to ensure sustainability is too slow. The need for adaptive management was emphasised including strengthening of area-based management tools such as Marine Protected Areas. The creation of mobile Marine Protected Areas, requiring a creative process, was also proposed.

Arctic Council should work in close connection with ICES and PICES about Central Arctic Ocean issues because science should be the foundation for Central Arctic Ocean management.

The Arctic Ocean seabed should not be forgotten in future management because there will be increased interest for the arctic seabed in the coming decades.

Other International Issues

Geopolitical tensions are increasing in the Arctic. The Decade could help foster international cooperation and limit tensions. Since there are so many different, and potentially conflicting, interests in the Arctic Ocean, it is important to get all actors to talk together at an equal level, not as member states and observers. Science diplomacy is critical, and Arctic science naturally lends itself to international cooperation due to the geography as well as the high costs of doing research.

International agreements take long time to develop, so we need other actions in addition.

Crosscutting Topics

The public and especially young people are a driving force for policy change. The Decade needs to be attractive to scientists and should be project driven. The Decade processes need to be transparent. The Decade needs to have calls to apply for project support to be attractive to scientists and coordinated calls between different countries on bilateral or multilateral basis. We need to foster a culture to focus on solutions and positive vision, not the environmental and mental grieve about catastrophic and apocalyptic scenarios.

Education

It is important to develop interdisciplinary Arctic Ocean research projects. The interdisciplinarity should be covered already in the education. We need education that translate expertise, so that different experts can communicate and understand one another. We need to learn and educate collaboration and increase trust among different disciplines and science institutions/educational organisations. By collaborating, we strengthen cross-countries communication and translate our knowledge to future people.

Young Scientists

The Decade should be a platform for more fellowships for young scientists. The future generation will work cross sectors with cross competences to achieve common goals. Young scientists should be more engaged and included in discussions at conferences and meetings (Arctic Frontiers Young program as an example).

Communication

Communication should be directed to specific audiences. Communication of science to policy makers is different than communication to the general public or schoolchildren. Communication from e.g. ICES to decisionmakers are good. Bringing ocean science to the classroom will have a long-term effect on children's knowledge and awareness of ocean issues. The Decade should be spread among lecturers, professors, educators, local communities, so that everyone could be involved in different ways.

Science should be communicated through various channels that reach the young generation; social media (Instagrâm, twitter), YouTube. Young people collect information from different sources than older people. For good communication professionals should be involved helping us to communicate knowledge, e.g. video makers, artists, actors, etc. Young people themselves know much about how to communicate to their own generation.

Narratives are important in communication. Scientists need to think how they communicate their work and need to improve their communication in general. The concept of value is of importance, science can provide information that influence the perception of value among the public and among decisionmakers.

Civil society is key to create a concrete plan of action that involves people across the segments of society, and in particular involve the younger generation through education and outreach and build on its enthusiasm. Citizen science could be one way of including communities in ocean literacy.

Funding

The different Arctic regions are limited by national funding. There is a missing international funding mechanism. One positive funding mechanism is the Belmont funding. A joint strategy, action plan and funding, which is building on and contributing to national initiatives, should be developed.

Annex

Programme

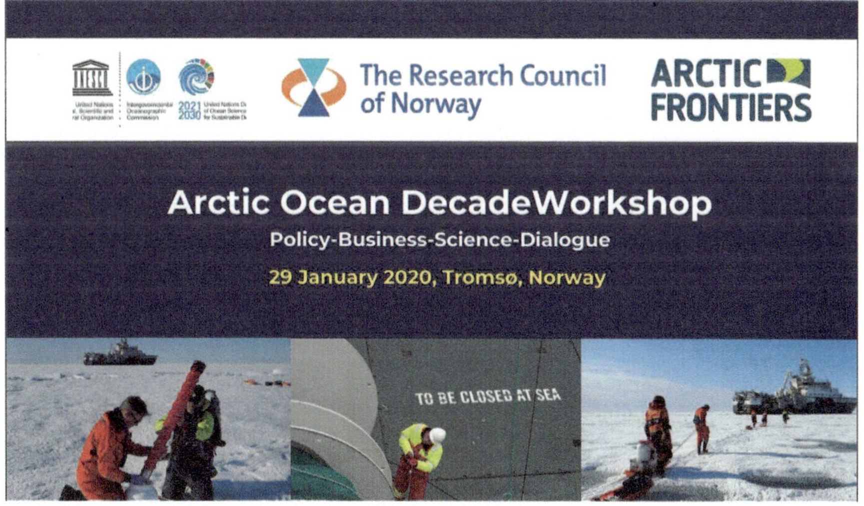

09.00	**Welcome Statements and Opening Remarks** **Fridtjof Unander**, Executive Director of the Division of Oceans, Energy and Sustainability, The Research Council of Norway **Ole Øvretveit**, Director of Arctic Frontiers **Ivet Petkova**, Arctic Frontiers Young **Peter Thomson**, UN Secretary-General's Special Envoy for the Ocean (video) Moderator: Christina Abildgaard, Director of Department for Ocean and Polar Research, the Research Council of Norway
09.30	**Introduction to UN Decade of Ocean Science for Sustainable Development** *By **Vladimir Ryabinin**, Executive Secretary Intergovernmental Oceanographic Commission of UNESCO*

(continued)

P. A. Berkman et al. (eds.), *Building Common Interests in the Arctic Ocean with Global Inclusion, Volume 2*, Informed Decisionmaking for Sustainability,
https://doi.org/10.1007/978-3-030-89312-5

	Introduction to group work *By Christina Abildgaard*
09.45	**Topic 1: A clean, healthy and resilient Arctic ocean** *Keynote talk by* ***Anne Christine Brusendorff****, General Secretary, International Council for the Exploration of the Sea (ICES)*
10.00	**Topic 2: A predicted ocean** *Keynote talk by* ***Heidar Gudjonsson****, Chair, Arctic Economic Council*
10.15	**Break Out Groups: Topic 1** Addressing the Decades Societal Outcome *A clean ocean* and *A healthy and resilient ocean*
11.30	**Break Out Groups: Topic 2** Addressing the Decades Societal Outcome *A predictable ocean*
13.00	**Lunch**
13.45	**Topic 3: A sustainably harvested and productive ocean** *Keynote talk by Senior Fellow* ***David Balton****, the Woodrow Wilson Center's Polar Institute*
14.00	**Break Out Groups: Topic 3** Addressing the Decades Societal Outcome *A sustainably harvested and productive ocean*
15.30	**Summing up and way forward** Summing up the group discussions with the working group chairs: **Marianne Kroglund**, Norwegian Environment Agency **Colin Moffat**, Scottish Government **Tore Furevik**, Bjerknes Centre of Climate Research **Elizabeth McLanahan**, National Oceanic and Atmospheric Administration **Jose Moutinho**, Atlantic International Research Centre in Portugal Way forward **Colin Stedmon**, Technical university of Denmark / Danish Centre for Marine Research **Vladimir Ryabinin,** Executive Secretary Intergovernmental Oceanographic Commission of UNESCO
16.15	End of Workshop
18.15–19.15	**Bringing the Arctic Ocean into United Nations Decade of Ocean Science** *Arctic Frontiers Side Event at Clarion Hotel the Edge, room: Arbeidskontoret* 1 Introduction by **Vladimir Ryabinin**, Executive Secretary IOC of UNESCO Panel discussion with: **Anne Christine Brusendorff**, General Secretary, International Council for the Exploration of the Sea (ICES) **Colin Moffat**, Chief Scientific Advisor Marine, Scottish Government **Hanna Kauko**, Association of Polar Early Career Scientists (APECS) **Paul Arthur Berkman**, Professor, Tufts University Moderator: **Peter Haugan**, Programme Director at Institute of Marine Research

Group Leaders and Rapporteurs

- Group 1: Chair: Colin Moffat, Scottish Government. Rapporteur: Anna Silyakova, UiT The Arctic University of Norway
- Group 2: Chair: Tore Furevik, Bjerknes Centre. Rapporteur: Anne Katrine Normann, Centre for the Ocean and the Arctic
- Group 3: Chair: Marianne Kroglund, Norwegian Environment Agency. Rapporteur: Mario Acquarone, AMAP
- Group 4: Chair: Elizabeth McLanahan, NOAA. Rapporteur: Jon L. Fuglestad, Research Council of Norway
- Group 5: Chair: Jose Moutinho, Atlantic International Research Centre. Rapporteur: Øyvind Lundesgaard, Norwegian Polar Institute

Participating Institutions and Countries

Aarhus University	Denmark
Akvaplan-niva	Norway
Alfred Wegener Institute	Germany
AMAP	International
Arctic Economic Council	Iceland
Association of Arctic Expedition Cruise Operators	Norway
Atlantic International Research Centre	Portugal
Bjerknes Centre	Norway
Bren School, University of California	USA
Centre for the Ocean and the Arctic, Nofima	Norway
Equinor	Norway
EurOcean	International
EuroGOOS	International
European Commission, DG Maritime Affairs and Fisheries	International
European Polar Board	International
Finnish Environment Institute	Finland
Fisheries and Oceans Canada	Canada
French National Research Agency	France
GenØk – Center for biosafety	Norway
GRID-Arendal / UNEP	Norway/ International
Hafenstrom	Norway
ICES	International
Indiana University / Lilly Family School of Philanthropy	USA/France
Institut Francais Norvege	France
Institute of Earth Sciences	France

(continued)

Institute of Ecology and Environment – National Centre for Scientific Research	France
Institute of Marine Research	Norway
IOC of UNESCO	UN Agency
Japan Agency for Marine-Earth Science and Technology (JAMSTEC)	Japan
Kola Science Centre of the Russian Academy of Sciences	Russia
Marine Research Centre at Lomonosov Moscow State University / The University of Edinburgh	United Kingdom
MET Norwegian Meteorological Institute	Norway
Ministry for Foreign Affairs	Iceland
Ministry of Foreign Affairs, Ocean Team	Norway
Ministry of Trade, Industry and Fisheries	Norway
Multiconsult	Norway
NAMMCO – North Atlantic Marine Mammal Commission	Norway
Nansen Environmental and Remote Sensing Center	Norway
National Institute of Aquatic Resources, Technical University of Denmark	Denmark
NILU – Norwegian Institute for Air Research	Norway
NMBU – Norwegian University of Life Sciences	Norway
NOAA, Office of International Affairs	USA
Norwegian Environment Agency	Norway
Norwegian Polar Institute	Norway
APECS	International
NTNU – Norwegian University of Science and Technology	Norway
Orca Research Ltd	Ireland
Plymouth Marine Laboratory	United Kingdom
Sabima	Norway
SAON	International
Scottish Government, Marine Scotland	United Kingdom
SINTEF Ocean	Norway
Swedish Meteorological and Hydrological Institute	Sweden
Swedish Polar Research Secretariat	Sweden
Technical university of Denmark / Danish Centre for Marine Research	Denmark
The Norwegian Coastal Administration	Norway
The Ocean Foundation	USA
Tomsk Polytechnic University	Russia
UiT The Arctic University of Norway	Norway
University of Bergen	Norway
University of Oslo	Norway
University of the Arctic – UArctic	International
Wilson Center	USA
Young friends of the earth Norway	Norway

Index

P. A. Berkman et al. (eds.), *Building Common Interests in the Arctic Ocean with Global Inclusion, Volume 2*, Informed Decisionmaking for Sustainability,
https://doi.org/10.1007/978-3-030-89312-5

J

K

W

Printed by Printforce, the Netherlands